MANUAL OF VERTEBRATE DISSECTION

Comparative Anatomy

SECOND EDITION

Dale W. Fishbeck
Youngstown State University (Emeritus)

Aurora M. Sebastiani
Youngstown State University (Emerita)

Morton Publishing Company
925 W. Kenyon Ave., Unit 12
Englewood, CO 80110

http://www.morton-pub.com

Book Team

Publisher:	Douglas N. Morton
Biology Editor:	David Ferguson
Production Manager:	Joanne Saliger
Production Assistant:	Desiree Coscia
	Patricia Billiot
Typography:	Ash Street Typecrafters, Inc.
Copyediting:	Carolyn Acheson
Cover Design:	Bob Schram, Bookends, Inc.

ISBN 10: 0-89582-748-4
ISBN 13: 978-0-89582-748-7

Printed in the United States of America

10 9 8 7 6 5 4 3 2 1

Preface

With this book, students, perhaps the first time, will become familiar with the core of disciplines encompassing the concepts of evolution, embryology, comparative morphology and histology, and physiology—all in the same course. This book provides a unique opportunity to learn about some tangible evidence supporting the concepts of evolution, demonstrated in the laboratory portion of this course in comparative anatomy. Students can "get their hands dirty" in performing the comparative study of a number of representative vertebrate animals. To set the stage for the study of vertebrates, students will examine the characteristics of the non-vertebrate chordates, because many members of this group exhibit the unique traits of chordates.

If we are to understand the evolutionary record of vertebrates, studying a selection of animals representing significant advances in this history is essential. The earliest vertebrates were jawless, and they lacked paired fins. The lamprey best represents this early condition. The dogfish shark is a representative gnathostome fish, as the morphology of its organ systems lays the groundwork for evolutionary changes to come. It is large enough to dissect and is available.

Because amphibians were the first tetrapods, examination of a representative of this group seems logical. The morphology of the caudate members is the least modified of the modern amphibian fauna. Although the mudpuppy possesses a number of specialized characters, its size and availability make it a reasonable choice. The cat probably is the best choice as a mammalian representative, again because it is readily available. Further, we are interested in our own morphology, and cats and humans have numerous similarities.

This book is designed to permit the instructor maximum latitude to plan a laboratory course that will enhance the objectives of the lecture, fit the student's needs, and meet a reasonable time schedule. The laboratory experience has been organized to permit either a systemic or an organismic study. Among the strengths of this book are the following:

- The major systems of each organism are covered in depth and in a logical sequence.
- Color photographs are the main source of illustration. Many are unique among comparative vertebrate anatomy manuals.
- Detailed dissection instructions for obtaining superior specimens for comparison are given throughout the text.
- Important anatomical terms are in boldface type.

NEW IN THE SECOND EDITION

A major new section—a discussion of the most significant events in the evolution of the comparative anatomy of vertebrates—was added. Copious illustration enhances the text. Some sections were re-organized, as follows, for a more logical presentation:

- Positional figures were included to aid in determining specific body regions illustrated in many photographs.
- Some dissection instructions were enhanced.
- The Glossary was expanded to include many more terms.

Acknowledgments

The authors express their sincere appreciation to David Ferguson, Biology Editor at Morton Publishing Company, for his assistance in the production of the second edition. We also acknowledge the contribution of Carolyn Acheson, copy editor. For the outstanding page layout, format, and design, we thank Joanne Saliger and her staff at Ash Street Typecrafters. We also thank Mike Schenk for the superior illustrations.

For their review and helpful suggestions in preparation of the final draft of the text, we thank Dr. Susan Hill, Michigan State University, and Dr. Clare Hays, Metropolitan State College of Denver. In addition, a number of suggestions for changes in the second edition by instructors of comparative anatomy were incorporated. In particular, Avery Williams and his students at Louisiana State University at Eunice were helpful.

Contents

Introduction   1

What is a Chordate?   1
  Classification   2
Terms: Anatomical and Directional   4
Suggested Equipment List   4
  Dissection Tools   5
  Other Equipment   5

**PART ONE
Phylum Hemichordata   7**

1 The Hemichordates   9

**PART TWO   Phylum Chordata—
The Non-Vertebrate
Chordates   13**

2 Subphylum Urochordata   15

3 Subphylum Cephalochordata   21

**PART THREE   The Evolution
of Vertebrate Systems:
An Overview   27**

4 Integumentary System   29

Scales   29
Hair   31
Claws   31
Horns, Antlers, and Other Derivatives   32
Glands   33
Pigment   33

5 Skeletal System   35

Skull   35
Chondrocranium   36
Splanchnocranium   36

Jaws   37
Teeth   38
Dermatocranium   41
Vertebral Column, Ribs, and Sternum   42
Limbs and Girdles   46

6 Muscular System   51

**7 Body Cavities
and Mesenteries   55**

**8 Digestive and
Respiratory Systems   57**

9 Urinary System   61

Urinary System   61
  Kidney   62
  Cloaca   62
  Urine Elimination   63
  Membranes and Ducts   63

10 Reproductive System   65

**11 Circulatory and
Lymphatic Systems   69**

Blood   69
Heart   70
Aortic Arches   74
Blood Vessels   76
Lymphatic System   78

**12 Nervous System
and Sense Organs   79**

Cells of the Nervous System   79
Brain   80
Meninges and Cerebrospinal Fluid   83
Spinal Cord   84
Spinal Nerves   84
Common Neuronal Pathways   85
  Reflex Arcs   85
  Ascending and Descending Pathways   85

v

Cranial Nerves   86
    Terminal Nerve (0)   87
    Olfactory Nerve (I)   87
    Optic Nerve (II)   87
    Oculomotor Nerve (III)   87
    Trochlear Nerve (IV)   87
    Trigeminal Nerve (V)   87
    Abducens Nerve (VI)   87
    Facial Nerve (VII)   87
    Vestibulocochlear Nerve (VIII)   87
    Glossopharyngeal Nerve (IX)   88
    Vagus Nerve (X)   88
    Spinal Accessory Nerve (XI)   88
    Hypoglossal Nerve (XII)   88
Autonomic Nervous System   88
Sense Organs   88
    Olfaction   89
    Vomeronasal Organ   89
Gustation   90
    Lateral Line System   90
    Equilibrium and Audition   91
Vision   92
    Lateral Eyes   92
    Median Eyes   94
    Thermoreceptors   94

13 Endocrine Systems   95
Hypophysis   95
Pineal Gland   96
Adrenal Gland   97
Thymus   97
Thyroid   97
Ultimobranchial Bodies   97
Parathyroid Glands   98
Pancreas   98
Ovaries and Testes   98
Placenta   98
Enteric Hormones   98

**PART FOUR Phylum Chordata:
Subphylum Vertebrata   99**

14 Lamprey   101
External Anatomy   101
Integumentary System   102
Internal Anatomy   106
Skeletal System   107
Muscular System   108
Digestive System   109
Respiratory System   110
Excretory System   110
Reproductive System   112
Circulatory System   113
Nervous System and Sense Organs   114
Larval Lamprey   114

**PART FIVE Phylum Chordata—
Subphylum Vertebrata—
Dogfish Shark   119**

**15 Dogfish Shark—External Anatomy
and Integumentary System   121**
External Anatomy   121

16 Dogfish Shark—Skeletal System   129
Axial Division   130
    Chondrocranium   130
    Splanchnocranium   132
Appendicular Division   138
    Pectoral Girdle and Fins   138
    Pelvic Girdle and Fins   138

17 Dogfish Shark—Muscular System   141
Skinning   141
Trunk or Axial Muscles   143
Pectoral Fin Muscles   143
    Extensor or Abductor   143
    Flexor or Adductor   143
Pelvic Fin Muscles   145
    Extensor or Abductor   145
    Flexor or Adductor   145
Dorsal Fin Muscle   146
Muscles of the Gill Arches or Their Derivatives   146
Branchiomeric Muscles   146
    Visceral Arch I (Mandibular Arch)   146
    Visceral Arch II (Hyoid Arch)   147
    Visceral Arches III–VII   147
Hypobranchial Muscles   150
Eye Muscles   151

**18 Dogfish Shark—Body Cavities
and Mesenteries   153**
Opening the Shark   153
Mesenteries   155
    Derivatives of the Dorsal Mesentery   155
    Derivatives of the Ventral Mesentery   157

**19 Dogfish Shark—Digestive
and Respiratory Systems   159**
Digestive System   159
    Accessory Digestive Organs   163
Respiratory System   166
    Internal Gill Slits   166
    Gills   167

20 Dogfish Shark— Urogenital System 169

Urinary System 169
  Kidneys and Ducts 169
Reproductive System 170
  Female 170
  Male 170

21 Dogfish Shark—Circulatory and Lymphatic Systems 175

Heart and Arteries 176
  Heart and Afferent Branchial Arteries 175
  Efferent Branchial Arteries, Dorsal Aorta, and Its Branches 176
Venous System 186
  Portal Systems 186
  Systemic Veins 188
Lymphatic System 189

22 Dogfish Shark—Nervous System and Sense Organs 191

Brain and Nerves 191
  The Brain: Dorsal View 191
The Brain: Ventral View 197
  Ventricles of the Brain 197
  Cerebrospinal Fluid Circulation and Meninges 197
  The Occipital Nerves, Hypobranchial Nerve, Spinal Nerves, and Spinal Cord 198
  Cervicobrachial and Lumbosacral Plexus 199
Autonomic Nervous System 199
Sense Organs 199
  Olfactory Apparatus 199
  Lateral Line System 199
  Auditory and Equilibrium Apparatus 200
  Eye and Extrinsic Eye Muscles 202

23 Dogfish Shark— Endocrine System 207

Hypophysis or Pituitary 207
  Pineal Gland or Epiphysis 208
  Thyroid Gland 208
  Thymus Gland 208
  Adrenal Gland 208
  Pancreas 209
  Ovaries 209
  Testes 209
  Other Endocrine Tissues 209

PART VI Phylum Chordata— Subphylum Vertebrata— Mudpuppy (*Necturus*) 211

24 Mudpuppy—External Anatomy and Integumentary System 213

External Anatomy 213
Integumentary System 214

25 Mudpuppy—Skeletal System 217

Axial Division 218
  Skull 218
  Vertebral Column 220
  Vertebral Regions 221
Appendicular Division 223
  Pectoral Girdle and Forelimb 223
  Pelvic Girdle and Hindlimb 223

26 Mudpuppy—Muscular System 225

Skinning 225
Axial Muscles 226
  Epaxial Muscles 226
  Hypaxial Muscles 226
Dorsal and Lateral Muscles of the Head 228
  Adductor Mandibulae Anterior 228
  Adductor Mandibulae Externus 228
  Depressor Mandibulae 228
  Branchiohyoideus 229
  Levatores Arcuum 229
  Dilatator Laryngis 229
Ventral Muscles of the Throat 229
  Intermandibularis 229
  Interhyoideus 229
  Sphincter Colli 230
  Geniohyoideus 230
  Genioglossus 230
  Rectus Cervicis 230
  Subarcuales 230
  Transversi Ventrales 231
  Depressores Arcuum 231
Ventral Pectoral Muscles 231
  Omoarcual 231
  Procoracohumeralis 231
  Supracoracoideus 232
  Pectoralis 232
Dorsal/Lateral/Medial Pectoral Muscles 232
  Pectoriscapularis 232
  Cucullaris 232
  Dorsalis Scapulae 232
  Latissimus Dorsi 232
  Thoraciscapularis 232
  Levator Scapulae 232
  Subcoracoscapularis 233

Muscles of the Forelimb   233
  Triceps Brachii   233
  Coracobrachialis   233
  Humeroantebrachialis   233
  Flexors and Extensors of the Forearm   234
Muscles of the Pelvic Girdle and Hindlimb   234
  Puboischiofemoralis Externus   234
  Puboischiotibialis   234
  Ischiofemoralis   234
  Puboischiofemoralis Internus   234
  Pubotibialis   234
  Ischioflexorius   235
  Ischiocaudalis   235
  Caudocrualis (or Caudopuboischiotibialis)   236
  Caudofemoralis   236
  Iliotibialis   236
  Ilioextensorius   236
  Iliofibularis   236
Flexors and Extensors of the Hindlimb   236

27 Mudpuppy—Body Cavities and Mesenteries   237
Body Cavities   237
Opening *Necturus*   237
Mesenteries   239
  Derivatives of the Dorsal Mesentery   239
  Derivatives of the Ventral Mesentery   241

28 Mudpuppy—Digestive and Respiratory Systems   243
Digestive System   243
  Alimentary Canal   243
  Accessory Digestive Organs   245
Respiratory System   245

29 Mudpuppy—Urogenital System   249
Urinary System   249
Kidney and Ducts   249
Reproductive System   250
  Female   250
  Male   250

30 Mudpuppy—Circulatory and Lymphatic Systems   255
Circulatory System   255
Heart   255
Arteries   258
  Ventral Aorta and Afferent Branchial Arteries   258
  Efferent Branchial Arteries, Dorsal Aorta and Its Branches   258
Veins   262
  Renal Portal System   262
Lymphatic System   266

31 Mudpuppy—Nervous System and Sense Organs   267
Nervous System   267
  Brain and Cranial Nerves   267
  Meninges and Ventricles of the Brain   267
  Telencephalon   268
  Diencephalon   268
  Mesencephalon   269
  Metencephalon   269
  Myelencephalon   270
Spinal Nerves and Spinal Cord   270
  Brachial Plexus   270
  Lumbosacral Plexus   271
  Autonomic Nervous System   271
Sense Organs   271
  Olfactory Apparatus   271
  Lateral Line System   271
  Auditory and Equilibrium Apparatus   272
  The Eyes   272

32 Mudpuppy—Endocrine System   273
Hypophysis or Pituitary   273
  Pineal Gland   273
  Thyroid Gland   274
  Parathyroid Glands   274
  Ultimobranchial Bodies   274
  Thymus Gland   274
  Suprarenal or Adrenal Glands   274
  Pancreas   274
  Ovaries   275
  Testes   275
Other Endocrine Tissues   275

PART VII Phylum Chordata— Subphylum Vertebrata—Cat   277

33 Cat—External Anatomy and Integumentary System   279
External Anatomy   279
Integumentary System   280
  Skin   280
  Epidermal Derivatives   281

34 Cat—Skeletal System   283
Axial Division   284
  Skull   284
  Anatomy of the Skull   285
  Surface Features of the Skull   296
  Cavities and Sinuses   296
  Cranial Foramina   296
  Mandible   297
  Vertebral Column   297
  Lumbar Vertebrae   300
Appendicular Division   302
  Pectoral Girdle and Forelimb   302
  Pelvic Girdle and Hindlimb   305

35 Cat—Muscular System 309

Skinning 309
    Preparing the Cat for Muscle Dissection 310
    Direction of Muscle Fibers 310

Superficial Thoracic Muscles 310
    Pectoantebrachialis 310
    Pectoralis Major 311
    Pectoralis Minor 311
    Xiphihumeralis 312

Abdominal Muscles 312
    External Oblique 312
    Internal Oblique 312
    Transversus Abdominis 312
    Rectus Abdominis 313

Superficial Back Muscles 313
    Clavotrapezius 313
    Clavobrachialis 313
    Acromiotrapezius 314
    Spinotrapezius 314
    Latissimus Dorsi 314

Deep Thoracic Muscles 315
    Serratus Ventralis 315
    Scalenes 315
    Transversus Costarum 315
    Intercostalis Externus 315
    Intercostalis Internus 316
    Transversus Thoracis 316
    Serratus Dorsalis 316

Lower Back Muscles: Lumbar and Thoracic 317
    Multifidus Spinae 317
    Longissimus Dorsi 317
    Spinalis Dorsi 317
    Iliocostalis 318

Muscles of the Neck 318
    Sternomastoid 318
    Cleidomastoid 318
    Sternohyoid 318
    Sternothyroid 319
    Thyrohyoid 319
    Stylohyoid 319
    Mylohyoid 320
    Geniohyoid 320
    Genioglossus 320
    Hyoglossus 320
    Styloglossus 320

Deep Neck and Back Muscles 320
    Rhomboideus Capitis 320
    Rhomboideus Cervicis and Thoracis 320
    Splenius 320
    Longissimus Capitis 320
    Longus Colli 322

Muscles of the Head 322
    Masseter 323
    Temporalis 323
    Pterygoideus Externus 323
    Pterygoideus Internus 323
    Digastric 324

Muscles of the Shoulder 324
    Supraspinatus 324
    Infraspinatus 324
    Teres Major 324
    Levator Scapulae Ventralis 324
    Acromiodeltoid 324
    Spinodeltoid 325
    Teres Minor 325
    Subscapularis 325

Muscles of the Upper Forelimb, or Brachium 325
    Coracobrachialis 325
    Epitrochlearis 325
    Biceps Brachii 326
    Triceps Brachii 326
    Anconeus 326
    Brachialis 326

Muscles of the Lower Forelimb, or Antebrachium 327
    Brachioradialis 327
    Extensor Carpi Radialis Longus 327
    Extensor Carpi Radialis Brevis 327
    Extensor Digitorum Communis 329
    Extensor Digitorum Lateralis 329
    Extensor Carpi Ulnaris 329
    Supinator 329
    Abductor Pollicis Longus 329
    Extensor Indicis 329
    Flexor Carpi Ulnaris 329
    Flexor Digitorum Superficialis 329
    Flexor Carpi Radialis 329
    Pronator Teres 329
    Flexor Digitorum Profundus 330
    Pronator Quadratus 330

Muscles of the Manus 330
    Lumbricales 330

Muscles of the Thigh 331
    Sartorius 331
    Gracilis 331
    Biceps Femoris 332
    Tenuissimus 332
    Caudofemoralis 333
    Semitendinosus 333
    Semimembranosus 333
    Adductor Femoris 333
    Adductor Longus 334
    Pectineus 334
    Iliacus 334
    Psoas Major 335
    Psoas Minor 335
    Tensor Fascia Latae 335
    Quadriceps Complex 335

Muscles of the Shank 336
    Tibialis Cranialis 337
    Extensor Digitorum Longus 337
    Peroneus Longus 337
    Peroneus Tertius 337
    Peroneus Brevis 337
    Popliteus 337
    Flexor Digitorum Longus 337
    Flexor Hallucis Longus 337

Tibialis Caudalis   338
Gastrocnemius   338
Plantaris   338
Soleus   339
Triceps Surae   339
Muscles of the Pes   339
Extensor Digitorum Brevis   339
Flexor Digitorum Brevis   339
Muscles of the Hip   339
Gluteus Maximus   339
Gluteus Medius   339
Pyriformis   340
Gluteus Minimus   340
Articularis Coxae   340
Gemellus Cranialis   340
Coccygeus   340
Obturator Internus   340
Gemellus Caudalis   340
Quadratus Femoris   340
Obturator Externus   340
Tail Muscles   340

36 Cat—Body Cavities and Mesenteries   343
Opening the Cat   343
Body or Coelomic Cavities   345
Mesenteries of the Thoracic Cavities   345
Mesenteries of the Abdominopelvic Cavity   345

37 Cat—Digestive and Respiratory Systems   355
Digestive System   355
Alimentary Canal   355
Salivary Glands and Ducts   355
Mouth or Oral Cavity   357
The Digestive Tube   360
Accessory Digestive Organs   361
Respiratory System   364

38 Cat—Urogenital System   371
Urinary System   371
Kidney and Ducts   371
Reproductive System   374
Female   374
Male   377

39 Cat—Circulatory and Lymphatic Systems   381
The Heart   381
External Anatomy   381
Internal Anatomy   382
Blood Vessels   386
Lymphatic System   408

40 Cat—Nervous System and Sense Organs   411
Brain   411
Meninges and Ventricles of the Brain   412
External Anatomy   412
Internal Anatomy: Sagittal Section of Brain   415
Spinal Cord   419
Spinal Nerves   419
Autonomic Nervous System   421
Sense Organs   424
Tongue   424
Olfactory Apparatus (Nose)   424
Auditory and Equilibrium Apparatus   424
Eye   426

41 Cat—Endocrine System   431
Pituitary Gland or Hypophysis   431
Pineal Gland   432
Thyroid Gland   432
Parathyroid Gland   433
Thymus Gland   433
Adrenal Glands   433
Pancreas   434
Ovaries   434
Testes   434
Other Endocrine Tissues   434

References   437

Glossary   439

Index   447

Introduction

To discover the evolutionary relationships among animals, we must examine the anatomy of a number of animals in detail. For example, to make some sense of the history of the development and function of an organ or a structure in a mammal, we may have to search for its possible origin among non-mammalian vertebrates. Perhaps what makes comparative anatomy most fascinating is that the functional organ in a vertebrate may not even resemble its homolog in its ancestor, but the relationship often can be demonstrated by carefully examining the morphology, embryology, histology, and physiology of ancestral structures.

To make this experience possible and meaningful, we will study and compare the anatomy of the lamprey, the dogfish shark, the mudpuppy, and the cat as important representative members of the subphylum Vertebrata. This selection of animals introduces the student to taxa that play or have played an important role in interpreting the developmental history of vertebrates. Each of these modern vertebrates chosen for our study is the result of millions of years of evolution, and all are adapted to their current environments.

- The jawless **lamprey** is an example of the earliest vertebrates in which we find a combination of ancestral vertebrate structures and some highly specialized structures to allow it to function as a viable member of the modern fish fauna.

- The **dogfish shark** is a gnathostome, or jawed vertebrate, in which we again see a combination of ancestral and derived structures. Much of the ancestral anatomy foreshadows anatomical conditions found among the tetrapods.

- The **mudpuppy** is a tetrapod, or four-legged vertebrate, whose anatomy is predominantly ancestral with specialization that allows it to function as a viable obligate aquatic vertebrate.

- The **cat** is a mammal that exhibits a body plan that is rather unspecialized and similar to our own. Most of us are intrigued by our history.

Notice that many of the characteristics of the vertebrates used in our comparative study are ancestral. Rarely, however, do we find an animal existing in an environment without some specialization. Therefore, in the lamprey and shark, for example, we see anatomy and physiology that are adapted to the conditions of their habitat.

WHAT IS A CHORDATE?

The relationships among animals have been, and remain, a major point of interest among biologists. Two distinct groups of animals have evolved: the protostomes and the deuterostomes. The protostomes include many of the most numerous and successful animals—mollusks (clams, snails, squids), annelids (earthworms, leeches) and arthropods (crayfish, lobsters, insects, ticks, mites, spiders). The deuterostomes are not as numerous, and include the echinoderms (starfish, brittlestars, sea urchins), the hemichordates (acorn worms, pterobranchs), and the chordates (urochordates, cephalochordates, vertebrates or craniates).

The major reason that protostomes and deuterostomes can be recognized is based on events that occur during embryonic development. Early cleavages in both groups lead to a more or less hollow ball of cells known as the blastula. It is the manner by which the next developmental stage, the gastrula, in each group is formed that distinguishes the two. Among protostomes, the gastrula is formed through an invagination of a wall of the blastula forming the primitive gut, the archenteron, leaving an opening called the blastopore. The blastopore in protostomes gives rise to the mouth of the animal. In contrast, in deuterostomes, the blastopore gives rise to the anus, and the mouth arises from a second invagination in an area of the gastrula that is destined to become the head of the animal (see Figure I.1). In addition, during development, protostomes and deuterostomes exhibit differences in cleavage patterns, germ layer and coelom (body cavity) formation.

All **chordates** possess certain characteristics at some time during their life that may persist as permanent adult structures or in a transitory embryological developmental sequence. This suite of characteristics includes:

1. a longitudinal supporting rod, the notochord

2. a dorsal, hollow nerve cord

3. pharyngeal slits

4. a subpharyngeal organ (endostyle or thyroid gland) that binds iodine

5. a postanal tail.

An old maxim among vertebrate zoologists emphasizes that "all vertebrates are chordates, but not all chordates are vertebrates." Two of the three groups of chordates—the urochordata and the cephalochordata—are not craniates.

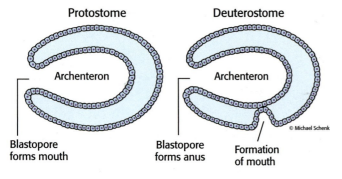

FIGURE I.1 Comparative gastrulation of protostomes and deuterostomes.

Do chordates have any relatives among the rest of the invertebrates? The answer seems to be "yes." Most biologists now consider a group of animals, the hemichordates, formerly classified with the chordates, to be a separate phylum (phylum Hemichordata). Of the definitive traits exhibited by members of the phylum Chordata, the hemichordates share only pharyngeal slits with them. The hemichordates produce a larva that is similar to the larva of echinoderms, thereby possibly providing a bridge between the phylum Chordata and the phylum Echinodermata. Our closest living relatives among non-chordate invertebrates are starfish, sea urchins, sea cucumbers, and their relatives!

Classification

For centuries scientists have dealt with relationships among organisms. Determining how closely organisms are related depends on the past evolutionary history of these groups. Currently, the two major classification systems are traditional, or evolutionary, and cladistic, or phylogenetic.

Evolutionary System

The evolutionary system of classification had its origin prior to Charles Darwin and remains dependent upon some of the philosophy and taxa of a famous Swedish natural historian of the 18th century, Carl von Linné. You probably know him better by his Latinized name, Carolus Linnaeus, often shortened to Linnaeus.

Our system of naming organisms is based largely on methods he proposed and established. He invented a system using two terms for each organism. The first, called the genus, refers to the generic name, usually a Latin noun or Latinized name and the second or the specific epithet, is usually a Latin adjective or similar word to describe the generic name. The system has become known as binomial nomenclature. This is the system we continue to use today.

In this system, only one organism can be assigned a binomial name—e.g., *Rana catesbiana*, the bullfrog. The generic name, *Rana*, can be used to name other frogs, and the specific epithet, *catesbiana*, can be used to describe other genera, but there can only be one *Rana catesbiana*. Because other frogs can share the generic and higher categories, a related, but distinct species, *Rana sylvatica*, the woodfrog, belongs to the same genus, family, and all other categories to which ranid frogs belong. But only one narrowly defined group of animals possesses the criteria that qualify it as a distinct species.

Further, Linnaeus and other naturalists of his time recognized that similar organisms were related to one another, and they grouped similar species together into higher categories that were increasingly inclusive, with broader criteria, and consequently more subjective. Later, taxonomists developed a scientific classification scheme in which organisms were placed in the following seven fundamental categories, listed in increasing order of objectivity:

> Kingdom
> > Phylum
> > > Class
> > > > Order
> > > > > Family
> > > > > > Genus
> > > > > > > Species

In an attempt to refine the system, super-categories and sub-categories have been added—subspecies, superorder, subphylum, and so forth.

Cladistic System

Proponents of the cladistic system, employed by phylogenetic systematists, consider it to be more objective, and possibly yielding a truer idea of the actual sequence of events in the evolutionary history of organisms. Two types of characteristics can be recognized in any taxon: those that are ancestral and those that are derived from ancestral traits.

All organisms in any taxon exhibit a number of characteristics or plesiomorphies that were inherited from a common ancestor. Ancestral characteristics that are shared with related organisms are called symplesiomorphies. An example among chordates, the group to which vertebrates belong, is the presence of a notochord at some time during the life cycle. Other than permitting us to separate chordates from other animals, it indicates nothing concerning the relationships among the chordates.

By contrast, apomorphies, characteristics derived from the ancestral condition, are useful in establishing relationships. In the phylogenetic classification, the more shared derived characteristics or synapomorphies between animals, the more likely they are to have had a recent common ancestor and to be closely related. The sharing of derived characteristics is what indicates relatedness among organisms, as these have arisen more recently than ancestral characteristics. Generally, relationships are determined by consideration of the number of shared derived characteristics among groups; and the greater the number of synapomorphies occurring between groups, the more closely related they are thought to be. This implies that they share a greater number of genes with one another than they do with distantly related groups. Figure I.2 shows the relationships of animals discussed in this book.

Why do we classify some vertebrates as mammals? Some of their characteristics are: hair, teeth specialized for a variety of food habits, mammary glands, etc. These characteristics distinguish mammals from other vertebrates, such as reptiles, fish, etc., and are synapomorphic, *i.e.*, they are unique to mammals. When we attempt to learn about the evolution of mammals, however, these characteristics are useless because all mammals possess them and are symplesiomorphic for mammals.

The phylogenetic system requires that all organisms within a taxon have a common ancestry and include all

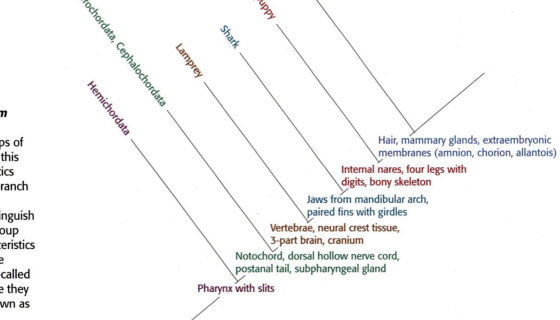

FIGURE I.2 *Cladogram* illustrating the possible phylogenetic relationships of animals encountered in this manual. The characteristics appearing below each branch are newly derived traits (apomorphies) that distinguish that group. Each new group exhibits the new characteristics and also shows all of the characteristics below it—called plesiomorphies. Because they are shared, they are known as symplesiomorphies.

Hemichordata

Urochordata, Cephalochordata

Lamprey

Shark

Mudpuppy

Cat

Hair, mammary glands, extraembryonic membranes (amnion, chorion, allantois)

Internal nares, four legs with digits, bony skeleton

Jaws from mandibular arch, paired fins with girdles

Vertebrae, neural crest tissue, 3-part brain, cranium

Notochord, dorsal hollow nerve cord, postanal tail, subpharyngeal gland

Pharynx with slits

descendants of the common ancestor, *i.e.,* they are mono-phyletic. This is not a requirement in the evolutionary system, although the assumption is that all organisms in a family, order, class, and so on, have a common ancestry. Evolutionary systematists place organisms in a common category based primarily on their possessing similar charac-teristics with the supposition that similar characteristics indicate a common ancestry. For example, the hagfish (a jawless relative of lampreys) and the dogfish shark are placed in the phylum Vertebrata.

In the phylogenetic classification system, the hagfish and dogfish shark are placed in the taxon Craniata, which refers to the fact that they have a braincase or cranium. In contrast to the evolutionary system, however, only the dogfish shark is classified as a vertebrate, as hagfish do not have vertebrae. Moyle and Cech (2000) have placed the hagfish in its own subphylum, the subphylum Myxini, phylum Chordata. More recent evidence suggests that hag-fishes and lampreys are probably monophyletic, however (Ota & Kuratani, 2007).

Sometimes evolutionary systematists place organisms that seem to have distinctive characteristics, such as birds and dinosaurs, in two widely separated groups: class Aves and class Reptilia, respectively. Fossil evidence suggests that a number of related dinosaurs shared feathers with birds, once thought to be a unique bird characteristic.

In addition, some lines of dinosaurs shared a number of other anatomical characteristics and probably some be-havioral traits with birds. Therefore, because birds and dinosaurs exhibit a considerable number of synapomorphies, phylogenetic systematists place birds and dinosaurs in the taxon Dinosauria.

Homologous or Homoplastic?

How do evolutionary systematists or phylogenetic systema-tists determine what information to use to construct a classification scheme? They look for inheritable similarities among the organisms with which they are dealing. Only the structures or organs inherited through a common ancestry are important. These characteristics are **homologous,** or exhibit homology. Homologous structures may or may not resemble each other.

Where did the middle ear bones (ossicles) of mammals originate? If we study their embryology, we discover that they develop in the same area from similar tissues that give rise to jaw suspension elements in fish. The ear ossicles of mammals—the stapes, incus, and malleus—are homologous with the hyomandibular, quadrate, and articular cartilages, jaw suspension elements, of sharks!

What is so difficult about determination of homologous characteristics? The challenge lies in the fact that sometimes structures and organs resemble one another or have similar functions but do not share a common ancestry. A classic example is the evolution of wings in three different groups of vertebrates—birds, reptiles, and mammals. Although the organs resemble one another and are used in the same fashion, they evolved independently and are not homologous but, rather, are homoplastic. On the other hand, many of the bones inside the wings of the three groups are homolo-gous. Forms of homoplasty include analogy, parallelism, and convergence.

During our discussion of the animals selected, we will use the traditional or evolutionary system because of its convenience.

TERMS: ANATOMICAL AND DIRECTIONAL

To comprehend dissection instructions and become a literate anatomist, you must understand and speak the language of anatomy. You are expected to pronounce and spell ana-tomical and directional terms correctly. To appreciate the biology of vertebrate anatomy, you must read the text, con-sult the illustrations and actually do dissections. **"Picture book dissection"** does not work!

General directional terms include **dorsal** (toward the back of the animal), **ventral** (toward the belly of the animal), **cranial** (toward the head), and **caudal** (toward the tail) (see Figure I.3). Just as often, among animals, **anterior** (meaning ahead or before) and **posterior** (meaning after or behind) are encountered in descriptive anatomy. Commonly, students are directed toward the **medial** (toward the midline) or **lateral** (toward the side) aspect of the animal. The midline is an imaginary line that extends directly down the middle of the ventral and dorsal surfaces. Frequently, you will encounter the directional terms **proximal** (next to or nearest the point of origin or attachment) and **distal** (some distance from the point of origin or attachment) (Figure I.3).

Often, planes of reference are important in understand-ing relationships of the morphology of organs, relationships among organs of a system within a body cavity, or relation-ships of organs and systems in a presented view. Referring to Figure I.3, a section parallel to the midline is a **sagittal** section. Thus, there are numerous sagittal sections, as long as you do not run out of animal. But there is only one **mid-sagittal** section, which passes exactly down the midline of the body. Figure I.3 shows a **transverse** or cross-section. Just as many sagittal sections are possible, so, too, numerous transverse sections may extend from the tip of the snout to the tip of the tail. Transverse sections are analogous to the slices of a loaf of bread, although usually much thinner. A **frontal** section is made along the entire length of the animal parallel to the belly and back. Numerous frontal sections also are possible.

SUGGESTED EQUIPMENT LIST

To produce well dissected specimens, good tools are neces-sary. The following list is suggested.

Dissection Tools

- 1 pair of fine point dissection scissors
- 1 scalpel handle, preferably No. 4
- Replaceable blades, preferably designated as 21–25
- 1 steel probe, preferably a Huber-Mall
- 2 pairs of straight forceps, one with medium points and one with fine points
- Dissecting pins

Other Equipment

- Safety goggles—strongly recommended
- Gloves—optional but strongly recommended
- Lab coat—optional but strongly recommended
- Small spray bottle to hold preservative fluid to prevent dehydration and deterioration of the specimens

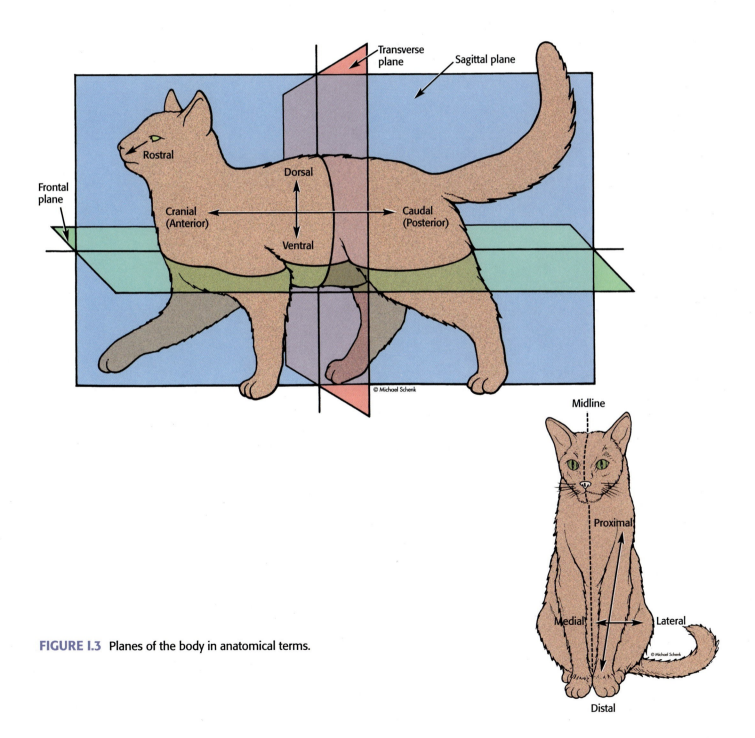

FIGURE I.3 Planes of the body in anatomical terms.

Phylum Hemichordata

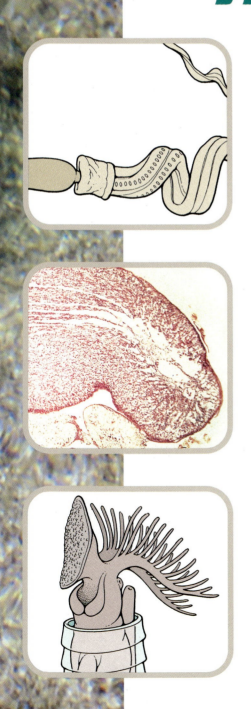

CHAPTER ONE

The Hemichordates

The hemichordates consist of a marine group of invertebrates that once were considered to be members of the phylum Chordata. Of the five diagnostic chordate characteristics, only the pharyngeal slits are present. A structure formerly thought to be the notochord, but now called the stomochord, seems to be an outpocketing of the primitive gut and does not have the typical histological composition of the notochord of chordates. The nervous system consists of a dorsal and ventral nerve cord, but the dorsal cord is not hollow. Further, no iodine-binding tissue has been found and no post-anal tail is present. Regardless, hemichordates seem to be related to the chordates and, because they possess larvae similar to those of the phylum Echinodermata (starfish, brittle stars, and sea urchins), they bridge the gap between the chordates and other invertebrates.

Hemichordates are recognizable as two classes quite different in appearance, the Enteropneusta and the Pterobranchia. The enteropneusts, also known as acorn worms, are sedentary, wormlike animals that burrow into the substratum of shallow waters.

Obtain available slides or plastic mounts and find the following structures.

The body of the acorn worm has three regions: the **proboscis**, the **collar**, and the **trunk** (Figure 1.1). The proboscis and collar are used during burrowing and locomotion. Movement is similar to wormlike invertebrates. The proboscis is anchored while the rest of the animal is pulled toward it. Some acorn worms feed by projecting their proboscis above the burrow. Cilia on the surface of the proboscis create currents that flow over the mucus covering the proboscis (Figure 1.2).

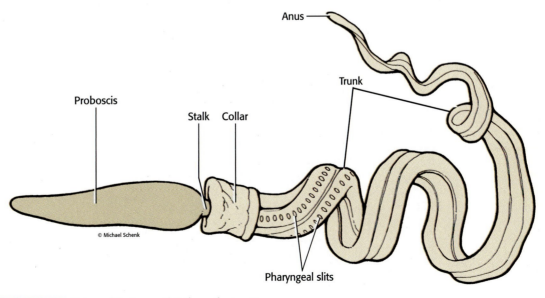

FIGURE 1.1 External features of *Balanoglossus*, Acorn worm.

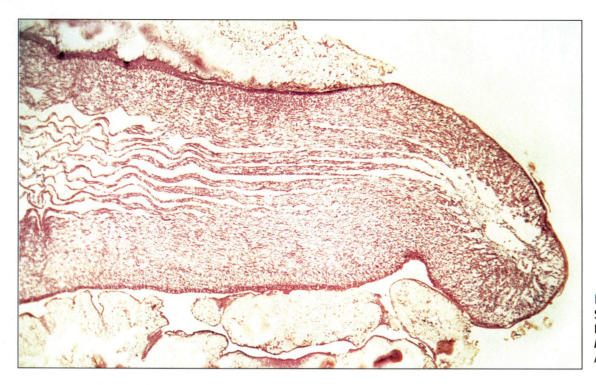

FIGURE 1.2
Sagittal section:
Proboscis of
Balanoglossus,
Acorn worm.

Microscopic organisms and particles in the water are trapped in the mucus and are directed into the mouth by the cilia. The **mouth** lies ventral and anterior to the collar and opens into the pharynx in *Balanoglossus* (Figure 1.3).

Other species ingest suspended material while in the burrow, digesting the organic material and passing the inorganic matter (e.g., sand) from their anus as a cast on the substrate, similar to earthworms. The **stomochord** projects into

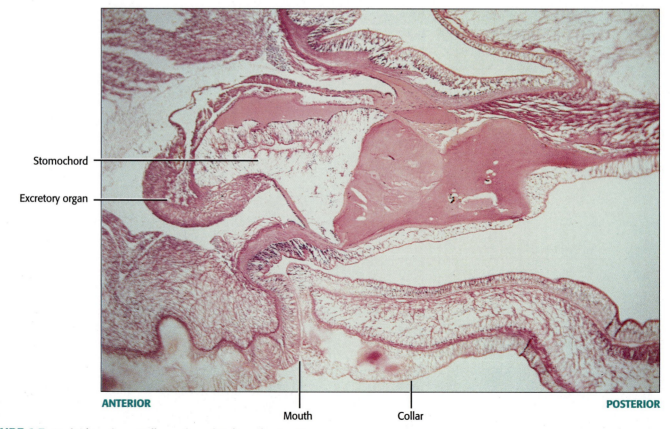

Stomochord

Excretory organ

ANTERIOR

POSTERIOR

Mouth

Collar

FIGURE 1.3 Sagittal section: Collar region of *Balanoglossus*, Acorn worm.

Pharyngeal pores

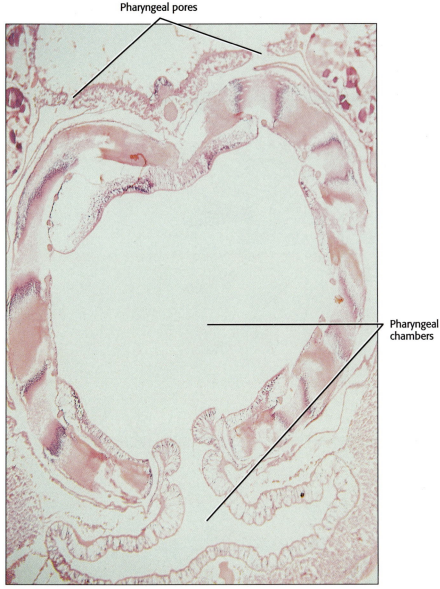

Pharyngeal
chambers

FIGURE 1.4 Transverse section: Pharynx of *Balanoglossus,* Acorn worm.

the cavity of the proboscis and is capped by a small organ that may be excretory in function (Figure 1.3).

The anterior end of the trunk is characterized by a large number of U-shaped internal pharyngeal slits that open to the outside through small, round, porelike openings, **pharyngeal pores**, allowing water to escape from the **pharynx** (Figure 1.4). Posterior to the pharynx is the simple **intestine** (Figure 1.5), where digestion and absorption take place and undigested matter is eliminated through the terminal **anus** (Figure 1.1).

The pterobranchs are a sessile group of hemichordates with a body plan similar to that of the enteropneusts. In contrast to enteropneusts, they are confined to a tubelike structure that may be part of a colony. The feeding mechanism consists of armlike structures called **lophophores**. Cilia on the lophophores assist in directing food toward the mouth (Figure 1.6).

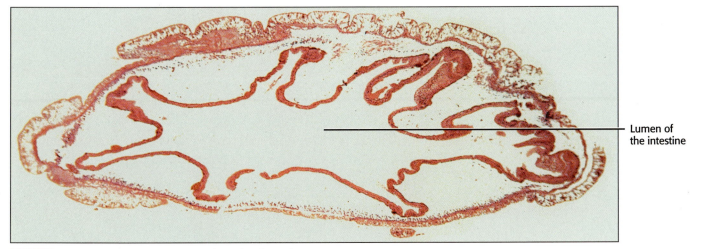

Lumen of
the intestine

FIGURE 1.5 Transverse section: Intestine of *Balanoglossus,* Acorn worm.

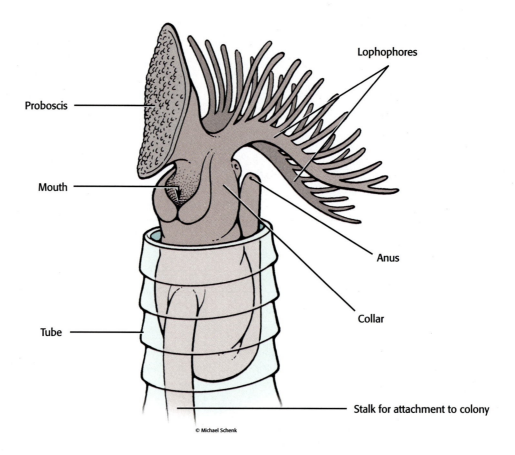

Proboscis

Lophophores

Mouth

Anus

Collar

Tube

Stalk for attachment to colony

© Michael Schenk

FIGURE 1.6 Pterobranch.

Phylum Chordata

The Non-Vertebrate Chordates

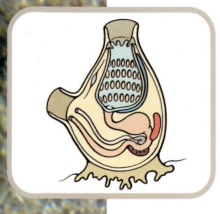

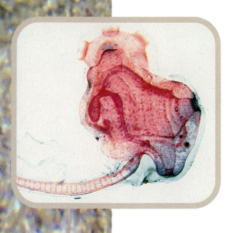

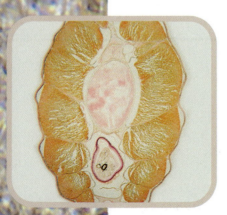

CHAPTER TWO

Subphylum Urochordata

Urochordates are marine non-vertebrate chordates consisting of three classes:

1. class Ascidiacea, the tunicates or sea squirts

2. class Larvacea

3. class Thaliacea.

The adult tunicate is sessile on substrates such as rocks and pilings. Tunicate adults are capable of producing new individuals asexually through the process of budding. Consequently, they often form aggregates of individuals. Dispersal is accomplished by a swimming larva. The adult form of the class Larvacea retains the basic morphology of the larval form and therefore is free-swimming. Members of the class Thaliacea are also **pelagic** and swim by means of jet propulsion. They also may be colonial. Because the tunicates are readily available, the following discussion is based on this group.

The adult tunicate, or sea squirt, is a sessile, filter-feeding, baglike organism with two openings (Figure 2.1).

Obtain the necessary slides to identify the following structures.

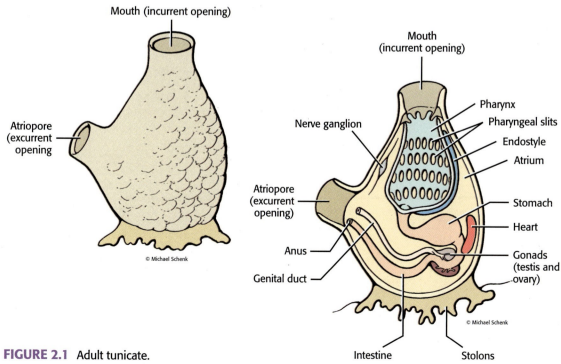

FIGURE 2.1 Adult tunicate.

The **mantle, or body wall,** contains elongated muscular fibers that aid in shortening the body during contraction (Figure 2.2). Surrounding the mantle is the loose-fitting outer covering, or **tunic.** Tunicin, a complex carbohydrate similar to cellulose found in higher plants, is a major component of the tunic. Cellulose is not a typical component of animal bodies.

The **mouth,** or **incurrent siphon,** leads into an extensive **pharynx** with **pharyngeal slits.** Along one side of the pharynx lies the thickened, deeply staining, ribbon-like structure, the **endostyle.** The pharynx leads into a short, narrow **esophagus,** which continues into a somewhat inflated **stomach** that opens into the narrow, elongated **intestine** ending with a terminal **anus** in the vicinity of the **atriopore** or **excurrent siphon** (Figure 2.2, Figure 2.3, and Figure 2.4).

Among the non-vertebrate chordates, the pharynx functions as a filter-feeding apparatus. Water currents, generated by movement of the cilia lining the inner surface and slits of the pharynx, are directed into the incurrent siphon. The water continues through the pharyngeal slits into the **atrium,** an extensive space between the pharynx and mantle, and out through the excurrent siphon (Figure 2.2). Microscopic organisms and organic and inorganic matter are contained within the water, trapped in mucus produced by the endostyle, and then moved by the cilia into the alimentary tract. There, digestion takes place and wastes are discharged through the anus.

The endostyle of urochordates also is capable of removing environmental iodine and synthesizing compounds containing the iodine in the form of mono-iodotyrosine and di-iodotyrosine. In cephalochordates and vertebrates, these two compounds combine to form thyroxine, which in vertebrates plays a significant hormonal role in controlling metabolism and tissue and sexual maturity.

In the adult tunicate, two definitive chordate characteristics—the notochord and the dorsal hollow nerve cord—are absent. The nerve cord is modified into a solid **ganglion** lying in the mantle between the siphons (Figure 2.2). Nerves extend from this ganglion to various parts of the body.

The circulatory system consists of a muscular **heart** (Figure 2.1), which contracts, propelling blood through spaces and some blood vessels. An interesting feature of the heart is that it regularly reverses its direction of contraction, causing a reversal in blood flow, perhaps ensuring that various areas of the organism are oxygenated adequately.

Reproduction is sexual in noncolonial tunicates, whereas asexual reproduction occurs primarily in colonial forms. Most tunicates are hermaphroditic, possessing both testes and ovaries, jointly called gonads, which lie near the stomach. The gonads shed gametes into **genital ducts** leading into the atrium near the anus, where they are carried into the surrounding water (Figure 2.5). In noncolonial tunicates, fertilization occurs in the water, whereas in most colonial species fertilization takes place in the body (Young, 1981).

The young tunicate hatches as a swimming tadpole-shaped larva and serves as an agent of species dispersal. During the larval stage, the urochordate

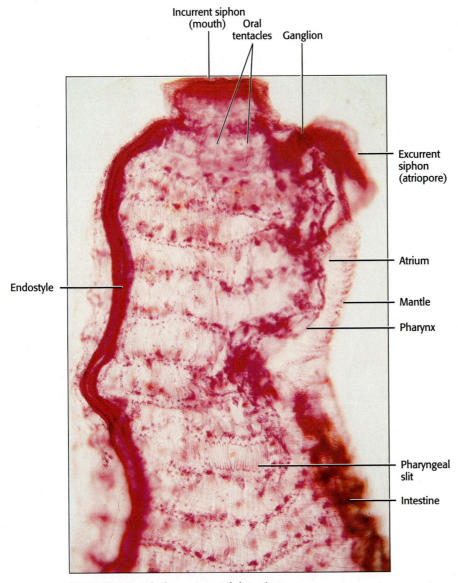

FIGURE 2.2 Whole mount: Adult tunicate.

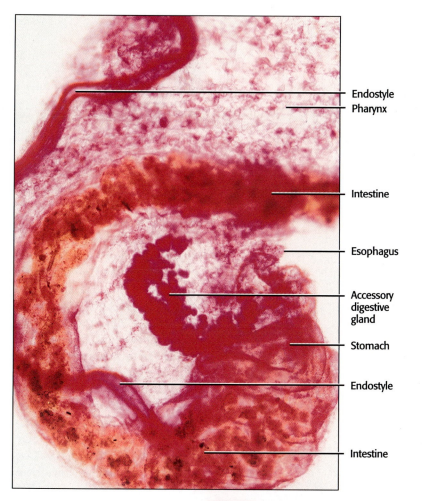

Endostyle
Pharynx

Intestine

Esophagus

Accessory
digestive
gland

Stomach

Endostyle

Intestine

FIGURE 2.3 Whole mount: Adult tunicate (digestive structures).

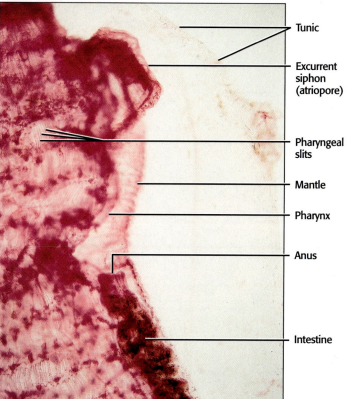

Tunic

Excurrent
siphon
(atriopore)

Pharyngeal
slits

Mantle

Pharynx

Anus

Intestine

FIGURE 2.4 Whole mount: Adult tunicate (anus/excurrent siphon).

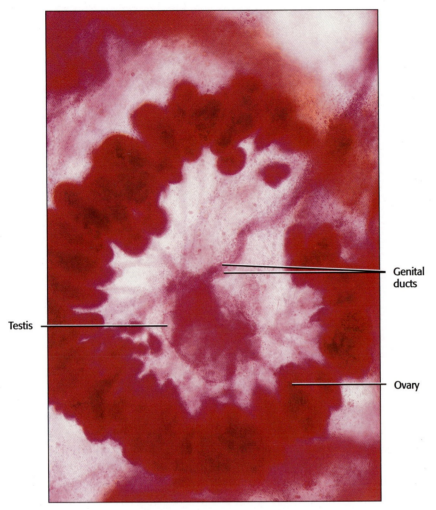

Testis

Genital ducts

Ovary

FIGURE 2.5 Whole mount: Adult tunicate (reproductive structures).

reveals its relationship to the rest of the chordates.

Obtain prepared slides of a tunicate larva. You may have to examine more than one specimen, as some may be lying at inconvenient angles on the slide.

In contrast to the adult, the larva is bilaterally symmetrical. It has a well-developed **notochord**, which appears as a series of boxlike cells extending from the tail tip to the body. The dorsal hollow **nerve cord** appears as a clear, tubelike structure dorsal to the notochord and extends from the tail tip well into the body, where it expands into a bubblelike "**cerebral vesicle.**" Muscle cells in the tail may mask the notochord and nerve cord, requiring careful focus (Figure 2.6, Figure 2.7, and Figure 2.8).

Note the two dark structures in the vesicle. One of these is the **statocyst,** which functions as an organ of balance, and the other is an **ocellus,** consisting of a pigmented cup-shaped photosensitive layer overlain by a number of clear, lenslike structures and functions as a photoreceptor (Figure 2.8 and Figure 2.9).

The larval anatomy consists of a **mouth (incurrent siphon),** an **atriopore**

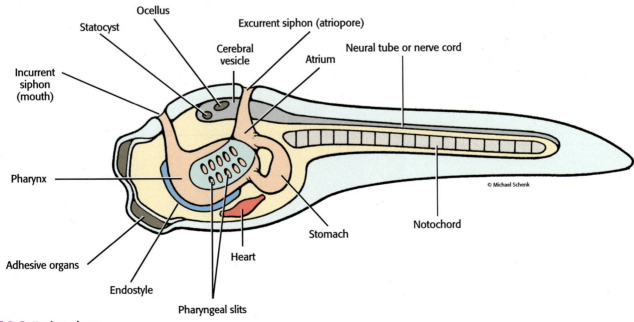

FIGURE 2.6 Tunicate larva.

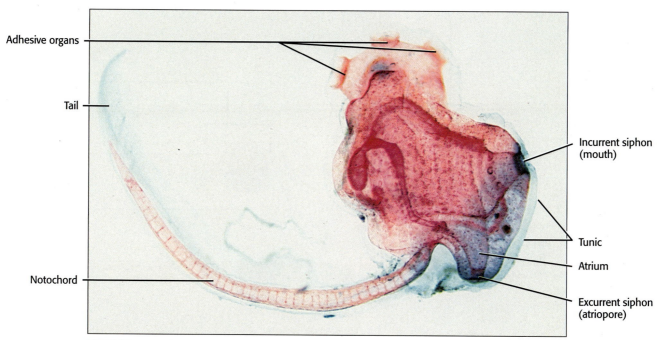

Adhesive organs
Tail
Notochord

Incurrent siphon (mouth)
Tunic
Atrium
Excurrent siphon (atriopore)

FIGURE 2.7 Whole mount: Tunicate larva.

(**excurrent siphon**), a **heart**, an **alimentary canal**, including a **pharynx with slits**, an **esophagus**, a **stomach**, and an **intestine**, terminating in an **anus** that empties in the vicinity of the excurrent siphon (Figure 2.6 and Figure 2.10). On the anterior end are **adhesive organs** (Figure 2.7 and Figure 2.8). You may be able to detect developing gonads, genital ducts, and the **cerebral vesicle**.

The larval stage exists for 24–48 hours. During this period the larva swims away from the parent(s) and surfaces because of a positive phototaxis and negative geotaxis. Shortly thereafter, the taxes reverse. The larva then seeks out a dark area at the bottom and attaches to an appropriate substratum with the adhesive organs. It resorbs its tail and metamorphoses into the sessile adult form (Young, 1981).

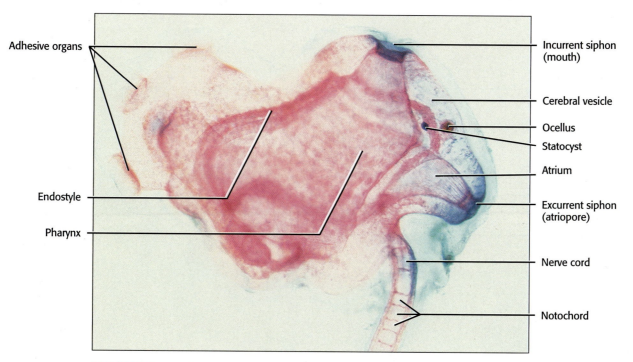

Adhesive organs
Endostyle
Pharynx

Incurrent siphon (mouth)
Cerebral vesicle
Ocellus
Statocyst
Atrium
Excurrent siphon (atriopore)
Nerve cord
Notochord

FIGURE 2.8 Whole mount: Tunicate larva (internal anatomy).

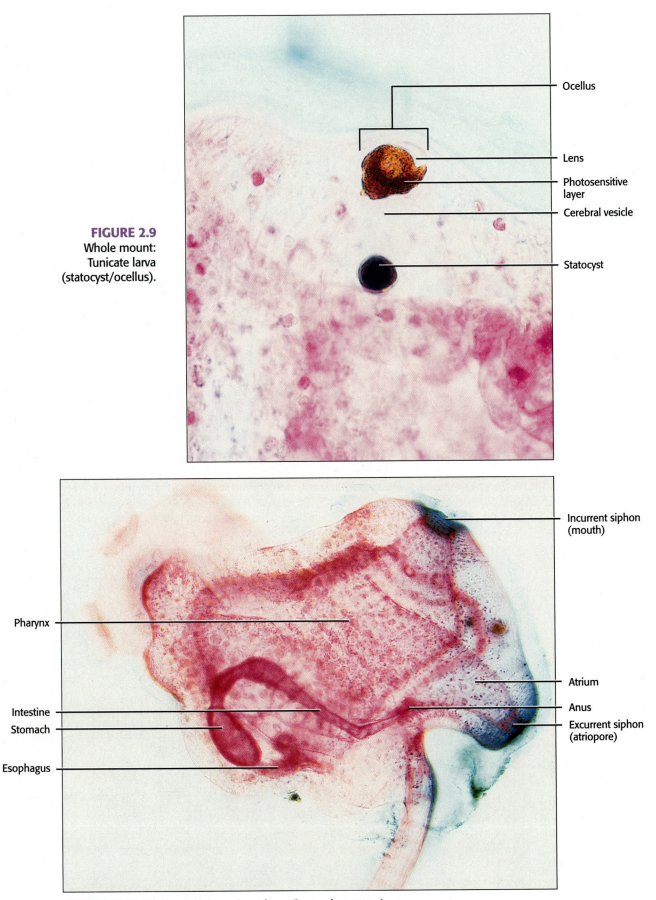

FIGURE 2.9
Whole mount:
Tunicate larva
(statocyst/ocellus).

Ocellus

Lens

Photosensitive
layer

Cerebral vesicle

Statocyst

Incurrent siphon
(mouth)

Pharynx

Atrium

Intestine

Anus

Stomach

Excurrent siphon
(atriopore)

Esophagus

FIGURE 2.10 Whole mount: Tunicate larva (internal anatomy).

CHAPTER THREE

Subphylum Cephalochordata

Cephalochordates are marine organisms that are widely distributed in shallow water on continental shelves. Although they are capable of swimming, they spend most of their adult life buried in the substratum with only their oral hoods projecting above the bottom. We we will examine in some detail Amphioxus, the common name of an excellent example of this group.

Obtain the appropriate slides, which might include a whole mount, representative cross-sections of various body regions, and male and female specimens. Several of these slides may be thick and can be damaged easily during careless use.

Among the non-vertebrate chordates, amphioxus resembles the overall vertebrate "fish" body plan the most closely. Along the dorsal surface of its lancet-shaped body runs a raised ridge, a **dorsal fin**, containing a linear series of **fin ray boxes**. A short **ventral fin**, similar in construction and extending between the atriopore and anus, is also present. The posterior end of the body terminates as a postanal **tail**. A pair of ventral **metapleural folds** extend along the ventral surface from the oral hood to the atriopore. In contrast to the fins, these do not contain supportive fin ray boxes (Figure 3.1 and Figure 3.2).

Striated muscle blocks, the **myomeres**, separated by sheets of connective tissue, the **myosepta**, extend almost along the entire length of the body (Figure 3.1 and Figure 3.2]. Forward body movement (swimming) is accomplished by an anterior to posterior serial contraction of the myomeres on one side of the body while those on the opposite side relax. Contraction and relaxation sides then are reversed. The resulting alternating sinusoidal curves move amphioxus through the water. If amphioxus has to move backward, the serial contractions begin at the animal's posterior end.

As an adult, amphioxus retains all of the diagnostic chordate characteristics—a notochord, a dorsal hollow nerve cord, a pharynx with slits, an endostyle, and a postanal tail. The **notochord** extends from the tip of the anterior end to the tip of the tail (Figure 3.1 and Figure 3.2). Its appearance is striated, and it functions as a longitudinal supporting structure that probably acts to stiffen the body, enabling amphioxus to burrow, and also acts as a force against which the body muscles can pull. The **nerve cord** lies dorsal to the notochord. In contrast to vertebrates, it is not expanded into a brain. Numerous small pigmented **photoreceptors** can be seen along the ventral portion of the cord.

An aggregation of these photoreceptors, the **eyespot**, is found at the anterior tip of the cord. In spite of its name, it does not function as an image-forming organ (Figure 3.2 and Figure 3.3). Light stimulating the photoreceptors elicits a negative phototaxis, causing the organism to respond by burrowing in the substratum.

The anterior end of amphioxus is known as an **oral hood** (Figure 3.1 and Figure 3.4). Its cavity is the **vestibule**. Associated with the oral hood are the anterior end of the pharynx and structures designed to monitor the particle size of both potential food and debris that could block the pharynx. **Buccal cirri**, extending from the edge of the oral hood, possess receptors, some of which are mechano-receptors, probably responsible for rejecting oversized particles (Figure 3.2).

Mucus is secreted by cells of **Hatschek's pit** in the anterior roof of the vestibule (Figure 3.4). Some biologists believe that this organ has some endocrine functions similar to the vertebrate hypophysis

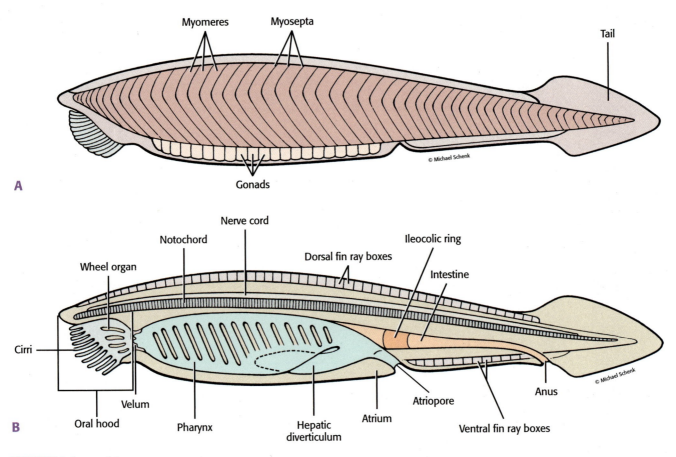

FIGURE 3.1 Amphioxus.

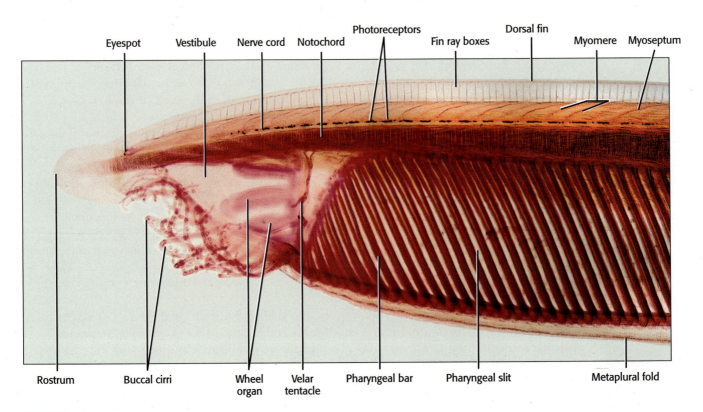

FIGURE 3.2 Whole mount: Amphioxus (anterior end).

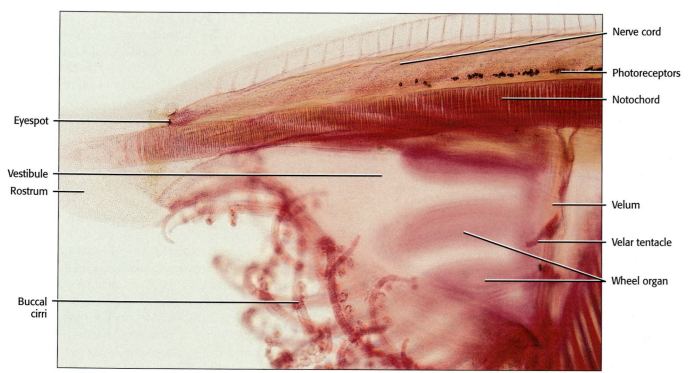

FIGURE 3.3 Whole mount: Amphioxus (anterior end—magnified).

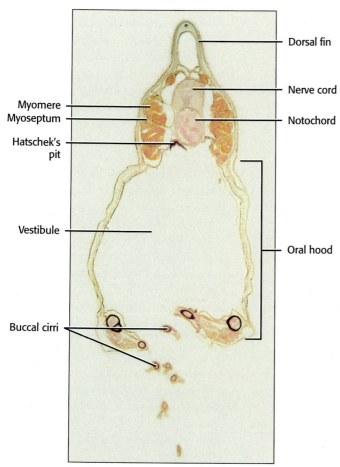

FIGURE 3.4 Transverse section: Amphioxus (oral hood).

(pituitary gland). The **wheel organ** (Figure 3.3), consisting of ciliated, fleshy, fingerlike projections, just posterior to Hatschek's pit, creates currents carrying mucous-entrapped particles and directs them into the pharynx. **Velar tentacles**, attached to a membrane, the **velum**, act as the final screening "barrier" through which particles pass into the pharynx (Figure 3.2 and Figure 3.3).

A voluminous **pharynx** with numerous **pharyngeal slits** is located posterior to the velum. A series of narrow rod-like structures, the **pharyngeal bars** (Figure 3.1, Figure 3.2 and Figure 3.5), support the walls of the pharynx. Note the small connecting **synapticulae** extending between the pharyngeal bars. The inner surface of the pharynx and slits is heavily ciliated (Figure 3.5).

The grooved, ciliated **endostyle** lies in the floor of the pharynx. It produces copious amounts of mucus and also synthesizes the iodine-binding proteins mono-iodotyrosine and di-iodotyrosine. In contrast to the urochordates, however, the endostyle further binds these compounds to produce tri-iodothyronine and thyroxine. The roof of the pharynx likewise is grooved and is known as the **epibranchial groove** (Figure 3.6). The mucus, secreted by the endostyle, traps suspended particles and is moved dorsally by the pharyngeal cilia into the epibranchial groove, where it is directed into the digestive tract.

The pharynx leads into the **midgut**. From the midgut projects a finger-like organ, the **midgut diverticulum**, on the right side of the body (Figure 3.1, Figure 3.7, and

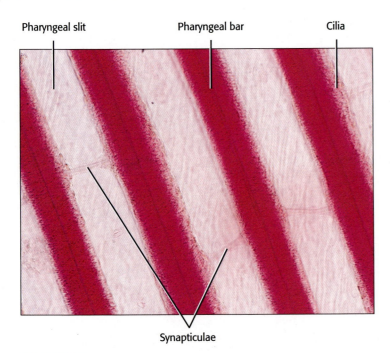

Pharyngeal slit   Pharyngeal bar   Cilia

Synapticulae

FIGURE 3.5 Whole mount: Amphioxus (pharynx magnified).

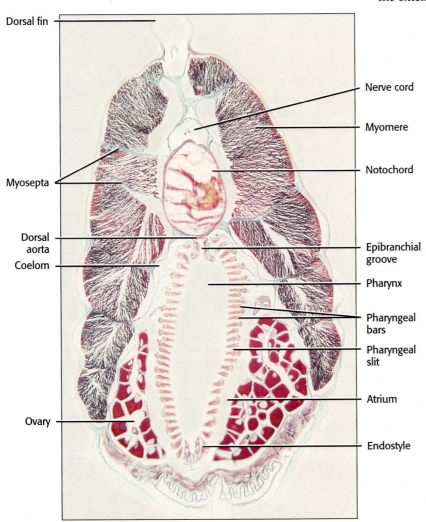

Dorsal fin

Nerve cord

Myomere

Notochord

Myosepta

Dorsal aorta

Coelom

Epibranchial groove

Pharynx

Pharyngeal bars

Pharyngeal slit

Atrium

Ovary

Endostyle

FIGURE 3.6 Transverse section: Amphioxus (female/pharynx).

Figure 3.8). This structure has been referred to as a "hepatic cecum" or "hepatic diverticulum," with the suggestion that this organ is similar to the vertebrate liver. Actually, some food circulates within the cavity of the diverticulum, where an enzymatic secretion (typical of the vertebrate pancreas) results in digestive activity. Perhaps the position of the diverticulum, similar in position to the liver of vertebrates, and the presence of glycogen and lipids, typically stored by liver cells, has contributed to its reference as a "hepatic" organ.

The **ileocolic ring**, a ciliated band at the posterior end of the midgut, moves mucus and food into the **hindgut**, where further digestion and absorption occur. Both extracellular and intracellular digestion occur in the gut. Undigested material is expelled through the **anus** (Figure 3.7, Figure 3.9, and Figure 3.10).

The **atrium**, a large space surrounding the pharynx and extending to the beginning of the *hindgut*, opens via the **atriopore** to the outside, anterior to the anus. The **coelom**, prominent in most vertebrates, is severely restricted anteriorly in amphioxus by the presence of the extensive atrium (Figure 3.6, Figure 3.8, and Figure 3.9). Posteriorly, the coelom is more prominent (Figure 3.10). Note its presence in some of the cross-sections. Water, entering via the pharynx, passes through the pharyngeal slits, continues posteriorly within the atrium, and finally is expelled through the atriopore. During the passage of water through the pharyngeal bars, some gas exchange occurs. Additionally, gas exchange occurs across the epithelium.

The closed, tubular circulatory system in amphioxus is similar to the vertebrate embryo, and the pattern of blood vessels is similar to that of the ancestral plan observed in early vertebrates. A major difference is the absence of a muscular pump or heart in amphioxus. Venous return is by way of a caudal vein, cardinal veins, and a sub-intestinal vein, with the blood emptying into a venous sinus in the approximate vicinity of the position of the vertebrate heart. From there, blood travels through the ventral aorta to the pharynx, dorsally through branchial vessels in the pharyngeal bars, and then into the paired dorsal aortae (anterior), continuing into the single dorsal aorta (posterior) supplying the tissues.

The excretory system in amphioxus is problematic. For a long time, biologists

Notochord

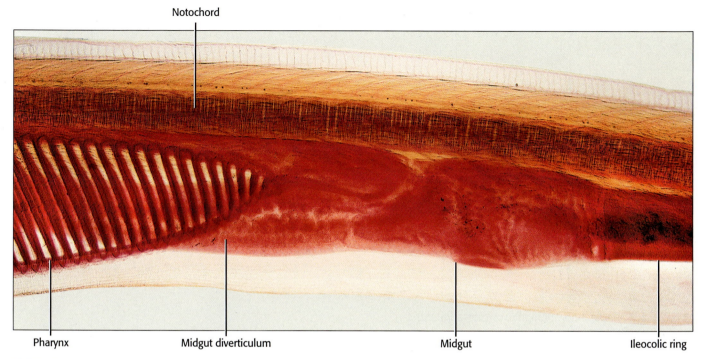

Pharynx   Midgut diverticulum   Midgut   Ileocolic ring

FIGURE 3.7 Whole mount: Amphioxus (midgut).

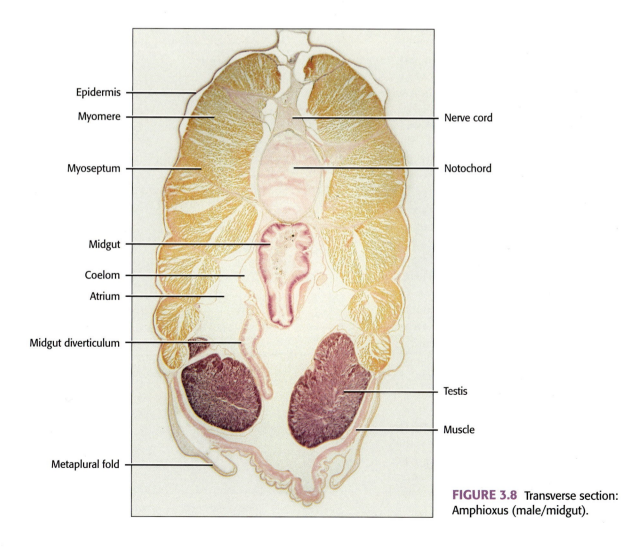

Epidermis

Myomere — Nerve cord

Myoseptum — Notochord

Midgut

Coelom

Atrium

Midgut diverticulum

Testis

Muscle

Metaplural fold

FIGURE 3.8 Transverse section: Amphioxus (male/midgut).

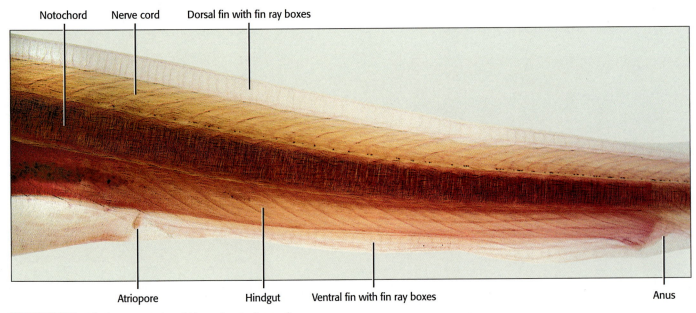

Notochord   Nerve cord   Dorsal fin with fin ray boxes

Atriopore   Hindgut   Ventral fin with fin ray boxes   Anus

FIGURE 3.9 Whole mount: Amphioxus (posterior end).

thought it was similar to that found among some invertebrates, such as the annelids. More recent research involving electron microscopy suggests that at least part of its anatomy may resemble podocytes of the vertebrate nephron and may function somewhat the same. Excretory wastes are dumped into the atrium and flushed through the atriopore.

In this subphylum the sexes are separate. In a mature amphioxus, the gonads virtually fill the body and, for that reason, only immature animals are used to make the whole mount slides you are using. The **ovaries** and the **testes** are hollow, laterally paired, serially repeated organs (Figure 3.6 and Figure 3.8). In cross-section, the ovary is recognizable, with its large eggs containing obvious nuclei and nucleoli. In contrast, the testes are oval organs appearing granular or striated. There is no system of ducts through which gametes are carried to the outside. The walls of the gonads burst, releasing mature eggs or sperm into the atrium. Gametes are carried into the environment by water flowing through the atrium and out the atriopore. Following release of the gametes, the walls of the gonads close and a new cycle of gamete development begins.

Fertilization and development occur externally. A larva resembling the adult amphioxus hatches from the egg and, after spending some time as a member of the plankton, settles on the substratum and metamorphoses into a relatively sedentary filter-feeding adult.

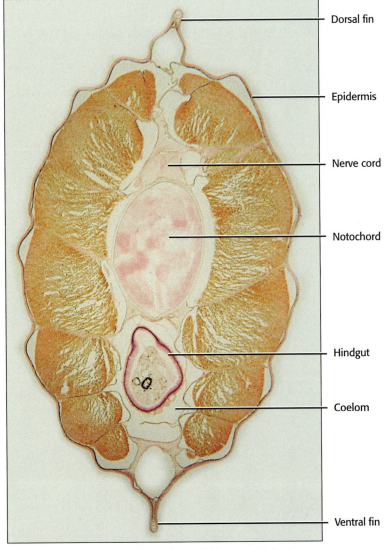

Dorsal fin

Epidermis

Nerve cord

Notochord

Hindgut

Coelom

Ventral fin

FIGURE 3.10 Transverse section: Amphioxus (hindgut).

The Evolution of Vertebrate Systems

An Overview

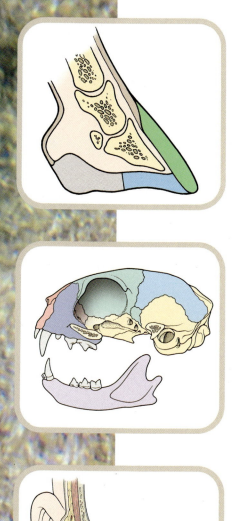

CHAPTER FOUR

Integumentary System

T he outer covering of organisms is called the integument or skin. The integumentary system serves a number of functions. The skin separates the animal from its environment, and an unbroken integument of vertebrates is an effective deterrent against the entry of bacteria and viruses. The skin also provides an effective barrier against unrestricted diffusion of water and ions between the organism and the environment.

Among vertebrates, the integument is a complex system consisting of an outer multilayered epidermis lying over a thicker dermis. Chromatophores, or pigment cells, secreting a variety of pigments, producing characteristic colors and patterns, are found in the integument.

Melanin, a common brown to black pigment in the skin of vertebrates, occurs in melanophores, a type of chromatophore. The integument also contains sensory receptors—tactile, temperature, pain, and so on, as well as blood vessels, muscles, and fat. The epidermis of vertebrates has given rise to a number of distinctive derivatives— scales, feathers, hair, sweat glands, sebaceous glands, scent glands, and so on.

SCALES

Scales are present within members of the major vertebrate groups. In the earliest vertebrates, the ostracoderms, external dermal armor in the form of plates was present over large areas of the body and dermal scales occurred in the tail region. At the base of the plates was a layer of bone, overlain with dentine, which was covered by a layer of enamel, derived from the epidermis, or in some vertebrates, enameloid, derived from the dermis. A great deal of controversy concerns the derivation of early surface layers. The surface has small projections called **denticles** (Figure 4.1A).

Among the ancestors of the chondrichthes the dermal armor apparently disappeared (was secondarily lost), leaving the surface denticles as placoid scales (Figure 4.1B).

Several types of scales are found within the bony fishes. The structure of the cosmoid scale is

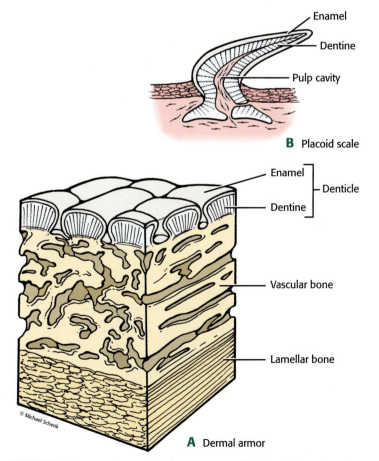

FIGURE 4.1 Primitive ostracoderm armor and derived placoid scale.

29

similar to the plate tissues found in ostracoderms, differing only in thickness and, according to some authorities, is found in *Latimeria*, a living sarcopterygian. In gars and the bichir a thick ganoid scale consists of a basal layer of bone covered with enamel or enameloid (Figure 4.2A). The shape of these scales is rhomboid, and in the gar forms a heavy, interlocked scaly covering.

Most teleosts possess thin dermal scales composed of acellular bone lacking dentine and enamel. They occur in two different configurations, cycloid and ctenoid (Figure 4.2B and Figure 4.2C). Ctenoid scales differ from cycloid scales in that they possess tiny spines or "teeth" or cteni at their posterior ends. Both are imbricated, or overlapping.

Ancestral amphibians continued the pattern of external dermal scalation. Modern amphibians universally have a moist, scaleless skin; however, some caecilians possess embedded dermal scales.

With the evolution of the amniotes, thick skins covered with epidermal scales appeared (Figure 4.3). A major feature of these scales is heavy keratinization of the epidermis. Among the squamate reptiles, snakes and lizards, the entire epidermis is shed (ecdysis) under the influence of hormones.

Reptilian scales play a critical role in mechanical and osmotic protection. In snakes, ventral scales or scutes play a part in locomotion. Turtles possess external shells that consist of an underlying bony component consisting of dermal and endochondral bone, covered by thick, horny epidermal plates. The fresh-water soft-shelled and marine leatherback turtles lack hard shells. Crocodilians have external epidermal scales underlain by dermal bony plates.

Within the turtles and crocodilians, ecdysis does not occur, but as the epidermis is worn away, it is restored gradually.

Modified epidermal structures, claws, appear for the first time among the reptiles. The jaws of some dinosaurs, birds, and turtles are covered with a keratinous beak-like structure. Among many of the dinosaur groups, epidermal scales were present and, in addition, greatly elaborated dermal plates—for example, among the stegosaurs, ceratopsians, and others. Reinforcing rib-like structures, the gastralia, are found in the ventral body regions of crocodilians, some lizards, and *Sphenodon*. They play a role in respiratory movements.

In contrast to reptiles, the skin of birds is thin and not heavily keratinized. Among modern vertebrates, birds are distinguished by the presence of feathers, a derivative of the reptilian epidermal scale. For decades, scientists considered feathers to be a unique characteristic of birds. Within the last few years, however, paleontologists have discovered a number of theropod dinosaurs that appear to have possessed feathers. The theropods include the infamous *Tyranosaurus rex*, as well as more lightly built forms that appear to include bird ancestors.

The original function of feathers has been a controversial topic for some time. One school adheres to the flight theory. Another proposes that feathers evolved parallel to the evolution of endothermism within birds. A third theory holds that feathers may have evolved to facilitate social behavior such as courtship and nesting. Feathers are shed at regular intervals, and the process (molting) is under hormonal influence.

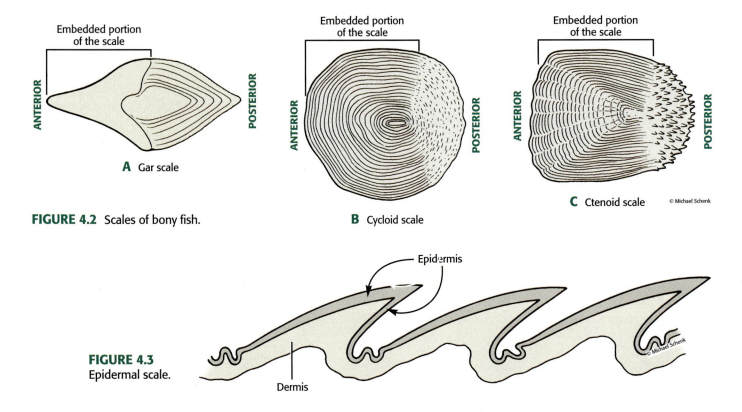

FIGURE 4.2 Scales of bony fish.

A Gar scale

B Cycloid scale

C Ctenoid scale © Michael Schenk

FIGURE 4.3
Epidermal scale.

Epidermis

Dermis

© Michael Schenk

Epidermal scales cover the legs and feet of birds. Birds are characterized by heavily keratinized beaks. Myriad adaptations for food gathering have evolved. Examples include sturdy beaks to crack seeds, found among seed-eaters (*e.g.*, cardinals); beaks with tooth-like strainers for sieving small aquatic organisms (*e.g.*, flamingos, ducks, and geese); and strongly recurved, razor-like beaks (*e.g.*, eagles, hawks, and their relatives), to cut and tear prey.

The integument of mammals, particularly the dermis, is thick. The epidermis is multi-layered, and its outer layers are highly keratinized, constantly shed, and replaced by underlying layers of cells originating in the mitotically active basal stratum germinativum.

Epidermal scales are retained in some mammals. Examples are the tails of rodents and opossums.

HAIR

Hair is a mammalian synapomorphy. These highly keratinized structures originate in the epidermis but sink into the dermis and are anchored there. In contrast to feathers, hair is not a derivative of epidermal scales. Hair in animals that possess scales, such as rodents, appears in clusters between the scales and may hint at the possible origin of hair. Among some of the synapsid reptile ancestors of mammals, hair may have occupied similar sites and perhaps played a role in mechanoreception. In addition, a heat conservatory role may have ensued (Maderson 1972).

The hair of mammals usually consists of an outer coat of coarse hairs, the guard hairs, and an inner coat of finer, "wooly" underfur next to the skin. Among mammals that possess very thick underfur, the pelage is known as fur. Hair sometimes appears in a modified form—for example, the quills of porcupines.

Hair patterns in mammals often have evolved as distinguishing features. Tufts of coarse hairs known as vibrissae, or whiskers, are positioned on the snout and serve as mechanoreceptors. Other examples are eyelashes, eyebrows, nasal hair, manes, tails, and axillary and pubic hair in humans. These provide functions such as protection, filtration, warding off insects, and social displays. Hair has been reduced in a number of mammals, elephants, rhinoceroses, hippopotamuses, and humans, and is almost nonexistent in cetaceans. In mammals, hair generally is shed once or twice a year. Perhaps you may have noticed the process in your pet or in horses, particularly in the spring.

CLAWS

Well-developed claws attached to the terminal phalanx appear among reptiles. They are derivatives of the stratum corneum. The outer, tough covering of claws, hooves, and nails is called the unguis and the underlying surface of these structures is known as the subunguis. A few modern amphibians, *e.g.*, the African Clawed Toad, *Xenopus*, possess claws, but almost universally, amphibians do not have claws.

In birds, the toes terminate in claws, which are modfied in some birds, *e.g.*, hawks, owls, and eagles, for predation (Figure 4.4A). Laterally compressed claws are common in a wide variety of mammals. Some are retractile, such as those

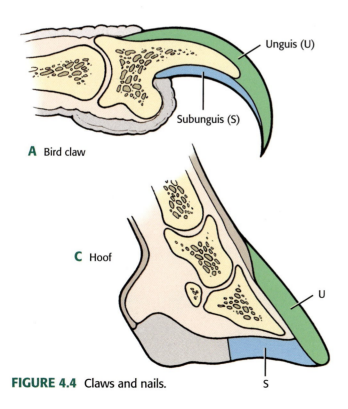

A Bird claw

Unguis (U)

Subunguis (S)

C Hoof

U

S

FIGURE 4.4 Claws and nails.

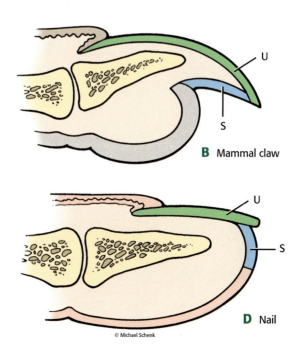

B Mammal claw

U

S

D Nail

U

S

© Michael Schenk

found in most cats, and others are not, such as those of dogs (Figure 4.4B). In primates, claws have become dorso-ventrally flattened nails (Figure 4.4D). Claws among the ungulates have evolved into enlarged structures known as hooves (Figure 4.4C). Claws and their derivatives serve a number of functions, some of which are traction and loco-motion, food gathering, self-defense, and digging.

HORNS, ANTLERS, AND OTHER DERIVATIVES

In many mammals a variety of cranial epidermal derivatives have evolved. Several distinct groups of mammals are recognized because they possess projections called horns and antlers. The non-shedding horns of sheep, cattle, goats, buffaloes, bison, and African antelopes consist of an inner

bony core covered by a thick, keratinized sheath (Figure 4.5). In contrast, the pronghorn "antelope," an American species, has a horn with a similar construction but sheds the sheath (Figure 4.5). The non-shedding horn(s) of the rhinoceros consists of a tightly compressed group of keratin fibers. Horns are found in both males and females and function primarily for defense and in social behavior (Figure 4.5).

An entirely differently-constructed head ornament is the antler (Figure 4.5). It consists of a dermal bony projection attached to the skull and is shed annually. When antlers first appear in the spring, they are covered with a plush-like hairy skin known as "velvet." During the ensu-ing months the velvet is worn off, leaving the exposed bone of the functional antler, which is lost during the winter months. With the exception of caribou, antlers are found

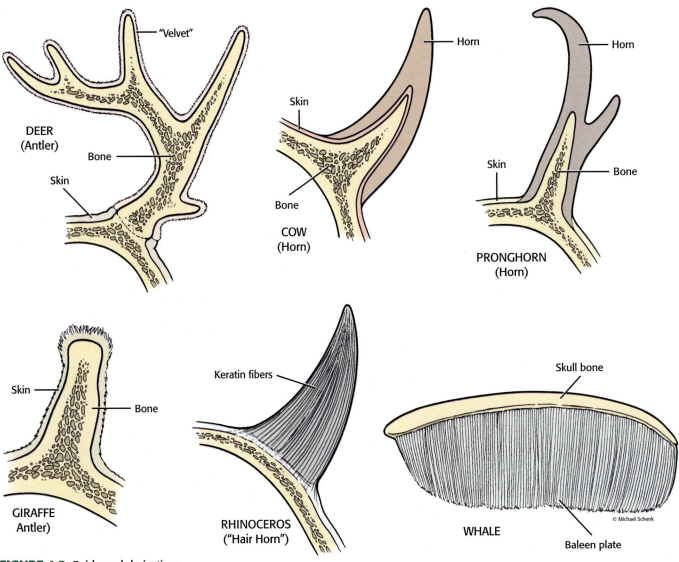

FIGURE 4.5 Epidermal derivatives.

only in males. Antlers generally are branched, with extra spikes being added each year.

Non-shedding giraffe "horns" are bony projections covered by skin. They actually are antlers and are found in both sexes. The functions of antlers are similar to those of horns.

The baleen of toothless whales represents yet another epidermal modification (Figure 4.5). Baleen consists of compacted keratin sheets that are suspended from the lateral aspects of the palate and are utilized in filtration of small crustaceans (krill) in the plankton.

GLANDS

The scaly skin of fish has evolved with numerous mucous glands, which produce large amounts of mucus. This compound protects the body surface from bacterial, viral, and fungal invasion and also has been discovered to facilitate the smooth flow of water layers over the body during locomotion.

Amphibian skin is unique in that it functions as a respiratory organ in addition to its other functions. Because respiratory surfaces must be kept moist to allow gaseous diffusion, mucous glands are found throughout the skin. Several groups of amphibians—for example, the poison dart frogs—produce virulent toxins from modified mucous glands.

With the evolution of the largely terrestrial amniotes, mucous glands were eliminated from the skin. Reptiles have few integumentary glands, and those present are a variety of scent glands found in various areas of the body. Their function is species identification and sexual attraction.

Bird skin, likewise, is largely devoid of glands. A notable exception is the uropygial gland found on the dorsal surface of the rudimentary tail. It secretes oil that is distributed by the bird over the feathers, waterproofing them.

Unlike reptiles and birds, the mammals possess several types of integumentary glands, of which two prominent examples are sweat and sebaceous glands. Sebaceous glands are oil-producing glands generally opening into the hair follicles and lubricating the hair and skin. Two kinds of sweat glands, apocrine and eccrine, are found in the skin of mammals. Apocrine glands are generally distributed over the body surface of most mammals and are associated with hairs. In the human they are concentrated in the axillary and ano-genital areas. Secretions of apocrine sweat glands are responsible for what humans refer to as a specific mammal's body odor.

Various scent glands are derived from modified apocrine glands and play a significant role in the social life of mammals. Commonly, the organic secretions of these glands are employed during territorial marking. Examples of information that can be elicited by sniffing the secretions of scent glands are sex, rank, and reproductive condition.

Eccrine glands are not associated with body hair but, instead, open directly onto the skin surface. Their secretion is an aqueous solution containing sodium chloride, urea, and some other waste compounds. In most mammals these glands are confined to the foot pads and muzzle. In primates, sweat glands are widespread over the surface of the body. Evaporation of sweat cools the body. Various types of modified sweat and/or sebaceous glands have become associated with the eye and ear of mammals, giving rise to lubricating secretions bathing the eye and ear wax in the outer ear canal.

In reptiles and birds whose diets include heavy salt loads —marine species—various glands, such as nasal, lacrimal, and oral, have become modified as salt-secreting organs.

Unique to mammals are mammary glands, upon which the class name is based. The evolutionary history of these glands is controversial. Some popular theories propose that these glands developed from apocrine sweat glands, sebaceous glands, or as Blackburn (1991) has suggested, that mammary glands arose *de novo* of dual origin from both.

Just think about drinking modified sweat with a little grease thrown in, the next time that you have a glass of milk! The milk is released from mammary glands from specific structures called nipples, or in the case of ungulates, from teats. These glands occur in both sexes and may extend along the entire abdominothoracic region or be confined to certain areas of the ventral body, such as, axillary, inguinal, or pectoral. In monotremes, the milk is secreted onto the surface of the skin and is licked from overlying hairs by the young. This may have been the original method of milk ingestion among ancestral mammals.

PIGMENT

A common integumentary inclusion among all vertebrates consists of chromatophores, or pigment cells. They may be found in the dermis and also in the epidermis. They are derived from neural crest cells, a type of embryonic cell found in craniates.

Melanin is a common black or brownish pigment found in melanophores of almost all vertebrates. Other types of pigments are the red pigment in erythrophores, yellow pigment in xanthophores, or usually crystalline guanine, a light-scattering colorless pigment found in iridophores, which produces an iridescent or silvery effect. The blue color in vertebrates is due to a scattering phenomenon, whereas green results from the presence of yellow pigment overlain by guanine, which produces a scattering phenomenon. The combination of yellow and blue associated with the scattering phenomenon yields green.

Because feathers and hair, found in birds and mammals, respectively, are epidermal derivatives, it should not be surprising that pigments in these two groups of vertebrates are found not only in the skin but also in these structures. By contrast, the pigments in fish, amphibians, and reptiles are found in the skin.

Pigmentation in vertebrates serves a number of functions, such as camouflage, warning, courtship, and sex identification. In some vertebrates rapid color or pattern changes occur under the influence of the endocrine and nervous systems.

Among some vertebrates coloration in specialized areas is due to the shallow distribution of blood vessels. Examples are the red faces and rumps of some monkeys, and apes and the cheeks and lips of humans.

CHAPTER FIVE

Skeletal System

Some type of skeletal support has evolved repeatedly in animals. A skeleton is necessary to lend shape to the body, as well as to provide a mechanical framework against which muscles can pull to produce movement of the body.

Among vertebrates, portions of the skeleton protect the brain and spinal cord and thoracic and pelvic organs. The skeleton houses hemopoietic (blood-forming) tissue and also is a depository for calcium and phosphorus, two elements that are essential in many chemical reactions in cells.

Among the earliest vertebrates, the ostracoderms, the skeleton consisted of an external articulated system of bony plates covering the anterior body region and heavy scales clothing the remainder (See Chapter 4). This external skeleton was produced in the dermis and, therefore, consisted of dermal bone. There was no internal bony skeleton or any paired fins. A functional notochord was the main internal axial support.

In later ostracoderms, both an extensive external dermal skeleton and an internal endochondral skeleton were present. Endochondral bone is formed within a cartilaginous facsimile or model of the bone. This combination of dermal and endochondral skeleton persisted in the placoderms, osteichthyes, sarcopterygii (tetrapod ancestral line), and tetrapods.

The general trend among vertebrates was increased emphasis of the endochondral skeleton, accompanied by a dramatic reduction of the dermal skeleton, especially among tetrapods. In modern fishes the thick scales became thin and overlapping, and they were lost in the tetrapods. Some dermal bones became associated with endochondral bones and remained as important components of the skull and portions of the shoulder girdle among the osteichthyes and tetrapods.

Tetrapod skeletons can be subdivided into two basic divisions:

1. the axial division, consisting of the skull, vertebral column, sternum, and ribs, and

2. the appendicular division, consisting of the skeletons of the appendages and their girdles.

Historically, in the tetrapods the two divisions have different emphases in function. Housed within cavities in spaces surrounded by much of the axial division are soft tissues, commonly called "vital organs." Among them are the brain, spinal cord, sense organs, heart, respiratory organs, and others. Much of the appendicular division functions during postural and locomotory activities. The surface of the entire skeleton serves as an attachment site for muscles. This is the primary function of the appendicular division.

SKULL

The skeleton of the head, better known as the skull, is a complex consisting of:

1. the endochondral chondrocranium, derived primarily from neural crest tissue, supporting the brain and its associated sense organs;

2. the endochondral splanchnocranium, derived from neural crest tissue, supporting the gills and giving rise to various structures associated with the head and neck;

3. the dermatocranium, derived from neural crest tissue and mesoderm of somite origin, forming in the skin and encasing elements of the chondrocranium, the splanchnocranium, and also giving rise to independent bones of the skull.

CHONDROCRANIUM

The cartilaginous chondrocranium, or shelf on which the brain and its associated sense organs rested, was already well developed in the ostracoderms, the earliest vertebrates to appear in the fossil record. The basic plan of development of the chondrocranium is similar in the embryos of all jawed vertebrates (Figure 5.1). Associated with the **notochord** is a pair of cartilaginous bars, the **parachordals**, and a second pair of bars, the **trabeculi**, which develop anterior to the parachordals. Three pairs of capsules—a nasal (olfactory) pair associated with the forebrain, an **optic** pair associated with the midbrain, and an **otic** pair associated with the hindbrain—contribute to the development of the chondrocranium. The trabeculi fuse at the cranial end forming the ethmoid plate leaving an opening in which the hypophysis rests. The parachordals fuse incorporating the cranial portion of the notochord into a basal plate. The nasal capsules unite with the ethmoid plate and the otic capsules unite with the basal plate. Eye movement precludes fusion of the optic capsules with the rest of the chondrocranium. They form the tough sclera of the eyes.

With the exception of the optic capsules, all of the components fuse to form a cohesive unit complete with walls and roof. The presence of cranial nerves and blood vessels results in characteristic foramina in the chondrocranium. The foramen magnum forms around the spinal cord extending from the hindbrain. A number of vertebral arches contribute to the formation of the occipital arch. Associated with the occipital region are occipital condyles, which articulate with the first vertebra.

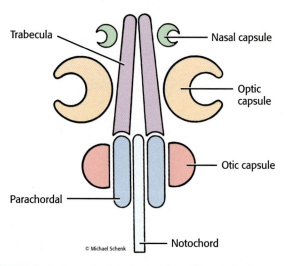

FIGURE 5.1 Embryonic chondrocranium (dorsal view).

Some portions of the chondrocranium of the agnatha—for example, the parachordals and sense organ capsules—are homologous with those of the gnathostomes and are recognizable as such. Other independent cartilages, however, are questionable.

SPLANCHNOCRANIUM

The skeleton of the gills is known as the splanchnocranium. Among the earliest vertebrates that were agnathous, the gills were numerous. Each pair of gills was supported by a visceral arch alternating with gill slits leading into gill pouches (Figure 5.2A). In even the most primitive of modern

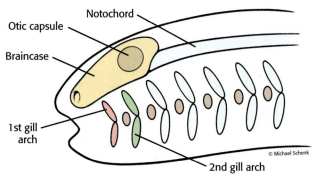

A Splanchnocranium of early jawless fish

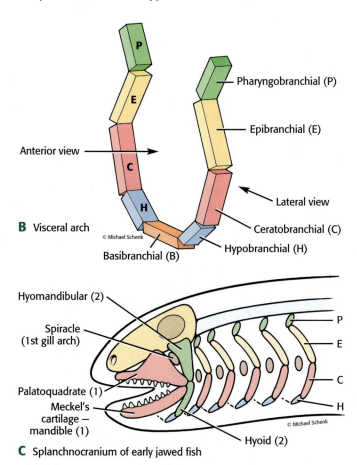

B Visceral arch

C Splanchnocranium of early jawed fish

FIGURE 5.2 Jaw development.

agnathous vertebrates, the number of gills has been reduced to seven, and this remains the basic number among vertebrates. Hence, the number of visceral arches is seven. Each of the visceral arches consists of four segments, a **pharyngobranchial**, an epibranchial, a **ceratobranchial**, and a **hypobranchial**, arranged in a W-shaped configuration and attached to a centrally located **basibranchial** (Figure 5.2B).

JAWS

Alfred Sherwood Romer, an eminent 20th-century vertebrate paleontologist, stated that the evolution of jaws in vertebrates was a monumental evolutionary event. With the evolution of jaws came the advent of carnivory, herbivory, omnivory, the ability to alter the environment, build nests, grasp mates, punish territorial interlopers, develop parental behavior, and so on. The vertebrate line probably would have languished or become extinct without jaws!

The earliest vertebrates were jawless, relying primarily on microscopic food supplies gleaned from the respiratory stream of water produced by muscular contraction of the pharynx. Two ancient lines of jawed vertebrates, the acanthodians and the placoderms, appear in the fossil record of the Ordovician and Silurian periods of the Paleozoic era, respectively.

For several generations the derivation of the jaw skeleton was contentious. One school proposed that the first pair of visceral arches gave rise to the palatoquadrate and Meckel's cartilage. Two other theories held that they arose from a more posterior pair or from supporting elements associated with a velum.

Recent research suggests that the origin lies in an entirely different direction. Cerny, *et al.* (2004) discovered that as neural crest cells migrate into the first pharyngeal arch, they form an upper "maxillary" and lower "mandibular" condensation. The intriguing fact is that both the palatoquadrate and Meckel's cartilage form from the "mandibular" condensation and the trabeculae cranii and part of the frontonasal process, elements of the neurocranium form from the "maxillary" condensation.

Among early jawed vertebrates such as the placoderms, the palatoquadrate was fairly firmly attached to the chondrocranium by means of ligaments, and articulation occurred between the posterior portions of the upper and lower jaws, the quadrate and the articular, respectively. The hyoid was not involved in jaw suspension. This type of suspension is known as autostylic (Figure 5.3A).

In other groups of fishes, early cartilaginous and bony fishes and acanthodians, a second type of suspension known as amphistylic, evolved. The palatoquadrate was associated with the chondrocranium via a number of articulating points, and the hyomandibula of the hyoid arch was employed as a jaw prop extending from the otic region of the chondrocranium to the posterior portion of the palatoquadrate (Figure 5.3B).

Among most modern chondrichthyes and osteichthyes a third type of jaw suspension, hyostylic, occurs (Figure 5.3C). The palatoquadrate is loosely associated with the chondrocranium, and the major suspension is via the

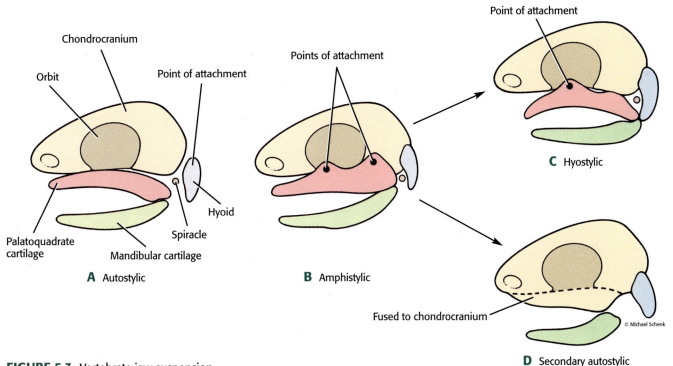

FIGURE 5.3 Vertebrate jaw suspension.

hyomandibula, enabling these fishes to project their jaws away from their crania. This facilitates tearing large pieces of flesh from prey, as in the case of sharks. By contrast, suction feeding, common among many bony fishes, is facilitated by a tube-like mouth made possible by the mobile hyomandibula.

Another version of the autostylic jaw suspension evolved among tetrapods (Figure 5.3D). The palatoquadrate became solidly associated or fused with the chondrocranium. The hyomandibula no longer played a role either as a jaw prop or as a suspensory element. Its position qualified it as an ideal candidate to act as a vibration or sound transmission structure from an outer membrane, the tympanum, to the inner ear. Vibrations and sound are easily transmitted in an aqueous environment but air is resistant to these transmissions. The hyomandibula was transformed into just such a transmission structure, the stapes in the middle ear. Among all vertebrates except the mammals, the point of articulation of the jaws occurs between the quadrate derived from the upper palatoquadrate, and the articular derived from the lower Meckel's cartilage.

The new jaw suspension, a mammalian synapomorphy, between the **squamosal** and **dentary**, released two other jaw elements, the **quadrate** and the **articular**, from their duties of jaw suspension among nonmammalian vertebrates. In mammals the quadrate was transformed into the **incus** and the **articular** into the **malleus**, both also associated with the middle ear (Figure 5.4). Within the mammalian middle ear the **stapes** articulates with the oval window of the inner ear and with the incus, which in turn articulates with the malleus, which abuts the inner surface of the tympanic membrane. Vibrations originating from airborne sound cause corresponding vibrations of the tympanic membrane, which are transmitted without degradation to the malleus-incus-stapes-oval window. Vibration of the oval window creates waves in the fluid-filled tubes of the cochlear canal and duct, which in turn causes deformation of hair cells within the duct, creating impulses that travel in the stato-acoustic nerve to the brain, where they are interpreted as sound.

TEETH

A distinguishing feature of the skull is the teeth. The earliest vertebrates were jawless and, therefore, toothless (edentulous). In modern jawless fish, the lampreys and hagfish, teeth are found on the tongue and in the mouth cavity. Because they are ectoderm in origin, they are not considered true teeth. A true tooth consists of tissues derived from mesoderm and ectoderm.

Although placoderms had jaws, they remained edentulous. The margins of the jaws often possessed bony plates

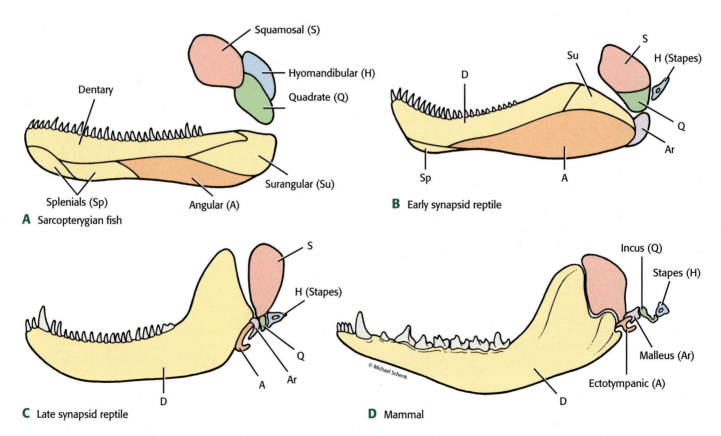

A Sarcopterygian fish

Squamosal (S)
Hyomandibular (H)
Quadrate (Q)
Dentary
Splenials (Sp)
Angular (A)
Surangular (Su)

B Early synapsid reptile

S
Su
H (Stapes)
D
Q
Ar
Sp
A

C Late synapsid reptile

S
H (Stapes)
Q
Ar
A
D

D Mammal

Incus (Q)
Stapes (H)
Malleus (Ar)
Ectotympanic (A)
D
© Michael Schenk

FIGURE 5.4 Evolution of the jaw articulation and ear ossicles of mammals. Lateral views. In the sarcopterygian fish, the articular (ar) is located medially and cannot be seen in a lateral view.

derived from dermal bone and served functions similar to teeth. The teeth of some acanthodians were associated with their jaws. Embryological and fossil evidence suggests that teeth evolved from the denticles of the dermal armor (Figure 5.5A). The anatomy of **denticles** is similar to the anatomy of teeth, with an inner **pulp cavity** containing blood vessels and nerves, surrounded by a layer of **dentine** and capped with **enamel**.

Teeth in bony fish are found not only along the jaw margin but also on the surface of the oral cavity, as well as on the inner surfaces of gill bars. The majority of bony fish can be classified as piscivorous, with teeth that are conical and similar in construction, a condition known as homodont. Some bony fish have specialized structures related to their food habits—*e.g.,* flat plates in lungfishes for crushing exoskeletons of invertebrates, and serrated triangular teeth of characins, a familiar member of which is the piranha.

The teeth of most sharks are serrated and triangular, occurring in a series of rows that continuously replace teeth that are lost in the functional set on the margin of the jaw. The chondrichthyes feed on a variety of prey. Many sharks are capable of eating large prey, and with their ability to protrude their jaws and shake their bodies from side to side, they are able to wrest large chunks of flesh from their prey. This serves to disable prey and also leads to great loss of blood, hastening its death. The largest sharks, the basking and the whale sharks, like the largest marine mammals, the whales, are filter feeders, living on small and microscopic invertebrates.

The invertebrate diet of skates and rays is correlated with their flat crushing plates. The males in many species of benthic rays possess cusped teeth, which play an important role in copulation, during which the male clamps his jaws on the pectoral fin of the female. Again, the largest ray, the

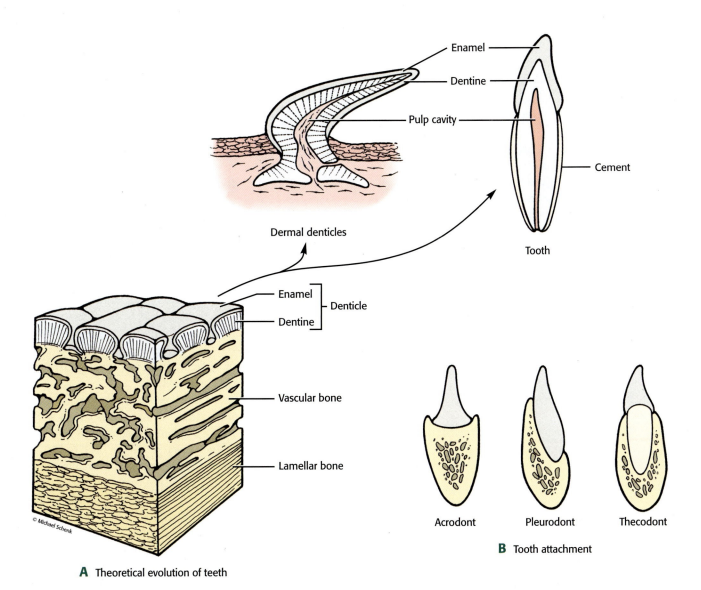

A Theoretical evolution of teeth

B Tooth attachment

FIGURE 5.5 Tooth evolution and attachment.

manta, like the largest whales and sharks, is a filter feeder that feeds on suspended invertebrates.

Distant relatives of cartilaginous fishes are the chimaeras. As invertebrate feeders that feed on mollusks and sea urchins, they have fused plates to deal with the resistant skeletons of their prey.

Various tooth attachments occur among the vertebrates. Two superficial attachments are acrodont, in which the tooth sits on the jaw rim and occurs among most fish, and *pleurodont,* in which the tooth is attached on the labial side of the jaw and is found in anuran amphibians, salamanders, and many lizards. In the third type of attachment, thecodont, the teeth are set into bony sockets and held in place by ligaments. This type is found in crocodilians, dinosaurs, primitive birds, and the therapsid reptiles, ancestors of mammals, and mammals (Figure 5.5B).

Early tetrapods possessed distinctively constructed conical teeth with elaborate folds and formerly were known as labyrinthodonts. Research has revealed that this type of tooth was found among early amphibians and amniotes and, therefore, retains little meaning. The crossopterygian fish from which the tetrapods evolved, however, also possessed this type of tooth. The modern amphibians have simple conical teeth with a weakened base, referred to as pedicillate teeth. Amphibian diets are mainly invertebrate.

The teeth of modern reptiles are basically conical. Among snakes, certain teeth had a tendency to evolve with grooves on the surface, which in some groups became completely enclosed within the tooth, giving rise to hypodermic needle-like teeth in conjunction with the evolution of venom-producing glands. In one genus of lizards, the Gila Monster, and its relative the Mexican Beaded Lizard, grooved teeth associated with venom-producing glands have evolved.

Like other vertebrates, the tooth design in dinosaurs reflected the diet of the group. Meat eaters possessed dagger-like, often serrated teeth, whereas plant eaters possessed peglike or platelike teeth, which offered a large surface area for grinding.

An interesting phenomenon is common to a number of reptiles: Modern turtles and some dinosaurs and birds have lost their teeth. It seems likely that the evolution of the beak that is peculiar to these groups occurred as the teeth were eliminated.

Beaks among birds generally reflect their food habits. Birds cannot use their anterior limbs for food handling but, rather, use their beaks to manipulate food. Therefore, a variety of sizes and shapes have evolved.

Tooth replacement among non-mammalian vertebrates is continual—i.e., as a tooth is lost, it is replaced. This condition, known as polyphyodont, continues throughout the animal's lifetime.

A distinguishing feature of mammals is their teeth. From the basic conical teeth of reptiles, four types of teeth evolved in mammal-like reptiles and their descendents.

The mammals gave rise to the heterodont dentition, in which tooth shape relates to function. Mammalian teeth are set into bony sockets, hence known as thecodont teeth. Most mammals are diphyodont, meaning that during their lifetime they have two sets of teeth. A juvenile deciduous set consists of incisors, canines, and premolars, which function for a varying amount of time in young mammals, then are replaced tooth for tooth with the addition of molars in adults as a permanent set. Molars are the only teeth in mammals that are not replaceable.

The teeth of carnivores evolved into anterior chisel-like incisors adapted for nipping; elongated, conical canines for stabbing and holding struggling prey; and molariform teeth, subdivided into premolars and molars for shearing and grinding. Further adaptation in carnivores occurs between the fourth upper premolar and first lower molar, permitting them to deal with tendons and ligaments as well as to crack bones. These teeth are called carnassial teeth.

Herbivores and omnivores employ large surfaces to grind vegetation. The upper canines and incisors of many herbivores (*e.g.,* bovines) are lost, creating a space known as a diastema. A thickened pad of tissue has replaced these teeth, and the lower teeth pressed against the pad permit a tearing function as the head is lifted. Therefore, the molariform teeth are much better developed than in the carnivores. Many mammals exhibit unique tooth specialization for feeding and reproductive functions—the long canines of walruses used to pry mollusks from the sea floor, the blade-like incisors of vampire bats to create wounds from which to feed, the canines of boars and male horses used during competition for females, etc.

Two types of whales have diverged in response to their food habits. The toothed whales possess teeth that are conical and undifferentiated or homodont, and they eat fish and other large prey. Because these whales have only a single set of teeth, the teeth are referred to as monophyodont. Baleen whales do not have teeth but, instead, have keratinous sheets of baleen growing from the epithelium lining the palate (see Figure 4.5). The baleen functions as a straining apparatus permitting baleen whales to consume crustaceans and other planktonic organisms.

The custom of indicating the number and position of mammalian teeth originated with systemetists who were interested in establishing a system of identification from skulls alone, *i.e.,* one only has to have the skull to identify most species of mammals. The presence of teeth in the skull facilitates identification, but even if they are missing, the identification often can be made. Hence, dental formulae were established.

A dental formula is a fraction in which the numerator represents the teeth in half of the upper jaw and the denominator represents half of the teeth in the lower jaw. Number and tooth position indicate tooth type in the formula, beginning with the incisors, *i.e.,* I/I, C/C, P/P, M/M.

The dental formula of primitive placental mammals was 3/3, 1/1, 4/4, 3/3. In bovine mammals (*e.g.*, the cow), the formula is 0/3, 0/1, 3/3, 3/3. Cats are represented by the formula 3/3, 1/1, 3/2, 1/1. The human dental formula is 2/2, 1/1, 2/2, 3/3. The dental formulae reflect food habits.

DERMATOCRANIUM

The earliest fossil agnatha appeared with a well-developed external skeleton produced in the skin. It consisted of thick anterior plates covering the head, the dermatocranium, enclosing the chondrocranium and splanchnocranium and, in some groups of "ostracoderms," extending to the pectoral region. Extending posteriorly from the anterior complex of plates was a covering of thick, non-overlapping scales.

The dermatocranium of later, yet primitive fish can be grouped into a related series of bones: (1) a skull roof covering the dorsal and lateral portion of the head, including the upper jaw, *e.g.*, the **parietals**, the **frontals**, the **nasals**, the **premaxillae**, and **maxillae**; (2) a group of bones covering the ventral surface of the chondrocranium, or palate, *e.g.*, **vomers**, **palatines**, **pterygoids**, **ectopterygoids**, and the central **parasphenoid** completing the primary palate; (3) groups of **opercular** and **gular** bones (branchiostegals) covering the respiratory region; (4) a series of bones encasing Meckel's cartilage, *e.g.*, the **dentary**, the **angular**, the **splenial**, and the **articular** forming the lower jaw (Figure 5.6A and Figure 5.7).

A series of bones associated with the pectoral girdle in fish, the **posttemporal** and a number of **cleithral** bones, unite the cranium with the rest of skeleton via the pectoral girdle (Figure 5.6 and Figure 5.17A).

With the evolution of the tetrapods, the skull became detached from the pectoral girdle as a result of the loss of the posttemporal and cleithral bones. For the first time, vertebrates developed necks, which allow more sophisticated head movements. During this revolution the opercular and gular bones that protected the delicate respiratory apparatus in the fishes and the fish ancestors of the tetrapods also disappeared because of the employment of lungs in the terrestrial environment. In addition, the number of anterior facial bones was reduced (Figure 5.6B and Figure 5.6C).

The primary palate persists among tetrapods; however, among some reptiles (*e.g.*, crocodilians and some lizards and all mammals), a shelf-like

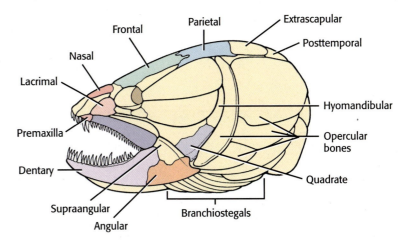

A *Amia*

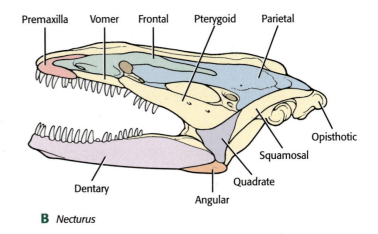

B *Necturus*

C Cat

FIGURE 5.6 Dermatocranium.

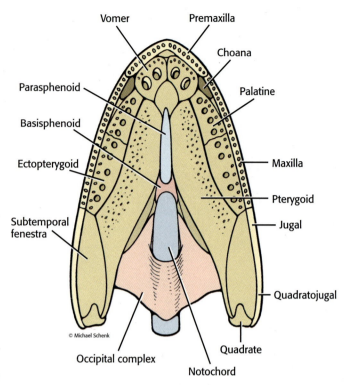

Vomer Premaxilla
Choana
Parasphenoid
Palatine
Basisphenoid
Ectopterygoid
Maxilla
Pterygoid
Subtemporal
fenestra
Jugal
Quadratojugal
© Michael Schenk
Quadrate
Occipital complex
Notochord

FIGURE 5.7 Palatal aspect of *Eusthenopteron*, an extinct sarcopterygian.

extension of the primary palate forms the secondary palate. This separates what is a multipurpose passage into a dorsal passage with a respiratory function from a ventral passage with a digestive function. The secondary palate allows crocodilans to breathe and drown their prey by projecting the tip of their snout above water. In vertebrates such as mammals, in which endothermism developed and somewhat lengthy food processing occurs in the oral cavity, the evolution of the secondary palate permits simultaneous respiration and feeding. Development of the secondary palate in mammals was accompanied by a concomitant posterior shifting of the position of the internal nares.

The skulls of fishes, amphibians, and early reptiles were without any breaks in the temporal region and are known as **anapsid** skulls. The weak jaw adductor musculature of fishes and amphibians and the stem reptiles remained confined to the inner surface of the skull, passing through the **subtemporal fenestra** and attaching to the posterior portion of the lower jaw (Figure 5.7). The action of the adductor muscles in these animals was a relatively simple jaw snap, impaling prey on their unspecialized conical teeth.

Later reptilian lines evolved openings in the temporal region, giving rise to the **synapsid** skull found in extinct mammal-like reptiles and mammals, the **diapsid** skull found in *Sphenodon*, crocodilians, and other archosaurs such as dinosaurs, and several types of a modified version of the diapsid skull found in squamate reptiles, lizards, snakes, and in the dinosaurian derivatives, birds. Another type of

probable modified diapsid condition known a **euryapsid** skull is found in the extinct ichthyosaurs and plesiosaurs (Figure 5.8). The evolution of temporal openings, or fossae, probably correlated with the subdivision of the jaw adductor musculature into further functional groups. The origin of these muscles tended to migrate to the outer surface of the skull. Correlated with the more complex musculature was an increase in the efficiency of mechanical food processing (mastication or grinding), such as we see in mammals.

VERTEBRAL COLUMN, RIBS, AND STERNUM

In the sister group of vertebrates, the cephalochordates, the only axial support is the notochord. The notochord occupied a similar position and served a similar function among the ostracoderms. In the living agnatha, hagfishes, and lampreys, the notochord remains the axial support. It continues to be important in the skeletal system of sturgeons, chimeras, etc., and also appears in the vertebrae and/or between the vertebrae of many other fishes. Furthermore, all early vertebrate embryos possess a notochord, which is found to be a vital organizer of the axis of the body, as well as an inducer of organ systems such as the nervous system.

Around the notochord evolved cartilaginous or bony coverings known as vertebrae. Among the agnatha, the hagfishes do not have vertebrae but the lampreys have isolated cartilaginous plates, axcualia, associated with the notochord.

In the gnathostomes, that portion of the vertebra directly surrounding the notochord is the centrum. The spinal cord lies in a vertebral canal dorsal to the notochord. The vertebral canal is created as the spinal cord is surrounded by cartilage or bone, forming the neural arch and spine. Lying below the notochord in the caudal region are the caudal artery and vein, which are surrounded by a hemal arch, terminating in a hemal spine.

The shape of the articulating surface of the centrum is often characteristic of a group of vertebrates. Centra with concave surfaces at either end are called amphicoelous (Figure 5.9A). Amphicoelous vertebrae are characteristic of fish, some amphibians, and a few reptiles. They are separated by remnants of the notochord or other types of tissue such as cartilage or connective tissue, and permit limited but smooth movement in any direction between adjacent vertebrae.

On the one hand, vertebrae with anterior concave and posterior convex articulating surfaces are known as procoelous (Figure 5.9B). On the other hand, opisthocoelous vertebrae possess anterior convex and posterior concave articulating surfaces (Figure 5.9C). Motion between these vertebrae is multidirectional except in the dorso-ventral plane. Many amphibians and early reptiles possess either of these types of vertebrae. Centra with flat articulating surfaces, acoelous vertebrae, are found among mammals

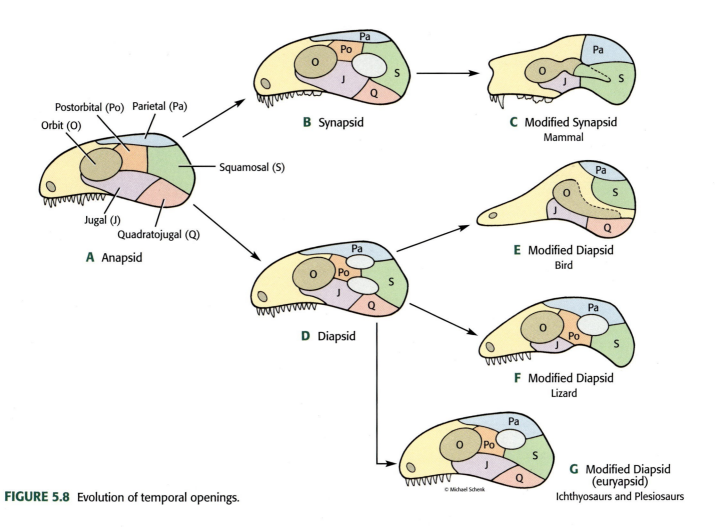

FIGURE 5.8 Evolution of temporal openings.

(Figure 5.9D). Movement between these vertebrae is somewhat limited but allows them to support increased weight.

A centrum with a saddle-shaped posterior end that articulates with an appropriately concave anterior end of an adjacent vertebra is known as a heterocoelous vertebra (Figure 5.9E). Motion includes lateral and vertical components but does not permit rotation. Heterocoelous vertebrae are present in the necks of birds. Intervertebral discs consisting of notochord remnants and/or cartilaginous connective tissue separate adjacent vertebrae and facilitate smooth movement between them.

Various processes project from the centrum. Among the tetrapods are dorsal processes, the zygopophyses. From the posterior edge of the neural arch project posterior zygopophyses (**postzygopophysis**) whose articulating surfaces point dorsally. These articulate with anterior zygopophyses (**prezygopophysis**), whose articulating surfaces point ventrally, and project from the anterior edge of the **neural arch** of an adjacent vertebra (Figure 5.10). Zygopophyses prevent rotational motion between vertebrae and strengthen the vertebral column enabling it to support appreciable amounts of weight without sagging.

Transverse processes occur on the centra and **neural arches** of many vertebrates. In general, **ribs** articulate with the transverse processes. Muscles also may attach to these processes. The placoderms and agnatha have no ribs. Ribs in the cartilaginous fishes are weakly developed intermuscularly and attached to the centrum at its base to the basapophysis, a short transverse process.

The ribs of bony fishes may develop at a number of sites, in the horizontal septum (dorsal ribs), subperitoneally (ventral ribs) and within myosepta between muscle bundles. Each of these ribs attaches by a single head to the centrum. The subperitoneal ribs attach to the **basapophysis** (see Figure 16.15).

With the evolution of the tetrapods, the **ribs** become two-headed or bicipital, the **capitulum** and the **tuberculum** (Figure 5.11 and Figure 5.12). The capitulum articulates with the centrum or a **parapophysis,** a short transverse process attached to the centrum; and the tuberculum articulates with the **diapophysis,** a more dorsal transverse process attached to the neural arch.

In the evolutionary history of vertebral columns, regional specialization has occurred. Among the bony fish,

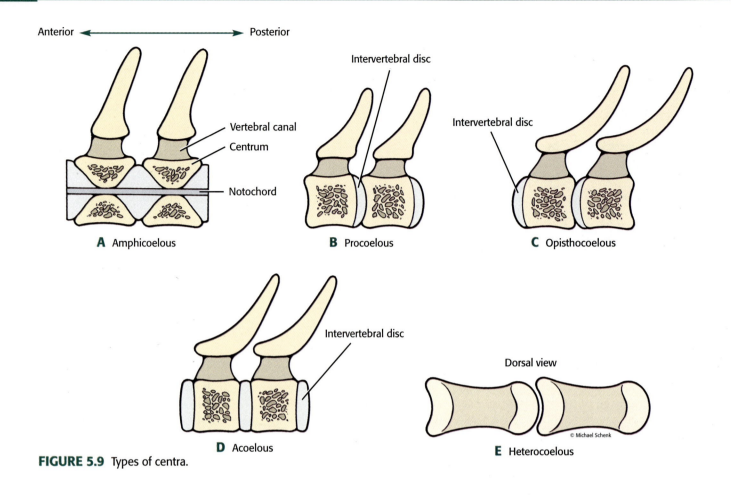

FIGURE 5.9 Types of centra.

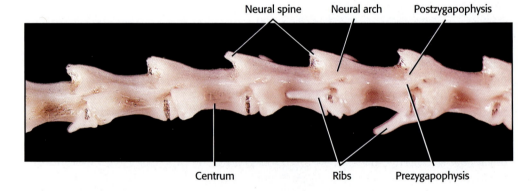

FIGURE 5.10 Trunk vertebrae: The anterior end is on the right.

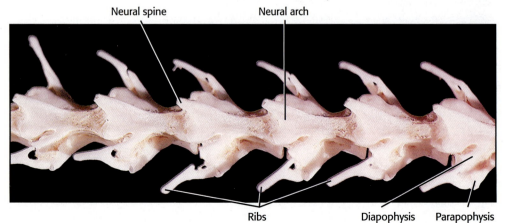

FIGURE 5.11 Trunk vertebrae: The anterior end is on the right.

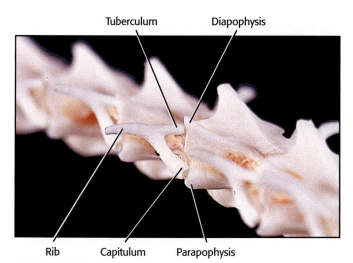

FIGURE 5.12 Trunk vertebrae with ribs.

Tuberculum   Diapophysis

Rib   Capitulum   Parapophysis

trunk and caudal vertebrae can be distinguished, with the first trunk vertebra modified to some extent for articulation with the skull. With the exception of the first trunk vertebra, the rest of the trunk vertebrae are basically similar. The caudal vertebrae are set apart by the presence of the ventral hemal arches enclosing the caudal artery and vein.

Among the earliest tetrapods, the first of the trunk vertebrae, the atlas, was modified with an articular surface to allow relatively free movement between the skull and the rest of the vertebral column. In amphibians, the atlas, representing a single cervical vertebra is present. Posterior to the atlas are a number of similar trunk vertebrae leading to a single sacral vertebra, followed by a number of caudal vertebrae. In the anuran amphibians, a number of posterior trunk vertebrae fuse, forming an elongate urostyle that, along with elongated ilia, forms a relatively rigid pelvic region that is important in resisting rotation or movement in frogs and toads, adapted for jumping and hopping.

With the evolution of amniotes, a second cervical vertebra, the axis, evolved, permitting greater freedom of movement of the head by permitting nodding and rotation of the head. The successive vertebrae in the neck, the cervicals, vary considerably in number and may be associated with short ribs. Following the cervicals are a number of trunk vertebrae, the thoracics, associated with ribs, which are important in respiration and protection of body organs.

In the region of the pelvic girdle another sacral rib appears, followed by a number of lumbar vertebrae. A varying number of caudal vertebrae make up the skeleton of the tail. This is the basic pattern found in reptiles. Among the birds the construction of the cervical vertebrae permits extremely mobile neck movements, which are important in feeding, nesting, etc. These are followed by fused thoracic vertebrae. A fused complex, the synsacrum, consisting of the last thoracic,

a number of lumbars, two sacrals, and a number of caudals, occurs in birds. The synsacrum is fused to the pelvic girdle. The extensive fusion found in the vertebral column of birds is correlated with stabilization of the skeleton, preventing rotation of the vertebral column during flight. In addition to a small number of free caudals, a fused group of caudals forms the pygostyle, forming a base for the tail feathers.

The five basic regions of the vertebral column—cervical, thoracic, lumbar, sacral, and caudal—also are found in mammals. Well developed atlas and axis occur in mammals. With rare exceptions, mammals have seven cervical vertebrae, a variable number of thoracic vertebrae, a variable number of lumbar vertebrae, three to five sacral vertebrae fused into a rigid structure, the sacrum, and a variable number of caudal vertebrae. Three to five caudal vertebrae fused into a structure, the coccyx, is found in humans and tailless apes.

Generally, in tetrapods, a midventral cartilage or bone, the sternum, developed. A great deal of variation occurs among them. The best-developed sterna are found among the birds and the mammals, with which the ribs become associated (Figure 5.13). In many tetrapods, important

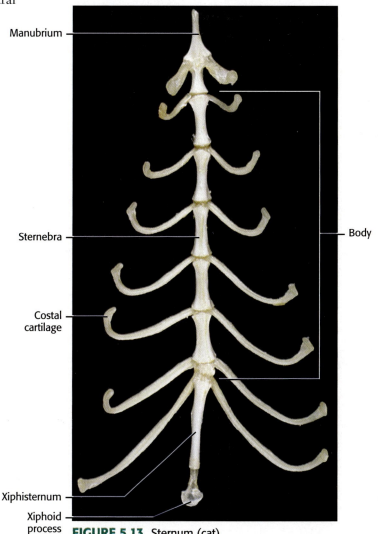

FIGURE 5.13 Sternum (cat).

Manubrium

Sternebra

Costal cartilage

Body

Xiphisternum

Xiphoid process

muscle groups, such as the pectoral muscles associated with pectoral limb attitude and locomotion, originate on the sternum. In amniotes, the sternum, along with the ribs, is involved in respiration. The sterna of birds and bats have a large midventral keel from which flight muscles originate.

LIMBS AND GIRDLES

While attempting to hold onto struggling prey with its jaws, a fish probably would be tossed around and subjected to destabilizing forces and would have little control over its body. If, however, projections (fins) evolved in strategic areas of the body, the fish could maintain some stability to deal with this problem. For this reason, many biologists believe that jaws and stabilizing fins coevolved.

As a flat projection (fin) is pressed against a dense medium such as water, the resistance of the water being displaced will cause an equal reaction in the opposite direction. For example, a right pectoral fin pressed down (adduction) on the water will cause the body to **roll** to the left, whereas a left pectoral fin pressed down on the water (adduction) will cause the body to roll to the right (Figure 5.14). Furthermore, as a fin moves through the water, a force known as lift occurs perpendicular to the surface of the fin.

To propel the body through water, most fish move their caudal fins (tail) back and forth in the horizontal plane.

Three major types of caudal fins (tails) arose among the ostracoderms.

1. The **diphycercal** caudal fin, in which the notochord extends straight to the tip.

2. The **hypocercal** caudal fin, in which the notochord projects into the lower lobe of the tail.

3. The **heterocercal** tail, in which the notochord projects into the upper lobe of the tail.

Modern fish possess the diphycercal tail, the heterocercal tail, or a modification of the heterocercal tail, the **homocercal** tail (Figure 5.15).

A problem with body displacement occurs, however, as the caudal fin is moved laterally through the water to produce thrust, because the head is attached firmly to the vertebral column, generating head movements opposite to those of the tail. As the caudal fin moves to the left, the head moves to the right, and as the caudal fin moves to the right, the head moves to the left. These actions result in **yaw**, a lateral deflection of the fish's progress through the water (Figure 5.14). If this deflection is uncorrected, the track of the fish will be quite erratic and inaccurate. To get an idea of the consequences of having no paired fins, watch a frog tadpole swim. It works for a tadpole that grazes on algae, but for a fish chasing lunch, it will not be effective.

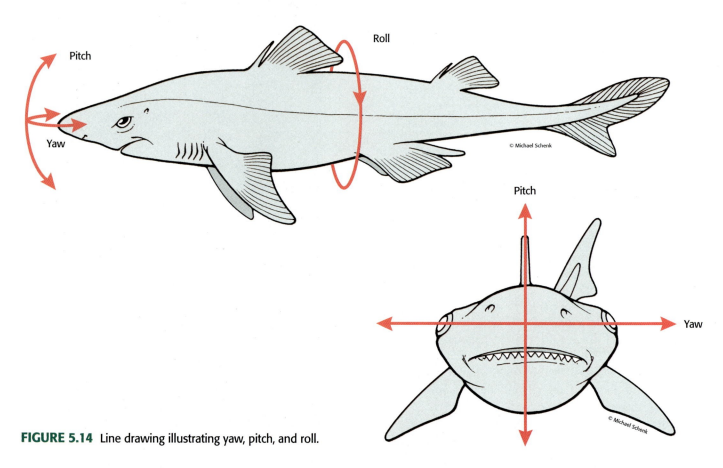

FIGURE 5.14 Line drawing illustrating yaw, pitch, and roll.

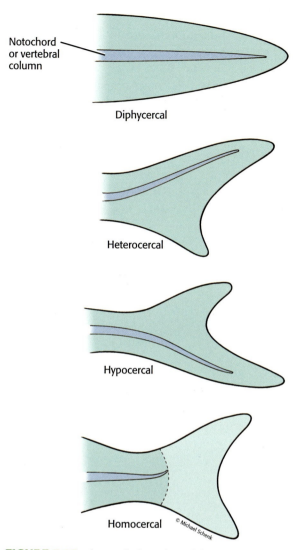

FIGURE 5.15 The evolution of caudal fins.

Yet other forces may also cause it to veer off course in different directions. **Pitch**, a course deflection occurring in the vertical plane, results in the anterior end moving up and down (Figure 5.14). A number of structures in fish (and other aquatic vertebrates) have evolved to counteract the problems of moving through water. Median dorsal fins control the yaw. The dorsal fin is single in some fish and double in others. In many fish, dorsal fins are collapsible, which actually induces yaw to promote tight turning circles.

Paired fins, the anterior pectorals and posterior pelvics, also evolved among fish. The pectoral fins effectively control pitch and **roll**. In addition, in conjunction with the flattened anterior ventral surface of the body in cartilaginous fish, the pectorals can supplement the lift generated on the fish's body, affecting its mobility.

If jaws were the most monumental event during the evolution of vertebrates, as Alfred Romer suggested, certainly another milestone was the evolution of paired limbs with digits. Although we now associate limbs with terrestrial

vertebrates, current fossil evidence suggests that jointed limbs probably evolved in an aquatic environment (Ahlberg and Milner, 1994; Coates and Clack, 1991). In addition, a number of characteristics of tetrapods evolved—for example, pectoral girdles detached from the skull, pelvic girdles attached to the vertebral column, mutually supporting vertebrae to withstand weight stress in the absence of a buoyant aquatic medium.

Even more recent evidence for these ideas has come to light with the discovery of *Tiktaalik roseae* , a fossil sarcopterygian fish, a sister group of tetrapods, from the late Devonian period. Along with its fish characteristics of scales, fin rays, and some skull anatomy are a shortened skull roof, mobile neck, and a functional wrist joint with a flexed shoulder and elbow, identified with tetrapods. An important adjunct to the evolution of terrestrial limbs was the development of flexions between the humerus and the radius and ulna (*i.e.,* the elbow), and between the radius and ulna and the digits (*i.e.,* the wrist) (Daeschler, Shubin, and Jenkins, 2006; Shubin, Daeschler, and Jenkins, 2006). Similarly, flexions between the femur and the tibia and fibula (*i.e.,* the knee), and between the tibia and the fibula and the digits (*i.e.,* the ankle) occurred in the posterior limb.

This combination of morphological features in fossils found in their relative geological surroundings suggests that early tetrapod ancestors lived along the margins of shallow water and probably were capable of augmenting their respiration with lungs (Daeschler, Shubin and Jenkins, 2006; Shubin, Daeschler and Jenkins, 2006). Although the changes we have just outlined permitted the continuation of an aquatic lifestyle, they permitted the exploitation of the terrestrial habitat with its developing invertebrate fauna and provided an environment free from competition with large predatory fishes, as well as avoiding predation by these same fishes.

Efficient movement of vertebrates on land is next to impossible with fins, despite the fact that fish such as mudskippers and their relatives, the gobies, manage to "walk" across tropical mudflats in pursuit of prey. Nevertheless, no terrestrial tetrapod retained fins. Jointed limbs with their attached muscles made it possible to hold the body in position, as well as move the body over land efficiently. Various important activities such as seeking shelter, procuring food, pursuing a mate, demonstrating parental behavior, defending a territory, and fleeing a predator were possible.

When the first land tetrapods appeared, they already had four limbs with digits. Although five digits once were thought to be the primitive number for vertebrates, in many of the earliest tetrapods, the number of digits far exceeded five. It was in the ancestral line of vertebrates that gave rise to the reptiles and their descendents that the number of digits became constant at five (Figure 5.16).

The pectoral anatomy and the limb construction of these early tetrapods remained similar to those of their fish ancestors (Figure 5.17). Both limbs possessed a single proximal bone, the **humerus**, in the anterior limb and the **femur** in the posterior limb, with a pair of distal bones, a **radius** and an **ulna**, in the anterior limb and a **tibia** and a **fibula** in the posterior limb. In addition, a caudal fin and fishlike scales appeared in the early tetrapods, belying their ancestry. Apparently these early "land" vertebrates occupied aquatic niches, probably using their caudal fins to propel them through the water.

Despite the close similarity of the pectoral anatomy of the ancestral fishes and tetrapods, there was a profound change in the relationship of the girdle to the cranial anatomy. Unlike the fish in which the skull is firmly attached to the pectoral girdle via the posttemporal, the postcleithrum, and supracleithrum, in the tetrapod, the girdle became detached from the skull and was suspended in a muscle sling (Figure 5.18A). Concurrently, a neck evolved, enabling the head to move independently from the rest of the skeleton.

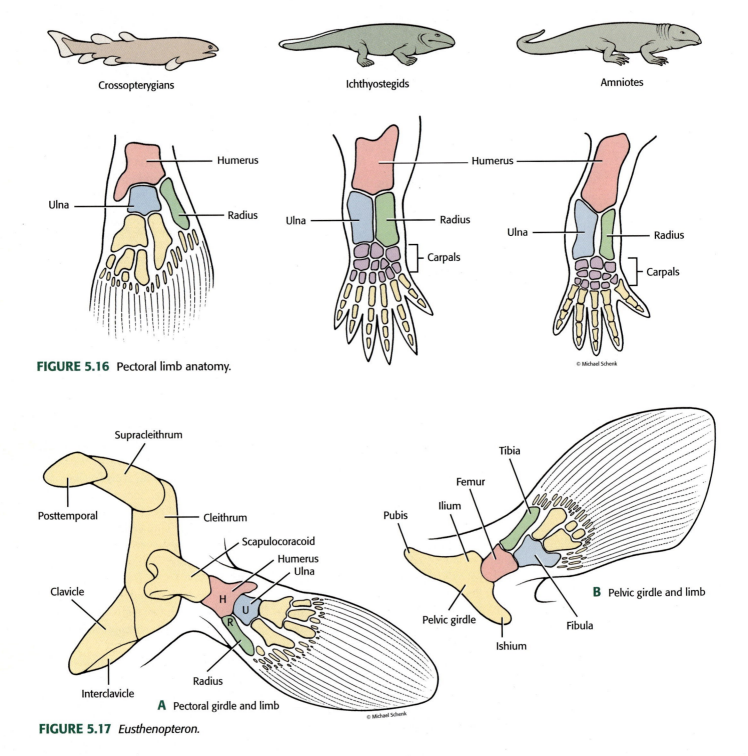

Crossopterygians

Ichthyostegids

Amniotes

FIGURE 5.16 Pectoral limb anatomy.

© Michael Schenk

FIGURE 5.17 *Eusthenopteron.*

A Pectoral girdle and limb

B Pelvic girdle and limb

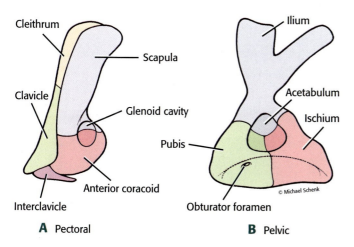

A Pectoral

B Pelvic

FIGURE 5.18 Early ancestral amphibian girdles.

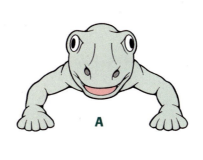
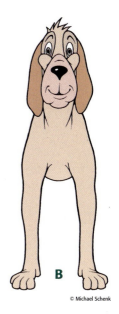

FIGURE 5.19 Comparison of primitive and advanced limb postures.

In contrast, the pelvic girdle became more substantial, consisting of three pairs of articulating bones: the **pubis** projecting anteriorly, the **ischium** projecting posteriorly, and the **ilium** projecting dorsally (Figure 5.18B). The pelvic girdle became associated more strongly with the vertebral column via attachment of the ilium to a sacral vertebra.

A sprawling limb position in early terrestrial tetrapods was similar in the anterior and posterior limbs. The humerus and femur were oriented parallel to the land, and the radius and ulna and tibia and fibula were oriented perpendicular to the humerus and the femur, respectively, with appropriate flexions at the wrist and ankle. A similar condition can be seen in modern salamanders (Figure 5.19A). This stance promotes a rotary motion around the shoulder and hip joint. This limb orientation is not conducive to the development of efficient and rapid locomotion but imparts a side-to-side motion similar to a fish swimming in water, which still is used by modern salamanders.

In vertebrates such as the carnivorous dinosaurs, birds, and most mammals in which we see rapid and agile locomotion, the position of the limbs changed. The limbs moved from the sprawled position toward the centerline of the body. Among early tetrapods support of body weight was related largely to the cross-section of the humerus and femur. The new orientation, however, shifted support of the weight of the body along the length of these two long bones. The ability of a bone to support weight is much greater along its length, permitting more efficient locomotion and also allowing an increase in body size. An increase in body size has its obvious advantages; *e.g.,* fewer predators are capable of chasing and subduing larger prey (Figure 5.19B).

The physics of locomotory systems adapted for various functions among vertebrates dictates the design of limbs. In fossorial mammals, for example, the humerus of the

pectoral limbs is stout and short, with a manus possessing sturdy claws that serve as an efficient digging tool. In mammals that are adapted for different types of locomotion, other trends evolved.

Adaptations in the structure of tetrapod feet are notable. Among mammals in which sure-footedness and stability are characteristic (*e.g.,* bears, beavers, and humans), the entire foot is walked on and is known as the **plantigrade** foot (Figure 5.20C). Among cursorial mammals two other designs have evolved. One is the **digitigrade** foot, in which the heel is pulled up off the substratum and the animal walks or runs on its toes. This design is found in animals such as dogs and cats (Figure 5.20B). The other design is the **unguligrade** condition, in which only the tips of the toes are in contact with the substratum (Figure 5.20A). Among these mammals the toes usually are covered with horny hooves. Unguligrade mammals such as goats, cattle, sheep, and antelopes, with the artiodactyl foot, walk or run on an even number of toes. Unguligrade mammals such as rhinoceroses and horses walk and run on perissodactyl feet with an odd number of toes.

If we compare the leg segments, *i.e.,* the upper arm and thigh, with the lower legs of mammals using the plantigrade foot, we find them to be of almost equal length and girth. In the legs of these animals, the muscle mass is fairly equally distributed between the upper and lower segments. If we make these same comparisons of mammals employing the digitigrade and unguligrade feet, we find that the lower portions of the limbs are considerably longer and much more slender. In these animals the muscle mass is concentrated near the body and upper portion of the leg. The distal or lower portion of the leg is lighter and longer than the proximal portion, permitting an extended stride and thereby resulting in more rapid locomotion.

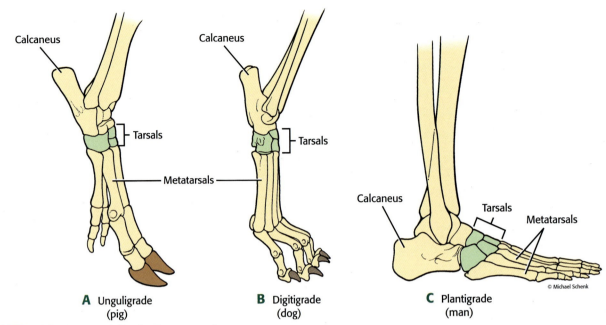

A Uniguligrade
(pig)

B Digitigrade
(dog)

C Plantigrade
(man)

© Michael Schenk

FIGURE 5.20 Foot specialization in mammals.

CHAPTER SIX

Muscular System

The function of muscle is contraction causing movement. Vertebrates have three types of muscle.

1. **Smooth or involuntary muscle** is associated with systems not involved in body locomotion, such as the digestive system (Figure 6.1A).

2. **Cardiac muscle** is found only in the heart and is involuntary in function (Figure 6.1B).

3. **Skeletal or striated muscle** is voluntary in function and generally is associated with visible body movements (Figure 6.1C).

Striated muscles pulling on various skeletal elements bring about body movements, including locomotion, feeding, and mating. Contraction of the tissues of the skeletal muscular system produces

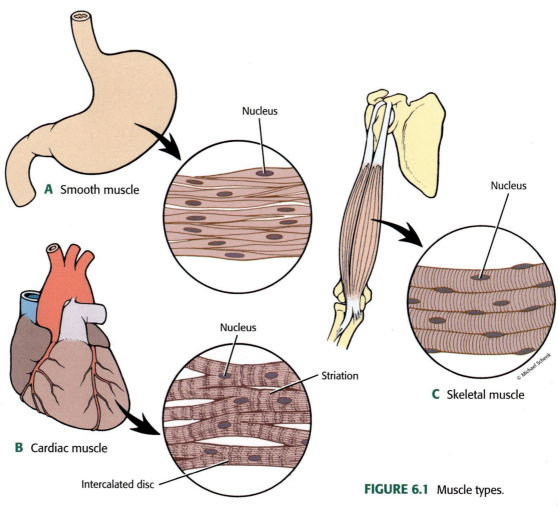

A Smooth muscle — Nucleus

B Cardiac muscle — Nucleus, Striation, Intercalated disc

C Skeletal muscle — Nucleus

© Michael Schenk

FIGURE 6.1 Muscle types.

51

externally visible body movements while some essential subtle, sophisticated, internal movements occur through the contraction of smooth muscle. Examples of the latter are food propulsion through the digestive tract, adjustment of blood vessel diameters to control blood volumes in various body regions, regulation of respiratory tube diameter, erection of feathers in birds and hair in mammals, and intrinsic eye functions such as dilation and constriction of the pupil. Cardiac muscle function is responsible for the contractions of the atria and ventricles of the heart.

The gross anatomy of a whole skeletal muscle includes the middle region called the **belly**, the less movable end known as the **origin**, and the more movable end called the **insertion** (Figure 6.2). In some muscles that are capable of several actions, the origin and insertion may be reversed during contraction. The points of origin and insertion are marked by the presence of dense connective tissue that anchors each muscle to a bone, to straplike tendons, or to other muscles, by means of flat, sheetlike aponeuroses.

Macroscopically, the structure of a striated muscle consists of a variable number of muscle cells (fibers), each encased in connective tissue, the endomysium, occurring in bundles called fasciculi and wrapped in connective tissue, the perimysium. Finally, groups of fasciculi surrounded by connective tissue, the epimysium, make up the whole muscle (*e.g.*, the Biceps brachii in tetrapods).

Microscopically, each of the muscle fibers contains two types of contractile proteins, actin and myosin, which are organized into regular and serially repetitive arrangements. This repetitive banding pattern is a prominent characteristic of striated muscle (Figure 6.1C).

Muscle shapes vary, depending primarily upon the arrangement of fibers within each muscle and its relationship with the tendon of insertion. An arrangement with which you probably are most familiar is known as **convergent**. The fibers in this arrangement are basically parallel but converge at either end of the muscle. Throughout the length of **strap** muscles, the fibers are parallel. Muscles whose architecture includes a straight, blunt origin and a convergent insertion are called **fan-shaped**. An oblique arrangement of fibers inserting into a tendon or tendons constitutes the pennate class of muscle architecture. Within this group are **unipennate**, in which the fibers insert into one side of a tendinous insertion, **bipennate**, in which the muscle fibers insert into both sides of a centrally located tendon, and **multipennate**, in which the muscle fibers insert into several tendons whose orientation may vary and may appear as combinations of the unipennate and bipennate subgroups (Figure 6.3).

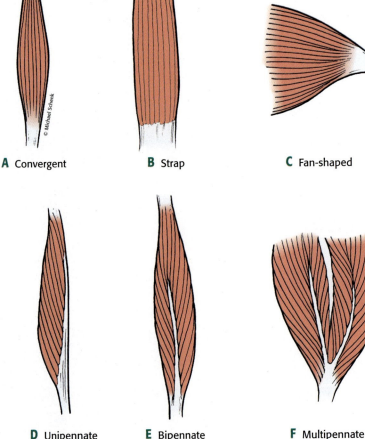

A Convergent   **B** Strap   **C** Fan-shaped

D Unipennate   **E** Bipennate   **F** Multipennate

FIGURE 6.3 Muscle architecture.

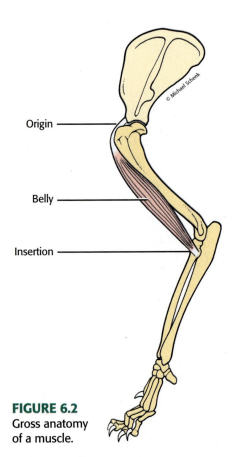

Origin

Belly

Insertion

FIGURE 6.2
Gross anatomy
of a muscle.

Muscles are capable of producing a variety of movements called actions. Movement of muscles associated with hinge joints (*e.g.,* the pectoral fin of a fish and the elbow of a tetrapod) produce actions known as flexion, causing reduction of the angle at the joint and extension, in turn causing an increase in the angle of the joint. When appendages or portions of appendages (*e.g.,* the pectoral fins of a fish and the digits of a tetrapod) are moved away from a midline reference point, the action is referred to as abduction. In contrast, movement toward a midline reference point is called adduction.

Movement of an appendage forward, producing an anterior action (*e.g.,* swinging a leg forward among tetrapods) is known as protraction, and the opposite action is known as retraction. In the living *Latimeria*, thought to be a somewhat distant relative of tetrapod sarcopterygian ancestors, the pectoral and pelvic fins exhibit these movements. Furthermore, the fins are protracted and retracted in the same primitive pattern as utilized by early tetrapods.

Rotation involves the movement of a portion of the body around a central axis, such as the head turning on the neck in mammals. A specialized action involving rotation of the radial head in the ulnar notch in mammals produces actions known as pronation and supination. When a mammal is standing, the manus is pronated or palm down; when grooming itself, however, the manus is palm up or supinated.

Individual muscles generally do not bring about actions by themselves. Most actions are the result of the combined efforts of several muscles. Muscles that affect the action directly are called prime movers. Prime movers usually are assisted by others, known as synergists. These muscles not only aid in bringing about the main action but also may stabilize the joint or portions of the skeleton involved in the action. They are known as fixators. Muscles whose actions oppose one another are called antagonists.

The axial or trunk and tail musculature of fish is arranged into a series of muscular segments, or myomeres, separated by sheets of connective tissue, the myosepta. Except for the agnathans, the myomeres are folded into a configuration resembling a complex W (Figure 6.4). Because the myomeres are attached to the vertebral column, alternating wavelike contractions on either side of the body, assisted by lateral tail movement, produce the motion typical of fish. Again with the exception of the agnathans, the myomeres are divided into a dorsal **epaxial** group and a ventral **hypaxial** group by a sheet of connective tissue called the **horizontal septum** (Figure 6.4). The epaxial musculature is

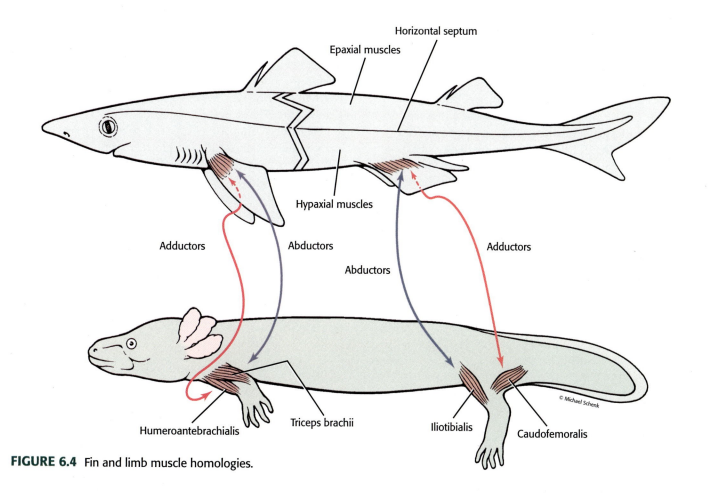

FIGURE 6.4 Fin and limb muscle homologies.

separated by the mid-dorsal vertical septum into left and right halves. The hypaxial musculature is subdivided into left and right halves by the linea alba.

The musculature of the throat and head region became specialized. Muscles associated with the gills in fishes are known as branchiomeric muscles. They function primarily in jaw and pharyngeal movements, therefore they are important during food procurement and processing. Some of these muscles are retained as jaw muscles but others have found other functions, especially in mammals, in which such muscles as the facial muscles, the trapezius group, and the laryngeal muscles were derived from the ancestral branchiomeric muscles.

The hypaxial or hypobranchial musculature of the throat became separated into a number of straplike muscles associated with respiration, depression of the mandible, and food handling. Some examples are the tongue muscles (e.g., various glossus muscles and muscles of the hyoid and thyroid region of the throat). The epaxial musculature in the head and shoulder region became specialized, affecting respiration and movement of skeletal elements such as the gill arches and scapula. Extensors or abductors associated with the pectoral and pelvic fins were derived from the epaxial musculature, and flexors or adductors of these fins evolved from the hypaxial musculature.

With changes in locomotion, respiration, feeding, and other functions as vertebrates moved from an aquatic to a terrestrial environment, the muscles associated with these functions became more complex. The muscles of the trunk and tail regions remained fairly simple and fishlike. Locomotion in terrestrial tetrapods relies almost exclusively on the appendages, promoting a dorso-ventral flexion and extension of the spine, instead of the lateral movement of the body in most fishes. Salamanders and some reptiles such as snakes and lizards continue to undulate, and their trunk and tail musculature retains a segmented anatomy.

The hypaxial muscles in tetrapods, however, have subdivided generally into a ventral rectus abdominis extending from the sternum to the pubis and a series of three thin layers: an outer external oblique, a middle internal oblique,

and a deep transversus abdominis covering the lateral aspect of the trunk and abdomen. A bandlike subvertebral muscle lying beneath the transverse processes acts as a flexor of the vertebral column.

With the evolution of a mobile neck and more flexible vertebral column, the epaxial muscles undergo a complex specialization in many amniotes, e.g., the reptiles and mammals. Commonly, the epaxial muscles differentiate into iliocostalis, spinalis, longissimus, and intervertebral. Further specialization occurs in the neck and head regions of many amniotes, reflecting the sophistication of neck and head motion in these tetrapods.

Girdle and limb musculature in the tetrapod, derived from the simple dorsal abductor and ventral adductor fin musculature in the fish ancestor, resulted in differentiation of discrete muscles capable of producing and sustaining terrestrial locomotion. Because the pectoral girdle is no longer attached to the skull but still must be stabilized to function as a brace and point of muscle attachment for pectoral appendage muscles, the girdle now is suspended in a muscle sling. In addition, because the pectoral girdle was freed from the skull, neck development with correspondingly increased head movements occurred in tetrapods.

The dorsal abductors of the fish pectoral fin gave rise to muscles such as the latissimus dorsi, deltoids, subscapulars, and teres muscles and the extensors of the brachium such as the triceps, supinator, and digit and manus extensors (Figure 6.4). The ventral adductors gave rise to muscles such as the pectoralis, supracoracoid becoming the spinatus muscles of the mammal, and the brachium flexors such as the biceps, humeroantebrachials, brachialis, various antebrachial muscles, e.g., flexor carpi ulnaris, and manus and digit flexors (Figure 6.4).

The dorsal abductors of the fish pelvic fin gave rise to muscles such as the iliotibialis, gluteals, rectus femoris, sartorius, pes, and digit extensors (Figure 6.4). The ventral adductors gave rise to a number of muscles such as the caudofemoralis, semimembranosus, semitendinosus, gracilis, biceps femoris, and flexors of the pes and digits (Figure 6.4).

CHAPTER SEVEN

Body Cavities and Mesenteries

T he body plans of many invertebrates and all craniates include an extensive body cavity lined with mesoderm, the coelom. The coelom permits the internal organs projecting into it to move independently from the body wall. This space is lined with a serous membrane, the **parietal peritoneum**, which, during embryonic development, converges as a double layer dorsally and ventrally, forming continuous mesenteries that diverge and wrap around any organs projecting into the coelom and suspend them within this cavity. This serous covering of suspended organs is known as the **visceral peritoneum**. The parietal layer secretes serous fluid into the cavity. This fluid functions as a lubricating medium, allowing frictionless movement of the organs (Figure 7.1).

Among invertebrate animals such as annelids, one can see the plan just described, in which organs are suspended by dorsal and ventral mesenteries. Among vertebrates, though, most of the dorsal mesentery remains as subdivided membranes that often are variously twisted and related to organs projecting into the cavity. The ventral mesentery, in contrast, is much reduced in vertebrates and remains as the membranes of the liver and the urinary bladder.

Subdivision of the coelom seems to have been important during the evolution of vertebrates, because it occurs in the most primitive of living vertebrates. As we will observe in the lamprey, the pericardial (heart) cavity already is separated from the pleuroperitoneal cavity by the **transverse septum**. This is similar to the shark condition (Figure 7.2). It may have been important to segregate the circulatory pump (heart) from the abdominal viscera to provide developmental and functional space.

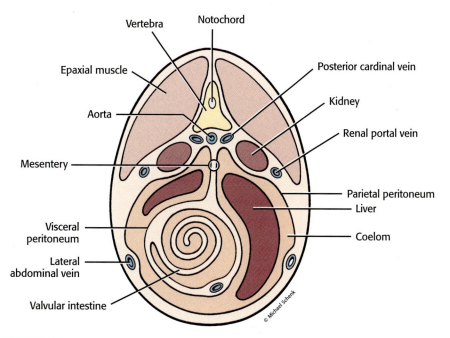

FIGURE 7.1 Body cavity with parietal and visceral peritoneum.

Cranial

Caudal

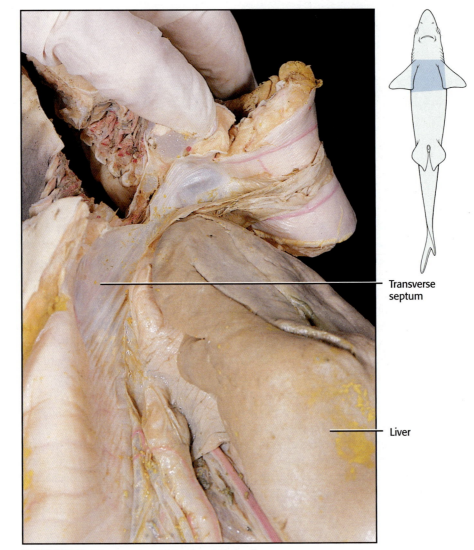

Transverse
septum

Liver

FIGURE 7.2 Transverse septum (ventral view).

Digestive and Respiratory Systems

8

The evolutionary history of the ancestral space known as the pharynx in chordates is an old one. In the non-vertebrate chordates, the pharynx with its slits functioned actively as a filtering apparatus and passively as a respiratory organ. Among vertebrates, we find that the muscular pharynx functions as an active respiratory and digestive passageway. In fish living in well-oxygenated water, water pumped through the pharynx and then past the gills facilitates the diffusion of oxygen and carbon dioxide across gill membranes.

Fossil evidence indicates, however, that even some of the earliest fishes occupied bodies of water in which oxygen tension was low and gill respiration was inadequate. Lungs were common in these early fishes, probably enabling them to gulp air, and thereby permitting them to supplement their aquatic breathing apparatuses. Because carbon dioxide is more soluble than oxygen, the gills remained essential to eliminate the carbon dioxide.

In early fishes the lungs were bilobed, ventrally located, and connected to the anterior gut. Members of this early group of vertebrates, sarcopterygian fishes, ancestors of the tetrapods, possessed ventral lungs as well as gills.

Gas-filled organs such as the lungs and swim bladders found in most modern bony fish not only can act as a supplementary respiratory organ but also affect the buoyancy of the animal, allowing it to more closely approach the density of the medium and conserving considerable energy. Generally, while lungs are ventrally located, swim bladders are located dorsal to the gut and act primarily as organs of balance.

The most primitive swim bladders retain a connection to the gut, the pneumatic duct. The fish possessing these are called physostomes. Some examples are goldfish, minnows, salmons, and herrings. These fish retain the capability to use the swim bladder as a supplement to the gills, and they are able to gas their swim bladders by gulping air and to degas by "burping" gas bubbles.

Other species of fish known as physoclists possess swim bladders that are not connected to the gut. Examples are perches, basses, and sunfish. Their swim bladder, in contrast to the physostomatous condition, is an organ of balance. The physiology of these bladders is very different from the physostomes. Gassing is accomplished by secretion of gases into one part of the bladder, and degassing occurs passively across the membranes of a second compartment of the bladder.

Why gas and degas the swim bladder? With the dorsal position of the swim bladder, a potential problem affects the homeostasis of the animal's position in its environment. In an aquatic medium the hydrostatic pressure varies with depth. For each 10 feet in depth, the pressure increases one atmosphere. For example, from 10 to 20 feet the pressure increases one atmosphere, and from 20 to 10 feet it decreases one atmosphere.

The properties of gases are affected by physical factors such as temperature and pressure. As the environmental hydrostatic pressure increases, the gases in swim bladders are exposed to increasing pressures. Therefore, the volume of the swim bladder decreases as the fish dives. The opposite is true as the fish ascends in the water column; the gases in the bladder expand. What does this mean for the fish? Without a means of adding gases to the swim bladder as the fish dives, the swim bladder would become smaller and smaller, causing the fish to descend precipitously and make it impossible for the fish to regain its original position. The opposite is true as the fish ascends; as the environmental pressure decreases, the gases expand and the ascent of the fish would be catastrophic, catapulting the fish through the surface, causing the swim bladder to explode or be forced out through the mouth.

In some physostomous fish the swim bladder acts as a sound production organ by modulated release of air from the bladder. In physoclists, sound is produced by rubbing bones or structures known as pharyngeal teeth together, and the sound resonates off the surface of the swim bladder. Other fish produce sounds by changing the configuration of the swim bladder through muscular action.

The cartilaginous chondrichthyes do not possess swim bladders. Another means of body density reduction has evolved. The pelagic species store low-density oils in their large livers, thereby reducing their body weight and permitting them to approach the density of their environment, which also conserves considerable energy.

Some of the early amphibians (*e.g., Acanthostega*) appear to have been largely aquatic and, like their fish ancestors, retained the ability to use gill respiration (Coates and Clack, 1991, Clack, 2005). To function, respiratory organs must remain moist. Obviously, gills were not an option for terrestrial vertebrates. Most of the later tetrapods abandoned gills in favor of pulmonary or lung respiration. Aerial respiration is much more effective than aquatic respiration, as air has a much higher oxygen level. In addition to pulmonary respiration amphibians utilize their skin as a respiratory organ. They pay a price for this option in that the skin must remain moist.

Various vertebrate structures or parts of structures that once were part of the respiratory apparatus in fishes are transformed into structures with new functions in tetrapods. Parts of visceral arches were incorporated into the hyoid apparatus that supports the tongue. The spiracular gill pouch became the auditory (eustachian) tube connecting the middle ear with the pharynx. Tissues originating from various gill pouches gave rise to the thyroid, thymus, and parathyroid glands.

Bordered by the lips and just anterior to the pharynx is the oral cavity. Two obvious features of the oral cavity are the teeth and the tongue. The vertebrate tongue begins as an immovable primary tongue, supported by elements of the hyoid arch and found in fishes and amphibians such as *Necturus*, which retain gills as adults. In anurans and most other salamanders, the tongue consists of a primary tongue and an anterior extension capable of being projected in a volatile fashion, resulting in prey adhering to it because of adhesive secretions. In later vertebrates, reptiles and mammals, lateral lingual folds are added, giving rise to a mobile muscular tongue. Some reptiles, such as chameleons, possess elongate, protrusible tongues used to capture prey. Most snakes and lizards have forked, protrusible tongues to which environmental chemicals adhere. The tongue is inserted into the vomeronasal apparatus, permitting odor detection. The tongues of mammals are modified in a number of ways.

1. Carnivores often possess spines used for rasping flesh from the surface of bones of their prey, as well as grooming.

2. Many herbivores possess long, mobile tongues, which are used to tear off vegetation by wrapping their tongue around it.

3. Several mammals that feed on ants and termites— *e.g.,* the aardvark, anteaters, and pangolins—possess protruscible, elongated, sticky tongues, which enable them to capture many prey at a time.

4. Birds possess mobile tongues with reduced lingual folds and often are modified for food gathering. Examples are the tongue of the woodpecker, which is elongated and barbed to facilitate extraction of arthropods from holes drilled into trees and other plants; and the tongue of hummingbirds, which is formed into a tubular organ used to transfer nectar from host plants.

Taste receptors, generally associated with the tongue, also are found in the pharynx or on the body surface of some vertebrates, such as catfish. The pharynx in tetrapods leads through the slit-like glottis into the larynx at the anterior end of the respiratory system. In mammals, the glottis is guarded by the epiglottis, which functions to close off the larynx during swallowing.

The pharynx also leads into the esophagus. Capturing of prey and cropping of vegetation and initial processing occur in the oral cavity. Food then passes through the pharynx (swallowing) into the esophagus, beginning its passage through the digestive system. The esophagus is specialized in some vertebrates. In certain birds the distal portion of the esophagus, the crop, is enlarged as a specialized area where food may be stored. This anatomy is found in seed-eating birds and permits softening and fermentation of the seeds. In some predaceous birds (*e.g.,* owls), the crop functions as a reservoir for indigestible parts of prey such as bones, feathers, and hair. This debris is formed into pellets that are regurgitated at roosting sites. These pellets have been used in studies of owl ecology.

A curious function in members of the Family Columbidae (*e.g.,* pigeons) is the production of a nutritious fluid known as "pigeon milk" by both sexes, fed to nestlings. Interestingly, the production of this fluid is stimulated by the hormone prolactin, the same hormone that stimulates milk production in mammals, as well as the "water drive" in salamanders that promotes movement of the sexes toward breeding ponds.

With the exception of the non-vertebrate chordates and the agnatha, in which the pharynx leads into the intestine, food enters the stomach from the esophagus. Most vertebrates have a J-shaped or U-shaped stomach. Exceptions occur in slim-bodied vertebrates such as snakes and salamanders, in which the stomach is straight. Three regions are typical of vertebrate stomachs: an esophageal region connected to the esophagus, followed by the fundic region, and terminating in the pyloric region. Each region is characterized by a typical mucosal lining.

Perhaps the stomach evolved to permit vertebrates to gorge on food, when available, making it possible to "weather" other periods when food was scarce. Later, mechanical processing and initiation of protein digestion evolved as additional stomach functions. Vertebrate stomachs notoriously secrete large amounts of hydrochloric acid, producing pHs of 2.0–3.0. Mucus secreted by cells in the stomach lining protect it from damage. Biologists believe that this capability to produce a strongly acidic environment evolved to kill live prey, as well as to reduce the concentration of bacteria in the food. A low pH is required to activate pepsinogen, the enzyme produced by the stomach that initiates protein digestion in the stomach.

Specializations of the stomach are found in various vertebrates. Among some fish, some reptiles (e.g., crocodilians), and birds, the stomach is specialized as a grinding organ called a gizzard. In crocodilians and birds the stomach is subdivided into a secretory section, the proventriculus, and a grinding portion, the gizzard. Gizzards typically are lined with a tough inner layer to avoid abrasion of the lining. Crocodilians and birds often swallow pebbles that lodge in the gizzard and serve as grinding surfaces because of the muscular contraction of the walls of the gizzard. Fossil evidence suggests that some dinosaurs probably possessed gizzards and also swallowed rocks that performed a similar function.

Among mammals, the proximal chambers of the ruminant (cud-chewing) mammalian stomach—the reticulum, rumen, and omasum—are lined with esophageal epithelium and apparently represent a considerable specialization for fermentation, mechanical processing, and early digestion of vegetation. These three chambers are reservoirs for a rich bacterial and protozoan mixture and the organisms are responsible for the chemical breakdown of various complex chemical compounds such as cellulose, a structural compound of plant cell walls, making energy available to the ruminant. Cellulase, the enzyme responsible for cellulose digestion, cannot be secreted by the ruminant but is produced by the resident bacteria. The distal compartment of the ruminant stomach is the abomasum, which produces the typical stomach secretions associated with other mammals. Common cud-chewing mammals include cows, sheep, antelope, deer, and camels.

In animals lacking a stomach, such as the lamprey, food moves from the pharynx into the intestine. Among vertebrates with a stomach, food is released slowly from the stomach into the small intestine through the pyloric valve.

Vertebrates show a tendency for regional specialization of the intestine. Extracting the greatest amount of energy

from foodstuff (i.e., maximizing digestion efficiency) has been achieved in different ways. Two solutions—increasing the area of the digestive and absorptive surface of the gut and slowing the speed of passage of food through the digestive system—have evolved among vertebrates.

An early mechanism in fish that utilized both solutions was a spiral valve in the intestine. Spiral valves are particularly characteristic of a number of primitive fish in addition to sharks. These include holocephalans, sturgeons, lungfish, and others. Among other vertebrates, the intestine lengthened and coiled. The turns of the valve greatly increased the surface area and also slowed the passage of food considerably. Lengthening ensured longer exposure of the food to digestive enzymes, and coiling avoided the requirement of elongate bodies to accommodate the longer intestine.

In many vertebrates with regional specialization, three regions appear, beginning at the intestinal side of the pyloric valve: the duodenum, the jejunum, and the ileum. In addition, the walls of the intestine were folded, creating plicae, and the mucosal surface (digestive and absorptive surface) of the intestine was modified by means of villi and microvilli, small finger-like processes. Both of these adaptations vastly increase the inner surface area without unduly increasing the outer diameter of the intestine. Among many teleosts, evaginations of the proximal portion of the intestine, pyloric caeca, evolved, thereby increasing the surface area. Almost all chemical digestion and absorption of digested proteins, fats, and carbohydrates occur in the small intestine of vertebrates. Most water absorption also takes place here.

Among fishes, the intestine is not differentiated into a small and a large region. With the advent of terrestriality among vertebrates, a large intestine becomes differentiated. In vertebrates (e.g., teleosts) that possess an unspecialized intestine, undigested material is released through the anus.

In tetrapods, undigested material passes from the small intestine, often through an ileocecal valve, into the large intestine. Many amniotes possess an intestinal cecum marking the beginning of the large intestine. A single dorsal cecum is found in many reptiles. In birds, a pair of ventral ceca occur. Among mammals, however, a single ventral cecum extends from the large intestine. In some mammals a smaller diameter evagination, the appendix, extends from the cecum. The evolution of a cecum and an appendix are further examples of increased surface area in a finite space.

The size of the cecum is correlated with food preference in mammals. Herbivores have a tendency to have a long cecum, and often an extensive appendix is associated with bacterial digestion of plant material. Conversely, carnivores have a shortened cecum and often no appendix. The remainder of the large intestine may be specialized as a colon and rectum. Additional water is absorbed in the large intestine. Resident colonies of bacteria in the large intestine metabolize the remnants of food, producing organic compounds and vitamins that are absorbed, as well as decomposing the remains, producing feces that are eliminated through the anus.

Most vertebrates possess a ventral area, the cloaca, at the posterior end of the body into which open the anus, the tubes of the urinary system, and the tubes of the reproductive system. Therefore, feces, urine, and eggs or sperm pass into the cloaca and to the outside through its opening. With the exception of the dipnoans and the coelacanth, teleosts and most mammals have lost the cloaca and, consequently, the digestive system terminates with an anus opening directly on the body surface. The reproductive and urinary systems also open on the body surface separately or through a common papilla posterior to the anus.

Two outgrowths develop from the embryonic gut of most vertebrates. A liver is common to all. It functions as a primary metabolic center of the body, including storage of energy in the form of glycogen, chemical interconversions of energy sources (*e.g.,* amino acids, fatty acids, glucose), storage of fat-soluble vitamins, detoxification of harmful chemicals, and synthesis of bile (a solution of hemoglobin residues, salts, cholesterol, etc.), which plays a role in fat digestion and absorption.

The second outgrowth in most vertebrates is a distinct, well-formed pancreas. Exceptions are the lamprey, dipnoans, and teleosts, which possess diffuse pancreatic tissue associated with the intestine. In some vertebrates the pancreas develops from a dorsal and a ventral lobe that fuse to form the adult organ. In those vertebrates with a discrete pancreas, it is glandular and consists of two distinct tissues.

1. An exocrine portion produces a solution of digestive enzymes and a buffer solution. These solutions are carried by a duct (or ducts) from the pancreas to the duodenum. The pancreatic enzymes play a major role in digestion of proteins, fats, and carbohydrates in the small intestine, and the buffering solution neutralizes the high acidity of the stomach contents.

2. Endocrine tissue (islets of Langerhans) produces two hormones, insulin and glucagon, which are involved in glucose metabolism. The endocrine tissue is not ducted, and the hormones leave the tissue via the circulatory system.

CHAPTER NINE

Urinary System

I n vertebrates, the urinary and reproductive systems are closely interrelated, anatomically and embryologically. Throughout the evolution of vertebrates, the urinary and reproductive systems in the embryo develop in close proximity and remain closely associated with one another in the adult.

URINARY SYSTEM

In spite of the interrelationships of the urinary and reproductive systems, the functions are worlds apart. In vertebrates the urinary system functions to maintain the homeostatic balance of fluids in intracellular and extracellular (blood, interstitial fluid, joint fluids, ocular fluids) compartments. Electrolytes, glucose, hormones, proteins, and other chemical substances are closely controlled. This balance is mediated through filtration, reabsorption, secretion, and elimination of excess chemicals above normal blood threshold levels.

Some of the most important excretory compounds that must be dealt with consist of nitrogenous waste, ammonia (NH_3), urea ($CO(NH_2)_2$), and uric acid ($C_5H_4O_3N_4$). These chemical compounds are synthesized using nitrogen that originates during deamination of amino acids occurring in protein digestion.

The availability of water to a vertebrate and its nitrogen excretory end product are closely correlated. Most freshwater teleosts, most larval amphibians, and adult aquatic amphibians excrete primarily ammonia, a condition known as ammonotely. Ammonia is extremely toxic, highly soluble in water, and requires copious volumes of water to excrete. Though each molecule of ammonia rids the body of only a single atom of nitrogen, it diffuses readily across the gill surface or is excreted by the kidneys.

Urea is less soluble in water than ammonia, is less toxic and can therefore be stored for variable periods of time. An advantage of being a ureotelic vertebrate is that urea can be stored and released as a concentrated solution, resulting in water conservation. In addition, two atoms of nitrogen are eliminated with each molecule of urea produced. As one might guess, the majority of terrestrial vertebrates, with the exception of most birds and reptiles, are ureotelic. Odd as it might seem, the chondrichthyes also are ureotelic and the marine teleosts are primarily ureotelic.

The hyperosmotic marine habitat contains a high concentration of ions and little water compared to the hyposmotic freshwater habitat, which contains a low concentration of ions and a high concentration of water. Water, following its concentration gradient, diffuses out of the marine fish and potentially causes dehydration, whereas water, again following its concentration gradient, diffuses into a freshwater fish, potentially causing dilution of body fluids. At the same time, ions diffuse across the gills into the marine teleost and chondrichthyes while they diffuse out of the freshwater teleost across the gills. Therefore, the marine teleosts and chondrichthyes are being "salted" while the freshwater teleosts are being "desalted."

To counteract these physiological problems, a number of evolutionary solutions have occurred.

1. Cartilaginous fishes synthesize and retain high concentrations of urea in their tissues, thereby raising ionic concentrations above that of the sea, and counteracting the negative flow of water from their bodies. The marine chondrichthyes function largely as a freshwater teleost because

water diffuses into the body across the gills as a result of retaining high concentrations of urea, raising its osmolality to a level slightly above that of sea water. Ions diffuse into the fish across the gills following the ionic concentration gradient.

2. Marine teleosts, to counteract their dehydration, drink seawater, excreting the salts, which neither the chondrichthyes nor the freshwater teleosts do.

3. Marine teleosts produce scant amounts of slightly hyposmotic urine, and freshwater teleosts and marine chondrichthyes produce copious amounts of hyposmotic urine.

4. Specialized cells, "salt cells," in the gills of marine and freshwater teleosts are capable of "pumping" sodium and chloride across the gill membranes—in the case of the marine teleost into the environment, and in the freshwater teleost into the fish. These processes occur against strong ionic concentration gradients and are expensive energetically. The gills of marine chondrichthyes, however, do not possess ionic pumps but, instead, eliminate highly hyperosmotic sodium chloride solutions secreted by the rectal gland.

5. Divalent ions consumed by marine teleosts are eliminated in the urine by the kidney.

Most birds and reptiles are uricotelic; they secrete uric acid. Of the three nitrogenous waste products, uric acid is the least toxic, the least soluble, and requires only small volumes of water to excrete. It generally is eliminated as a semisolid paste. Further, four atoms of nitrogen are eliminated with each molecule of uric acid.

The hagfish, mentioned previously, is unique in that its body fluids are isosmotic with sea water. Its only ionic problem is the elimination of divalent ions via the kidney.

Kidney

The kidney is the primary organ of the urinary system. It is composed of functional units known as nephrons. A nephron consists of a filtering cup-shaped structure known as a Bowman's capsule, into which fits a capillary called the glomerulus. Leading from the capsule is the renal tubule. Bowman's capsule, in conjunction with the glomerulus, acts as a filter of the blood. With the exception of the formed elements and most proteins, the filtrate reflects the chemical composition of the blood of the vertebrate. The filtrate then flows down the length of the renal tubule during which it is subject to chemical changes dependent upon the concentration of fluid surrounding the tubules and the physiological condition of the vertebrate. Both processes of active and passive transport occur.

Differentiation of the adult vertebrate kidney is diverse. From a ridge of nephric tissue in vertebrate embryos, a paired series of filtering organs, the glomeruli form, beginning at the cranial end and proceeding caudally. Accompanying this serial formation is a pair of tubes, the archinephric ducts (pronephric ducts) that drain the glomeruli.

Many biologists believe that the ancestral condition of the kidney was the holonepros, a primitive segmented kidney that probably extended from the cranial to the caudal end of the coelom. Each segment was drained individually by a pair of tubules that connected to paired archinephric ducts.

In all modern vertebrates the first embryonic kidney to develop is the cranial pronephros. In most vertebrates the pronephros is ephemeral and not functional, although the archinephric duct persists. The pronephros is functional in the larvae of agnatha, some fishes, and amphibians. Some biologists suggest that it may be somewhat functional in adult hagfishes and some teleosts. These animals may have other functional kidney tissue, however.

The archinephric duct continues to develop and induces the serial formation of the nephrons of the next embryonic kidney, the mesonephros, in the middle of the nephric ridge and caudal to the position of the pronephros. The mesonephros is the functional kidney of the embryos and larvae of all vertebrates. Among anamniotes the archinephric duct continues to induce nephron formation caudal to the mesonephros to give rise to the adult kidney, the opisthonephros.

The kidney in amniotes is derived from nephric tissue posterior to the mesonephros and is induced by a "new" structure, the ureteric bud originating from the posterior end of the archinephric duct. A break exists between the mesonephros and cranial end of the metanephros. The ureteric bud further differentiates into collecting ducts and ureters.

Cloaca

At the distal end of the urogenital system of all vertebrate embryos develops a space known as the cloaca, into which nitrogenous waste, reproductive cells, and intestinal waste products are released. A horizontal shelf develops in the cloaca of many fish, segregating the dorsal urodeum from the ventral coprodeum. Urine and reproductive cells are released into the uroderm and feces into the coprodeum.

Most adult vertebrates retain a cloaca. A major exception is the teleost fish, in which urine, reproductive cells, and feces are eliminated through two separate openings on the surface of the body. Other piscine exceptions include the lamprey and the ratfish (chimaera), a relative of elasmobranchs. Mammals, with the exception of the monotremes, comprise another noted group in which the openings of the urinary, reproductive, and digestive systems are separate.

Urine Elimination

How do vertebrates solve the problem of urine elimination? Among freshwater fishes urine is discharged constantly. In some fish the distal ends of the archinephric ducts or accessory ducts are enlarged, which some biologists recognize as a "urinary bladder." Whether it actually functions as a urine storage organ is debatable.

An organ of life-conserving convenience, the urinary bladder has evolved among terrestrial vertebrates. The major function of the urinary bladder is the storage of urine. Can you imagine a terrestrial vertebrate creating a urine trail during its daily routine? Predators would be ever "grateful." Not all terrestrial vertebrates possess urinary bladders, however. Major exceptions are the majority of birds and many reptiles, which eliminate urine and feces as a pasty mixture. The amphibian bladder evolved as a ventral evagination of the cloaca. Dehydration occurring among many amphibians is counteracted by diffusion of water from the urine into blood vessels. This is dependent upon the hyposmotic urine relative to the hyperosmotic plasma.

The urinary bladder in amniotes is derived from the basal portion of the allantois, an extraembryonic membrane. This "new" development in amniotes appears to have evolved from an extension of the cloacal bladder of the amphibians. In addition, among many amniotes a tube leading from the bladder, the urethra, develops from the allantois.

Membranes and Ducts

Three other extraembryonic membranes are found in amniotes: a yolk sac, the amnion, and the chorion. The yolk sac is a ventral, ancestral bag containing the nutritive yolk. It is found in all embryonic vertebrates. The amnion and chorion, along with the allantois, are innovations found among amniotes. The original function of these membranes is controversial at present. A popular explanation, however, is that they evolved at about the same time as the shelled (cleidoic) egg, to support and shelter the developing embryo.

In male vertebrates the archinephric duct usually is "parasitzed" by the reproductive system to function as an avenue for sperm transport. In most male anamniotes an accessory urinary duct has evolved. In male amniotes, the kidney is drained by the ureter, an innovative solution to the drainage problem. Further, nephric tubules in males often are converted to sperm transfer routes or, in some instances, carry both urine and sperm. With some exceptions, in anamniote females the archinephric duct continues to drain the kidney.

CHAPTER TEN

Reproductive System

The sex of a vertebrate embryo may be determined genetically at the time of fertilization, as in some snakes, some lizards, all birds, and all mammals. Among turtles, crocodilians and other lizards, sex is determined by environmental factors such as moisture and temperature. Until discovery of a species of viviparous lizard, *Eulamprus tympanum*, in which sex likewise is temperature dependent, with high temperatures producing males, it was thought that this could occur only in oviparous vertebrates. Females of this viviparous lizard, by regulating their temperature during gestation, are able to influence the sex of their broods depending upon sex ratios in their population (Robert and Thompson 2001)!

Among some vertebrates, for example, some marine teleosts, sex reversal may depend upon population dynamics, *e.g.,* a die-off of either sex. In other fishes, sex changes are sequential; *e.g.,* individuals are female, and under the influence of a balance of hormones, later become male, or vice versa.

Irrelevant to the nature of sex determination, the gonads exhibit no morphological differences during early development, and both sexes begin with an indifferent stage. The gonads develop from tissue medial to the kidney. The primitive germ cells that give rise to future sex cells migrate from the yolk sac into the developing gonad. Under the influence of hormones stimulated by a gene or environmental factors, the gonad differentiates into a testis or an ovary.

The functions of the ovaries are the production of oocytes or ova and sex hormones promoting "femaleness" and stimulating secretions by accessory glands in the female reproductive system. The functions of the testes and the accessory glands of the male reproductive system include production of semen (sperm and associated glandular fluids) and sex hormones. Male hormones are important to maintain "maleness" and continued stimulation of semen production.

For example, in mammals a sex determination gene on the Y chromosome of the male stimulates the secretion of testosterone to masculinize the testis. In the absence of the "male" gene and stimulation of the surge of testosterone, the female gonad develops into an ovary. In vertebrates the ovaries remain and function within the body cavity. The testes in male, non-mammalian vertebrates likewise remain and function in the body cavity. In most therian mammals the testes reside in an external pouch, the scrotum. In other therians the testes remain and function within the body cavity, *e.g.,* Cetacea. In still others the testes descend into the scrotum only during the reproductive period, *e.g.,* Lagamorpha.

In most vertebrates the oviduct forms lateral to the archinephric duct and is present in both embryonic males and females. Among the chondrichthyes and some amphibians, the oviduct forms as the result of a longitudinal fission of the archinephric duct. In males the duct regresses and is either lost or is present as a vestigial organ, *e.g.,* in *Necturus*. In females the oviduct persists to convey female sex cells.

A great deal of variation occurs in specialization of the oviduct. This often is correlated with the method of fertilization; whether external or internal. In most anamniotes, such as fishes and amphibians that are aquatic and oviparous, fertilization is external and the oviduct remains separate and relatively unspecialized. In teleosts, however, the oviducts are fused and open via a genital papilla. Among female amphibians the lining of the oviduct coats the fertilized eggs, and they are temporarily stored in the enlarged distal ovisacs.

In the chondrichthyes where fertilization is internal, the distal portion of the oviduct often is enlarged, forming a uterus. Near the proximal end of the oviduct has evolved a shell gland that deposits albuminous coats and a proteinaceous shell around the fertilized egg. In viviparous species the shell is temporary and subsequently disappears. The embryos of viviparous species settle in the uterus and sometimes are attached by means of a yolk sac placenta.

Most amniotes—for example, the reptiles, birds, and monotreme mammals—are oviparous, laying shelled (cleidoic) eggs. Development generally is terrestrial, and fertilization is internal. The large, yolky eggs are enveloped in albuminous coats in the proximal ends of the oviduct, and shells of various chemical composition are added by shell glands associated with the distal end of the oviduct. The reproductive tracts remain separate. In female reptiles the right oviduct is larger, but in birds only the left oviduct and shell gland are functional, the right oviduct having degenerated (Figure 10.1A, Figure 10.1B, and Figure 10.1C).

A great deal of variability within the anatomy of the oviducts has evolved among therian mammals. In the marsupials the distal ends of the oviducts, the uteri, have become enlarged but remain separate. It is in the uteri that fetuses develop. The uteri extend by means of cervices into the paired vaginae, which coalesce into a vaginal sinus opening into the urogenital sinus, which in turn opens to the outside. Paired vaginae have evolved in concert with the divided penis (bifid) of the male (Figure 10.2).

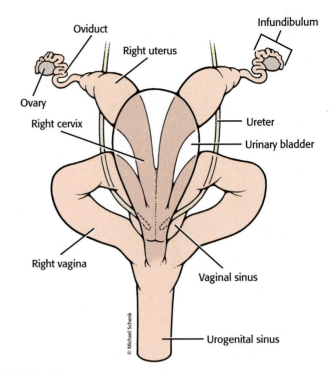

FIGURE 10.2 Female marsupial reproductive system.

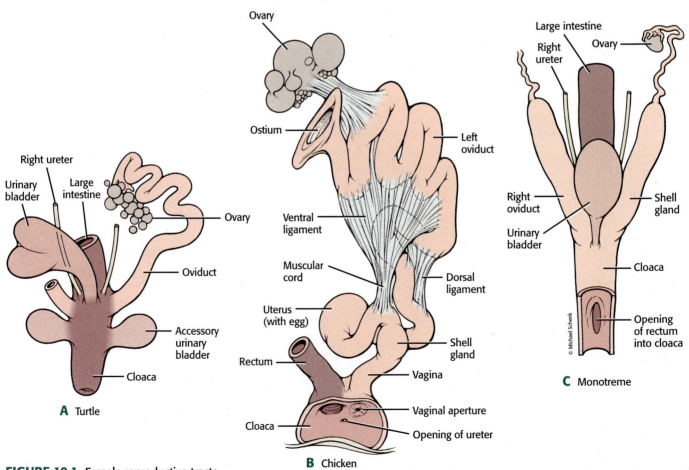

FIGURE 10.1 Female reproductive tracts.

Within the reproductive tracts of female eutherian mammals, the uteri and vaginae have fused to produce single organs, but belying their dual origin. The most primitive condition, a duplex uterus, consists of the uteri opening separately into the vagina via separate cervices and is found in many rodents, elephants, some bats and rabbits and their relatives (Figure 10.3A). Fusion of the distal ends of the uteri that open into the vagina by means of a single cervix distinguish a bipartite uterus (Figure 10.3B), found in most carnivores and many ungulates. A greater fusion of the distal ends of uteri opening into the vagina via a single cervix is the bicornuate uterus found in several carnivores, ungulates, whales, shrews, moles, and some bats (Figure 10.3C). A simplex uterus, consisting of completely fused

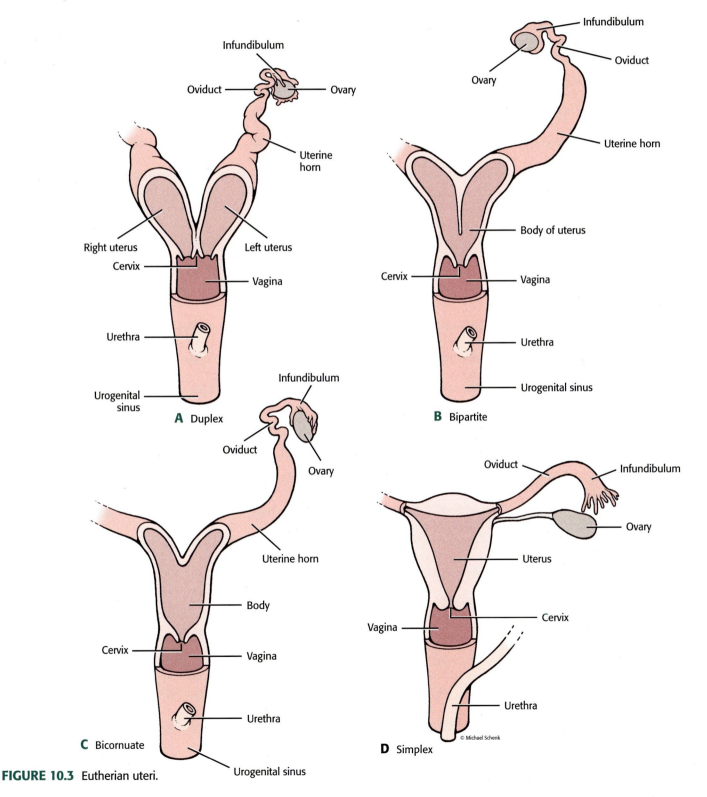

FIGURE 10.3 Eutherian uteri.

uteri and opening into the vagina by means of a single cervix, is found in armadillos and most primates, including humans (Figure 10.3D).

The uterus has evolved in conjunction with internal fertilization and serves as a space within which the developing young spend varying amounts of time. In viviparous species embryonic tissues and the lining of the uterus form a connection known as a placenta. Within amniotes, various combinations of extraembryonic membranes can be involved in the formation of the placenta.

How do the female gametes make their way from the ovary into and through the reproductive tract? In the cyclostomes the ovary wall ruptures, releasing the ovum into the coelom. The ovum then is transported from the body cavity through genital pores into the urogenital sinus, and ultimately is extruded through the genital papilla into the nest.

Within the gnathostomes the story is similar in that the ovary wall ruptures and the egg is swept into the coelom, but in these vertebrates the eggs then enter the often inflated proximal end of the oviduct, the ostium tubae. Exceptions are the teleosts that produce large numbers of ova, which if released into the coelom, could compress the viscera and impair the function of internal organs. Enclosed within the ovary is a small volume of coelom, from which lead the oviducts to their fused ends, opening through a genital pore or papilla, resulting in the direct expulsion of the ova. These "oviducts" represent extensions of the ovary and are not homologous with those of other vertebrates.

Most frogs and toads, on the other hand, practice external fertilization, involving the male clasping of the female with subsequent extrusion of the eggs by the female into water while the male releases sperm over the eggs. Internal fertilization requires a means of transferring sperm from the male into the reproductive tract of the female. Most urodeles practice internal fertilization. They transfer sperm by producing sperm packets on a pedestal, which the female picks up with her cloaca. As a result of this method of internal fertilization, various forms of courtship behavior have evolved to induce the female to pick up the sperm packets. On the one hand, in a highly specialized group of underground-dwelling amphibians, the caecilians, males possess a protrusible cloaca to facililtate sperm transfer to the female cloaca.

In many other vertebrates copulatory organs have evolved in association with internal fertilization. Among some teleosts (*e.g.*, the guppy) the anterior edge of the anal fin forms a gonopodium along which sperm are transferred into the female reproductive tract. In the chondrichthyes the pelvic fins have evolved into claspers, which are dorsally grooved organs that permit insertion into the female cloaca for sperm transfer. Typically, the contralateral clasper is used during this transfer.

Penises appear for the first time in male reptiles and comprise two different types. In the first type, occurring among lizards and snakes, the penis is bifurcated, known as a hemipenis, and lies inverted in a pouch on either side of the cloaca. During copulation only one of the hemipenes is everted and inserted into the cloaca of the female. In successive matings the hemipenes are alternated.

In crocodilians and turtles the penis lies in the floor of the cloaca and consists of a pair of erectile, spongy, vascular columns, the corpora cavernosa, capped by a glans penis. During sexual excitement the corpora cavernosa are infused with blood, causing them to swell and enclose the central groove along which sperm are transferred during copulation with the female. *Sphenodon*, a primitive reptile, has no penis, and sperm is transferred by direct cloaca-to-cloaca contact, a condition that may represent the primitive condition for amniotes.

Most birds transfer sperm via the "cloacal kiss," an apposition of the male and female cloaca. Among ostriches, rheas, emus, cassuarys, ducks, geese, and some other primitive birds, a penis has been retained and is involved during copulation.

The monotreme condition is similar to male reptiles except that the dorsal groove is enclosed permanently in the corpora cavernosa. The glans penis is bifurcated to accommodate the dual vaginal condition of the female.

Therian penises develop from a genital tubercle and consist of a corpus spongiosum that surrounds the urethra and a pair of erectile columns of tissue, the corpora cavernosa. The tip of the corpus spongiosum is enlarged into a glans penis. Typically, the penis is external and is enclosed in a fold of skin called the prepuce or foreskin. Within male marsupials, the glans is bifurcated in conjunction with the vaginal condition of the female. During sexual excitement, erectile tissues are infused with blood, causing erection of the penis, which then is used as a copulatory organ to facilitate sperm transfer.

A homologous organ, the clitoris, develops in female amniotes. In crocodilians, turtles, and birds that have a penis, the female has a clitoris. In therian mammals, the clitoris consists of paired erectile columns, the corpora cavernosa, but does not contain the urethra. During sexual excitement these erectile tissues become engorged with blood, bringing about erection.

CHAPTER ELEVEN

Circulatory and Lymphatic Systems

T he circulatory system of vertebrates consists of a network of continuous tubes connected to a muscular pump, the heart, constituting a closed system. This system, as its name implies, is responsible for the transportation, distribution, and circulation of various substances, *e.g.,* nutrients, gasses, hormones, and metabolites.

Although the circulatory system is continuous, it can be examined as individual components: (1) the blood, (2) the heart and arteries, and (3) the veins, arterioles, and venules that are generally connected by capillary beds.

BLOOD

Components of the blood are the fluid plasma and circulating elements, erythrocytes, leukocytes, and thrombocytes. Non-mammalian vertebrates possess nucleated blood cells. Most mammals, on the other hand, have enucleate erythrocytes but, similar to other vertebrates, their leukocytes are nucleated. In lieu of nucleated thrombocytes, mammals possess cell fragments known as platelets. Erythrocytes play an important role in gas transport. Leukocytes play a role in defensive and immune reactions against foreign proteins introduced into the body. Thrombocytes and platelets play a vital role in blood clotting.

Blood vessels transporting blood away from the heart are called arteries, and those transporting blood toward the heart are called veins. Most arteries carry oxygen-rich blood, and most veins transport oxygen-poor blood. There are exceptions among vertebrates. In most fish the exceptions are afferent branchial arteries, which carry oxygen-poor blood away from the heart to the gill region. In tetrapods pulmonary arteries carry oxygen-poor blood to the lungs and pulmonary veins carry oxygen-rich blood back to the heart.

A major function of capillaries is the diffusion of gases, water, and any dissolved substances between the circulatory system and tissues in which they occur. Under normal physiological conditions, blood elements and proteins occurring in the blood are too large to pass across the capillary wall and normally do not contribute to the overall movement of ions and water across the capillary cell membrane. Because the fluid (hydrostatic) pressure in the arterial end of capillaries exceeds that in the venous end, water has a tendency to diffuse out of the arterial end into surrounding tissues. Differences in osmotic pressures at the two ends of the capillaries, partially as a result of the diffusion of water out of the arterial end, lead to higher osmotic pressures within the venous end of the capillary, resulting in water diffusion into the venous end. Any fluid remaining in the tissues is returned to the circulatory system by way of lymphatic vessels.

A source of confusion often arises in studying the distribution of blood vessels associated with appendages. Be aware that the same vessel may have different names as it passes through various areas of the appendage. For example, in mammals the subclavian artery becomes the axillary as it passes through the armpit or axilla, and then becomes the brachial artery as it proceeds distally along the appendage.

HEART

The vertebrate heart lies in a **pericardial cavity** lined with a serous membrane, the **parietal pericardium**, which then continues over the heart surface as the **visceral pericardium**. Presumably, the ancestral vertebrate heart consisted of four linearly arranged chambers (Figure 11.1). The walls of all four of these chambers had typical cardiac muscle. From posterior to anterior, they were the **sinus venosus**, the **atrium**, the **ventricle**, and the **conus arteriosus**.

The heart of cyclostomes and modern chondrichthyes is constructed similarly to the ancestral heart (Figure 11.2). A difference between the ancestral condition and these fishes is that the chambers are realigned into an **S-shape** (Figure 11.3). Strategically placed valves between the chambers

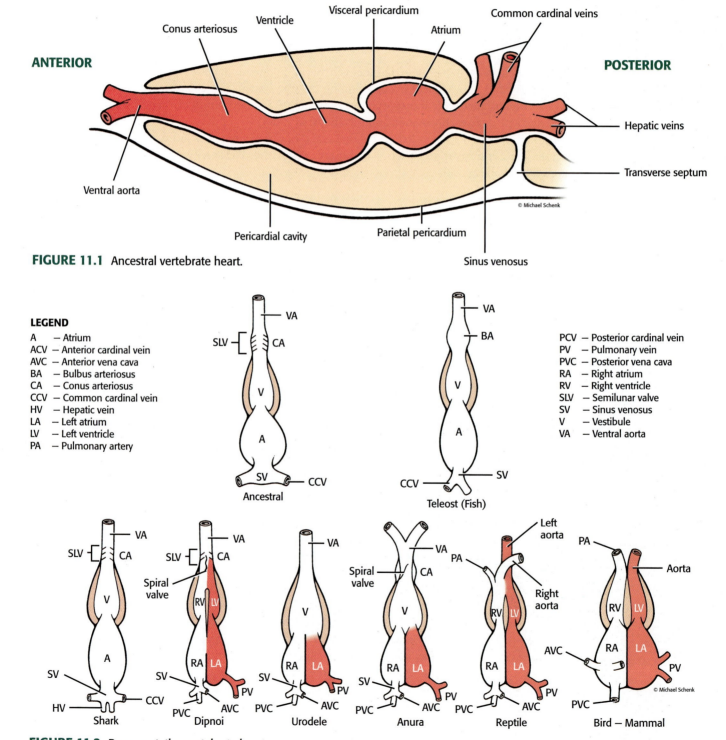

FIGURE 11.1 Ancestral vertebrate heart.

LEGEND

A   — Atrium
ACV — Anterior cardinal vein
AVC — Anterior vena cava
BA   — Bulbus arteriosus
CA   — Conus arteriosus
CCV — Common cardinal vein
HV   — Hepatic vein
LA   — Left atrium
LV   — Left ventricle
PA   — Pulmonary artery

PCV — Posterior cardinal vein
PV   — Pulmonary vein
PVC — Posterior vena cava
RA   — Right atrium
RV   — Right ventricle
SLV — Semilunar valve
SV   — Sinus venosus
V   — Vestibule
VA   — Ventral aorta

FIGURE 11.2 Representative vertebrate hearts.

ensure unidirectional blood flow. Modern teleosts have lost the conus arteriosus, and an elastic bulbus arteriosus devoid of muscle occurs in its place (Figure 11.2).

In almost all fish blood circulation depends upon a *single pump circuit*, in which all of the blood passing through the heart is oxygen-poor (Figure 11.4). From the heart, blood moves through the afferent branchial arteries to the gill lamellae, where gaseous exchange occurs in enclosed capillaries. From the capillaries, the blood flows from the gills via the efferent branchial arteries into the dorsal aorta. Arteries branching from the aorta transport oxygen-rich blood to the tissues. These vessels branch into capillaries in the tissues, and the blood returns to the heart through veins. This circulation is similar to the pattern found in the lamprey, first seen in Amphioxus.

Among some fish such as lungfish that live in tropical waters where oxygen levels may vary considerably, not only are gills present, but also lungs similar to those found in terrestrial vertebrates. The anatomy and function of the respiratory and circulatory systems in lungfish probably gives us a good idea of the way in which these systems evolved in tetrapods.

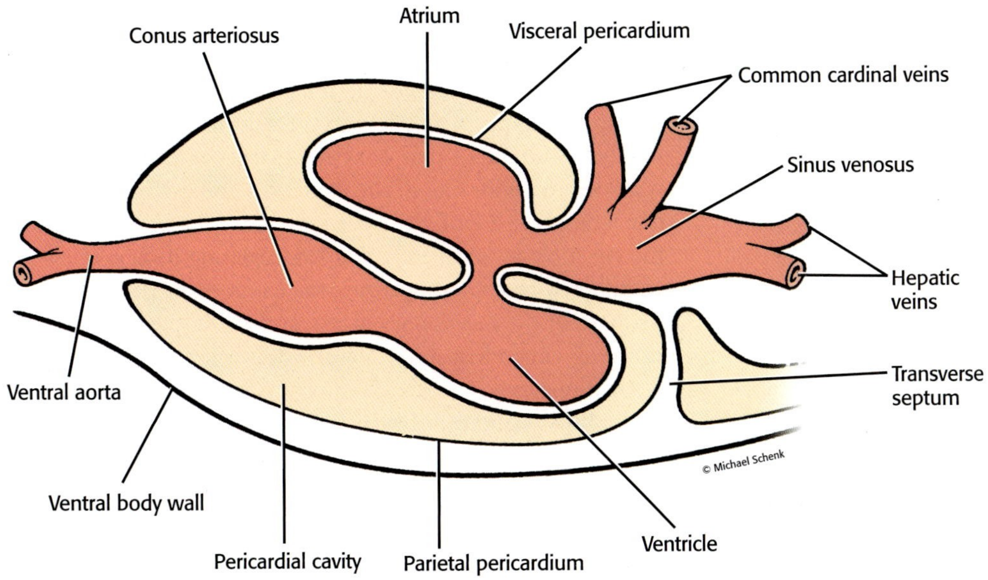

FIGURE 11.3 Shark heart: S-shaped configuration.

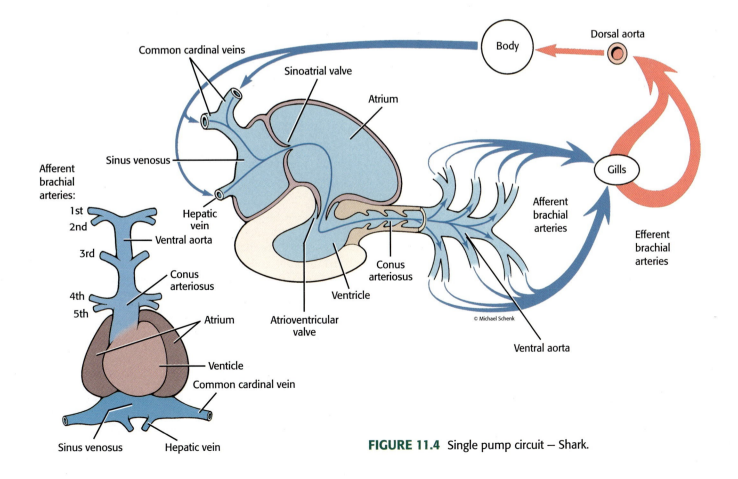

FIGURE 11.4 Single pump circuit – Shark.

The heart of the lungfish has evolved into an organ that is at least partially efficient at maintaining separation of oxygen-rich and oxygen-poor blood (Figure 11.2). The atrium is divided by an incomplete septum into left and right halves. In addition, an incomplete interventricular septum of the ventricle along with spiral folds in the conus arteriosus continue to prevent the admixture of the oxygen-rich and oxygen-poor bloodstreams. This condition likely foreshadows the ancestral tetrapod pattern.

In well-oxygenated water, lungfish utilize a typical fish respiratory pattern. Oxygen-poor blood enters the sinus venosus and is pumped consecutively into the atrium, ventricle, and conus arteriosus. With the appearance of lungs, pulmonary veins evolved and carry blood to the left side of the heart (atrium). Under normal oxygen tensions these fish utilize their branchial (gill) circulation in a fishlike manner. When oxygen levels drop, the pulmonary respiratory circuit is activated to augment gill circulation. During normal branchial circulation the blood is largely prevented from entering the pulmonary (lung) circulation by strategically positioned valves, while blood under hypoxic conditions and pulmonary use is channeled away from the branchial vessels, again through the action of valves. This reduces the probability of large amounts of oxygen from diffusing out of the body as would happen during hypoxia.

Neoceratodus, the Australian lungfish, retains a full complement of gills. *Lepidosiren*, the South American lungfish, and *Protopterus*, the African lungfish, have lost the gills on gill bars three and four. During aerial respiration blood travels directly into the dorsal aorta through the vessels of these gill bars. Experimental evidence indicates that *Lepidosiren* and *Protopterus* are obligate aerial breathers in addition to aquatic respiration, which continues to provide some oxygen and is necessary to eliminate CO_2. The Australian lungfish, however, is a facultative pulmonary breather.

Circulation in tetrapods is based on a *double pump circuit* (Figure 11.5). Among modern tetrapods lungs are present and aerial respiration is obligatory. The adult amphibian heart consists of a divided atrium with a left and a right chamber. The ventricle remains undivided, but the configuration of the heart keeps the oxygen-rich and oxygen-poor streams of blood fairly well separated. A spiral fold originating in the conus arteriosus maintains the relative separation of these streams. A much-abbreviated ventral aorta divides into a pair of carotid, systemic, and pulmocutaneous arches. Oxygen-rich blood is directed into the systemic circulation of the head and body, and oxygen-poor blood is directed toward the pulmocutaneous circulation to the lungs and skin.

However, *Necturus*, an aquatic amphibian, exhibits a combination of fish and tetrapod circulation (Figure 11.5A). Oxygen-poor blood returns to the heart from the body and enters the right atrium. Oxygen-rich blood returns from the

gills, passes through the pulmonary artery of the lungs, and enters the left atrium via the pulmonary vein. Both bloodstreams are pumped into the single ventricle, where some mixing occurs. From there, the oxygen-poor blood is pumped to the gills and skin, where oxygen and carbon dioxide diffusion occurs. The oxygen-rich blood, by contrast, is pumped from the ventricle to various body regions.

The circulatory system of reptiles resembles that of amphibians. The heart is three-chambered, with a ventricle that is partially divided by an interventricular partition consisting of complex folds (Figure 11.2). The conus arteriosus is subdivided into a pair of systemic trunks, each of which is associated with one of the ventricles and a pulmonary trunk associated with the right ventricle. Among crocodilians the ventricle is completely divided, resulting in a four-chambered heart. An opening, the foramen of Panizza, at the base of the pulmonary artery and aorta, permits blood shunting dependent upon whether the animal is terrestrial or is submerged in water. Because crocodilians function both above water and below water the foramen permits blood to be shunted in response to the pulmonary pressure. During underwater activity the pulmonary pressure is great, lungs are nonfunctional and great volumes of blood do not enter the pulmonary circuit.

The four chambers of the endotherms, birds and mammals, consist of two atria and two ventricles, representing the ancestral single atrium and single ventricle, completely subdivided by septa (Figure 11.2). The sinus venosus is incorporated into the wall of the right atrium and acts as a pacemaker for the heart, and the conus arteriosus probably is incorporated into the ventricles and bases of the pulmonary and systemic aortas. Oxygen-rich blood enters the left atrium, is pumped past valves into the left ventricle and then past valves into the aorta and then on to various parts of the body. Oxygen-poor blood enters the right atrium, is pumped past valves into the right ventricle and then past valves into the pulmonary aorta and into the lung circulation. The two streams are completely separated in the heart (Figure 11.5B).

Fetal development in placental mammals occurs in close internal contact with the lining of the female reproductive tract, where the placenta develops. The relationship between the fetus and the female is parasitic; the fetus is the parasite on the female host. The body of the female performs various physiological functions and, therefore, some fetal organs (*e.g.*, the lungs) are largely nonfunctional and fetal circulation to those organs is modified. Across capillaries in the placenta, gaseous exchange, nutrition, waste disposal, and the like take place.

Fetal blood, carrying carbon dioxide, metabolic wastes, and so forth, is transported to the placenta by way of the umbilical arteries. Blood laden with oxygen, nutrients, hormones, and other substances, is returned from the placenta via the umbilical vein to the liver, where most of the blood

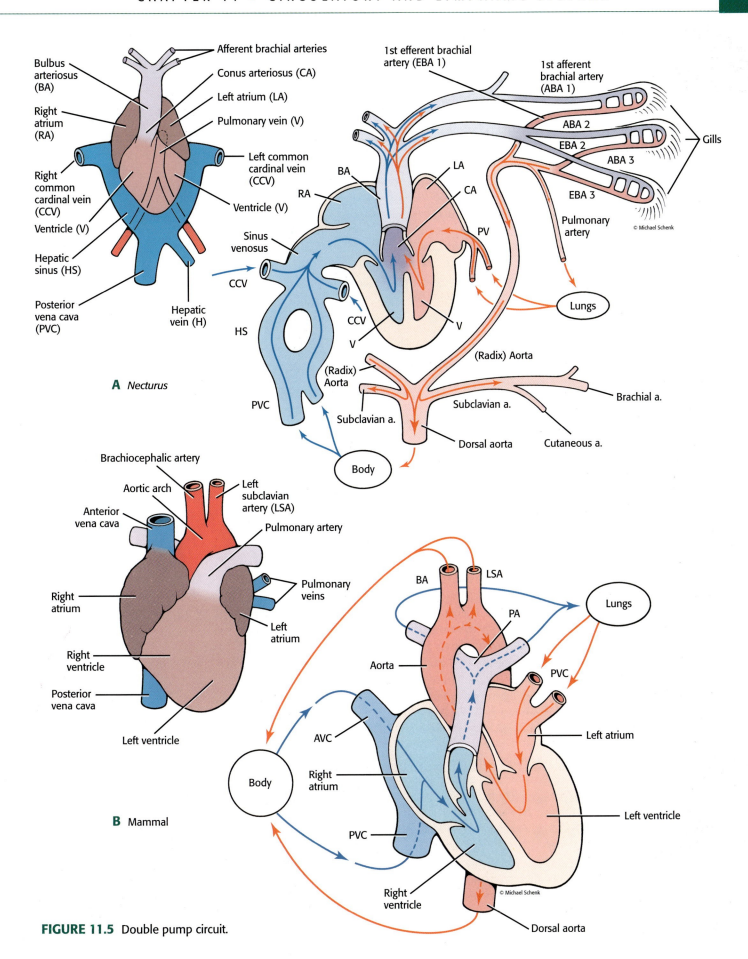

A *Necturus*

B Mammal

FIGURE 11.5 Double pump circuit.

is shunted directly into the posterior vena cava via the ductus venosus. This oxygen-rich blood returns to the right atrium of the heart, where it is shunted to the left atrium through an opening in the interatrial septum, the foramen ovale. From there it is pumped into the systemic circuit. Other venous blood returning via the anterior vena cava to the right atrium is pumped to the right ventricle and through the pulmonary trunk, where the ductus arteriosus connects to the aorta, and most of the blood is shunted away from the nonfunctional lungs into the systemic circulation.

All fetal structures are related directly to the fact that the fetus is basically dependent on the female for its existence. At birth, the fetal circuits are no longer functional and become remnants that usually are identifiable in the adult circulatory system. Under normal circumstances, contraction of the ductus arteriosus, separating the pulmonary artery from the aorta in the fetus, results in a tendinous band of tissue, the ligamentum arteriosum. The ductus venosus is transformed into the ligamentum venosum, and the umbilical vein becomes the round ligament of the liver. The basal portion of the umbilical arteries continue to supply the urinary bladder, whereas those portions between the urinary bladder and the umbilicus are constricted at birth and become the lateral umbilical ligaments of the border of the mesentery of the urinary bladder. The interatrial valve is forced against the foramen ovale by the abrupt change in blood pressure in the left atrium at birth. In a short time the edges are sealed through tissue growth, and a depressed oval area, the fossa ovalis, remains.

AORTIC ARCHES

We can learn a great deal by examining the circulatory systems of vertebrate embryos because during development the embryo repeats its evolutionary history in a general sense. Therefore, the development of the circulatory system of the upper torso, in particular, the fate of the aortic arches associated with the primitive gill arches, can be ascertained. Embryos among the gnathostomes almost invariably have six aortic arches, which develop sequentially from the anterior end of the embryo. The aortic arches traverse the gill bars and connect the paired ventral aortae with paired dorsal aortae.

Among sharks the number of aortic arches is related to the number of functional gill pouches. In most modern sharks the number of aortic arches is five, but some primitive sharks possess more, six or seven (Figure 11.6A). In a modern shark, the dogfish, the ventral portion of the first aortic arch disappears and the dorsal part gives rise to a spiracular artery. Ventral portions of arches two through six give rise to the afferent branchial arteries that supply blood to the gills. Collector loops derived from the rest of the aortic arches give rise to pretrematic and posttrematic arteries, which form efferent branchial arteries connecting to the dorsal aorta.

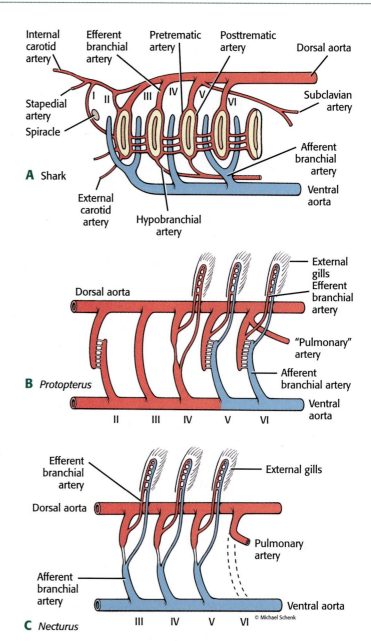

FIGURE 11.6 Adult aortic arch derivatives of the shark, *Protopterus* and *Necturus*.

Arteries associated with the head are the carotids. The common carotid is derived from the paired ventral aortae between the third and fourth arches. Two branches of the common carotid are (1) the external carotid artery, an extension of the ventral aorta, which supplies blood to structures of the ventral head region, and (2) the internal carotid artery, derived from the paired dorsal aortae, which supplies blood primarily to the brain.

Within the bony fishes, aortic arches 1 and 2 disappear (Figure 11.7). With the exception of the pre- and posttrematic arteries, the arterial scheme is similar to the shark. In the lungfishes (dipnoi), aortic arches 2–6 are retained and a pulmonary artery arises from the sixth aortic arch,

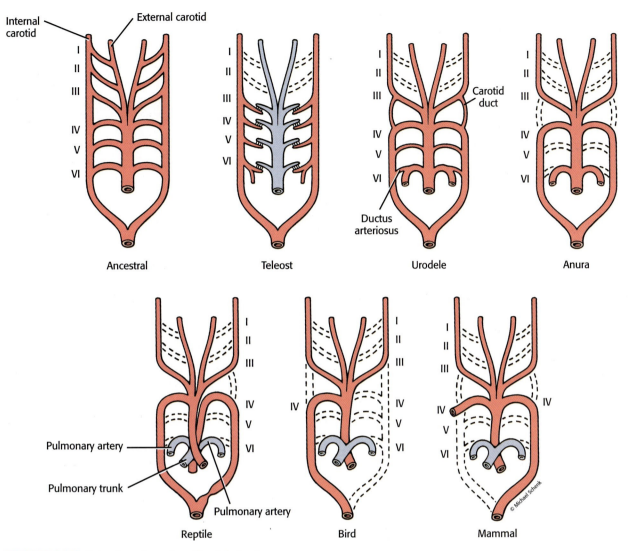

FIGURE 11.7 Fate of aortic arches (frontal view).

forecasting the condition among the tetrapods. Further, the third and fourth aortic arches in *Protopterus* and *Lepidosiren* function as passageways between the ventral and dorsal aorta (Figure 11.6B). *Neoceratodus*, however, retains functional gills associated with arches 2–5.

Among all tetrapods aortic arches 1 and 2 disappear. In amphibians the fourth aortic arches form the systemic arches, giving rise to the abdominal dorsal aorta and delivering blood to the posterior portion of the body (Figure 11.7). Anura lose the fifth arch. The urodeles retain the fifth and sixth arches. The sixth arch in amphibians gives rise to the pulmocutaneous artery, which transports blood to the lungs and skin in the adult.

Why deliver blood to the skin? Recall that the skin in amphibians is a supplementary respiratory organ. The ductus arteriosus connects that part of the sixth aortic arch dorsal to the pulmonary artery to the systemic arch. It is retained in urodeles but is lost in anura. The carotid duct, occurring between aortic arches 3 and 4, likewise is retained

in urodeles but not in anura. The anterior arterial scheme of amphibians is similar to that of the fishes. The neotenic condition of *Necturus* is accompanied by retention of external gills with a circulatory system similar to the lungfish condition (Figure 11.6B and Figure 11.6C).

The basic reptilian condition is similar to that found in the anurans. The third arches contribute to the internal carotids. The fourth arches remain as the systemic arches. The fifth are eliminated, and the sixth arch gives rise to the pulmonary arteries (Figure 11.7).

The fate of the aortic arches of birds and mammals, although not related, is similar, with the exception of the systemic arch configuration. The fourth right arch persists in birds, whereas the fourth left arch is present in mammals (Figure 11.7). In fetal mammals the ductus arteriosus, a dorsal extension of the sixth aortic arch, connects the left pulmonary and the aorta, thus bypassing the non-functional lungs.

BLOOD VESSELS

Blood vessels consist of the arteries, veins, and capillaries.

Arteries

Anteriorly, the primitive arterial pattern consists of paired dorsal aortae and paired ventral aortae connected by six pairs of aortic arches (Figure 11.8). Just posterior to the paired dorsal aortae, a single vessel, the dorsal aorta, traverses the length of the body and terminates in a caudal artery. Along the entire length of the dorsal aorta(e), paired segmental vessels originate.

The dorsal aortae persist as the internal carotid arteries, whereas the ventral aortae give rise to the external carotid arteries that supply cranial tissues. The subclavian arteries evolved from a pair of segmental vessels in the region of the anterior limbs as they made their appearance in the ancestral vertebrate. The iliac arteries, in turn, evolved from a pair of segmental arteries in the region of the posterior appendages when they appeared in the ancestral vertebrate.

Paired segmentals became associated with the kidneys and gonads as well as various paired glands (*e.g.*, the adrenals).

In addition, unpaired visceral arteries emanate from the abdominal region of the dorsal aorta. Arising anterior to posterior, the unpaired visceral arteries consist of a celiac artery and several mesenteric arteries. A common pattern in adult vertebrates is three—a celiac, a cranial mesenteric, and a caudal mesenteric artery.

Veins

The primitive venous drainage consisted of (1) anterior, posterior, and common cardinal veins; (2) caudal, subintestinal, and vitelline veins; and (3) lateral abdominal, iliac, and subclavian veins. The cardinal veins drained the anterior and posterior body. The caudal, subintestinal, and vitelline veins drained the tail, the abdominal organs, and the yolk sac. The lateral abdominal, the iliacs, and the subclavians drained the lateral body wall and anterior and posterior appendages, respectively (Figure 11.9).

Venous circulation in the shark exhibits the basic gnathostome condition (Figure 11.10). Two sets of major paired veins, the **anterior cardinals** drain the cranial portion of the body, and the **posterior cardinals** drain the trunk.

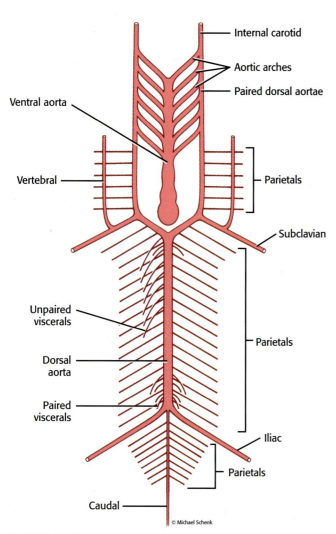

FIGURE 11.8 Ancestral arterial pattern of jawed vertebrates (frontal view).

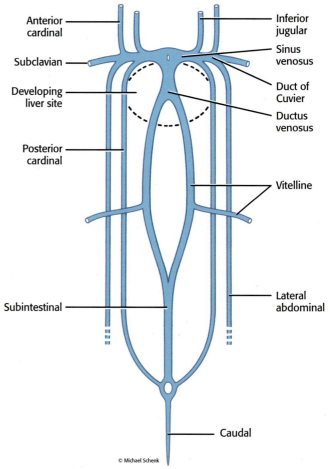

FIGURE 11.9 Ancestral venous pattern of jawed vertebrates (frontal view).

The cardinals unite to form a pair of common cardinal veins, which empty into the **sinus venosus**. Drainage of the ventral portion of the head is accomplished by the **inferior jugular veins**, which also empty into the common cardinal.

In the amniotes the anterior cardinals are known as internal jugular veins, draining the brain and internal structures of the head. Lateral portions of the anterior cardinals form the external jugulars, draining the external parts of the head. Among most vertebrates, with the evolution of the tongue, the inferior jugulars become the linguals, draining the tongue. In amniotes the common cardinal veins become known as the paired anterior vena cavae. In some

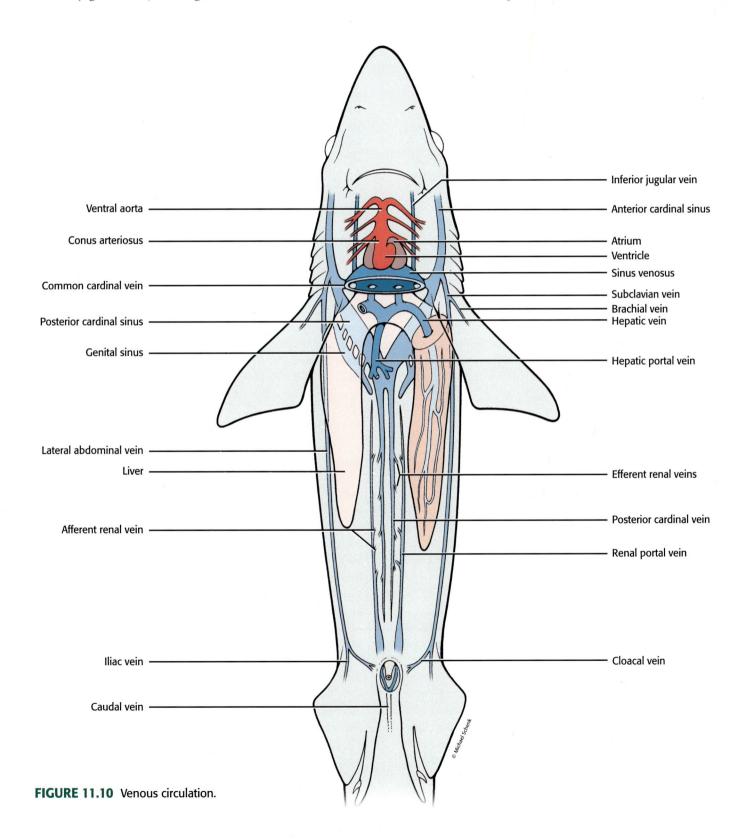

FIGURE 11.10 Venous circulation.

mammals (*e.g.*, cats and humans), the right anterior vena cava persists and is joined by the right subclavian vein to form the right brachiocephalic vein. The left brachiocephalic vein crosses the midline of the body and joins the right brachiocephalic. That portion of the left anterior vena cava posterior to the point at which the left crosses over is eliminated, leaving a remnant draining the heart called the coronary sinus.

Drainage of the posterior portion of the body among vertebrates is variable. The posterior cardinal veins of the ancestral vertebrate that drain the posterior body from the cloaca, the lateral body wall, and the reproductive and urinary organs join the common cardinal veins. Among the fishes the caudal part of the posterior cardinals switches its allegiance to the lateral portions of the kidneys, becoming the renal portal system. The caudal vein draining the tail joins the renal portal veins. In sharks the renal portal veins empty into the posterior cardinal veins.

Renal portal systems persist among the tetrapods through the birds. With the exception of some reptiles and birds, large volumes of blood flow through kidney tissues, utilizing the renal portal system. In these reptiles and all birds the blood is shunted directly through the kidneys into the posterior vena cava. A renal portal system is absent in mammals.

Among the craniates, specialized blood transport routes, called portal systems, between two organs are found. A portal system is defined as a capillary bed at either end of the system, connected by a vein. Examples are the hepatic portal system extending between the liver and digestive organs, the renal portal system extending between the posterior portion of the body and the kidney, and the hypophyseal portal system between the hypophysis and the hypothalamus.

Postcardinal veins among adult tetrapods generally have been eliminated. In mammals, however, a remnant of the right postcardinal remains as the azygous vein, and in some mammals a remnant of the left postcardinal remains as the hemizygous vein, respectively, each draining the intercostal spaces. The hemizygous joins the azygous, which in turn drains into the anterior vena cava.

Among the tetrapods and lungfish, drainage of the posterior body is accomplished by a "new" vessel, the posterior vena cava. Within ancestral vertebrates, not only was the posterior body drained by the posterior cardinals but also by vessels called subcardinal veins. Portions of these vessels give rise to the single posterior vena cava. In mammals the supracardinal veins also contribute to formation of the posterior vena cava. Among non-mammalian tetrapods that possess renal portal systems, the posterior vena cava extends from the kidneys to the heart, passing through the liver and emptying into the sinus venosus.

Among primitive gnathostome fish such as the shark, the lateral body is drained by a pair of **lateral abdominal veins**, the anterior paired appendages are drained by the **subclavian veins,** and the posterior appendages are drained by the **iliac veins.** Each, in turn, empty into the **lateral abdominal veins** at their respective levels (Figure 11.10). Among the tetrapods the subclavians empty into the anterior vena cava. Generally, in amphibians the two lateral abdominal veins fuse, forming a single ventral abdominal vein. In reptiles such as turtles and crocodiles, paired lateral abdominal veins persist. In adult birds and mammals, lateral abdominal veins are not present.

Among vertebrate embryos whose venous architecture resembles the ancestral condition, two paired veins, the vitelline and the subintestinal veins, associated with the yolk sac and intestinal tracts, give rise to the universal hepatic portal vein. The hepatic portal vein drains the viscera, carrying blood laden with most nutrients, waste products, toxins, etc. This vein subsequently enters the liver and forms sinusoids. The sinusoids coalesce into one or more hepatic veins, which empty into the sinus venosus or into the posterior vena cava and subsequently into the heart.

Lymphatic System

Although most fluid in the circulatory system remains confined within vessels, water and ions are exchanged in capillary beds between tissues and capillaries at both the arterial and venous ends. At the arterial end, water and ions diffuse into the tissues, and water and ions from the tissues diffuse into the capillaries at the venous end. This exchange is not 100% efficient; usually a net volume of fluid is left in the tissues at the venous end. Over time, if allowed to accumulate, this excess fluid would cause osmotic problems for the tissues.

A lymphatic system evolved in vertebrates to get the excess fluid back into the circulatory system. In most vertebrates the lymphatic system consists of a branching system of vessels beginning from closed, blind capillaries, unconnected to arterial vessels, that empty into veins. Therefore, high pressures, generally found in arteries, cannot influence fluid flow in lymph vessels. Low pressure in the lymphatic system promotes slow lymph flow within lymphatic vessels. Lymph ultimately empties into veins.

Lymph vessels are well developed in the gut and are the site of diffusion of digested fats from the small intestine. Some vertebrates have a series of pulsatile lymph hearts that aid in moving the lymph within the vessels. In amphibians, large lymphatic sinuses lie under the skin. Water from these sinuses diffuses out into the tissues during periods of desiccation, to minimize damage to the animal.

In mammals, small masses of lymphatic tissue (*e.g.,* lymph nodes, lymph glands, tonsils) are associated with lymphatic vessels. Within this tissue, lymphocytes filter and remove bacteria, viruses, and other foreign protein.

CHAPTER TWELVE

Nervous System and Sense Organs

T he nervous system is one of the most complex and least understood of any system in animals. The complexity of this system is reflected not only in the anatomy but also in its physiology. Consider that this system, often in conjunction with portions of the endocrine system, is the controlling center for the homeostatic well-being of the body. The nervous and endocrine systems initiate, moderate, and coordinate all body activities.

Classically, the brain and spinal cord are grouped together as the central nervous system, and the cranial nerves, spinal nerves, and autonomic nerves are known collectively as the peripheral nervous system.

CELLS OF THE NERVOUS SYSTEM

Cells of the nervous system are specialized for the transmission of information and the insulation and isolation of individual cells and bundles of cells. Neurons are the cells responsible for transmitting action potentials. A neuron consists of a cell body containing the nucleus and the center of synthesis of materials needed for the function of the cell. A receptive process or processes, the dendrite(s), transmits an impulse to the cell body and ultimately to the single axon, where a space, the synapse, exists between it and its effector (a muscle fiber or a gland) or another neuron. An axon usually branches into telodendria that may synapse with a number of effectors or other neurons. Axons also are known as nerve fibers.

Three different types of neurons are:

1. The multipolar neuron: possesses many dendrites and a single axon.

2. The bipolar neuron: possesses a single dendrite and axon separated by the cell body.

3. The pseudounipolar neuron: also possesses a single dendrite and axon; however the cell body is offset laterally (Figure 12.1). Embryologically, this neuron originates as a bipolar neuron whose two processes fuse, forcing the cell body to lie laterally.

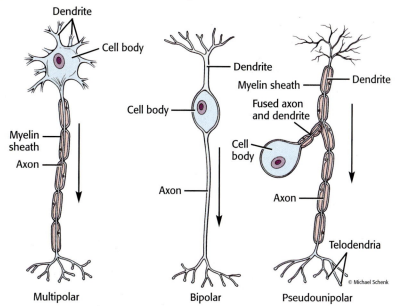

FIGURE 12.1 Types of neurons.

Another classification system of neurons is defined by the presence or absence of a fatty wrapping of myelin. Axons of myelinated neurons are wrapped by a series of individual Schwann cells; hence, there are gaps, nodes of Ranvier, between successive cells. Impulse transmission is much more rapid in myelinated fibers because the action potential is transmitted electrically across the myelin from node to node, where depolarization and repolarization occur only in the gaps. The generation of action potential brought about by depolarization and repolarization of the entire length of the fiber membrane through the exchange of sodium ions in unmyelinated fibers is a slow process compared to its counterpart in myelinated fibers, in which much of the fiber is not involved in this process.

The functional system of neuron classification is either sensory or motor and refers to the type of information transmitted by that neuron. In the somatic sensory portion of the nervous system, impulses including information such as odor, vision, temperature, shape, size, texture, and the like are transmitted from peripheral receptors, which are specialized dendritic endings, via a dendrite to the central nervous system.

In the visceral sensory portion of the nervous system, impulses are transmitted from receptors in the internal organs, carrying information of the conditional status of the organ. For example: Does your stomach ache? Are you hungry? Are you thirsty?

Motor neurons transmit impulses from the central nervous system via the peripheral nervous system to effectors such as striated locomotory muscles in the somatic motor portion of the nervous system or smooth or cardiac muscle or glands in the visceral motor portion of the nervous system. Interneurons occurring only in the central nervous system transmit impulses from sensory neurons to motor neurons (Figure 12.6 and Figure 12.7).

The telodendria of axons are not in direct contact with other neurons or effectors but are separated by a space called a synapse. Impulses are transmitted across this space by chemicals known as neurotransmitters. The neurotransmitters are stored in the presynaptic endings of the synapse and are released upon arrival of an action potential. These chemicals become associated with receptor sites on the postsynaptic ending, causing the generation of an impulse in the postsynaptic cell (neuron, muscle cell, or gland cell). Examples of common neurotransmitters are acetylcholine, norepinephrine, and serotonin. Neurotransmitters may be excitatory or inhibitory, and some may be either, depending upon where they occur.

Several other types of cells of the nervous system known as glial cells make up large volumes of tissue found in the central nervous system. Oligodendroglia are involved in myelination of central nervous fibers. Astrocytes function in metabolic activities. Microglia are the garbage detail of the central nervous system; *i.e.,* they are phagocytic. Also, all of the glial cells act as the central nervous system connective tissue, anchoring structures such as blood vessels, neurons, etc. Schwann cells are glial cells occurring in the peripheral nervous system and are responsible for myelination of the peripheral nerve fibers.

BRAIN

Characteristic of the most complex of any known animal nervous system, that of vertebrates, is a dorsal hollow nerve cord with a greatly elaborated anterior end known as a brain, related to the organization of a body plan with a well-defined head. Invertebrate animals often are said to possess a brain, but it consists of a mass of nervous tissue called ganglia and generally is capable of behavior associated only with stereotypic activities.

Early in the evolution of vertebrates, among the ostracoderms and the lamprey, the brain consisted of three basic regions, the **prosencephalon**, the **mesencephalon,** and the **rhombencephalon** (Figure 12.2). Obviously, this pattern

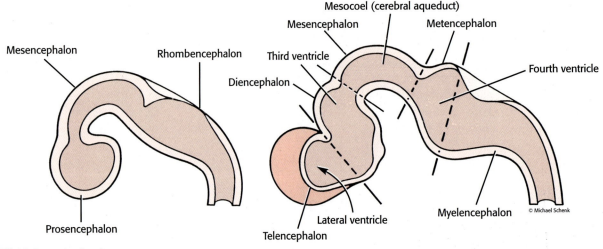

FIGURE 12.2 Brain development.

probably had been evolving for some time before the first identifiable vertebrates appeared. During the early embryological development of the nervous system of vertebrates, the same three primitive divisions of the brain differentiate from the enlarged anterior end of the neural tube: the anterior prosencephalon or forebrain and the posterior rhombencephalon or hindbrain, separated by the mesencephalon or midbrain. As differentiation progresses, the mesencephalon remains intact but the prosencephalon subdivides into the telencephalon and the diencephalon, and the rhombencephalon divides into the metencephalon and myelencephalon, yielding a five-part brain (Figure 12.2).

The **telencephalon** further differentiates to form the olfactory region, consisting of the **olfactory lobes** and **bulbs**, and the **cerebrum**. The **diencephalon** remains undivided. The **mesencephalon** is organized as dorsal **optic lobes** and ventral **cerebral peduncles**. The **metencephalon** differentiates into the **cerebellum**, and the **myelencephalon** becomes the **medulla oblongata** (Figure 12.3).

Tissues of the nervous system are classified as gray matter and white matter. Nerve cell bodies, dendrites, and unmyelinated axons appear gray, and myelinated axons appear white. Groups of myelinated neurons or fibers are

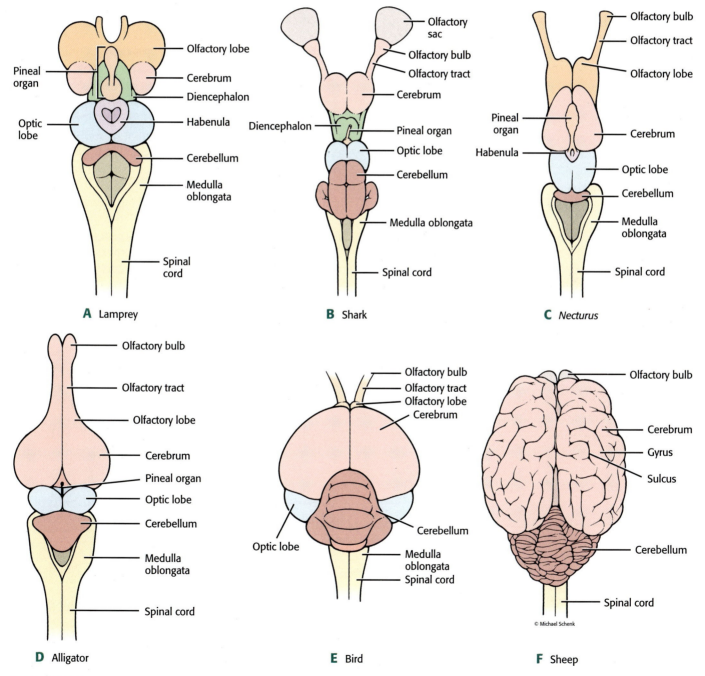

FIGURE 12.3 Representative vertebrate brains (dorsal view).

called nerves. The primitive or basic pattern of these tissues in the central nervous system is a core of gray matter, nerve cell bodies, dendrites, and unmyelinated axons, surrounded by a layer of white matter or mylinated nerve fibers. This is the condition found in the brains and spinal cords of anamniote vertebrates.

The cerebrum in the brain of fish, amphibians, and reptiles is organized largely as an olfactory brain. The major function, then, is to detect environmental chemicals. More recent studies, however, indicate that many other functions are centered here. Olfaction among birds has largely been eliminated, with exceptions among carrion-eating birds, which have keen senses of smell. Many mammals retain highly developed senses of smell.

With the evolution of the reptiles at the base of the amniote ancestry, we begin to see elaboration of the pallial region on the surface of the cerebrum giving rise to the cortex. Among birds and mammals the volume of the pallium becomes greatly enhanced and leads to a major change in cerebral organization. The pallium consists primarily of nerve cell bodies, dendrites and unmyelinated axons, and hence is gray matter. This marks a major change in the relationship of gray and white matter in the brain, *i.e.*, gray on the outside and white on the inside. It also leads to the marked increase in cerebral volume, particularly among mammals.

The brains of fishes and amphibians and most reptiles are basically linear. The brain of birds and mammals, however, undergoes flexure. The cerebrum comes to overlie the diencephalon and mesencephalon, and the telencephalon and diencephalon are angled with respect to the rest of the brain (Figure 12.2).

Within the confines of the skull, the space allotted to brain development is limited. In some mammals the solution is one often seen among structures confined within a finite space: The surface was thrown into complex folds resulting in convexities, **gyri**, separated by concavities, **sulci**, making it possible to "cram" large numbers of cells into the thin, outer gray layer of nervous tissue (Figure 12.3F). In birds and mammals the cerebrum has become greatly enhanced as the major integration center for body control and the voluntary activity initiation center (Figure 12.3E and 12.3F).

The diencephalon is characterized by a ventral hypothalamus, lateral thalami, and a dorsal epithalamus. From the surface of the epithalamus, the pineal and parietal bodies are projected. They function in regulating activities based upon circadian and circannian cycles. Some evidence suggests that the pineal and parietal organs may represent a relict pair of dorsal eyes. The thalamus of the diencephalon is the center of all sensory relays from the caudal regions of the central nervous system to the cerebrum, as well as a basic awareness interpretive center, *e.g.*, for pain and pleasure. Residing in the hypothalamus are centers for homeostasis through coordination of the autonomic nervous system, which control functions such as temperature regulation, feeding, sexual behavior, and so on. Both excitatory and inhibitory centers are present.

The brainstem consists of the mesencephalon and medulla oblongata in non-mammalian vertebrates, and includes the pons in mammals. The roof of the mesencephalon of anamniotes is the tectum, which is configured into a pair of rounded optic lobes. Optic lobes continue to be evident in reptiles and birds. Upon these lobes are projected a point-for-point set of visual data and it is here that visual interpretation is accomplished.

Caudal to the optic lobes is a pair of lobes receiving auditory impulses originating with the amniotes. Although clusters of neurons occur in a similar position in fishes, they are not obvious. In mammals the optic lobes and the auditory lobes form the corprorora quadrigemina. These two regions are the sites of visual and auditory reflexes and the relay of visual and auditory impulses to the proper cerebral sensory interpretive areas. The floor of the mesencephalon consists primarily of two thick collections of fibers, the cerebral peduncles, which carry impulses between the various portions of the central nervous system. The floor of the mesencephalon in fishes and tetrapods is a center where various locomotory movements, such as the characteristic movements of the paired appendages, are initiated.

The medulla oblongata grades into the spinal cord and consists of groups of fibers that carry sensory impulses from the body to the brain and also motor impulses from the brain to the body. Residing in the medulla are the control centers for a number of visceral functions such as heart beat, heart rate, respiration, and other visceral activities.

The cerebellum is found at the caudal end of the brain and often overlying the medulla oblongata (Figure 12.3). This is the second portion of the brain where we find gray matter overlying a central core of white matter. Its appearance is superficially similar to the cerebrum. The cerebellum of non-avian and non-mammalian vertebrates is generally smooth but is large in active animals such as sharks and many bony fishes. The cerebellum of vertebrates such as birds and mammals is large, and the surface is thrown into sulci and gyri, thereby increasing the surface area. In the cerebellum we find huge numbers of synapses with the fibers of neurons transmitting impulses from all parts of the body and brain. In the cerebellum are found centers for equilibrium, spatial position, and locomotion.

In some birds and in mammals ventral bulges, the pons, sit at the base of the cerebellum (Figure 12.4). Impulses are transmitted from the cerebrum to the cerebellum in the pons. In addition, impulses are transmitted between the hemispheres of the cerebellum.

Because the vertebrate brain and spinal cord developed from a hollow nerve cord, it should not be too surprising to discover that the central nervous system is not solid but, instead, has cavities. The cavities of the brain are identified

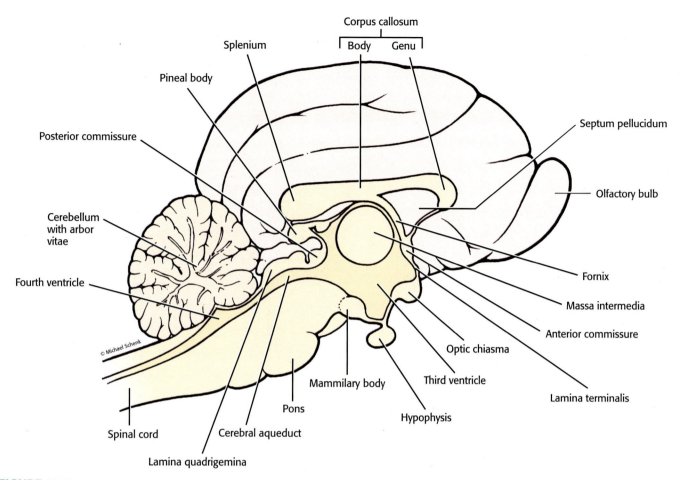

FIGURE 12.4 Brain: Left sagittal section.

as ventricles and traditionally are numbered using Roman numerals, while the constricted cavity of the spinal cord is known as the central canal. Lining the central nervous system is a modified squamous epithelium called the ependyma. With the bony fish as the only exception, ventricles I and II of the cerebrum are found in the left and right halves of the cerebrum, respectively. The bony fish possess a single large ventricle. Connecting ventricles I and II with ventricle III, located in the diencephalons, are narrow canals called the foramina of Monro. In anamniotes the mesencephalon contains a conspicuous cavity connected to the third ventricle. In amniotes this cavity is constricted, forming the aqueduct of Sylvius, which leads into ventricle IV in the medulla oblongata. Confluent with ventricle IV is the narrow central canal of the spinal cord.

MENINGES AND CEREBROSPINAL FLUID

The central nervous system is surrounded by a series of enveloping membranes, the meninges. Fishes possess a single vascularized covering referred to as the **primitive meninx** (Figure 12.5A). In amphibians and reptiles this membrane differentiates into two layers—an outer tough **dura mater**

that adheres to the inner membrane of the overlying bone, and an inner vascularized (**pia-arachnoid**) layer closely associated with the central nervous system, separated by a **subdural space** (Figure 12.5B). In birds and mammals the meninges consist of an outer tough dura mater, a delicate inner vascularized **pia mater** separated by an inner, cobwebby layer called the **arachnoid**. A space, the subdural space, occurs between the dura mater and the arachnoid, and a more extensive space, the **subarachnoid** space occurs between the arachnoid and the pia mater (Figure 12.5C and Figure 12.5D).

A series of complex highly vascularized membranes, choroid plexi (singular plexus), are associated with the ventricles. A choroid plexus is a combination of a vascularized meninx and the ependyma, the epithelial lining of the central nervous system.

Cerebrospinal fluid is secreted by the plexi. The cerebrospinal fluid circulates within the inner spaces of the central nervous system of vertebrates. Some fluid in anamniotes and reptiles probably bathes the outer surfaces. In birds and mammals it circulates within the ventricles and central canal, escaping through pores in the central nervous system into the subarachnoid space, where it continues to circulate and bathe the outer surface of the brain and spinal cord.

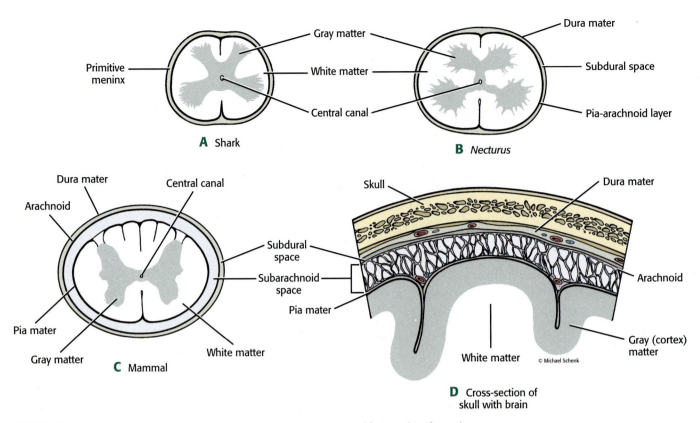

FIGURE 12.5 Cross-sections of vertebrate central nervous systems with associated meninges.

The cerebrospinal fluid drains by way of the circulatory system. The function of the cerebrospinal fluid is to ensure that the delicate nervous tissue remains moist and to dampen mechanical injuries. In addition, the fluid provides chemical homeostasis. Further, in mammals it provides a measure of buoyancy, reducing the weight of the brain.

SPINAL CORD

More primitive in construction is the spinal cord of vertebrates. It extends from the brain, passes through the foramen magnum, and continues through the vertebral foramen of the spinal column, which consists of vertebrae. Among the cyclostomes the spinal cord appears entirely gray because they either do not have myelinated fibers or these fibers cannot be distinguished. Among the jawed fish the inner core consists of **gray matter** containing the dendrites and nerve cell bodies of motor neurons and interconnecting interneurons, surrounded by a layer of **white matter** consisting of nerve fibers (Figure 12.5A). With the amniotes the gray matter tends to be configured into an H-shape. Another trend among tetrapods is for the outer surface to be indented, forming sulci or fissures dorsally and/or ventrally (Figures 12.5B and 12.5C).

All nerve tracts, both ascending or sensory and descending or motor, are represented in the white matter of the spinal cord. Tetrapods show a trend toward enlargement of spinal segments associated with the appendages—the cervobrachial of the anterior appendages and the lumbosacral of the posterior appendages. These are related to the greater complexity of the musculature associated with terrestrial locomotion.

SPINAL NERVES

Associated with the spinal cord in vertebrates are nerves of the peripheral nervous system. The function of fibers found in spinal nerves is either sensory or motor. Further, each may be somatic or visceral.

In the sister group of the vertebrates, the cephalochodates (*e.g.,* amphioxus), what appeared to be alternating dorsal and ventral roots of the nerves associated with the nerve cord turns out to be dorsal nerves alternating with striated muscle fibers connecting the nerve cord with the myomeres. In the lamprey the spinal nerves consist of alternating intersegmental dorsal spinal nerves carrying chiefly somatic and visceral sensory neurons and a few visceral motor neurons, and ventral spinal nerves carrying only somatic motor neurons. The majority of the visceral motor neurons are carried in a cranial nerve. This organization of the spinal nerves seems to be the basic blueprint for vertebrates.

Paired, fused, segmental spinal nerves associated with the innervation of skin and its structures, and muscles of individual body segments, are found among all vertebrates. The dorsal and ventral roots of each segment fuse to form a spinal nerve. All somatic sensory fibers enter the spinal cord through the dorsal root. The ventral root carries all somatic motor fibers. On the one hand, amniotes follow the rules, with all sensory fibers entering the spinal cord through the dorsal root and all motor fibers exiting the spinal cord through the ventral root. Anamniotes, on the other hand—similar to teenagers, who sometimes are re-luctant to follow rules—have many visceral motor fibers leaving the nerve cord through the dorsal root while some continue to be associated with the ventral root. Generally, the number of spinal nerves is correlated with the number of body segments.

COMMON NEURONAL PATHWAYS

We might ask, "How are impulses transmitted in the nervous system?" Among all vertebrates three circuits are found: reflex arcs and ascending and descending pathways.

Reflex Arcs

The functional pathway of the nervous system is the reflex arc (Figure 12.6). The simplest reflex arc consists of a two-neuron circuit—(1) a sensory neuron originating in the periphery of the body entering the spinal cord through the dorsal root of a spinal nerve that synapses with (2) a motor neuron originating in the spinal cord, exiting through the ventral root of the same spinal nerve and causing a response by innervating appropriate muscles. Most circuits, however, are much more complex, in which interneurons within the spinal cord are interposed between the sensory and motor neurons, permitting impulses to be transmitted to other segments of the cord as well as to the brain and opposite side of the body (Figure 12.7). These complex circuits enable decisions to be made within the nervous system, and whether the original stimulus requires an action other than a simple avoidance reaction to a noxious stimulus.

Ascending and Descending Pathways

Sensory impulses travel from the body periphery through a sensory neuron residing in a spinal nerve and enter the spinal cord through the dorsal root ganglion, where it may synapse with interneurons and motor neurons. From the point of entry, this information may be carried by either the original sensory neuron or, more commonly, by interneurons to higher centers in the nervous system. This part of the circuit constitutes an example of an ascending pathway (Figure 12.7).

Two events may result from the arrival of this sensory information:

1. A simple reflex action may result from transmission of impulses through interneurons to motor neurons in the same segment to effectors in the affected area.

2. If the stimulus is more complex or stronger, the higher centers of the central nervous system will initiate im-pulses that are carried in a descending pathway to the affected segment or segments, as well as to segments on the opposite side of the body, where they synapse with motor neurons that travel in spinal nerves to effectors, bringing about appropriate action (Figure 12.7).

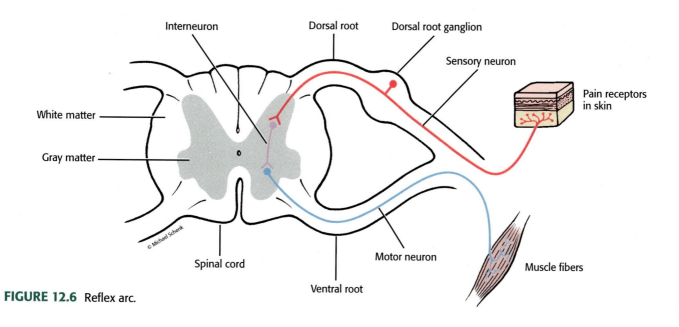

FIGURE 12.6 Reflex arc.

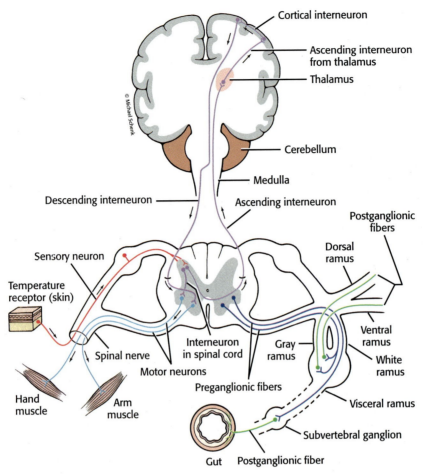

FIGURE 12.7 Common neuronal pathways including the sympathetic division of the nervous system.

CRANIAL NERVES

Originating from the vertebrate brain are 10–12 pairs of cranial nerves that generally innervate glands, muscles, and organs at the anterior end of the animal. The exception is the vagus nerve, which innervates most of the visceral organs. Cranial nerves are associated with various regions of the brain and pass through specific foramina of the vertebrate skull.

For several decades, biologists have adhered to the theory of cranial nerve origin advanced by Goodrich (1918, 1930). Goodrich suggested that the head was segmented originally and the cranial nerves were derived from the roots of spinal nerves innervating those segments. More recent reevaluation of existing and new evidence suggests that the organization of the head may not be a simple serial segmentation with subsequent evolution of the cranial nerves from portions of their respective spinal nerves, but instead may be represented by fewer segments with some cranial nerves derived by contributions from several sources (Northcutt, 1990, 1993).

A common scheme to deal with the cranial nerves is to identify three or four different categories depending on their source of derivation or their major function—sensory, special sensory, motor, and mixed. Although some cranial nerves are considered motor, they carry somatic mechanoreceptors called proprioceptors, which transmit information concerning muscle and tendon tension and joint position to regions of the brain, making it possible to control body position and movement.

In the following discussion, cranial nerves are identified as a part of the dorsal series or as a part of the ventral series. Designation of dorsal or ventral cranial nerves is predicated upon their resemblance to the spinal nerves of the lamprey. In the lamprey the dorsal and the ventral nerves are separate —the dorsal nerves carrying sensory neurons and the ventral nerves carrying motor neurons. Some cranial nerves are combinations of sensory and motor neurons and are designated as mixed nerves. Cranial nerves 0, I, II, and VIII are associated with the sense organs of the head and are designated as special sensory cranial nerves.

Terminal Nerve (0)

This special, somatic sensory nerve was discovered subsequent to the naming of the other 12 nerves. A great deal of controversy surrounds the function of this nerve. Some evidence points toward a reproductive function—*e.g.,* in some vertebrates gonadotropic releasing hormone (GnRH) is released and carried to various regions of the brain, retina, and nasal structures by way of the vascular system, perhaps coordinating reproductive and gonadal development. Further, vasomotor fibers of the terminal nerve project to the nasal mucosa (Demski, 1993).

Olfactory Nerve (I)

This special somatic, sensory nerve originates in the nasal olfactory epithelium of the nasal sac and terminates in the olfactory bulb of the brain. Because the nasal sac often is located near the olfactory bulb, it is a short cranial nerve. The olfactory bulb and the olfactory tract originate in the embryo as evaginations of the brain. Chemosensory impulses involved in smelling are transmitted from the epithelium to the brain.

Optic Nerve (II)

Cranial nerve II is also a special somatic sensory nerve extending from the retina through the optic chiasma and continuing to the brain as the optic tract. During embryonic development the eye originates as an evagination of the brain, the optic vesicle, which then invaginates to form the retina. The modified sensory cells of the retina—the rods and cones—synapse in the retina with interneurons whose fibers pass through a nerve tract commonly called the optic nerve. Nerve tracts are found only in the brain; hence, what is referred to as the optic nerve consists of the fibers of interneurons synapsing in the retina with the modified sensory neurons known as the rods and cones.

In many vertebrates all of the fibers of the interneurons cross over from one side to the other in the optic chiasma and continue to various areas of the brain. In some vertebrates, a few of the fibers of the interneurons pass through the chiasma on the same side as they originate. In mammals half of the interneurons of each eye pass through the chiasma, remaining on the same side as their origin, giving rise to binocular vision.

Oculomotor Nerve (III)

The oculomotor nerve is well named because it carries somatic motor fibers that innervate the dorsal rectus, medial rectus, ventral rectus, and ventral oblique muscles of the eye. In addition, in most vertebrates it carries visceral motor fibers that are distributed to the ciliary and iris muscles. This cranial nerve is part of the ventral series.

Trochlear Nerve (IV)

This somatic motor cranial nerve also innervates an eye muscle—the dorsal oblique— and is part of the ventral series.

Trigeminal Nerve (V)

Three branches—the ophthalmic, the maxillary, and the mandibular—make up this large mixed nerve, one of the dorsal series. All branches carry somatic sensory fibers; and in addition, the mandibular carries somatic motor fibers.

1. The ophthalamic may be divided into a deep and superficial branch, associated with the skin of the dorsal and lateral head regions, including the eye and nasal region.

2. The maxillary branch carries sensory impulses from the skin of the lateral and upper jaw regions, as well as the teeth of the maxilla, palate, upper lip, and surrounding area.

3. The mandibular branch carries sensory impulses from the teeth of the mandible, lower lip, ear region, and skin of the cheek and lower jaw. In addition, the mandibular branch innervates primarily jaw muscles, and in mammals, also a tiny muscle of the malleus, the tensor tympani, which evolved from the same muscle group as the jaw muscles. Recall that the malleus, one of the ear ossicles in mammals, evolved from an early ancestral jaw articulating bone. The mandibular branch also carries general somatic sensory sensations in the anterior third of the tongue.

Abducens Nerve (VI)

The abducens is a third somatic motor nerve that innervates an eye muscle, the lateral rectus. It is one of the ventral series.

Facial Nerve (VII)

The facial nerve, one of the dorsal series of cranial nerves, is a mixed nerve associated with the head and neck regions. Cranial nerves V and VII often are coupled closely with one another and may originate from the same point, such as in the shark. Sensory impulses from the head and oral cavities are carried by the facial nerve, along with motor impulses carried to muscles of the hyoid arch. In anamniotes this nerve contributes to the lateral line system, innervating structures such as the ampullae of Lorenzini and the lateral line organs associated with the head and mandible. The facial nerve carries general sensory impulses from the posterior two-thirds of the tongue and gustatory impulses from the taste buds of the anterior two-thirds of the tongue.

Vestibulocochlear Nerve (VIII)

This special sensory nerve innervates the inner ear. It consists of two branches, one of which innervates the organs of equilibrium and the other the organs of equilibrium and the organs of hearing.

Glossopharyngeal Nerve (IX)

A small mixed nerve, the glossopharyngeal, a member of the dorsal series, carries both sensory and motor fibers. As its name suggests, the glossopharyngeal innervates the pharynx and posterior portion of the tongue. Along with C.N. VII, it carries gustatory impulses from the taste buds to the brain. The motor fibers innervate muscles of the third visceral arch. A branch of this nerve contributes to the lateral line innervation at the back of the head.

Vagus Nerve (X)

This mixed nerve, the longest of the cranial nerves, has the most extensive distribution of any of them. It courses from portions of the head, where it innervates muscles of the last four visceral arches and serves as the major respiratory nerve in fishes, to a wide variety of visceral organs—*e.g.,* the heart, lungs, if present, and digestive organs. Along with V, VII, and IX, it is part of the dorsal series of cranial nerves. Among the aquatic anamniotes a large branch of the lateral line system receives mechano-sensory impulses from ampullae situated along the body. The vagus nerve also carries gustatory impulses from the pharyngeal region posterolateral to the tongue.

Spinal Accessory Nerve (XI)

This small nerve, a member of the dorsal series, is found in amniotes and some amphibians and is purely motor. It was derived from the vagus nerve and probably includes fibers derived from anterior occipitospinal nerves. This nerve innervates muscles of the pharynx and larynx and some shoulder muscles.

Hypoglossal Nerve (XII)

This ventral cranial nerve carries motor impulses to muscles of the tongue in amniotes. In anamniotes the hypobranchial nerve, derived from several occipitospinal nerves, innervates hypobranchial musculature and appears to have given rise to the hypoglossal.

AUTONOMIC NERVOUS SYSTEM

Although treated as a distinct entity, this part of the nervous system is not divorced from the rest of the nervous system. The autonomic nervous system of vertebrates is associated with other components of the peripheral nervous system— some cranial nerves and spinal nerves in general—and has a motor function. This part of the peripheral nervous system innervates the viscera, including the heart and blood vessels, glands, smooth muscle, and skin.

As opposed to somatic motor circuits in which motor fibers extend from the spinal cord to effectors, autonomic motor circuits always involve a peripheral ganglion and two motor fibers in the circuit. Therefore, autonomic circuits have a preganglionic neuron and a postganglionic neuron.

The preganglionic nerve cell body resides in the spinal cord, and its nerve fiber synapses with the postganglionic neuron in a collateral ganglion such as the subvertebral ganglion (Figure 12.7). The nerve cell body of the postganglionic neuron resides in the collateral ganglion, and its nerve fiber travels through a splanchnic nerve and innervates an effector.

In tetrapods the autonomic system is divided into sympathetic and parasympathetic divisions whose effects are antagonistic. Much of our knowledge concerning these two divisions in tetrapods has been gleaned from mammalian anatomy and physiology. The distribution of these divisions can be ascertained from their anatomical names. The sympathetic division is known as the thoracolumbar and consists of a pair of chains of connected ganglia. The craniosacral division, also called the parasympathetic division, is located in several cranial nerves and the sacral region of the spinal cord. In general, the parasympathetic division innervates and controls so-called vegetative activities such as digestion, elimination of urine and feces, decrease of the heart rate, constriction of bronchial musculature, gland secretion, sexual activity, and so forth.

Conversely, the sympathetic division innervates and controls many of the same activities but in an antagonistic fashion—*e.g.,* dilation of the bronchial musculature, increase of heart rate, inhibition of digestion, inhibition of urination and defecation, inhibition of the function of the genitalia, and so on. Structures of the skin such as sweat glands, arrector pili muscles (hair erectors), and cutaneous blood vessel muscles are innervated only by the sympathetic division. The architecture of the two divisions also differs: The craniosacral division is not incorporated into the spinal nerves; however the sympathetic division is incorporated, in a somewhat complicated manner (Figure 12.7).

The adrenal medulla is unique in its innervation because it is innervated only by preganglionic neurons of the sympathetic system. The adrenal medulla, however, actually functions similar to postganglionic ganglia and secretes noradrenalin similar to postganglionic fibers. In contrast to the postganglionic neurons in which the noradrenalin is released from the terminal end of the axon, the adrenal medulla releases its secretions into the bloodstream, enhancing the effects of the sympathetic division.

The autonomic system of most amniotes functions in a fashion similar to mammals. Anamniote systems are less well defined and tend to be diffuse.

SENSE ORGANS

Survival in a constantly vacillating environment presents a number of problems that vertebrates must solve. Sense organs have evolved to perceive the presence and changes in chemicals, electrical currents, fluid currents, pressures, light energy, sound and equilibrium, temperature, and so on. This wide variety of stimuli is detected by receptors in and on the body and transformed into electrical impulses carried by sensory neurons to the brain for interpretation and integration, permitting appropriate responses by the vertebrate.

Olfaction

This sense is dependent upon detecting chemicals in the environment, whether aquatic or terrestrial. Detection of odorous chemicals depends heavily upon receptors that reside in moist epithelia. Vertebrates are capable of detecting large numbers of odors—in the thousands. It is not clear how many receptors are responsible for this ability. From research in this area, however, receptors probably are able to detect more than one odor. Odor may involve not only detecting a certain class of chemical, but also the strength of the chemical stimulus and its processing in the central nervous system. The impulses generated in the receptors are transmitted via the olfactory nerve, cranial nerve I.

To process odorous chemicals, craniates must move their aqueous or gaseous environment across moist olfactory epithelia. Among craniates, the lamprey and the hagfish are unique with a single nostril. Associated with the nostril is an olfactory sac, which is exposed to water-borne odorous chemicals. In the chondrichthyes and osteichthyes, which possess paired bilateral incurrent and excurrent nostrils leading into olfactory sacs, water currents are drawn into the incurrent nostrils across the olfactory epithelium by pumping or suction and out through the excurrent nostrils.

Among the fish relatives of the tetrapods, the excurrent nostril became associated with the oral cavity and gave rise to the choanae or internal nares. The nasal cavity of tetrapods developed between the external nostril and the choana, with olfactory epithelium restricted to the upper portion of the cavity. During normal respiration odorous chemicals enter the nasal cavity and flow over the olfactory epithelium, which is kept moist by mucus. Mammals in particular are capable of forcefully inhaling or sniffing the air to enhance their ability to smell odors (Figure 12.8).

Vomeronasal Organ

Tetrapods possess a vomeronasal organ, a chemoreceptor once thought to be associated with olfaction (Figure 12.9). The vomeronasal organ is found in the roof of the mouth as an invaginated pocket. Many biologists, however, now believe that the vomeronasal organ and its function are separate from the olfactory apparatus. The structure of its epithelium and its neural pathways are discrete from those employed during olfaction with the impulses carried by the vomeronasal nerve to the accessory olfactory bulbs lying beside the large olfactory bulbs.

In squamate reptiles the forked tongue is employed to sample the chemicals of its environment, and then is projected into the paired pockets of the vomeronasal organ (Figure 12.9A). Following food trails or potential mates are examples of its use. Many tetrapods produce pheromones,

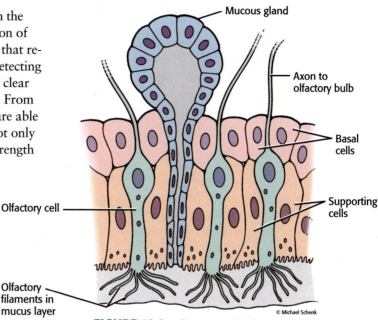

FIGURE 12.8 Olfactory epithelium.

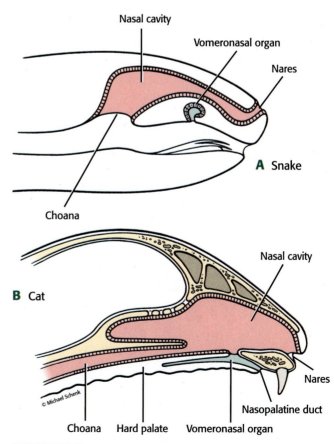

FIGURE 12.9 Vomeronasal organs.

chemicals that are recognizable by conspecifics. Pheromones are chemicals that convey territorial boundaries, mating condition, health and other aspects of social behavior.

Male mammals commonly use the vomeronasal organ to detect the sexual condition of females (Figure 12.9B). Many male mammals sense the sexual condition of females by elevating their upper lip and drawing in pheromones released by the female, which are directed over the epithelium of the vomeronasal organ (the Flehmen behavior).

GUSTATION

The distribution of taste buds in vertebrates is variable. Most *gustation* receptors are confined to the pharyngeal region as well as the tongue. In tetrapods the tendency is for taste buds to be confined to the tongue and posterior surface of the pharynx. In many fishes taste buds are distributed over the entire surface of the body. Bottom-feeding fishes often have evolved sensory barbels associated with the head, and taste buds are particularly numerous on their surface. Innervation of the gustatory receptors in vertebrates from fish to mammals is accomplished by branches of cranial nerves VII, IX, and X, the facial, glossopharyngeal and vagus nerves, respectively (Figure 12.10).

Lateral Line System

The lateral line system of vertebrates appears to be an ancient one. Despite the discussion of cranial nerves VII, IX, and X, in which a branch of each of these nerves was assigned to the lateral line system, more recent research has revealed at least six separate lateral line nerves (Northcutt, 1990). They originate from six cranial embryonic placodes and are equivalent to cranial nerves. The inner ear was derived from the cranial portion of the lateral line. The embryological development of both is similar. A lateral line system is found in fish, larval amphibians, and adult aquatic salamanders such as *Necturus*.

The basic structure of the lateral line system is the neuromast (Figure 12.11). It consists of a supporting cell in which are rooted a number of hair cells. Projecting from the free surface of each hair cell is a kinocilium and a number of microvilli (stereocilia), which are embedded in a gelatinous dome-shaped structure, the cupula. Hair cells are paired with their kinocilia, oriented 180 degrees from each other. Each neuromast contains a number of these pairs. Primitively, neuromasts are found as free-standing organs, but among most fish they occur in canals embedded below the skin. Among larval amphibians and *Necturus*, however, the neuromasts are free-standing.

The function of the neuromast is to detect water displacement. As a stream of water causes deformation of the cupula (*i.e.*, bends it in one direction or another), the kinocilium likewise is bent. A kinocilium bent in one direction causes an increase of electrical discharge, and the bending of its paired member causes a decrease in discharge. The brain interprets a comparison of these respective hair cell outputs as relative water displacement. With this system a vertebrate can determine vital information such as its position in a school, presence and size of objects, such as rocks, plants, prey or movement and speed of prey, and so on.

Another very old system consisting of ampullary organs is capable of electroreception. A swollen ampulla embedded below the skin is connected to the surface by a tube opening in a pore. Modified neuromasts lacking microvilli and a cupula but possessing a kinocilium occupy the ampullae. The tube wall is nonconductive and filled with an electroconductive gel. Therefore, only environmental electrical potentials are detected. Thus, the fish's detection system is not confused by its own body activities. Cartilaginous fishes have clusters of ampullary organs on the head, known as the ampullae of Lorenzini (Figure 12.12). These remarkable organs can detect electrical currents. Electrical currents are normal outputs of resting muscle, and even prey covered by a substance such as sand can be located. Electrical currents generated by chemicals moving in a stream also can be

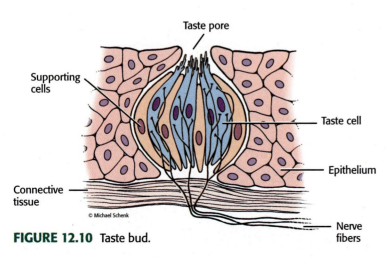

FIGURE 12.10 Taste bud.

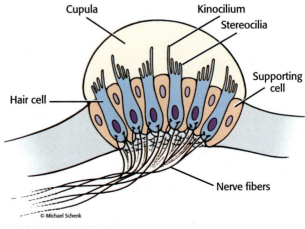

FIGURE 12.11 Neuromast.

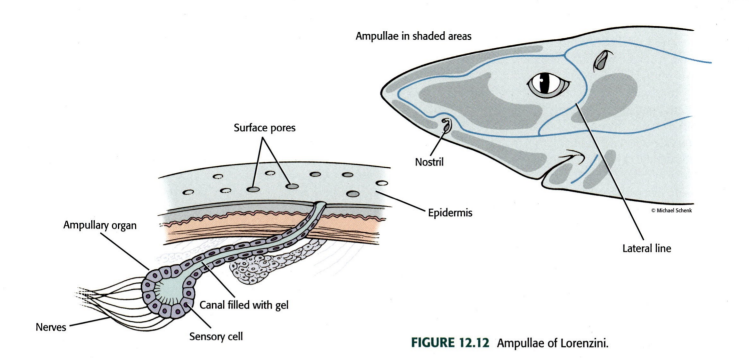

FIGURE 12.12 Ampullae of Lorenzini.

detected, making it possible for a shark to navigate by body orientation relative to the current. Minor temperature and salinity changes also can be sensed.

A number of vertebrates, the lampreys, primitive ray-finned fishes, chondrichthyes, the coelacanth, lungfish, and larval amphibians are electro-sensitive. The organs of detection are ampullary organs derived from tissue similar to the lateral line.

Electroreceptors apparently have re-evolved among some teleosts and in the monotreme mammals, the echidnas and the platypus. The morphology of the organs, however, is different.

Equilibrium and Audition

The inner ear apparatus of vertebrates likely was derived from a portion of the lateral line system. Interference from signals monitored by the lateral line system probably necessitated the sensory portion of the "new" ear to be isolated from the rest of the lateral line. Therefore, this portion of the lateral line sank beneath the surface, perhaps giving rise to the ability to maintain body position in the environment, or a sense of equilibrium. Among vertebrate embryos an ectodermal otic placode arises on the surface in the region of the hindbrain. It sinks below the surface, forming an otic vesicle from the membranous labyrinth, which develops into the semicircular ducts.

With the exception of the "cyclostomes," vertebrates possess three ducts, two of which are oriented more or less in the vertical plane, and the third is horizontal. Hagfishes have a single duct, and the lamprey possesses two.

An ampulla occurs at the end attached to an enlargement of the labyrinth, the utriculus. Sensory hair cells, the cristae,

occur in the ampullae. A second swelling of the labyrinth and continuous with the utriculus, the sacculus, occurs ventral to the utriculus. Groups of sensory hair cells, the maculae, occur in the utriculus and the sacculus. Associated with maculae are groups of calcium carbonate crystals. The lagena, an outpocketing of the sacculus, occurs in most vertebrates. The lagena in mammals becomes the coiled cochlea.

The membranous labyrinth is filled with the fluid endolymph. This delicate membranous system resides in a space within the skull that mirrors the shape of the system. This space is filled with a fluid perilymph. Movement of the endolymph in response to body and head movement distorts the hair cells, making it possible for the brain to determine position and control the appropriate response of muscles to maintain body position in the environment.

Hearing, or audition, among vertebrates varies from the detection of water displacement in fishes to the recognition of sound by terrestrial vertebrates. Some fish are capable of hearing underwater. One of the best studied groups is the ostariophysii, to which belong minnows, carp, suckers, catfish, and characins. This ability is mediated through a series of bones associated with the anterior vertebrae, Weberian ossicles, that extend between the swim bladder and the membranous labyrinth. Underwater sound causes waves of water displacement whose frequency parallels that of the sound. The swim bladder vibrates sympathetically with the water waves, causing the ossicles to transmit the frequency to the inner ear. The brain interprets the vibrations as sound.

Adequate sound transmission in air is associated with a membrane capable of vibrating in concert with sound waves impinging upon it. Just when this membrane evolved is conjectural, but the middle ear and tympanum seem to have evolved a number of times. Transmission of sound from the tympanum across the middle ear to the inner ear requires a mechanical means.

With the advent of an autostylic jaw suspension among early tetrapods, the hyomandibular acted as a jaw prop, extending between the otic capsule and quadrate of the upper jaw, placing it in a serendipitous position to be incorporated into the middle ear as the stapes. This condition remains adequate for tetrapods from the amphibians through the birds.

With the evolution of mammals, a series of three sound-transmitting ossicles appears. The stapes, derived from the hyomandibular jaw element, persists, and two more jaw-articulating bones contribute to the series. The incus was derived from the quadrate, and the malleus from the articular. The stapes, which abuts the oval window of the inner ear, forms a synovial joint with the incus, which in turn forms a synovial joint with the malleus that abuts against the tympanum (Figure 12.13). Impulses from the statoacoustic apparatus are carried by cranial nerve VIII.

VISION

With the exception of animals that have secondarily lost their sense of vision, such as some burrowing or cave-dwelling species and some deep-sea fishes, all vertebrates have bilateral, image-forming eyes.

Lateral Eyes

Developmentally, the eye begins as a hollow, lateral evagination of the diencephalon, the optic vesicle. As it grows toward the surface of the head, it maintains a connection with the diencephalon, the optic stalk. The vesicle invaginates, forming a two-layered optic cup. The inner layer becomes the visual portion of the retina, containing the visual cells, the rods and cones, and various other cells. The outer layer becomes the pigmented portion of the retina. The stalk becomes the optic nerve or tract.

As the optic cup approaches the skin surface, it induces the formation of an ectodermal placode, which then becomes the hollow lens vesicle. The vesicle becomes the solid lens through proliferation of the inner cells of the vesicle. The mesenchyme surrounding the developing eye gives rise to the choroid coat, the sclera, and the cornea. The iris and ciliary apparatus is derived from the edges of the optic cup. The cavity from the cornea to the iris is called the anterior chamber, and the cavity between the iris and the lens is called the posterior chamber.

Aqueous fluid, similar in composition to the blood plasma, is secreted by the ciliary body into the posterior chamber. It then flows into the anterior chamber, from which it diffuses into the bloodstream. The vitreous humor occupying the space between the lens and the retina appears to be secreted by the retina (Figure 12.14).

There are a number of similarities between the construction of a camera and the eye. At the front of the system of a camera that uses film is a refractive lens that bends light as it passes through the lens, focusing the image on a photosensitive film. The image on the film is latent until it is developed and fixed; hence, your images (pictures) can be seen.

At the anterior surface of the eye is the cornea, a transparent portion of the sclera, which in most vertebrates acts as a strong initial light refractor. In fishes the refractive index of the cornea is close to that of its environment, and very little refraction occurs. Therefore, fish corneas are more convex than corneas of terrestrial vertebrates. Light then passes through the aqueous humor with little refraction, and then passes through the lens, where most of the rest of light bending or refraction is accomplished. It then passes through the vitreous humor, where again little refraction is accomplished, and finally strikes the retina, where the photoreceptors, the rods and cones, are stimulated.

The rods are most sensitive to light and are responsible for sight in subdued light; they produce images in shades of gray. Cones, conversely, are stimulated by bright light and are responsible for images in color. The impulses generated in the rods and cones are transmitted to other cells in the visual pathway via the optic nerve to the brain, where the image is projected upside down and the brain then repositions the image as it is viewed by the vertebrate. How the brain accomplishes this feat is as yet unknown.

In the camera the diameter of the lens aperture is controlled by a diaphragm, dependent upon the distance and brightness of the object. In the eye the aperture, the pupil, is regulated by the iris, which is stimulated by the amount of light received, which in turn is dependent upon the distance of the object from the eye. As an object approaches the eye, the amount of light reflected from the object increases. To project a sharp image on the retina, the light passing through edges of the lens must be reduced because otherwise it will produce a distorted image with fuzzy edges. The shape of the pupil varies from elliptical (*e.g.*, snakes, cats, and sharks), to circular in most vertebrates, to horizontally oblong in some mammals (*e.g.*, bovines and whales).

The curvature of the lens in the camera is fixed, which explains why one needs a number of lenses to photograph objects at various distances—although zoom lenses have reduced the number required previously. Accommodation, the process of changing the focal length of the lens of the eye, is accomplished either by changing the shape of the lens or by changing the position of the lens.

Obviously, the eyes of aquatic vertebrates are exposed to water continuously and the cornea is not subject to drying. Therefore, fish do not shed tears because they do not have tear glands. In contrast, the eyes of terrestrial vertebrates are exposed to the air, which constantly causes evaporation of moisture from the surface of the cornea. Tear glands produce a solution that bathes the eye.

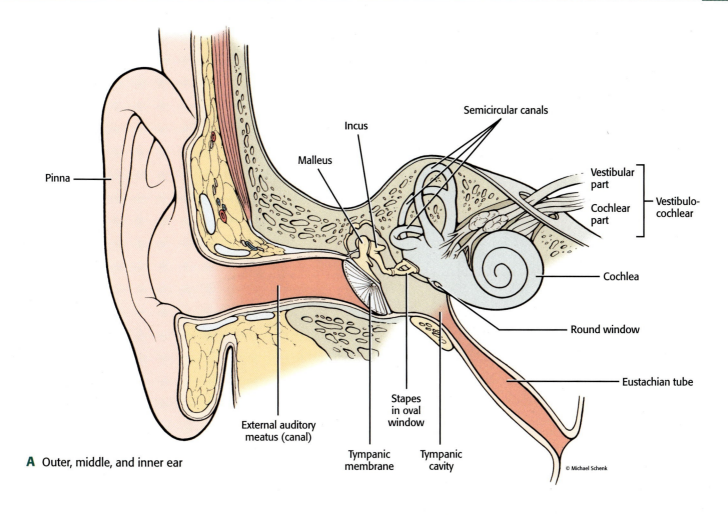

Pinna

Incus

Malleus

Semicircular canals

Vestibular part

Cochlear part

Vestibulo-cochlear

Cochlea

Round window

Eustachian tube

Stapes in oval window

External auditory meatus (canal)

Tympanic membrane

Tympanic cavity

© Michael Schenk

A Outer, middle, and inner ear

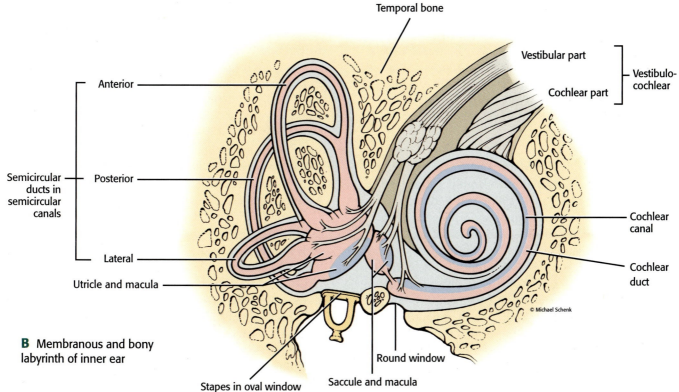

Temporal bone

Vestibular part

Cochlear part

Vestibulo-cochlear

Anterior

Semicircular ducts in semicircular canals

Posterior

Lateral

Utricle and macula

Cochlear canal

Cochlear duct

© Michael Schenk

B Membranous and bony labyrinth of inner ear

Stapes in oval window

Saccule and macula

Round window

FIGURE 12.13 Human ear.

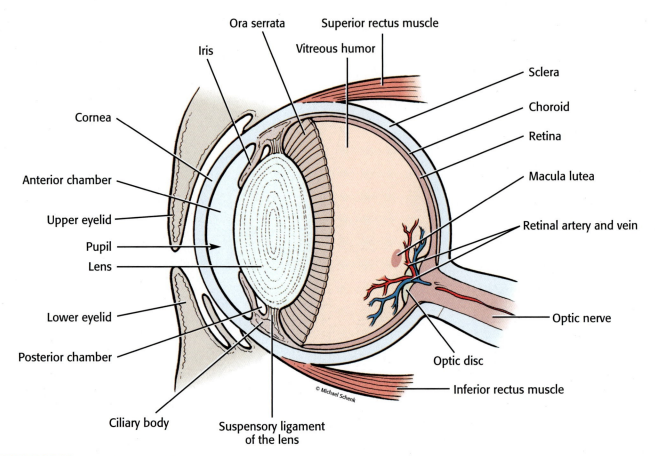

FIGURE 12.14 Sagittal section of human eye.

To protect the surface of the eye and to spread the solution of the tear glands, upper and lower eyelids have evolved in tetrapods. The ability to move the eyelids varies among the tetrapods. Some sharks have immovable folds of skin referred to as eyelids. In some sharks and in most tetrapods, a translucent third eyelid, the nictitating membrane, occurs in the medial corner of the eye, serving a protective function.

Median Eyes

Median eyes are widely distributed among all vertebrate classes. The anterior organ is known as the parietal eye, and the posterior organ as the pineal eye. Both originate in embryos as dorsal evaginations of the diencecphalon. Nearly all vertebrates possess a pineal eye (or organ). In some vertebrates, such as lampreys and some lizards, both eyes are present. In lampreys, the parietal lies beneath the pineal eye. Further, in some vertebrates, photoreceptors resembling the cones of lateral eyes, a pigmented retinal layer and a lens, may be present. Both seem to be photosensitive.

The parietal eye, in some lizards that shuttle back and forth between sunlight and shade to regulate their body temperature, has been implicated in controlling the duration of photoperiods, thereby signaling the animal when to move from one environment to another. The pineal eye also functions as a photoperiod-sensitive organ in some animals. Evidence now suggests that activity periods, circadian cycles, or 24-hour internal biological rhythms and seasonal gonadal activity based on circannual rhythms, are affected by median eyes. Perhaps the median eyes evolved as a bilateral pair of eyes with one moving anterior becoming the parietal organ, and the other posterior becoming the pineal organ.

Thermoreceptors

Some snakes are capable of detecting subtle temperature differences between the environment and the body temperature of available endothermic prey, birds and mammals. A group of American snakes, the crotalids, among which are the infamous rattlesnakes, evolved with a pair of pits between the nostrils and eyes, which are sensitive to small changes in infrared radiation, approximately .001–.003°C. Crotalid snakes prey almost exclusively on endothermic vertebrates and are nocturnal when environmental temperatures are generally cooler.

This group of snakes is capable of striking prey accurately at a distance in the dark. Shallow pits capable of detecting infrared radiation are found surrounding the mouth in pythons and boas. These are not as sensitive as those of the pit vipers. All of these thermodetectors are innervated by the trigeminal nerve, cranial nerve V.

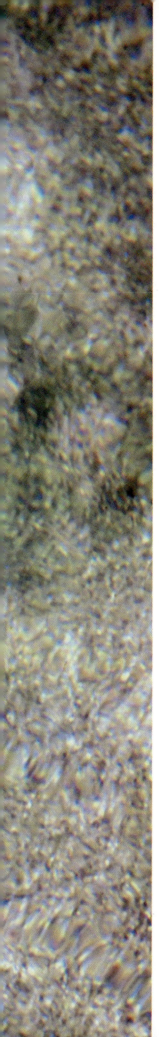

Endocrine Systems

13

Although the endocrine organs are known collectively as a system, in many cases they do not physically contact one another. The function of these organs, often known as glands, is coordinated exquisitely when operating normally. When the organs are not, the result often is a dysfunctional vertebrate.

Secretions of these ductless glands, called hormones, act as regulatory chemicals, often interacting among themselves to control a complex series of cellular activities. Hormones are transported in the blood of the circulatory system to their target tissues, where they are attracted to specific receptor sites of these target cells, mediating their characteristic activities. Inconspicuous endocrine tissues sometimes are found in the tissues of other organs (*e.g.,* the small intestine). The hormones produced by these endocrine tissues are essential in coordinating the physiology of other organs, including the pancreas and the gallbladder, among others.

HYPOPHYSIS

The hypophysis, or pituitary, attached to the hypothalamus of the diencephalon, is a small gland but plays a major role in influencing a wide variety of other endocrine glands and functions of the body. The hypophysis has an interesting embryonic origin, in which the adenohypophysis arising as an outpocketing of the primitive embryonic mouth, the stomadeum, and the neurohypophysis as an evagination of the floor of the diencephalon. The adenohypophysis detaches from the mouth when it comes into contact with the floor of the brain, and becomes associated with the neurohypophysis. The adenohypophysis consists of a pars distalis and the pars intermedia. The pars intermedia is absent in birds and a few mammals. A pair of thin strips from the adenohypophysis, the pars tuberalis, is found in most tetrapods and a few fish. Its function is unknown (Figure 13.1).

The pars distalis of the adenohypophysis secretes several hormones.

- Thyroid stimulating hormone (TSH) is released into the bloodstream and carried to the thyroid, where it stimulates the thyroid gland to synthesize thyroid hormone. Thyroid hormone affects most metabolic functions and promotes sexual and tissue maturation.

- Adrenocorticotropic hormone (ACTH) stimulates the adrenal cortex to synthesize adrenal cortex hormones. Among them are the glucocorticoids, which affect cellular metabolism, the mineralocorticoids, which promote vital electrolyte homeostasis, thereby affecting blood pressure, and androgens, which affect sex drive and masculinity in both sexes.

- Follicle stimulating hormone (FSH) stimulates the development of ovarian follicles in females and spermatogenesis in males.

- Luteinizing hormone (LH) stimulates the conversion of the ruptured follicle to form a corpus luteum, which secretes progesterone and estrogens in the female, and in the male stimulates the interstitial cells of the testes to secrete male sex hormones, called androgens. A corpus luteum is found among those vertebrates, which retain developing embryos within their reproductive tracts for a period of time, giving birth to young—for example, therian mammals, some snakes, some sharks, and some lizards.

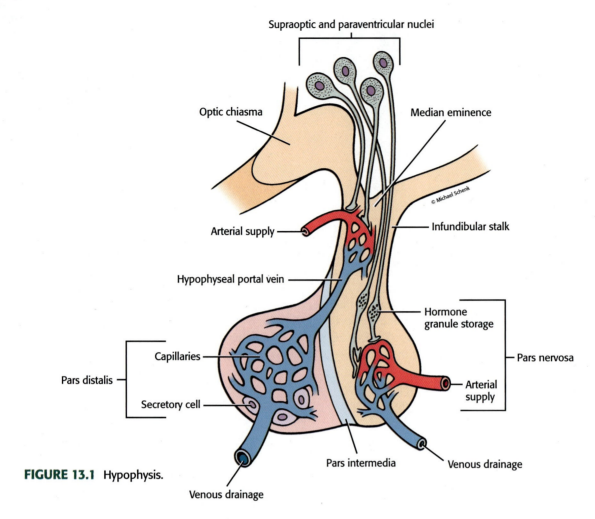

FIGURE 13.1 Hypophysis.

- Growth hormone (GH) stimulates a variety of metabolic activities, especially protein synthesis, leading to growth of the body.
- Prolactin (PRL) is widespread among vertebrates, stimulating a variety of activities such as the movement of salamanders to breeding ponds, nesting and the care of eggs and hatchlings in birds, and production of a secretion known as "pigeons' milk" in the crop of both sexes of doves and pigeons and the production of milk in female mammals.
- Melanocyte stimulating hormone (MSH) is utilized in vertebrates in different ways. In vertebrates possessing a pars intermedia, MSH affects the distribution of melanin in melanocytes that are present in the skin. In animals lacking a pars intermedia, MSH is synthesized in the pars distalis.

Synthesis and secretion of the hormones of the adenohypophysis generally are controlled by releasing hormones produced in the hypothalamus and transported via the hypophyseal portal system (Figure 13.1).

As opposed to the adenohypophysis, the neurohypophysis does not synthesize hormones but, rather, stores hormones produced by nuclei in the hypothalamus (Figure 13.1). The hormones, arginine vasotocin (arginine vasopressin VP in mammals) and oxytocin (OXY), are transported along axons of neurons that reside in the hypothalamus and released into the neurohypophysis, where they are stored and subsequently released in response to appropriate stimulation by the hypothalamus.

The basic function of vasotocin is to stimulate the reabsorption of water by the kidney tubules, thereby affecting the blood pressure. Oxytocin causes the contraction of smooth muscle in the reproductive system (*e.g.*, in oviducts of reptiles and birds) and the uteri of mammals, promoting egg laying and the expulsion of fetuses. In addition, it stimulates the release of milk in the mammary glands of mammals triggered by the suckling behavior of the newborn.

PINEAL GLAND

The small pineal gland present in vertebrates originates as an evagination of the roof of the diencephalon (Figure 12.3). The cells of this gland secrete *melatonin*. This hormone, being antagonistic to MSH, affects pigmentation.

The concentration of melatonin in the blood exhibits a diurnal and seasonal periodicity, which may affect photo-periods, perhaps influencing daily sleep/awake patterns and other diurnal phenomena as well as seasonal reproductive activities.

ADRENAL GLAND

In fish the "adrenal gland" is not really a compact gland but instead consists of scattered masses of tissue that generally is of two distinct types—interrenal tissue and chromaffin bodies—which are homologous with the adrenal cortex and adrenal medulla in amniotes. In all vertebrates the two tissues have separate origins. The adrenal cortex is derived from mesoderm, and the medulla develops from neural crest tissue.

In lampreys and cartilaginous fish the interrenal tissue and the chromaffin bodies are separate masses often in contact with dorsal blood vessels and lie between the kidneys. In bony fish both tissues are found on the ventral surface of the kidneys. In amphibians the two tissues are intermingled into a diffuse adrenal gland on the ventral surface of the kidney.

A compact, organized gland is found among amniotes with chromaffin tissue identified as the adrenal medulla and a cortex. Only in mammals is the medulla surrounded by a cortical layer, and even mammals show some variability. The adrenal glands in amniotes lie near the anterio-medial aspect of the kidneys.

Hormones secreted by the chromaffin tissue include norepinephrine and epinephrine, which are released during times when an animal encounters a stressful situation, thereby reinforcing the effects of the sympathetic nervous system, producing the "flight or fight" response. Among the effects of secretion of these hormones are increased release of glucose from the glycogen supply in the liver to support increased muscle contraction; and relaxation of smooth muscle affecting organs such as the bronchi and trachea, which enables the animal to receive increased volumes of oxygen.

The interrenal or adrenal cortex secretes steroid-based hormones of two basic classes—the glucocorticoids and the mineralocorticoids. The major glucocorticoids are corticosterone and cortisol, which affect carbohydrate and protein metabolism and are anti-inflammatory. The major mineralocorticoid, aldosterone, affects the regulation of electrolytes and thereby affects water balance and ultimately controls blood pressure.

Small amounts of sex hormones are produced in the cortex of both sexes. An effect of overproduction in females is the enhancement of male characteristics such as the appearance of a beard in the bearded lady.

THYMUS

The thymus, a lymphoid organ, arises from the epithelial lining of gill pouches. In the lamprey the thymus develops from all seven of the pouches. As we examine other groups of vertebrates, we find that the thymus develops from fewer and fewer pouches, and the thymus in many mammals, including humans, arises from the third pouch.

Cells from developing bone marrow in the embryo invade the thymus and give rise to T-lymphocytes (T cells), which travel in the circulatory system to organs of the lymphatic system, where they proliferate and continue to be produced throughout the lifetime of the animal. The thymus generally is active during juvenile stages and tends to regress in the adult. T cells are part of the immune system that usually acts against foreign cells such as bacteria, fungi, and viruses, and they are the culprits in transplanted organ rejection.

In many young birds, a dorsal, cloacal evagination, the bursa of Fabricius, functions as a site of B lymphocyte maturation. This function is similar to that of the thymus.

THYROID

Cells of the ventral lining of the pharynx of all vertebrates are capable of fixing iodine by bonding it to the amino acid, tyrosine, resulting in the formation of thyroxine (T_4) and triidothyronine (T_3). In amphioxus and the larval lamprey a ventral groove, the endostyle in both, is likewise capable of binding iodine.

The structure of the thyroid varies from a loose aggregation in bony fish to a more organized, lobed organ lying in the ventral neck region. In some fish it may migrate to other organs, such as the kidney.

The thyroid gland is organized into contiguous follicles with epithelial walls that synthesize a protein-rich colloid consisting of thyroglobulin, a combination of protein and iodinated tyrosine. Under the influence of TSH from the hypophysis, the epithelium releases T_3 and T_4, which enter the circulatory system to target cells throughout the body. This produces myriad effects, including metabolism, growth and development, metamorphosis, proper development and maturation of the integument, reproductive system, molting in amphibians, reptiles, mammals, and so on.

ULTIMOBRANCHIAL BODIES

Ultimobranchial bodies develop from the epithelium of the last pair of pharyngeal pouches in all gnathostomes and, with the exception of mammals, persist as discrete glands. The hormone *calcitonin*, synthesized in the ultimobranchial bodies, promotes the retention of calcium in bones. In mammals, cells from the ultimobranchial bodies migrate into the thyroid gland, lying between the follicles as parafollicular cells. The parafollicular cells produce calcitonin with the same function as the hormone produced by ultimobranchial bodies.

PARATHYROID GLANDS

With the exception of the fish, all vertebrates possess parathyroid glands, which develop from pharyngeal pouches. In most tetrapods the parathyroid glands are located in the neck region near the thyroid gland, or in mammals may be embedded in the thyroid gland. The cells of this gland secrete parathyroid hormone, which causes calcium to be released from the skeleton, thereby raising the calcium titer of the blood. Notice that the actions of parathyroid hormone and of calcitonin are antagonistic.

PANCREAS

The pancreas is an organ found only in vertebrates, although in amphioxus, cells that secrete digestive enzymes similar to those of the pancreas are found in the wall of the hepatic diverticulum. Both exocrine tissue, which is responsible for secretion of digestive enzymes that are capable of digesting most food categories, and endocrine tissue, which secretes the hormones glucagon and insulin that regulate carbohydrate metabolism, are found in vertebrates.

Agnathans do not possess a well-formed pancreas but have cells embedded in the intestine. In lungfishes and elasmobranches, however, a well-defined pancreas is present. The pancreas in many other fishes is identifiable but loosely organized. In tetrapods the pancreas is a distinct organ that may be composed of several lobes, the main portion of which consists of ducted exocrine tissue, while the endocrine tissue is organized into ductless islets.

OVARIES AND TESTES

In addition to the production of female reproductive cells, the ovaries produce estrogens, small amounts of androgens and progesterones, and in mammals, relaxin. Estrogens stimulate the development of the reproductive system and secondary sex characteristics and their maintenance. The function of progesterones is only well known in mammals where it plays a role in ovulation and preparation of the reproductive tract prior to ovulation and fertilization, as well as maintenance of the lining of the uterus during pregnancy. Relaxin functions just prior to *parturition,* relaxing the pelvic symphysis and associated ligaments, permitting passage of the fetus through the birth canal. The androgens influence the sex drive.

Testes, in addition to producing sperm, secrete androgens and small amounts of estrogens. Testosterone, the main androgen, stimulates development of the reproductive system, sexual behavior, and secondary sex characteristics.

Notice that female and male hormones are produced in both sexes. The reason is that these hormones are steroid-based and share most biosynthetic steps.

PLACENTA

In mammals the placenta produces a number of hormones, including estrogens, progesterone, gonadotropin, and relaxin. All of these hormones promote the continuation of gestation and ensure the health of the fetus and birth.

ENTERIC HORMONES

The enteric group of hormones plays a role in the function of various digestive organs. Food in the stomach stimulates the secretion of gastrin by cells of the pyloric portion of the stomach, stimulating the secretion of HCl, which promotes the conversion of the inactive stomach proteinase pepsinogen to the active form pepsin. As the pH falls (becomes more acidic), cells in the pyloric portion of the stomach secrete somatostatin, which inhibits acid secretion.

Partially digested food, known as chyme, is released slowly through the pylorus into the duodenum, where its presence stimulates the lining of the duodenum to release secretin, which is transported via the vascular system to the pancreas, where it stimulates the release of a watery solution of salts and bicarbonate ions into the duodenum, thereby buffering the chyme, raising its pH. The digestive enzymes of the intestine function most efficiently at a higher pH.

The release of the solution containing digestive enzymes, proteases, a lipase and an amylase, of the pancreas is stimulated by cholecystokinin, formerly known as two separate hormones, pancreozymin and cholecystokinin. In addition, cholecystokinin slows the release of chyme from the stomach and stimulates the gallbladder to contract, which empties bile, produced by the liver, into the intestine. Bile emulsifies fats, increasing the surface area by creating smaller globules of fat from large globules, thereby increasing the efficiency of lipid digestion.

An intestinal hormone formerly known as enterogastrone but now recognized to consist of a hormone complex, inhibits the secretion of HCl in the stomach, stimulates the release of insulin in the presence of appropriate glucose levels, and stimulates digestive tract motility.

Phylum Chordata

Subphylum Vertebrata

As members of the phylum Chordata, vertebrates exhibit the characteristics—a notochord, dorsal hollow nerve cord, pharyngeal slits, etc.—that set chordates apart from all other animals. As adults, they may not exhibit some of these distinguishing traits, but these characteristics nevertheless appear sometime during their life cycle. In addition, vertebrates have synapomorphic traits that distinguish them from all other chordates. A few of these characteristics include: vertebrae protecting the dorsal nerve cord; a brain associated with specialized sense organs and protected by a braincase; a closed circulatory system with a ventral, chambered, muscular heart; hemoglobin contained within erythrocytes; glomerular kidneys; dentine; enamel; and bone. Members of this distinguished group of vertebrate chordates include the lamprey, dogfish shark, mudpuppy, and cat.

The earliest vertebrates, commonly called ostracoderms, appear in the fossil record more than 500 million years ago during the Cambrian period. Although this early ostracoderm fossil material occurs only as scattered fragments of bone, an examination of these fragments reveals a distinct organization. Bone is found only among vertebrates and consists of fibrous connective tissue impregnated with calcium salts. Mature bone contains a high percentage of these salts in the form of hydroxyapatite crystals. These crystals, aligned in parallel with the forces of stress occurring in the bone, greatly increase its weight-bearing capabilities.

Later fossils indicate that early vertebrates possessed an extensive external bony armor. A characteristic of most of these heavily armored fish was a small ventral mouth that was not supported by jaws. For this reason, these vertebrates are classically placed in the class Agnatha (without jaws).

Lampreys, jawless descendants of ostracoderms, are marine and fresh water fish. Hagfishes are a marine group of jawless organisms that superficially resemble lampreys and are probably related to them (Ota and Kuratani, 2007).

Hagfishes are burrowing marine inhabitants that eat various invertebrates and have the curious habit of eating the inner organs of dead or dying fish, leaving a husk. Their elongate cylindrical bodies facilitate their burrowing habit and allow them to enter host fish. Their ability to produce large amounts of mucus has led to a common alternative name, slime-hag.

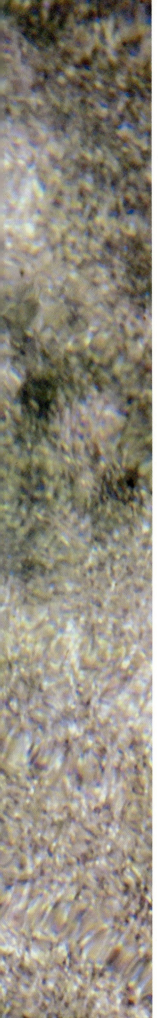

CHAPTER FOURTEEN

Lamprey

14

Lampreys, unlike hagfishes, are not confined to the marine habitat. Some are strictly fresh-water species. *Petromyzon marinus*, a lamprey that is predominately marine but is capable of invading fresh water, has devastated the Great Lakes fisheries. All lampreys, marine and freshwater, begin their life cycle with a freshwater larval stage. Lamprey larvae spend a number of years as burrowing filter feeders in their home stream, metamorphose into juveniles, migrate downstream, and eventually enter the marine environment or return to a freshwater habitat, such as Lake Erie. They probably spend one or two years feeding as a parasite on other fish or large marine mammals.

During the feeding period, the lamprey attaches itself to its host by means of its sucker, rasps a hole in the body wall, and feeds on body fluids and tissue debris. Occasionally lampreys have attempted to attach to human swimmers in the Great Lakes. Following the feeding period, the juvenile metamorphoses into a reproductive adult and returns to its home stream to build nests and spawn. The adult dies after spawning. Some strictly freshwater species produce normal feeding larvae that as juveniles do not feed but, rather, metamorphose directly into adults, spawn, and die; there is no parasitic stage.

Because most biologists consider the lamprey to be the most primitive living vertebrate, a careful dissection of the lamprey will introduce you to this fascinating animal group. The lamprey is actually a mosaic of ancestral and specialized structures. Many of its characteristics, such as the oral area and respiratory system, have become specialized for a parasitic feeding behavior.

EXTERNAL ANATOMY

The body of the lamprey is elongate and cylindrical, flattening laterally and tapering posteriorly (Figure 14.1 and Figure 14.2). Median fins occur as **dorsal** and **caudal fins**, but the lamprey has no paired pectoral or pelvic fins.

The skin is scaleless, in contrast to its ostracoderm ancestors, which were encased in a heavy bony armor. Embedded in the skin are numerous mucous glands that are detectable only microscopically.

At the anterior or head end, note the pair of lateral image-forming **eyes** typical of vertebrates. On the dorsal surface of the head, identify the single **nostril**. The **pineal organ**, an evagination of the roof of the diencephalon region of the brain, occurs just posterior to the nostril (Figure 14.3). In some vertebrates, this structure has been identified as an organ that detects light and may affect the distribution of skin pigments and reproductive behavior. It probably represents half of a pair of dorsal eyes that allowed early vertebrates to survey the environment above them. Some vertebrates, such as lizards, exhibit the other half of this primitive pair, the parietal organ.

A series of pores, representing the **lateral line system**, occur on the head in close association with the eyes and buccal funnel (Figure 14.4). The pores lead into canals, where receptor cells capable of detecting water currents are located.

Adaptations for the unique feeding behavior of the lamprey are found in the oral region. The **mouth** lies within a circular, sucker-like **buccal funnel**. At the edge of the funnel is a fringe of **buccal papillae**. The inner lining of the sucker is ornamented with a species-specific pattern of **toothlike organs** (Figure 14.5). These are not true vertebrate teeth, as they are of ectodermal origin, in contrast to the teeth of other vertebrates that are derived from mesoderm and ectoderm. A tongue with sharp "teeth"

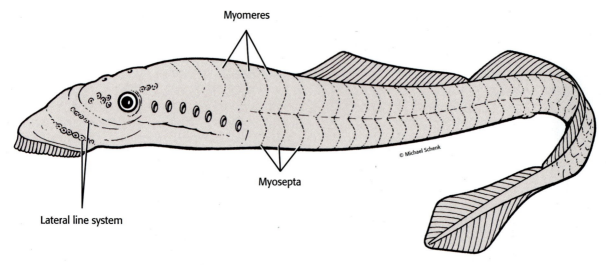

FIGURE 14.1 External features of Lamprey.

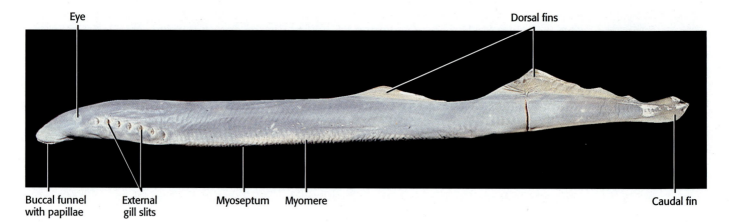

FIGURE 14.2 External features of Lamprey.

protrudes into the oral cavity. It is employed as a rasp to create holes in the surfaces of host animals. Because of the suction capability of the sucker, the lamprey remains attached to the host while consuming blood, tissue fluids, and tissue debris.

A series of seven pairs of **gill slits** open externally along the lateral aspect of the body, beginning just posterior to the eyes (Figure 14.2). During feeding, the flow of respiratory water in lampreys is tidal; it is pumped in and out through the gill slits. This is in sharp contrast to the normal unidirectional flow maintained by other fish in which water flows into the mouth and out through the gill slits. When not feeding, lampreys breathe similar to other fish.

If we consider the lamprey's dilemma, we will have no difficulty in realizing that its breathing pattern complements its parasitic feeding behavior. If it releases its suction on its host to allow respiratory water to flow into its mouth, guess what will the host do? Therefore, when feeding, it maintains its position with its oral sucker and it pumps respiratory water in and out of its gill slits.

At the posteroventral end, a **cloacal aperture** can be identified. A cloaca is a space into which urine, fecal material, and gametes are released. This seems to be an ancestral characteristic of many vertebrates, such as sharks and mudpuppies. In mature specimens, a small conical structure, the **urogenital papilla**, projects from the cloacal aperture. Urine and gametes are released through this papilla (Figure 14.6).

INTEGUMENTARY SYSTEM

All vertebrates possess an outer covering commonly called the skin. It typically consists of an outer multilayered epidermis overlying the dermis that, in turn, is connected to a subcutaneous layer. This organ system functions in a number of ways: protection against mechanical injury, bacterial invasion, and, in some instances, predation. Among aquatic vertebrates, osmoregulation is yet another important function. Amphibian skin functions as a respiratory organ.

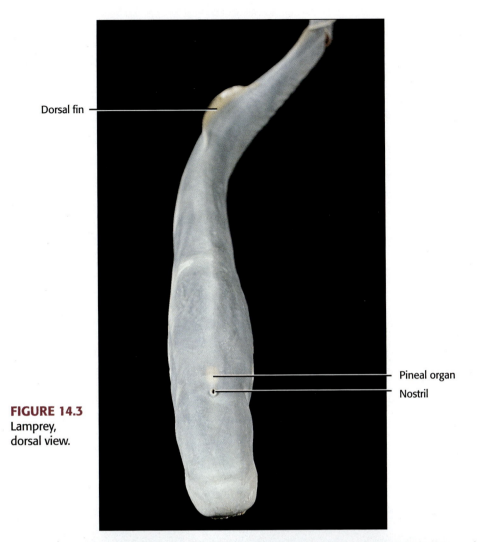

Dorsal fin

Pineal organ

Nostril

FIGURE 14.3
Lamprey,
dorsal view.

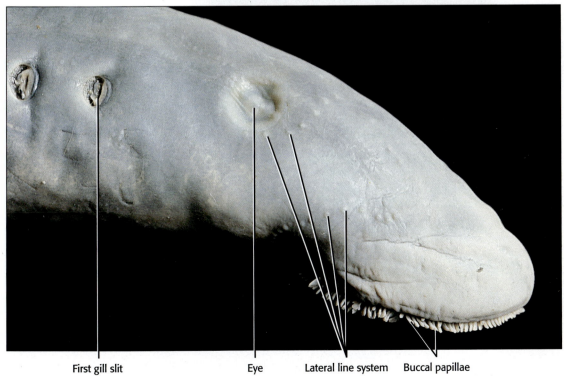

FIGURE 14.4
Lamprey,
anterior end.

First gill slit   Eye   Lateral line system   Buccal papillae

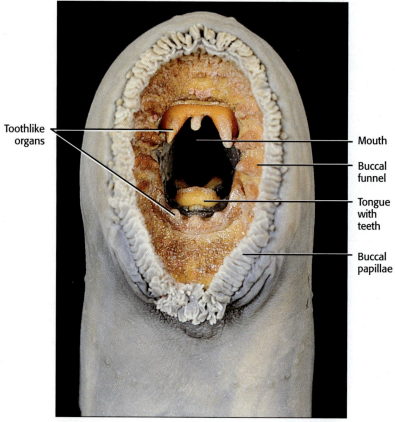

FIGURE 14.5 Lamprey, oral region.

Lamprey skin is scaleless. You will notice that a lamprey's skin is smooth if you pass your hand over it. The outermost layer of the epidermis is a cuticle that appears striated (Figure 14.7). Scanning electron microscopic (SEM) studies indicate that the striations are microridges (Figure 14.8 and Figure 14.9). Three distinct cell types are identifiable in the epidermis:

1. Large, elongated, binucleate club cells containing fibrous protein extend from the basal membrane to approximately halfway into the epidermis.

2. Large granular cells are located in the upper half of the epidermis.

3. A small third type of cell is scattered among the other two. As in most other fish, these cells produce copious amounts of mucus and other secretions.

The dermis, much thinner than the epidermis, consists of fibrous connective tissue and pigment cells. Underlying the dermis is a subcutaneous layer consisting of fat cells and other connective tissue, blood vessels, and muscle (Figure 14.8).

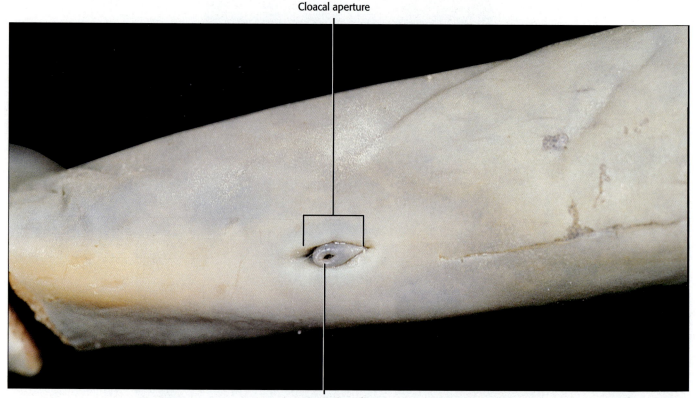

FIGURE 14.6 Lamprey, cloacal region (ventral view).

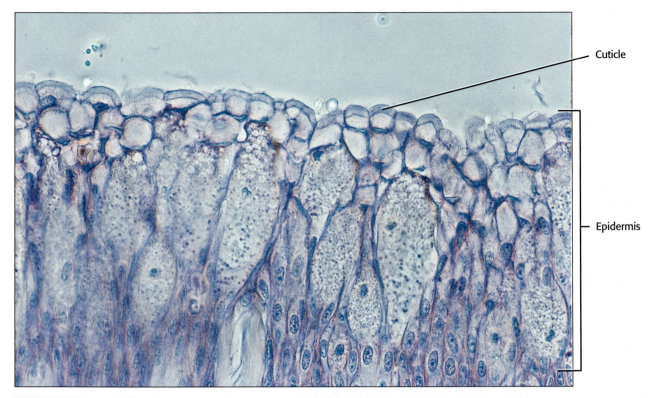

FIGURE 14.7 Sagittal section: Lamprey integument with cuticle.

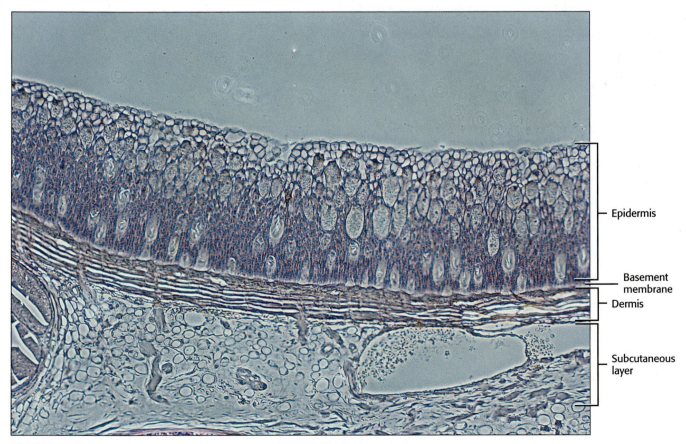

FIGURE 14.8 Sagittal section: Lamprey integument on underlying tissues.

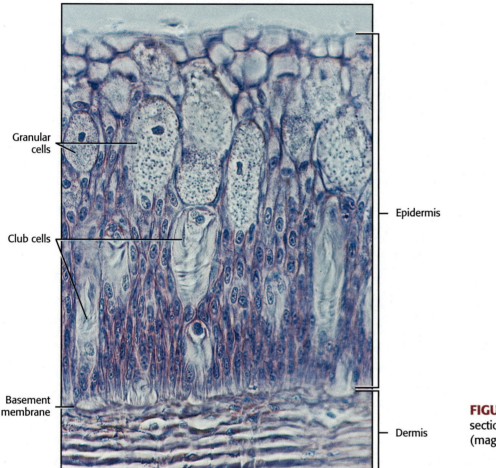

Granular cells

Club cells

Basement membrane

Epidermis

Dermis

FIGURE 14.9 Sagittal section: Lamprey integument (magnified).

INTERNAL ANATOMY

The lamprey cannot be dissected in the same fashion as one dissects the shark, the mud puppy, and the cat. A major problem with dissecting the lamprey is that most of the circulatory vessels are embedded in tissue, making it difficult to appreciate its primitive circulatory system. A more effective and efficient manner of studying the lamprey is through the use of midsagittal and transverse (cross) sections of mature specimens. To gain a thorough understanding of the anatomy of the lamprey, both types of sections have to be studied.

 When making the sections, one student must hold the specimen while a second student makes the cuts. Both students must take great care in making the cuts. The most likely person to receive an injury is the "holder."

Your instructor will direct you to complete one or more of the following:

1. **Midsagittal Section**

   With a sharp kitchen knife, beginning the cut at the dorsal edge of the sucker and taking care to slice as closely as possible through the midline of the dorsal surface of the lamprey, cut through the body to produce two equal halves. This cut should pass through the middle of the nostril, pineal eye, dorsal ridge, and fins.

    Be sure to use a "sawing" or cutting motion. You cannot simply press down on the knife and expect it to cut.

2. **Transverse Sections**

   With a sharp kitchen knife, make the following series of cross-sections (Figure 14.10):

   (a) Through the lateral and pineal eyes

   (b) Through the fourth gill slit

   (c) Through the body approximately 0.5 cm posterior to the seventh gill slit

   (d) Through the body approximately 4.5 cm posterior to the seventh gill slit

   (e) Through the body midway between the seventh gill slit and the cranial edge of the anterior dorsal fin

   (f) Through the body between the anterior and posterior dorsal fins.

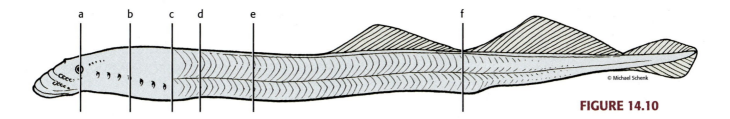

FIGURE 14.10

SKELETAL SYSTEM

The skeletal system consists of a notochord and numerous cartilages associated with oral, nervous, respiratory, and circulatory structures. The main longitudinal skeletal support of an adult lamprey consists of an unconstricted **notochord** extending from the tail tip to just below the ventral surface of the brain (Figure 14.11 and Figure 14.12).

A number of associated but un-united cartilages surround the brain and sense organs forming the cranial skeleton or chondrocranium. The oral skeletal elements include a supporting **lingual cartilage**, important during the

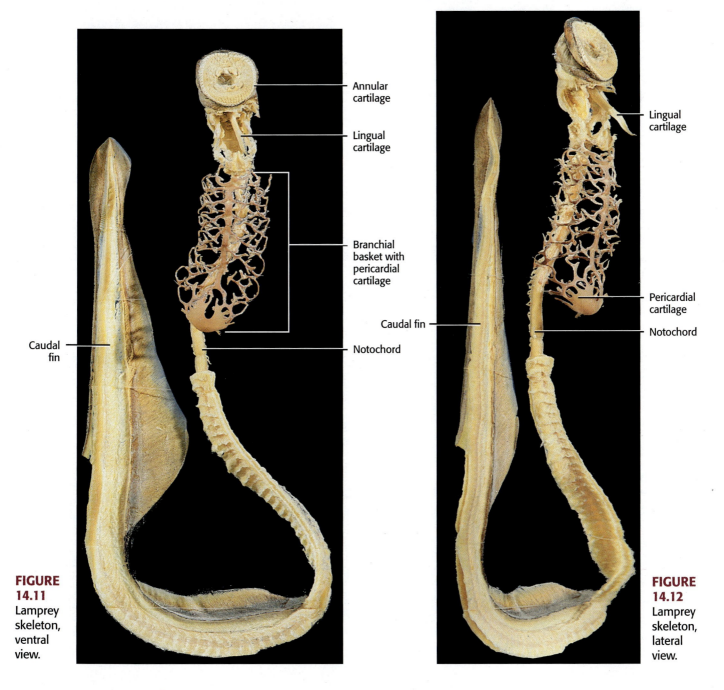

FIGURE 14.11
Lamprey skeleton, ventral view.

FIGURE 14.12
Lamprey skeleton, lateral view.

movement of the rasping tongue in the early feeding stages, and an **annular cartilage** with other cartilaginous skeletal elements supporting the oral sucker (Figure 14.11 and Figure 14.12).

Surrounding the gills and heart is an irregularly shaped **branchial basket**. During your inspection of both the sagittal and cross-sections of the specimens, you will notice irregular pieces of cartilage because the branchial skeleton consists of numerous interconnected individual bars of cartilage. In the lamprey, the branchial skeleton is lateral to the gills, in contrast to gnathostome fish, in which it is not basketlike and is medial. The posterior portion of the branchial basket encompasses the heart as the **pericardial cartilage** (Figure 14.11 and Figure 14.12). Many biologists believe that the gill supports of the lamprey and gnathostome fish are not homologous.

MUSCULAR SYSTEM

Body (axial) musculature consists of a series of blocks, the **myomeres**, that extend along either side of the lamprey. Muscle fibers of the myomeres are oriented longitudinally with their ends abutting the myosepta. The myomeres are folded into a complex W-shaped configuration and separated by sheets of connective tissue, the **myosepta** (Figure 14.13). The myomeres interlock and produce smooth,

alternating contractions on either side of the fish, facilitating its movement. In contrast to all jawed vertebrates, the lamprey does not possess a horizontal septum, the sheet of connective tissue that divides the axial musculature into epaxial (dorsal) and hypaxial (ventral) areas, nor does it possess ribs.

The musculature in the oral region has become specialized. A **circular muscle** associated with the sucker is employed during release of the host (Figure 14.14). Attachment of the sucker to the host is promoted by contraction of a muscle inserted into the tongue, thereby causing

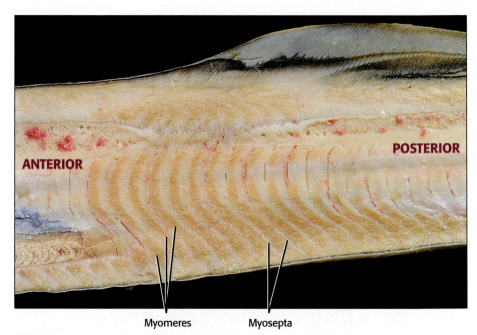

FIGURE 14.13 Sagittal section: Lamprey myomeres and myosepta.

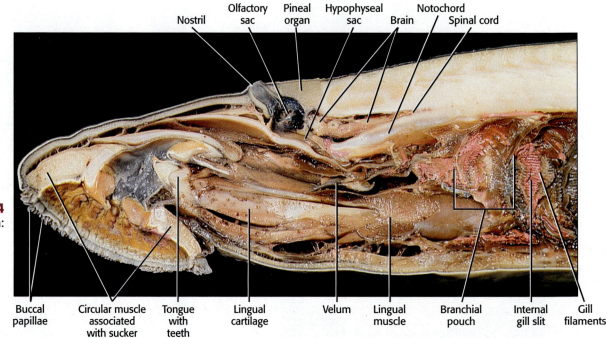

FIGURE 14.14
Sagittal section: Lamprey, anterior end.

Buccal papillae   Circular muscle associated with sucker   Tongue with teeth   Lingual cartilage   Velum   Lingual muscle   Branchial pouch   Internal gill slit   Gill filaments

Nostril   Olfactory sac   Pineal organ   Hypophyseal sac   Brain   Notochord   Spinal cord

suction. The tongue is controlled by other muscles that "rock" it dorsoventrally, enabling the lamprey to create a seeping wound in the side of a host vertebrate.

DIGESTIVE SYSTEM

As in all vertebrates, the digestive system of the lamprey is an alimentary canal—a mouth located at the anterior end and an anus at the posterior end. A body cavity seems to have been a prerequisite for any animal group that has attained an appreciable body size, such as annelids, arthropods, and vertebrates. The body cavity permits the internal organs and body wall to move independently. This cavity, known as a coelom, has walls lined with parietal peritoneum. Various organs projecting into this space are covered by the visceral peritoneum, a continuation of the cavity lining. The coelom in vertebrates is subdivided by a **transverse septum** into a ventral **pericardial cavity** and a postero-dorsal **pleuroperitoneal cavity**. Both are already present in the lamprey (Figure 14.15, Figure 14.18, and Figure 14.23).

In many other vertebrates, the pleuroperitoneal cavity is subdivided into a number of spaces separated by membranes, cartilage, and/or muscle that house specific organs of the respiratory, digestive, and other systems. Associated with these cavities are suspensory membranes, the mesenteries, that may be associated in a complex way with organs projecting into the cavity. In the lamprey, mesenteries are present but not well developed.

Recall that the lamprey leads a parasitic life during which it consumes body and tissue fluids, as well as tissue debris. This diet is one that seems to be digested easily with little reason to store food, and, therefore the digestive system is much simpler than that of other vertebrates. Clearly, any animal must present food to the digestive system for processing. The lamprey procures food by attaching the sucker to the host. The peripheral **buccal papillae** probably serve a sensory function and reinforce the suction. Notice the epidermal "teeth" adorning the surface of the **tongue** and the inner surface of the oral sucker (Figure 14.5 and Figure 14.14). These assist in producing the wound needed to facilitate feeding. The teeth in the funnel also aid in maintaining the attachment. A pair of **oral glands**, located ventrolaterally to the base of the tongue, secrete an anti-coagulating agent (Figure 14.16). This prevents coagulation of the host's blood and promotes a seeping wound.

The **mouth** (Figure 14.5 and Figure 14.14), located above the tongue, leads into a **pharynx**, which in the

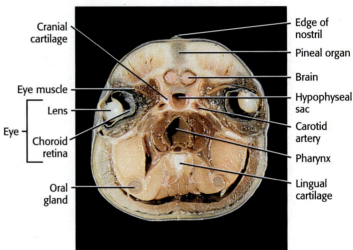

FIGURE 14.16 Transverse section: Lamprey, through lateral and pineal eyes.

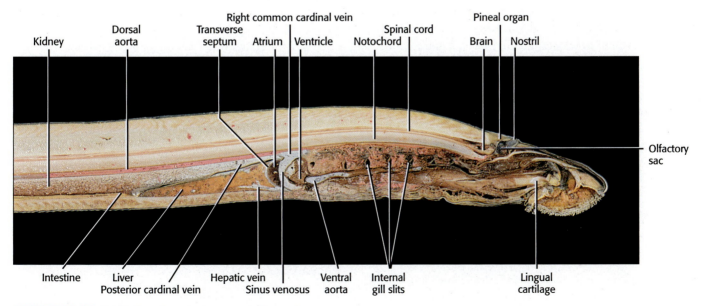

FIGURE 14.15 Sagittal section: Lamprey, anterior end.

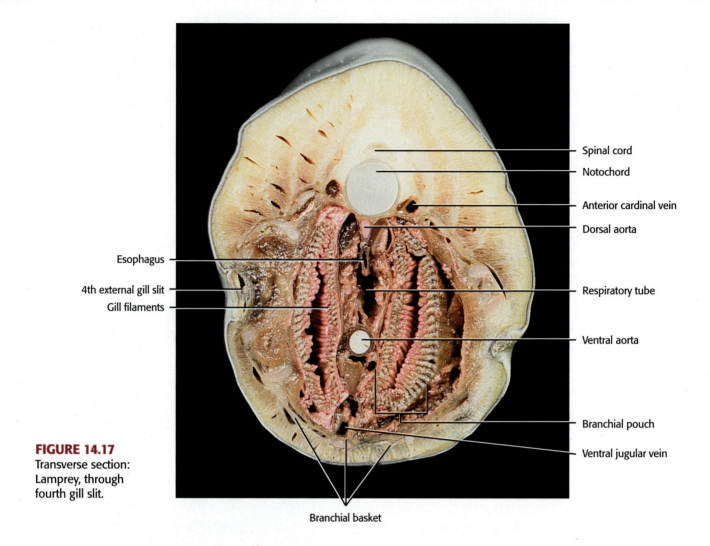

Spinal cord

Notochord

Anterior cardinal vein

Dorsal aorta

Esophagus

Respiratory tube

4th external gill slit

Gill filaments

Ventral aorta

Branchial pouch

FIGURE 14.17
Transverse section: Lamprey, through fourth gill slit.

Ventral jugular vein

Branchial basket

lamprey has been subdivided into a dorsal **esophagus** and a blind ventral **respiratory tube** (Figure 14.16 and Figure 14.17). This anatomical arrangement is unique to the lamprey and appears after metamorphosis into an adult. The esophagus leads into the straight **intestine**, terminating in an **anus**, opening into the **cloaca** (Figure 14.19). A spiral fold, extending along the inner surface area of the intestine, increases the effective digestive surface area. A stomach is not present.

Although a formed pancreas is not visible, isolated cells in the **liver** and the walls of the intestine in the vicinity of the liver produce chemicals similar to those secreted by the pancreas (Barrington, 1945). The **liver** is distinguishable; the gallbladder and bile duct disappear at metamorphosis. Observe that the liver abuts the posterior surface of the transverse septum (Figure 14.15).

RESPIRATORY SYSTEM

The morphology of the respiratory system in the lamprey is very different from that of other fish. The **velum**, located at the anterior end of the blind **respiratory tube**, shuts off the tube, preventing food from entering the respiratory system. The tube is pierced by seven **internal gill slits,** each of which leads into a **branchial pouch,** which, in turn, leads to the outside through an **external gill slit** (Figure 14.14, Figure 14.15, and Figure 14.17).

During feeding, water moves passively through the external gill slits into the branchial pouch by means of elastic recoil of the branchial basket following exhalation. As the water passes over the **gill filaments,** located within the pouches, gaseous exchange occurs (Figure 14.17). Water is actively expelled by means of muscular contraction of the branchial pouches through the external gill slits. When the lamprey is not feeding, respiratory water takes a more normal fish pathway, proceeding through the mouth into the respiratory tube, through the internal gill slits into the branchial pouches, and out through the external gill slits. During respiration, the velum closes off the esophagus.

EXCRETORY SYSTEM

Kidneys in lampreys consist of a pair of laterally flattened flaps that project ventrally into the posterior half of the pleuroperitoneal cavity. A small, collapsed tube, the **archinephric duct**, runs the length of the free border of the kidney. The ducts, carrying urine, converge in the vicinity of the cloaca into a urogenital sinus that drains externally through the **urogenital papilla.** Using a transverse section containing the kidney, with a dissecting needle carefully probe the very edge of the border of the kidney to observe the duct (Figure 14.18 and Figure 14.19).

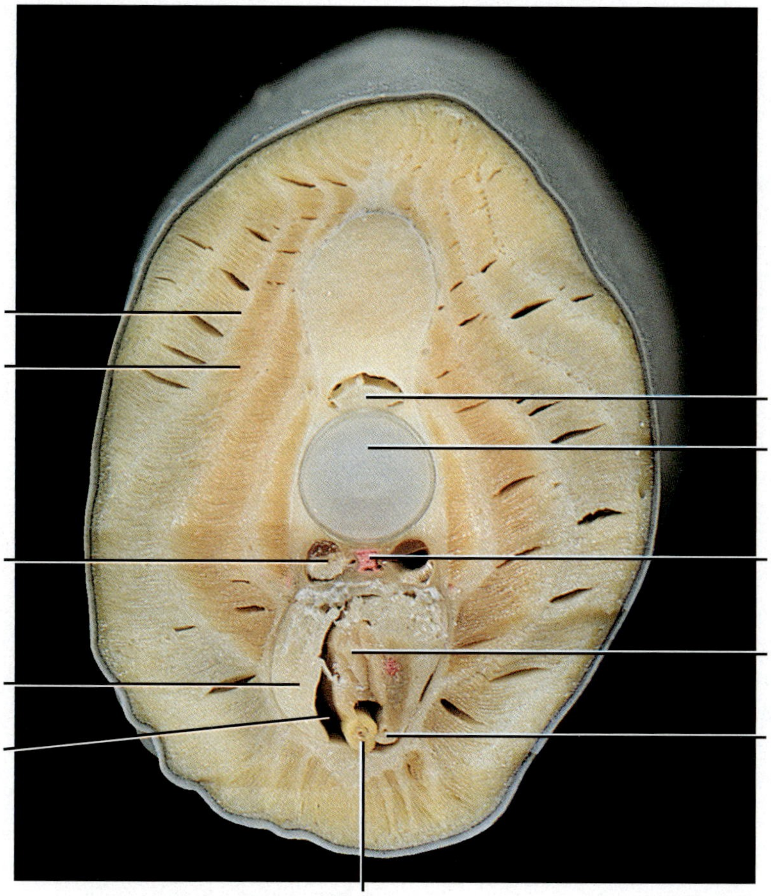

FIGURE 14.18
Transverse section: Lamprey, midway between seventh gill slit and cranial edge of anterior dorsal fin.

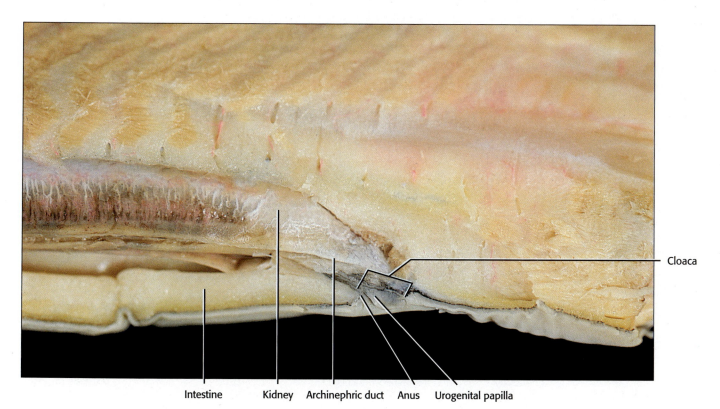

FIGURE 14.19 Sagittal section: Lamprey, through cloaca.

REPRODUCTIVE SYSTEM

In the lamprey, the sexes are separate—a condition known as dioecious. The adult ovary or testis is a single organ located medial to the kidneys in the pleuroperitoneal cavity. Embryologically, the gonads develop from paired primordia that fuse to form the adult structure.

In a mature specimen, the **ovary** appears as a large, centrally located organ crammed with numerous, creamy or yellowish, small spheres, the developing ova (Figure 14.20 and Figure 14.21). In contrast, the **testis** appears as smooth, tan folds (Figure 14.18). Because genital ducts are absent, the gametes in both sexes are discharged into the pleuroperitoneal cavity, from which they exit through genital pores in the walls of the urogenital sinus and out through the urogenital papilla into the freshwater environment.

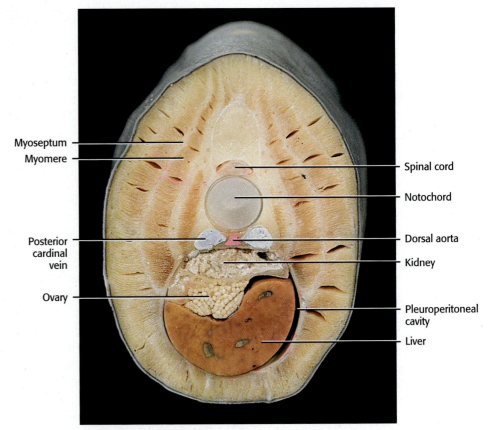

FIGURE 14.20 Transverse section: Lamprey, 4.5 cm posterior to seventh gill slit.

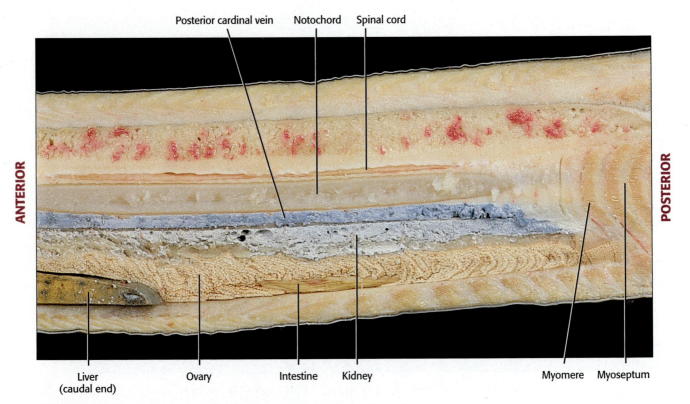

FIGURE 14.21 Sagittal section: Lamprey, mid-body.

CIRCULATORY SYSTEM

Circulation among vertebrates is closed (blood circulates entirely within vessels) and has evolved so blood transported in veins normally flows toward the heart, whereas blood in arteries is pumped away from the heart.

 To view many of the following blood vessels, refer to the midsagittal and cross-sections.

Blood from the tail region returns via a single **caudal vein**, lying ventral to the **dorsal aorta**, that empties into paired **posterior cardinal veins** as it enters the body cavity (Figure 14.22 and Figure 14.23). Muscles, kidneys, and gonads are drained by the posterior cardinals, located on either side of the dorsal aorta (Figure 14.18). A pair of **anterior cardinal veins**, also on either side of the dorsal aorta, drain the anteriodorsal region of the body (Figure 14.17).

As the left anterior and posterior cardinal veins approach the heart, they cross medially to join the **right common cardinal vein (duct of Cuvier)** formed by the union of the right anterior and posterior cardinal veins (Figure 14.15). Blood from the right common cardinal vein empties into the

sinus venosus. From the anterioventral region, blood drains through the **ventral jugular vein** into the sinus venosus (Figure 14.17).

Drainage of the intestine is accomplished by a hepatic portal system. A portal system consists of two capillary beds connected by veins. The hepatic portal system transports blood laden with products of intestinal digestion, as well as blood from the pancreas and spleen, to the liver. In the liver, this venous circulation breaks up into venous sinuses, which coalesce into a **hepatic vein** that empties into the **sinus venosus** of the heart (Figure 14.15).

As the muscular pump of the circulatory system, the heart, moves blood through the blood vessels. The pattern of the chambers of the heart in ancestral vertebrates consists of a posterior sinus venosus, a thin-walled atrium, a thick-walled ventricle, and a conus arteriosus, arranged linearly. A pair of bicuspid valves separate the sinus venosus and atrium (sino-atrial valve) and the atrium and ventricle (atrio-ventricular valve). A series of simple cup-shaped or semilunar valves guard the entrance into the ventral aorta. All of these valves regulate the volume of blood and promote a unidirectional flow of blood through the heart.

Like the heart of other vertebrates, the lamprey **heart** is surrounded by the pericardial cavity, a subdivision of the coelom. Also, as in other vertebrates, the pericardial cavity is delimited by a serous membrane, the parietal pericardium, that is continuous with the visceral pericardium, the outer layer of the heart.

In the lamprey the **sinus venosus** is tubular and lies between the **atrium** on the left side of the animal and the

FIGURE 14.22 Transverse section: Lamprey, between anterior and posterior dorsal fins.

Caudal fin-ray

Spinal cord

Notochord

Caudal artery

Caudal vein

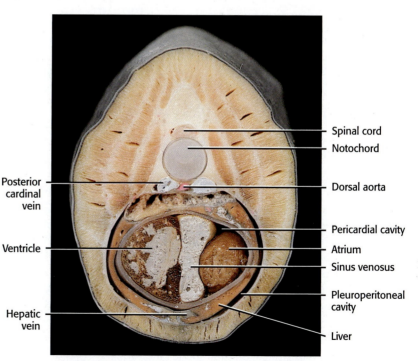

Posterior cardinal vein

Ventricle

Hepatic vein

Spinal cord

Notochord

Dorsal aorta

Pericardial cavity

Atrium

Sinus venosus

Pleuroperitoneal cavity

Liver

FIGURE 14.23 Transverse section: Lamprey, 0.5 cm posterior to seventh gill slit.

ventricle on the right side of the animal (Figure 14.15 and Figure 14.23). The peculiar position of the sinus venosus in the lamprey is found in no other vertebrate. A conus arteriosus, present in most vertebrates, is not present in the lamprey, and the **ventral aorta** is connected directly to the ventricle. The normal valves are present between the sinus venosus and the atrium (**sino-atrial valve**) and the atrium and the ventricle (**atrio-ventricular valve**).

Blood returns to the sinus venosus by way of the **right common cardinal**, **ventral jugular**, and **hepatic vein**. From the sinus venosus it is pumped into the atrium, and from the atrium into the ventricle. From the ventricle, blood is pumped past the semilunar valves and into the **ventral aorta**, which branches into **afferent branchial arteries**, transporting blood to the gills, where gas exchange occurs in capillaries (Figure 14.15 and Figure 14.17).

From the capillaries, blood laden with oxygen is carried away from the gills by the **efferent branchial arteries** into the dorsal aorta. Blood then is transported by **carotid arteries** to the tissues of the anterior portion of the body while the posterior body and tail of the lamprey receive blood through the **dorsal aorta** and **caudal artery**, the continuation of the dorsal aorta in the tail (Figure 14.16 and Figure 14.22).

Blood circulation in the lamprey illustrates what is known as a *single pump circuit* in vertebrates. In addition to the lamprey, almost all fish have this type of circulation. All of the blood returned to the heart is oxygen-poor, and this is the only blood that is pumped through the heart. From the heart, the blood is pumped to the gills. Oxygenation occurs within gill capillaries, from which blood is transported to the dorsal aorta and out to the body tissues by way of a system of arteries. Capillary beds in the tissues are the site of exchange of gases, nutrients, hormones, vitamins, and so on. From the capillary beds, veins transport blood back to the heart.

NERVOUS SYSTEM AND SENSE ORGANS

The **brain**, although exhibiting the typical anatomy of a vertebrate—consisting of a cerebrum, diencephalon, mesencephalon, medulla oblongata, and cerebellum—is difficult to study because of its small size. Ten pairs of cranial nerves originating from the brain are associated with various regions of the body. The brain is continuous with the **spinal cord**. Both lie above the **notochord** (Figure 14.14 and Figure 14.15). Spinal nerves have evolved uniquely among lampreys, in that the dorsal and ventral roots of each are not united as in most other vertebrates but, instead, emerge as alternating roots.

The functions of internal organs are controlled by the autonomic nervous system. It is not subdivided into a sympathetic and parasympathetic division in the lamprey.

A single dorsal **nostril**, located anterior to the brain, opens into the **olfactory sac**, enabling the lamprey to smell.

A **hypophyseal sac** is associated with the olfactory apparatus and facilitates olfaction by moving water in and out of the nostril and thereby over the olfactory epithelium (Figure 14.14 and Figure 14.16).

The **pineal organ,** just posterior to the nostril, functions as a light-detection organ. It controls various activities that are sensitive to photoperiodicity (Figure 14.3, Figure 14.14, and Figure 14.16).

The lidless, **lateral eyes** are capable of forming images. They possess a recognizable **lens** and pigmented **choroid-retinal layer** (Figure 14.2 and Figure 14.16).

The **lateral line system** consists of sensory organs at the ends of blind tunnels. It is distinguished externally by pores concentrated anteriorly near the eyes and buccal funnel (Figure 14.4). This primitive sensory system allows the lamprey to detect water movement.

Lampreys have two semicircular ducts, located posterior to the eyes and found in the inner ear, which are involved in detecting changes in body position and equilibrium. They are not easily seen.

LARVAL LAMPREY

After the female deposits the eggs and the male fertilizes them in fresh water, the adults die. Young lampreys, known as ammocoetes larvae, hatch after a short embryogenesis of about a week. The larvae then spend a number of years buried, with the exception of the head, in the substratum of the stream in which they hatch. The head projects above the substratum to facilitate filter feeding.

In many respects, ammocoetes bear a striking resemblance to amphioxus, but it differs in a number of significant ways (Figure 14.24, Figure 14.25, and Figure 14.26).

 To observe the anatomy of the larval lamprey, obtain a whole mount and representative cross-section slides as suggested by your instructor. Because these slides are thick mounts, be careful, particularly when switching to high power.

Like amphioxus, ammocetes have an **oral hood** with **oral lobes** (oral tentacles) that function as sensory organs and large particle strainers. Just posterior to the oral lobes is a pair of muscular flaps, the **velum,** that move water into the **pharynx** (Figure 14.24, Figure 14.25, and Figure 14.27).

Furthermore, the muscular pharynx is capable of expansion and contraction, causing volume changes and drawing a stream of water into it. The contraction phase propels the water across the **gill filaments** of the **gill pouches** and out through the external **gill slits** (Figure 14.28 and Figure 14.29A). Active use of the pharynx as a muscular pump represents a major divergence from the passive use of the pharynx as a filter-feeding apparatus by the non-vertebrate

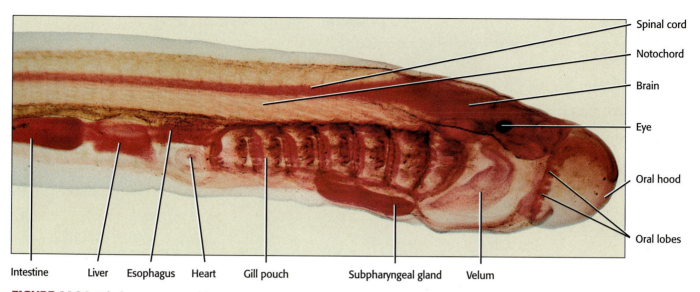

FIGURE 14.24 Whole mount: Larval lamprey.

Labels: Spinal cord, Notochord, Brain, Eye, Oral hood, Oral lobes, Intestine, Liver, Esophagus, Heart, Gill pouch, Subpharyngeal gland, Velum

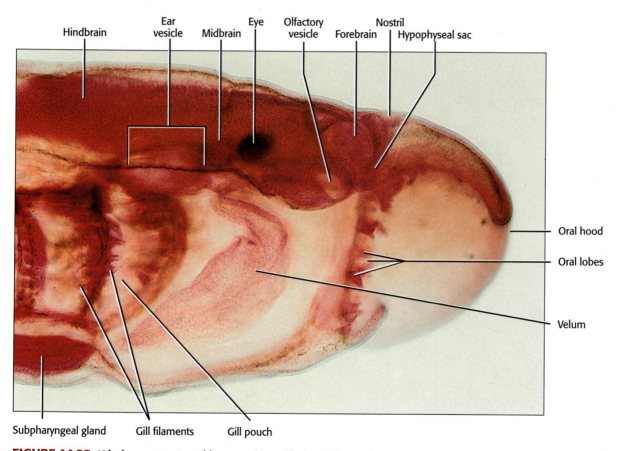

FIGURE 14.25 Whole mount: Larval lamprey (magnified anterior end).

Labels: Hindbrain, Ear vesicle, Midbrain, Eye, Olfactory vesicle, Forebrain, Nostril, Hypophyseal sac, Oral hood, Oral lobes, Velum, Subpharyngeal gland, Gill filaments, Gill pouch

chordates. This change was present already in ostracoderms. Note that this mechanism not only achieves respiration but potential food also enters the pharynx in the "inhaled" stream of water. Recall that in the urochordates and cephalochordates the propulsive structures that cause water to flow through the pharynx are cilia.

The **subpharyngeal gland,** an organ that probably is homologous with the endostyle of the non-vertebrate chordates, lies ventral to the pharynx (Figure 14.29A and Figure 14.29B). Some of its tissue is capable of fixing iodine, producing chemical compounds similar to thyroid hormones, and in the adult is transformed into the thyroid gland. In

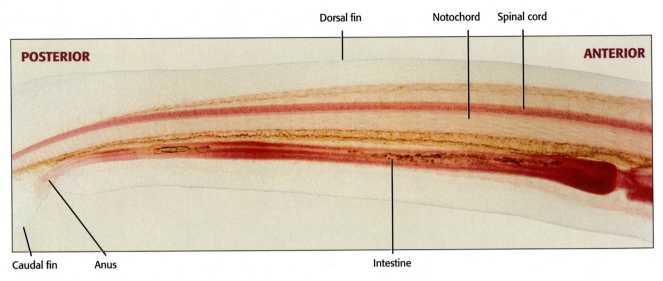

FIGURE 14.26 Whole mount: Larval lamprey, posterior end.

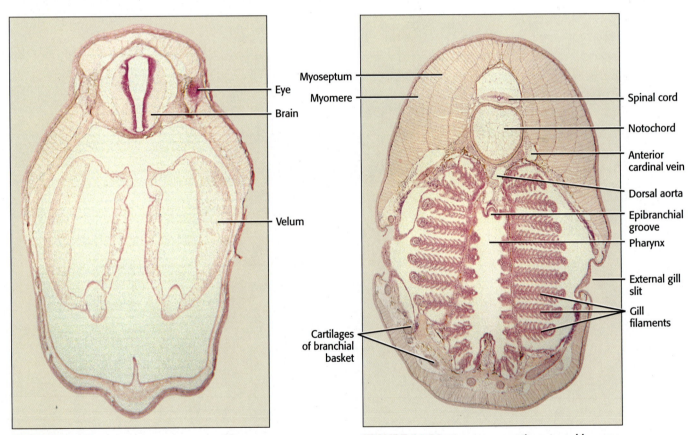

FIGURE 14.27 Transverse section: Larval lamprey, velum.

FIGURE 14.28 Transverse section: Larval lamprey, pharynx.

ammocetes, a major function of the subpharygeal gland is the production of mucus, which captures microscopic organisms and other particles brought in with the stream of water. Additional mucus is secreted by the pharyngeal lining. The mucus strands then move into the **esophagus** and into the **intestine**. Undigested material is discharged through the **anus** (Figure 14.24, Figure 14.26, and Figure 14.30).

Three organs not found in amphioxus but present in ammocetes are a **liver**, a **gallbladder**, and a **heart** (Figure 14.31). The heart sits just posterior to the pharynx. The liver is anterior to the intestine and posterior to the heart. Notice the developing **ventral aorta** (Figure 14.31). The gallbladder appears as a clear bubble closely associated with the liver.

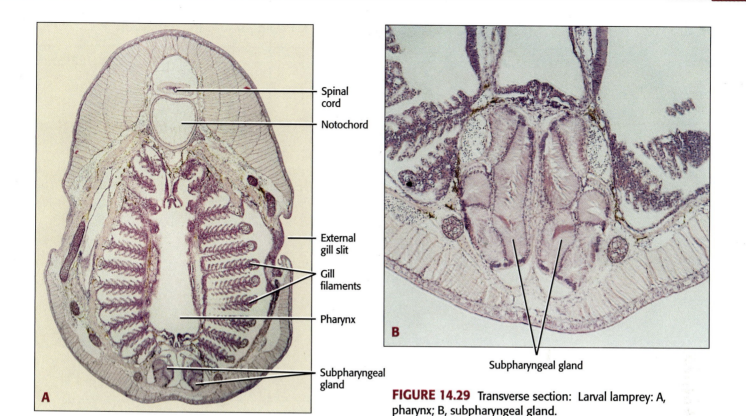

Spinal cord

Notochord

External gill slit

Gill filaments

Pharynx

Subpharyngeal gland

B

Subpharyngeal gland

FIGURE 14.29 Transverse section: Larval lamprey: A, pharynx; B, subpharyngeal gland.

Observe the tubules in the developing pronephric **kidney,** located dorsal to the heart (Figure 14.30). The pronephros is the first kidney to develop from the nephric tissue above the heart in the embryos of all vertebrates. This kidney functions in the larvae of some vertebrates. The **pronephric tubules** stimulate the development of the archinephric ducts that extend to the cloaca and serve as the primitive drainage tubules. The adult lamprey, however, possesses an opisthonephric kidney that develops from nephric tissue posterior to the pronephros.

The anterior portion of the nervous system of ammocetes is radically different from amphioxus and consists of a three-part brain, the **forebrain,** the **midbrain,** and the **hindbrain.** The **spinal cord** extends posteriorly from the hindbrain. Several developing sense organs are associated with the brain. An **olfactory vesicle** lies lateral to the forebrain. The **eye,** appearing as a black spot, is lateral to the midbrain. The clear oval **ear vesicle** lies lateral to the hindbrain. The single dorsal **nostril,** appearing as a small bump anterior to the forebrain, leads into the

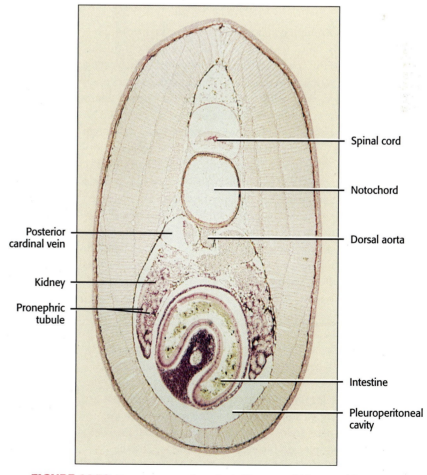

Spinal cord

Notochord

Dorsal aorta

Posterior cardinal vein

Kidney

Pronephric tubule

Intestine

Pleuroperitoneal cavity

FIGURE 14.30 Transverse section: Larval lamprey, kidney and intestine.

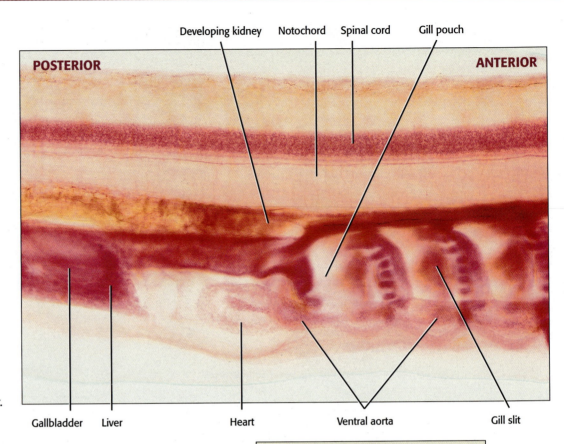

FIGURE 14.31 Whole mount: Larval lamprey, magnified heart and liver.

hypophyseal sac developing ventral to the brain (Figure 14.24 and Figure 14.25).

Just ventral to the spinal cord is the **notochord,** which retains its early chordate function as the major longitudinal supporting rod. The notochord, whose diameter is two to three times that of the spinal cord, stains less densely and is striated (Figure 14.31).

Similar to amphioxus, the body of ammocetes is serially muscular. Muscles are divided into muscle blocks, or **myomeres,** separated by sheets of connective tissue or **myosepta** (Figure 14.28). Although the specimen was treated chemically to make superficial tissues transparent and permit underlying organs to be seen, one still can see myomeres near the dorsal surface of the animal. Pigment cells or melanophores occur in the skin. Toward the posterior third of the body, a **dorsal fin** continuous with the **caudal fin** can be observed (Figure 14.24, Figure 14.26, and Figure 14.32).

Following its freshwater existence, the larva of a marine species leaves the burrow in the substratum that it may have occupied for three to seven years. It metamorphoses into a juvenile and migrates to the sea to feed. During this metamorphosis many organs are transformed into recognizable adult structures, *e.g.,* the parasitic mouth complex, "horny" teeth, rasping tongue, opisthonephic kidney, and maturing of the sense organs. Larvae of fresh-water species may metamorphose into juveniles that are parasitic and feed, *e.g.,* some species of *Lampetra* and *Ichthyomyzon,* while other species of the same genera metamorphose into reproductive adults, do not feed, but reproduce and die (Trautman, 1981; Scott and Crossman, 1979; Eddy and Underhill, 1974; Hubbs and Lagler, 1958).

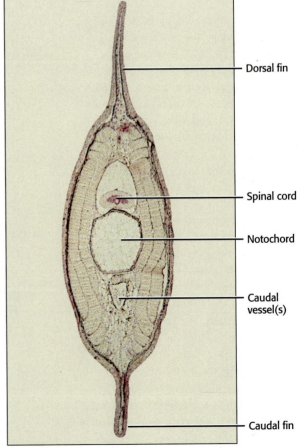

FIGURE 14.32 Transverse section: Larval lamprey, posterior end.

Phylum Chordata/Subphylum Vertebrata

Dogfish Shark

Two groups of jawed fishes, the acanthodians and the placoderms, appeared in the fossil record of the Paleozoic era. The placoderms, a group of fish exhibiting an extensive, external dermal, bony armor, became an important part of the marine and freshwater fish fauna during the Paleozoic era. This is a controversial group. Some paleontologists think that the placoderms were probably ancestral to gnathostomes. Others credit them with the ancestry of the Chondrichthyes, while still others consider them to be a fascinating group of vertebrates, which were a dead end, not related to any other modern group of vertebrates. Most paleontologists accept the acanthodians as the sister group of the bony fish or Osteichthyes.

For the following reasons, the dogfish shark, belonging to the genus *Squalus*, subclass Elasmobranchii and class Chondrichthyes, will be studied to illustrate the fish level of vertebrate organization:

1. Although specialized in a number of ways, the shark exhibits many ancestral vertebrate characteristics.

2. Sharks are relatively large, readily available from supply houses, and easy to dissect because their organ systems are large.

3. Removing the skin with its reduced scales is easier than removing the skin of heavily scaled bony fish.

 Obtain a shark, and observe the structures in Chapter 15.

Dogfish Shark — External Anatomy and Integumentary System

15

EXTERNAL ANATOMY

Water, with its great density and viscosity, has played a significant role in selecting for a streamlined body among aquatic vertebrates. Strong selection pressures to reduce body friction as the animal moves through the water have been exerted on aquatic vertebrates, resulting in a fusiform or teardrop-shaped body. With this conformation, energy expenditure for locomotion is reduced greatly.

If you run your hand, either gloved or ungloved, from posterior to anterior over the skin of the shark, you will experience a sensation similar to running your hand over sandpaper. What you are feeling are the spines of the posteriorly projecting **placoid scales** of the skin (Figure 15.1). The shape and the posterior orientation of the spines function to reduce the turbulence along the body surface and to facilitate movement through a thick, sticky medium. Notice that the surface of the body at the anterior end is compressed dorsoventrally, suggesting a wedge (Figure 15.2). This adaptation seems to play a role in creating lift at the anterior end of the shark.

Many of the specializations observed among modern sharks are related to jaws and food procurement. The modern shark, with its **rostrum**, **subterminal mouth**, and **protrusile jaws** equipped with a battery of shearing **teeth**, is capable of tearing out chunks of flesh from the bodies of large prey (Figure 15.3). The head-shaking behavior that accompanies the ability to protrude the upper jaw away from the skull makes it a formidable predator on the one hand. On the other hand, the same protrusile jaws enable the shark to pluck food items delicately from the seafloor.

To propel the body through water, most fish move their caudal fins (tail) back and forth in the horizontal plane. In the shark, a **heterocercal caudal fin** is employed to produce the thrust that results in forward movement (Figure 15.4). A heterocercal tail has a large dorsal lobe and a smaller ventral lobe with the vertebral column projecting

FIGURE 15.1 Placoid scales.

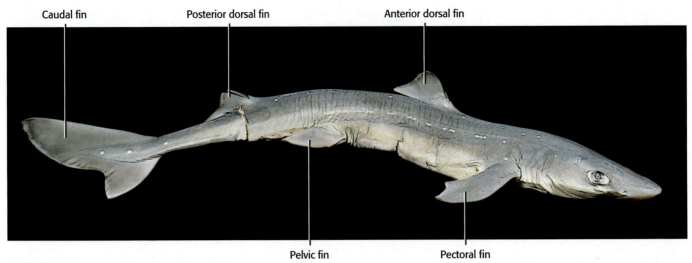

Caudal fin   Posterior dorsal fin   Anterior dorsal fin

Pelvic fin   Pectoral fin

FIGURE 15.2 External features of the shark.

into the dorsal lobe. Notice the lateral **keel** lying anterior to the tail. In swiftly swimming fish, the keel acts as a device, reducing the pressure required to move the tail laterally

through the water. As a result, tails with keels can be moved faster through the water. Notice the supporting fin rays in all of the fins.

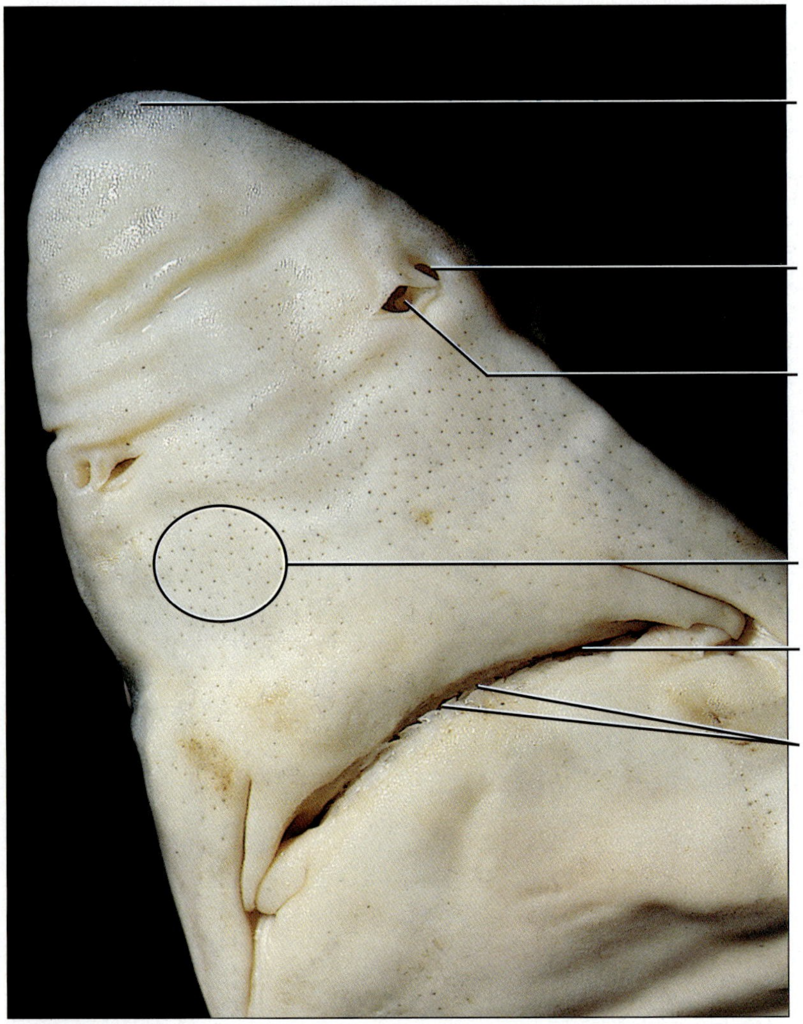

Rostrum

Incurrent aperture

Excurrent aperture

Openings of ampullae of Lorenzini

Subterminal mouth

Teeth

FIGURE 15.3 Ventral surface of anterior end of shark.

In contrast to the pectoral fins, the pelvic fins do not stabilize the course of locomotion. In the male, the pelvic fins are modified as sperm-transfer organs known as **claspers** (Figure 15.5). Obviously, the presence or absence of claspers will enable you to identify the sex of your specimen. Female pelvic fins are unmodified (Figure 15.6).

On the dorsal and ventral surface of the rostrum reside some highly sensitive sense organs, the **ampullae of Lorenzini** (Figure 15.3, Figure 15.7) of the **lateral line system** (Figure 15.3, Figure 15.7, and Figure 15.8). They also are found among the skates and rays, relatives of sharks. These organs function as sophisticated electroreceptors, capable of detecting electrical potentials of less than .01 microvolt. Electrical currents commonly are produced across cell membrane surfaces such as those of fish muscle. Even at rest, tissues produce currents, and the ampullae of sharks allow them to detect prey.

Because temperature affects electroreceptors, the ampullae also are sensitive "thermometers." Furthermore, as ions flow past each other in a solution, e.g., in

FIGURE 15.4 Heterocercal caudal fin and lateral keel.

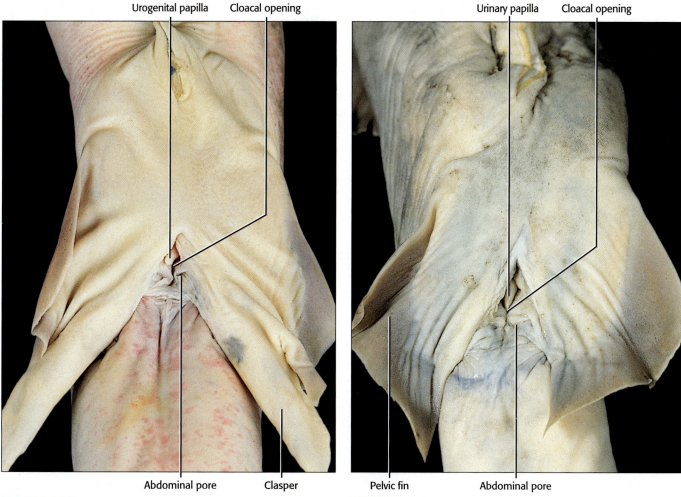

FIGURE 15.5 Male shark.

FIGURE 15.6 Female shark.

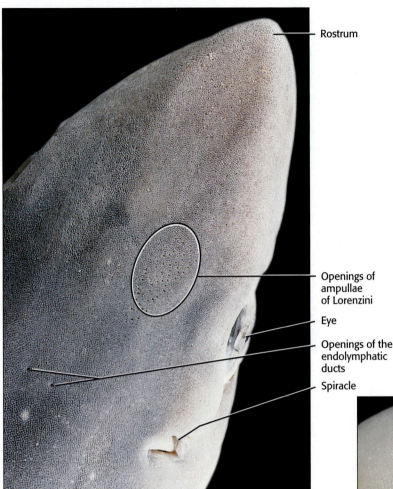

Rostrum

Openings of ampullae of Lorenzini

Eye

Openings of the endolymphatic ducts

Spiracle

FIGURE 15.7 Dorsal surface of anterior end of shark.

a **nasal flap** (Figure 15.8). As the fish swims, water enters the incurrent aperture, flows over the olfactory epithelium located on the surface of the **olfactory lamellae** and exits through the excurrent aperture. Detection of waterborne chemicals is exquisitely developed in sharks, giving them the ability to detect concentrations as low as 1 part in 10 billion (Pough et al., 2005). Obviously, this olfactory sensitivity in conjunction with the ability to detect subtle electrical currents generated by resting muscle tissue of potential prey gives the shark a tremendous edge as a predator.

The shark has paired, **lateral eyes**, typical of most vertebrates. Immovable **upper** and **lower eyelids** are present (Figure 15.9). A movable membrane is demonstrable in sharks as they approach for a bite of prey. The next time you watch one of the shark programs on TV, notice its function.

Posterior and somewhat dorsal to the eye, locate the circular opening of the **spiracle**, the modified first gill slit, between the jaws and hyoid

the ocean, they generate small currents well within the detectable range of the ampullae of Lorenzini. Some biologists believe the ampullae serve as a positional or navigational tool. Ampullary organs are found in a number of other fish and in aquatic amphibians.

> To demonstrate the presence of these organs, lying well beneath the skin, gently press on the surface of the snout and observe mucus expressed from the pores opening through the skin surface.

The lateral line system appears as an extensive system of pores arranged over the surface of the head and extending along the lateral surface of the body. The lateral line system functions as a group of coordinated mechanoreceptors, allowing the shark to detect water currents and pressure changes produced by objects in its environment. This system is common to most fish and many aquatic amphibians.

The paired **nostrils** are located in the ventral surface of the rostrum. They consist of two openings, a lateral **incurrent aperture** and a medial **excurrent aperture**, separated by

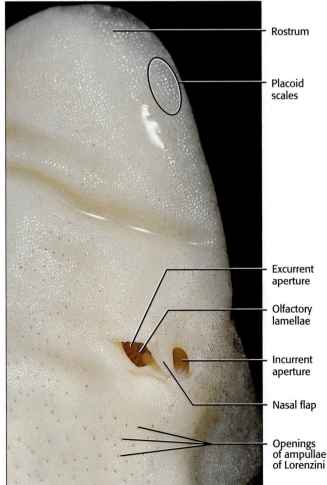

Rostrum

Placoid scales

Excurrent aperture

Olfactory lamellae

Incurrent aperture

Nasal flap

Openings of ampullae of Lorenzini

FIGURE 15.8 Nostrils of shark.

Spiracle

Upper eyelid

Eye

Lower eyelid

Nostril
(incurrent aperture)

FIGURE 15.9 Eye of the shark.

Spiracle

Fifth gill slit

First gill slit

FIGURE 15.10 Spiracle and gill slits (anterior end to the left).

arch. Five typical **gill slits** are located laterally just anterior to the pectoral fin (Figure 15.10). Respiratory water during nonfeeding periods exits through these openings. During feeding, water can be drawn in through the spiracle and expelled through the gill slits. The flattened rays and skates that spend a great deal of time on or in the substratum draw in respiratory water through the dorsal spiracle and pump it out through the ventral gill slits, thereby avoiding

injury of the delicate gill filaments by sand and debris that would occur if water were inhaled through the mouth.

Look for the tiny openings of a pair of **endolymphatic ducts** between the spiracles (Figure 15.7). In cartilaginous fishes the endolymphatic ducts, associated with the inner ear apparatus and involved in balance, open to the outside.

Similar to the lamprey and most other vertebrates, the shark also possesses a **cloaca**, a common space into which the digestive, urinary, and reproductive systems open. Identify the opening of the cloaca, located ventrally just posterior to the pelvic fins. Locate the **urogenital papilla** in the male, through which both sperm and urine pass (Figure 15.5). In the female, only urine passes through the homologous structure, the **urinary papilla** (Figure 15.6). Locate the **abdominal pores** just lateral and posterior to the cloaca (Figure 15.5). The function of these openings is conjectural. They may have served to allow the gametes to escape into the environment in early ancestors, or possibly to allow the escape of excess fluid from the pleuroperitoneal cavity.

INTEGUMENTARY SYSTEM

 Obtain available slides of shark skin and observe the skin and scale anatomy.

Typical of vertebrates, shark skin has a relatively thin outer **epidermis** underlain by a thicker **dermis** (Figure 15.11). The epidermis is a stratified epithelium with a basal layer of cells, the stratum germinativum, which gives rise to the epidermis through the process of mitosis. Among the epidermal cells are scattered mucous cells, cells that produce toxins associated with fin spines, and sensory cells of various types. The major constituent of the dermis is connective tissue. Melanophores, affecting the color of the skin, are present in the upper layers of the dermis. In addition, a number of inclusions (blood vessels and nerves) are located in the dermis.

Remnants of the ancestral bony armor are present in the form of **placoid scales**, embedded in the dermis (Figure 15.12 and Figure 15.13). Embryologically, the scales are derived from both the epidermis and the dermis in a complex process. The placoid scale consists of a **bony base** on which sits the bulk of the scale, consisting primarily of a bonelike material, **dentin**. Overlying the dentin of the caudally projecting **spine** of the scale is a thin layer of **enamel**, another bonelike material. The **pulp cavity**, lined by odontoblasts, contains blood vessels and nerves that maintain the living condition of the scale. Odontoblasts give rise to new dentin material. (See page 39, Figure 5.5.)

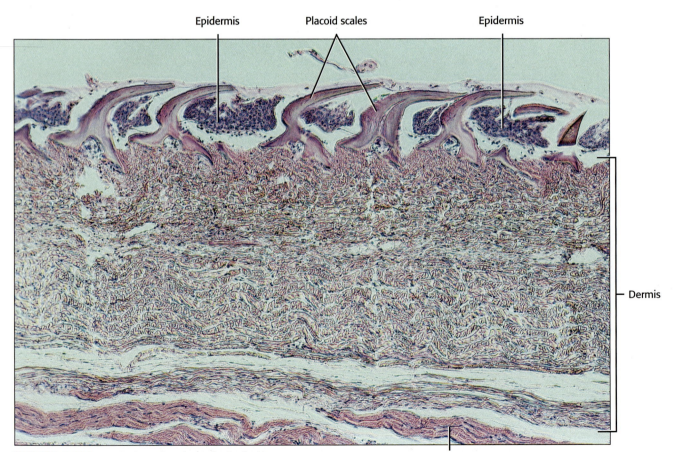

FIGURE 15.11 Sagittal section through shark skin.

Epidermis   Placoid scales   Epidermis

Dermis

Muscle tissue

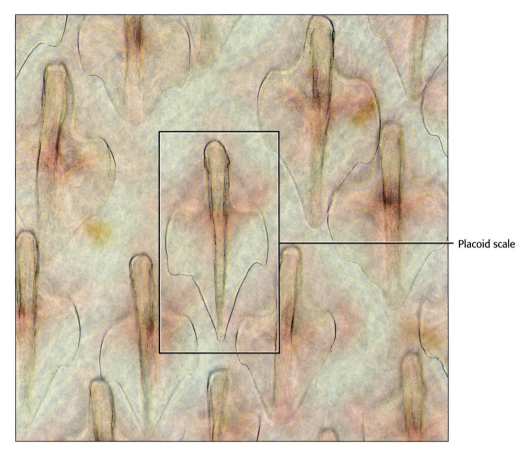

Placoid scale

FIGURE 15.12 Whole mount: Placoid scales.

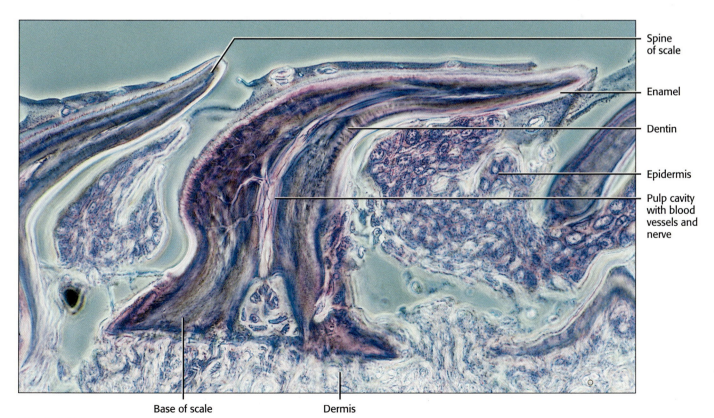

Spine of scale

Enamel

Dentin

Epidermis

Pulp cavity with blood vessels and nerve

Base of scale

Dermis

FIGURE 15.13 Sagittal section: Placoid scales.

Dogfish Shark — Skeletal System

The chondrichthyes are known as cartilaginous fishes because sharks, rays, skates, and their distant relatives, the Holocephali, have cartilaginous skeletons. In some sharks, portions of the skeletons are calcified, but calcified cartilage is not bone.

In spite of its delicate appearance, cartilage imparts great flexibility and permits bending of the body that would not be possible with inflexible bone. Sharks are capable of executing tight turns in their environment that are impossible for most bony vertebrates to duplicate. Typically, as in other vertebrates, the skeleton provides support and protection of the nervous system and other soft organs, as well as providing the surfaces for muscle attachment to produce body movement.

The skeleton can be conveniently subdivided into axial and appendicular divisions (Figure 16.1 and Figure 16.2). The axial skeleton consists of the chondrocranium, the splanchnocranium, and the vertebral column and ribs (Figure 16.3). We include the median fins, the dorsal fins, and the caudal fin in the axial division. The appendicular skeleton consists of the pectoral girdle and its associated pectoral fins and the pelvic girdle and its associated pelvic fins (Figure 16.1).

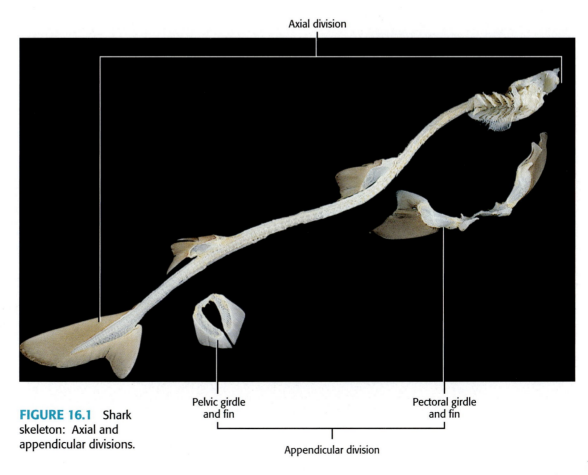

FIGURE 16.1 Shark skeleton: Axial and appendicular divisions.

FIGURE 16.2 Shark skeleton.

AXIAL DIVISION

Chondrocranium

Although somewhat specialized, the chondrocranium, or braincase, of the shark helps us interpret the anatomy of other vertebrates. The shark's chondrocranium is elongated, consists of a single piece of cartilage with no visible joints, and houses the brain and sense organs. During the shark's embryology, several areas of cartilage fuse together to form a braincase with its associated sense organ capsules. (See page 36, Figure 5.1.)

Similar developmental activities occur among all vertebrates, resulting in a similar complex, housing the brain and associated sense organs. One of the advantages of studying the shark is that it has no obscuring dermal bones over the chondrocranium or braincase as we find in "bony" vertebrates.

 Obtain available skeletal material and study the anatomy of the chondrocranium of the shark.

A unique characteristic of the **chondrocranium** is the anterior projecting **rostrum** overhanging the mouth and giving the shark its identifiable portrait (Figure 16.3). Lateral to the rostrum lie the **nasal capsules**, containing the olfactory apparatus. Posterior to these and opening laterally are the **eye orbits**, housing the eyeballs. The **preorbital process**, the **postorbital process**, and the **supraorbital crest** form a protective eye cup. At the posterior angles of the chondrocranium are the **otic capsules**, containing the semicircular ducts.

Concentrate on the dorsal aspect of the chondrocranium (Figure 16.4). Reidentify the rostrum with its **precerebral cavity**, in which resides a jellylike material. At the posterior end of the precerebral cavity is the precerebral fenestra opening into the cranial cavity. Note the **rostral fenestrae** opening ventrally on either side of the **rostral keel**. The dorsal body of the chondrocranium forms the roof of the braincase, extending from the anterior edge of the rostrum to the posterior ear region.

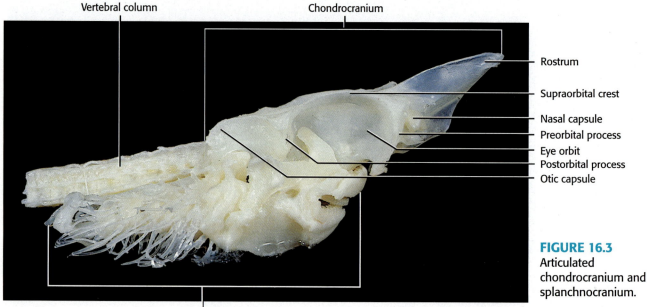

Vertebral column

Chondrocranium

Rostrum

Supraorbital crest

Nasal capsule

Preorbital process

Eye orbit

Postorbital process

Otic capsule

FIGURE 16.3
Articulated chondrocranium and splanchnocranium.

Splanchnocranium

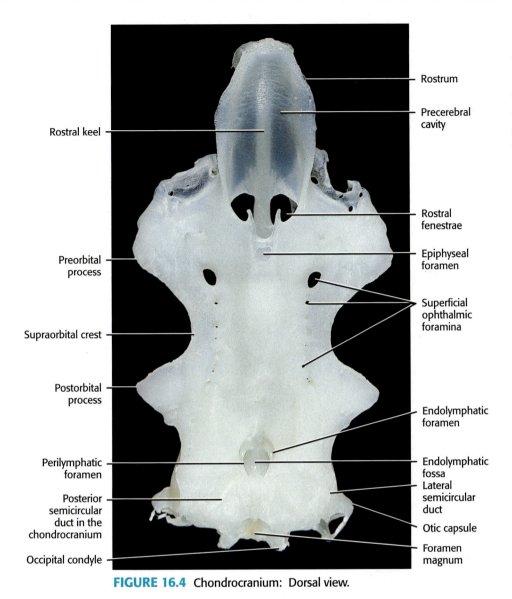

FIGURE 16.4 Chondrocranium: Dorsal view.

nerves (cranial nerve X), occur just lateral to the occipital condyles. The **glossopharyngeal foramina** lie in the otic capsules lateral to the vagus foramina. Through these openings pass the glossopharyngeal nerves (cranial nerve IX). In the roof of the otic capsules one sometimes can identify the groove associated with the lateral semicircular duct (Figure 16.4).

Turn the chondrocranium over and identify the following in the posterior, anterior, and ventral views (Figure 16.5, Figure 16.6, and Figure 16.7). Orient yourself by reidentifying the rostrum, the rostral keel, the rostral fenestrae, preorbital processes, and nasal capsules. In a wet preparation, the openings into the nasal capsules, the **external nares**, can be identified. A beamlike area extending between the orbits is punctuated posteriorly by the **basitrabecular processes**. Posterior to the processes is an expanded **basal plate**. Piercing the median aspect of the anterior portion of the basal plate is the **carotid foramen**, through which pass the internal carotid arteries.

Just posterior to the rostrum is the **epiphyseal foramen**, through which the epiphysis (pineal organ or eye) projects. Note the series of openings, the **superficial ophthalmic foramina**, paralleling the supraorbital crest, through which pass branches of the superficial ophthalmic nerve, portions of the trigeminal and facial nerves (cranial nerves V and VII, respectively). A median depression, the **endolymphatic fossa**, houses the smaller **endolymphatic foramina**, through which pass the endolymphatic ducts, and the larger **perilymphatic foramina**, through which pass the perilymphatic ducts, in its lateral walls. Both of these ducts are associated with inner ear function. The single, large foramen lying posterior to the endolymphatic fossa is the **foramen magnum**, through which passes the spinal cord (Figure 16.4 and Figure 16.5).

Just ventral and lateral to the foramen magnum are the **occipital condyles**, which articulate with the first trunk vertebra. The **vagus foramina**, through which pass the vagus

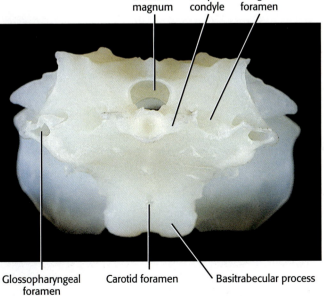

FIGURE 16.5 Chondrocranium: Posterior view.

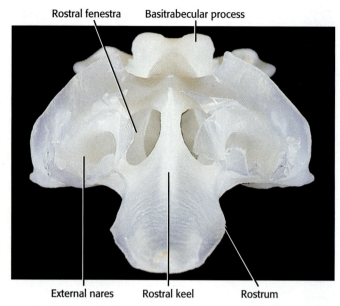

Rostral fenestra   Basitrabecular process

External nares   Rostral keel   Rostrum

FIGURE 16.6 Chondrocranium: Anterior view.

Rostrum

Rostral keel

Nasal capsule

Rostral fenestra

Superficial ophthalmic foramina

Basitrabecular process

Carotid foramen

Basal plate

Occipital condyles

Glosso-pharyngeal foramen

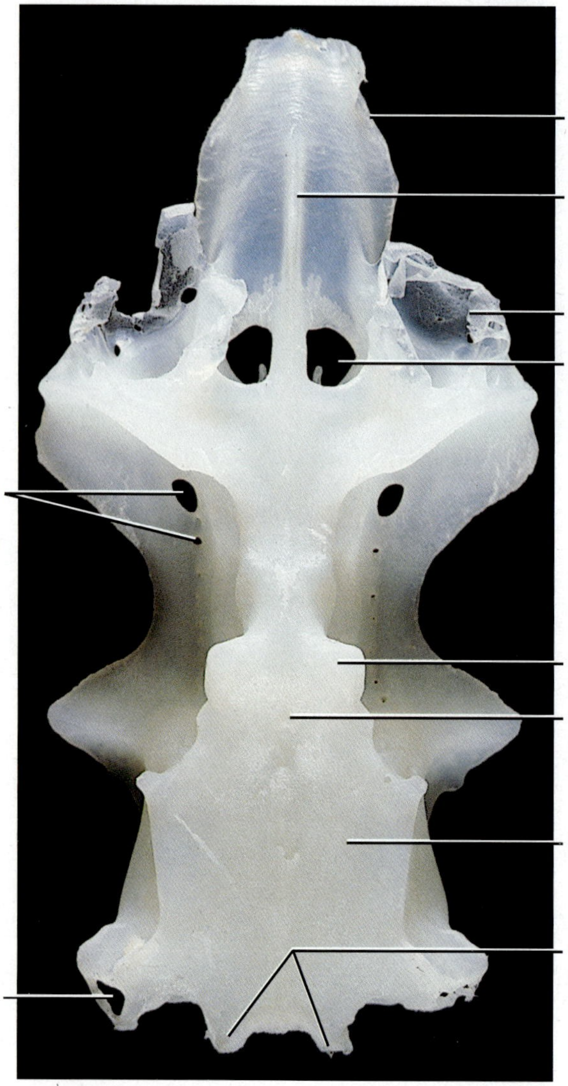

FIGURE 16.7 Chondrocranium: Ventral view.

In the lateral view, identify the following foramina in Figure 16.8. All but the hyomandibular foramen are located within the eye orbit. At the anterior end of the orbit is the large **optic foramen**, through which passes the optic nerve (cranial II), and at the posterior end is another large foramen, the **trigeminofacial foramen**, through which pass the trigeminal and the facial nerves (cranial nerves V and VII).

The small **trochlear foramen**, through which passes the trochlear nerve (cranial nerve IV), occurs just ventral to the supraorbital crest. A small **oculomotor foramen**, through which passes the oculomotor nerve (cranial nerve III), is located approximately 2–3mm anterior to the midline of the anterior border of the trigeminofacial foramen. The **hyomandibular foramen**, through which passes the hyomandibular nerve, a branch of the facial nerve (cranial nerve VII), pierces the otic capsule. Anterior to the hyomandibular foramen, identify the foramen allowing passage of the optic artery, and the **abducens foramen**, through which passes the abducens nerve (cranial nerve VI).

Splanchnocranium

Visceral Arches

As indicated in the discussion of the chondrocranium, the shark's skeleton is not obscured by overlying dermal bony elements. For this reason, the shark's **splanchnocranium**, or visceral skeleton, may provide us with an insight into the possible ancestral condition in vertebrates. The splanchnocranium of sharks consists of two anterior pairs of modified visceral arches followed by five pairs of unmodified visceral arches known as **branchial arches** (Figure 16.9, Figure 16.10, Figure 16.11, and Figure 16.12). (Also see page 36, Figure 5.2.) Because the anatomy of the five pairs of branchial arches is unmodified and similar, we will examine them first. The branchial arches, which support the gills, are located between the gill slits. Each half of the pairs consists of four components—the **pharyngobranchial**, the **epibranchial**, the **ceratobranchial**, and the **hypobranchial**—united ventrally by a **basibranchial**. (See page 36, Figure 5.2B.)

Two sets of projections emanate from the gill arches. The **gill rays** strengthen the branchial septa, walls of connective tissue that separate adjacent gill pouches. The branchial septa, on which the gills occur, extend to the lateral surface of the body. Gill slits occur between the septa. The second set of projections, the **gill rakers**, extend from the medial surface of the visceral arches and

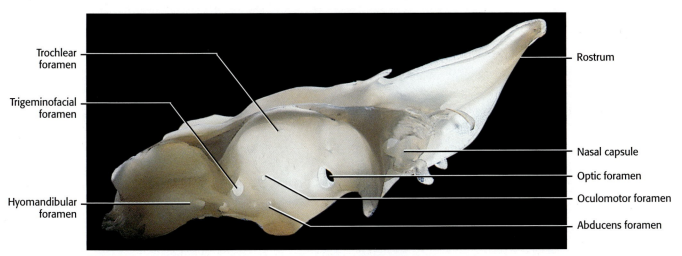

FIGURE 16.8 Chondrocranium: Lateral view illustrating foramina.

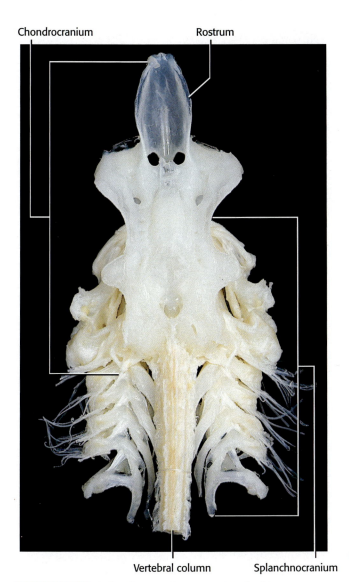

FIGURE 16.9 Splanchnocranium associated with chondrocranium: Dorsal view.

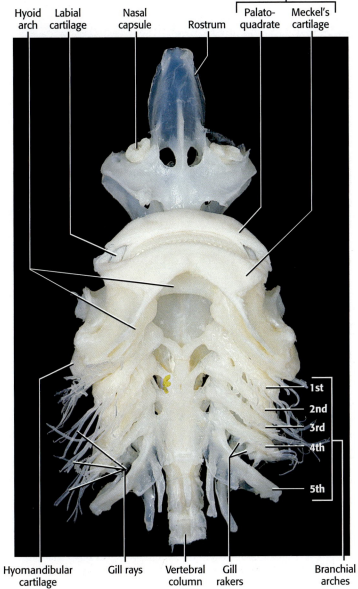

FIGURE 16.10 Splanchnocranium associated with chondrocranium: Ventral view.

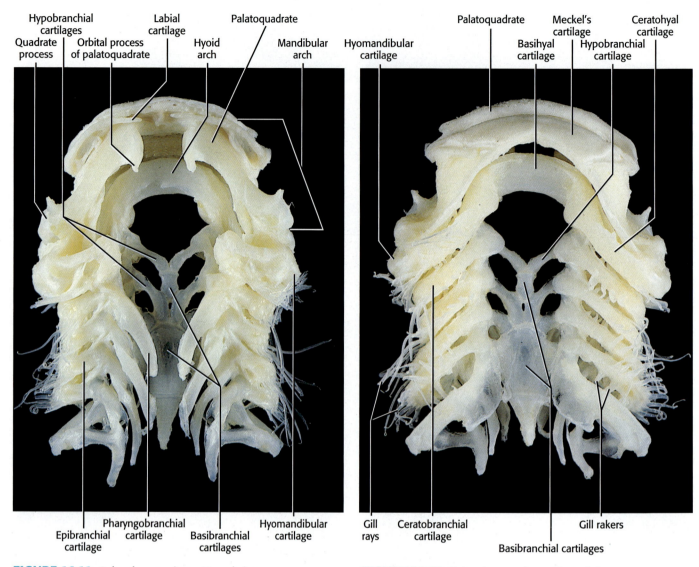

FIGURE 16.11 Splanchnocranium: Dorsal view.

FIGURE 16.12 Splanchnocranium: Ventral view.

function as straining devices, preventing food from entering the respiratory chambers (Figure 16.10 and Figure 16.12).

Mandibular Arch

Many comparative anatomists consider the mandibular arch to be the modified first visceral arch, but a great deal of controversy exists concerning the jaw skeleton (Figure 16.3, Figure 16.11, and Figure 16.12). The skeleton of the upper jaw, the **palatoquadrate** (Figure 16.10 and Figure 16.12) is thought to have been derived from the paired epibranchial elements of the arch, whereas the lower jaw skeleton, **Meckel's** or **mandibular cartilage** (Figure 16.10 and Figure 16.12), is thought to have been derived from the paired ceratobranchial elements. (See page 36, Figure 5.2.)

Identify the **orbital process** of the palatoquadrate (Figure 16.11) projecting dorsally into the eye orbit. The large laterally projecting **quadrate process** articulates with the posterior end of Meckel's cartilage. Teeth are attached to

the rims of both jaws and will be covered in more detail in the discussion of the digestive system (Chapter 19). On a well cleaned specimen, one may be able to distinguish a pair of **labial cartilages** (Figure 16.10 and Figure 16.11), projecting from the mandibular cartilages and overlying the palatoquadrate. In spite of their appearance and position, labial cartilages are not thought to be part of the visceral arch complex.

Hyoid Arch

Often identified as the second modified visceral arch in this series is the **hyoid arch** (Figure 16.10 and Figure 16.11). (Also see page 36, Figure 5.2.) Paired **hyomandibular cartilages,** the epibranchial elements of the hyoid arch, extend from the otic area of the chondrocranium and articulate with the **ceratohyals** (Figure 16.12), the ceratobranchial elements of the hyoid arch, in the angle of the jaws. The ceratohyals abut the posterior ends of the palatoquadrate and

Meckel's cartilage and extend ventrally, where they articulate with the **basihyal cartilage** of the hyoid arch.

In sharks, jaws are loosely connected to the rest of the skeleton. They are not connected directly to the chondrocranium but, rather, are held in place by small ligaments to the palatoquadrate. The hyomandibular cartilage creates a loose jaw association, known as hyostylic suspension (hyostyly). (See page 37, Figure 5.3C.) Hyostyly permits the fish to protract its jaws, allowing it to sink its teeth deeply into its prey. This, combined with head shaking, allows the shark to rip big chunks of flesh from its prey. Hyostyly also permits a shark to pluck food items delicately from the substratum (Moss, 1984).

Vertebral Column Part of the elegance of the graceful movements we associate with sharks can be attributed to the construction of the vertebral column (Figure 16.1). This construction permits the great flexibility necessary for these movements. Vertebral regions in sharks are not extensive, and only two types of vertebrae, **trunk** and **caudal**, make up the shark's vertebral column (Figure 16.13).

> Study prepared specimens, or ask your instructor to assist in the preparation of your own specimen.

Each vertebra possesses a cylindrical **biconcave (amphicoelous) centrum** (Figure 16.14). Although, as adults, sharks do not have a functional notochord, they do have remnants of this ancient chordate characteristic in their anatomy. During embryological development of the vertebrae, individual vertebral cartilages develop around and constrict the **notochord**, resulting in a thin central strand in the centrum (body) and gelatinous masses between adjacent vertebrae. These gelatinous notochordal remnants between the vertebrae act in a way similar to ball joints between the concave ends, allowing for great flexibility of the vertebral column.

In caudal vertebrae, a **vertebral** or **neural arch**, forming the **vertebral canal** that cradles the spinal cord, extends dorsally from the centrum and terminates in a **neural spine** (Figure 16.13 and Figure 16.14). Triangular **dorsal intercalary plates** (Figure 16.14) complete a continuous roof over the spinal cord. Note the **foramen** in each intercalary plate for passage of the **dorsal root of a spinal nerve** and the **foramen** in each neural arch

for passage of the **ventral root of a spinal nerve** (Figure 16.15). Extending ventrally from the centrum is the **hemal arch**, terminating in a **hemal spine**. Between adjacent hemal arches, small foramina for the passage of blood vessels may be identified. The **hemal canal** is subdivided by a thin cartilaginous shelf into a dorsal region cradling the caudal

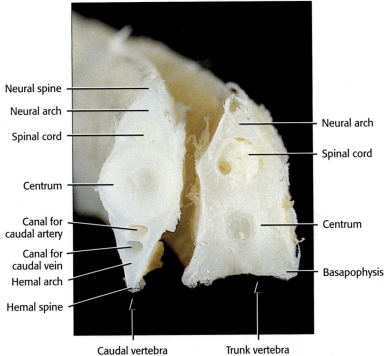

Neural spine
Neural arch
Spinal cord
Centrum
Canal for caudal artery
Canal for caudal vein
Hemal arch
Hemal spine

Neural arch
Spinal cord
Centrum
Basapophysis

Caudal vertebra   Trunk vertebra

FIGURE 16.13 Trunk and caudal vertebrae: Anterior view.

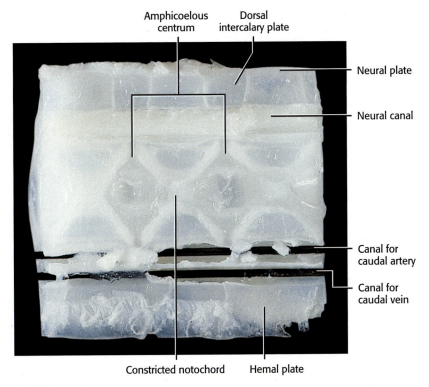

Amphicoelous centrum   Dorsal intercalary plate

Neural plate

Neural canal

Canal for caudal artery

Canal for caudal vein

Constricted notochord   Hemal plate

FIGURE 16.14 Caudal vertebrae: Lateral view.

artery and a ventral region housing the caudal vein (Figure 16.13 and Figure 16.14).

The basic morphology of trunk vertebrae is similar to that of the caudal vertebrae (Figure 16.13). An exception is the absence of the hemal arches. In their anatomical position, short lateral extensions, the **basapophyses**, project from the ventrolateral surface of the centra. Some comparative anatomists consider the basapophysis to be homologous with the basal portions of the hemal arches. Short cartilaginous **ribs** articulate with the basapophyses (Figure 16.2 and Figure 16.18). In addition, very small triangular **ventral intercalary plates** occur between the basapophyses (Figure 16.16). Again, notice the respective foramina in the **dorsal intercalary plates** and neural arches (Figure 16.15).

Median Fins The **caudal fin** is of the **heterocercal** type. This type of tail has two lobes, a larger dorsal lobe and a smaller ventral lobe. Notice that the vertebral column tilts upward into the upper lobe. Skeletal elements in the form of elongated hemal and neural arches support the tail, with fibrous **ceratotrichia** or **fin rays** extending into and stiffening the fin web (Figure 16.1, Figure 16.17, and Figure 16.18).

In the shark, there are two **dorsal fins**, an **anterior** and a **posterior**, each preceded by a protective **spine**. A large proximal cartilaginous **basal pterygiophore** and a series of distal **radial pterygiophores** support each of these fins. Fin webs are stiffened by ceratotrichia (Figure 16.19).

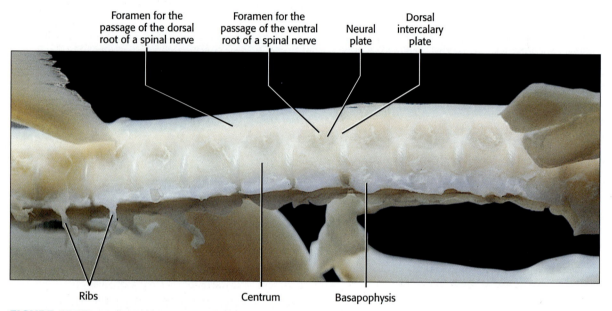

Foramen for the passage of the dorsal root of a spinal nerve   Foramen for the passage of the ventral root of a spinal nerve   Neural plate   Dorsal intercalary plate

Ribs   Centrum   Basapophysis

FIGURE 16.15 Trunk vertebrae: Lateral view.

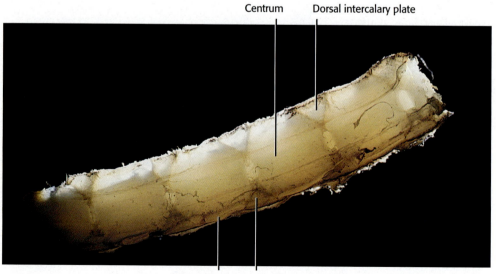

Centrum   Dorsal intercalary plate

Basapophysis   Ventral intercalary plate

FIGURE 16.16 Trunk Vertebrae: Lateral view.

FIGURE 16.17
Caudal fin and relationship of vertebral column.

Vertebral column   Caudal fin showing ceratotrichia

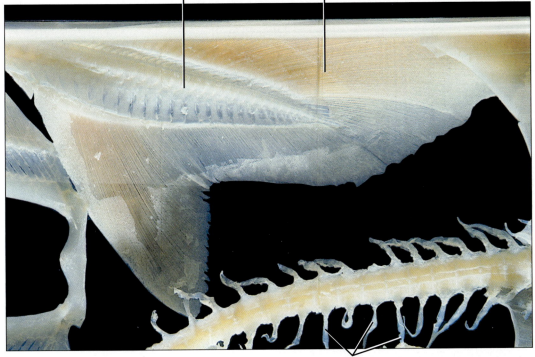

Ribs

FIGURE 16.18
Caudal fin illustrating ceratotrichia.

Basal pterygiophore   Spine

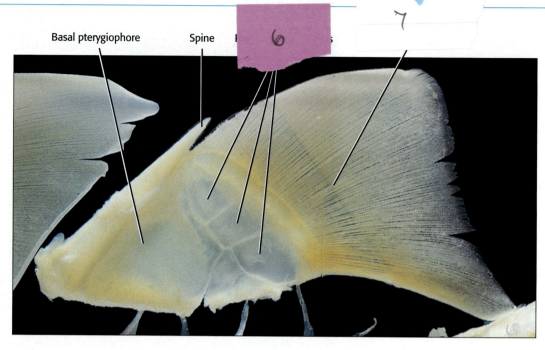

FIGURE 16.19
First dorsal fin:
Anterior end
to the left.

APPENDICULAR DIVISION

Pectoral Girdle and Fins

The **pectoral girdle**, consisting of a U-shaped complex often called a **scapulocoracoid bar,** lies just posterior to the last gill slit (Figure 16.20A). The ventral portion of the U is the **coracoid bar,** and the dorsal, distal part is the **scapular cartilage** (Figure 16.20B). A dorsolateral smooth surface, the **glenoid surface,** is present on the scapular cartilage to accommodate articulation of the pectoral fin (Figure 16.21). A small, attenuated, medially curved structure, the **supra-scapular cartilage** (Figure 16.20B), articulates dorsally with the scapular cartilage, completing the U.

The paired pectoral fins articulate with the scapular cartilages (Figure 16.21). The proximal end is supported by a series of three **basal pterygiophores**—an anterior **pro-pterygium,** a middle **mesopterygium,** and the posterior **metapterygium.** Articulating with and radiating from the basal cartilages are rows of **radial pterygiophores,** forming the major support of the fin. In the fin web, notice the numerous **fin rays,** or **ceratotrichia.**

Pelvic Girdle and Fins

Just anterior to the cloaca lies the **pelvic girdle** (Figure 16.22). Not well developed in fish, it consists of a short, horizontal **puboischiadic bar** with small **iliac processes** projecting from its anterior ends. A smooth **acetabular surface** occurs on the posterolateral surface of the bar.

The paired **pelvic fins** articulate with the pelvic girdle at the acetabular surface (Figure 16.22). Only two **basal pterygiophores**—the anterior **propterygium** and an elongated **metapterygium**—support the fins proximally. **Radial pterygiophores** extend from the basal cartilages. The fin web is supported by **ceratotrichia.**

In contrast to the pectoral fins, pelvic fins in sharks are sexually dimorphic (different in males and females) (Figure 16.23). In males, a sperm transfer organ, the **clasper,** supported by cartilages derived from radial pterygiophores, is obvious and is a handy way to sex sharks externally. Identify the lateral **spine** (Figure 16.22).

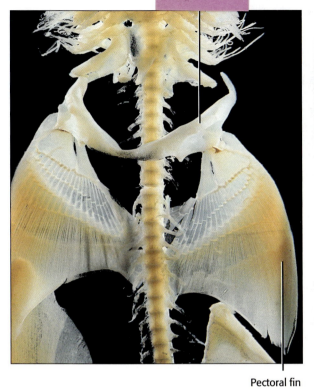

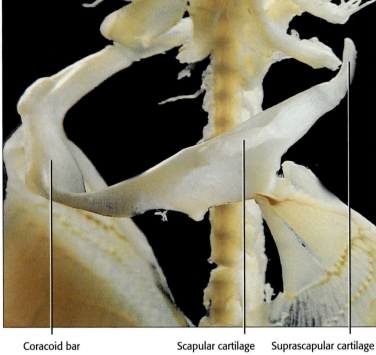

Pectoral fin

Coracoid bar   Scapular cartilage   Suprascapular cartilage

FIGURE 16.20A Pectoral girdle and fin.

FIGURE 16.20B Scapular cartilage and coracoid bar of pectoral girdle.

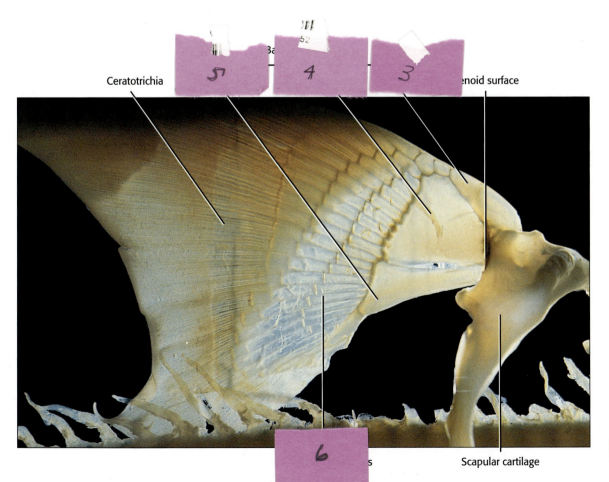

Ceratotrichia   3a   52   enoid surface

Scapular cartilage

FIGURE 16.21
Left pectoral fin.

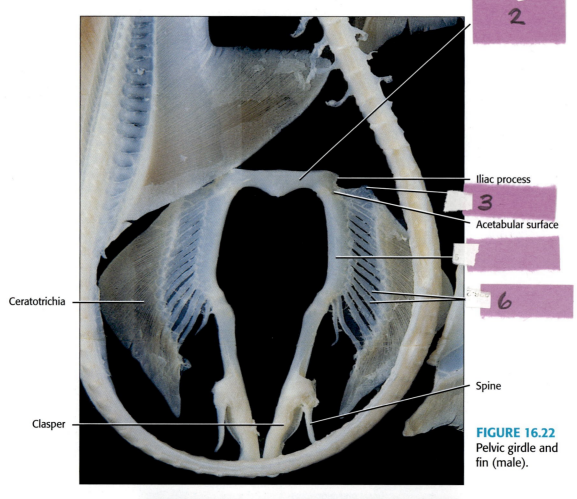

Iliac process

Acetabular surface

Ceratotrichia

Spine

Clasper

FIGURE 16.22
Pelvic girdle and
fin (male).

FIGURE 16.23
Pelvic girdle
and fins (male: top;
female, bottom).

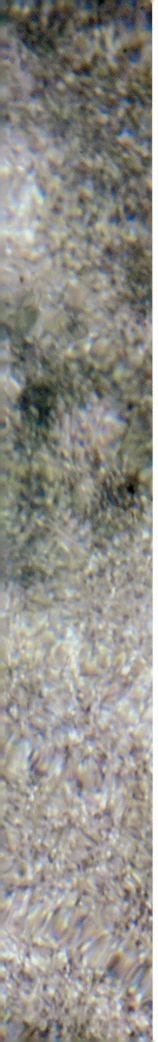

CHAPTER SEVENTEEN

Dogfish Shark — Muscular System

17

SKINNING

To study the muscles, the skin must be removed, which is easier said than done. Shark skin adheres tightly to underlying muscle and connective tissue.

> You must exercise great care during the skinning process, as many muscles can be destroyed or pulled out of place. The skinning process will take several hours to produce a superior specimen for muscle study. Don't hurry!

Place the shark on its ventral surface. With an injected shark, it probably will be best to begin skinning from the cut surface made by the supply house through the posterior portion of the trunk to inject latex into the circulatory system. Skin from ventral to dorsal—that is, skin the belly first and work toward the dorsal surface. Skin to the tip of the snout. Try to conserve the ampullae of Lorenzini that adhere to the undersurface of the skin. Take care to conserve the ducts leading from the pores. Carefully skin the entire head.

> A good dose of common sense and keen observation during the skinning process will greatly improve your chances of producing a usable specimen.

Use the tip of a steel probe or a flattened instrument to separate the skin from the muscle. A small spatula with rounded edges, found in some chemistry laboratories, is ideal. Work slowly, using a sharp scalpel to separate the tissue from the skin *only* when using the probe becomes impossible. Areas that will be particularly difficult to separate are the jaws, gill slits, and spiracle.

When you have finished skinning the shark, you will have removed the skin from the dorsal and ventral surfaces of the body from the tip of the snout to the posterior cut through the trunk. Skin the dorsal and ventral surfaces of the pectoral fins, beginning at the distal end of the muscle mass and proceeding toward the body to avoid detaching the muscles from the body. Follow this procedure for all fins. Skin the lateral surface of one of the dorsal fins.

Skin the dorsal and ventral surfaces of the pelvic fins, extending slightly dorsal to the fins to expose the origin of fin musculature associated with the trunk muscles in that area (Figure 17.1). In the male the siphon lies just below the ventral skin, and it should be left on the surface of the ventral muscle (Figure 17.2).

To see the association of trunk musculature with the heterocercal tail (Figure 17.3), carefully separate the skin overlying that area, using a spatula. Skin approximately 7.5 to 10 cm anterior to the tail, ending where the vertebral column terminates. It is not necessary to skin the tail lobes.

> Once the skin has been removed, the shark is susceptible to dehydration and rotting. Therefore, we suggest that you keep the specimen moist with a fluid preservative, and also covered when you are not actively working with it.

"Picturebook dissection" without reading the text of this manual leads to only partial understanding of comparative vertebrate anatomy!!!

Epaxial
muscle

Horizontal
septum

Hypaxial
muscle

FIGURE 17.1 Skinned shark.

Siphons

FIGURE 17.2 Male pelvic fins with siphons skinned.

Epaxial muscle   Muscle insertion tendons

Hypaxial muscle   Horizontal septum

FIGURE 17.3 Tail with associated muscles.

TRUNK OR AXIAL MUSCLES

The basic design of the musculature of sharks, as illustrated by the muscles of the trunk, consists of serially arranged **myomeres** (muscle blocks) separated by **myosepta** (sheets of connective tissue). The trunk muscles are separated into left and right halves by connective tissue, the **middorsal vertical** and a **midventral vertical septum.** Ventrally, a thin longitudinal line of connective tissue, the **linea alba,** meets the mid ventral vertical septum. The lateral musculature is separated by a sheet of connective tissue, the **horizontal septum,** dividing it into dorsal **epaxial** and ventral **hypaxial** muscles. In the vertebrate embryo, trunk myomeres often can be allied with specific vertebrae, but often, in the adult, the relationship of individual myomeres becomes blurred and complicated as a result of elaborate folding. Therefore, individual trunk myomeres may become associated with several vertebrae, making smoother locomotory movements possible (Figure 17.3, Figure 17.4, Figure 17.5, and Figure 17.6).

Muscle fibers extend longitudinally from one myoseptum to another within the myomeres. Folding of the myomeres results in a zig-zag appearance, with some muscle fibers maintaining their longitudinal orientation and others exhibiting an oblique orientation. This description is applicable to the trunk muscles. A great deal of specialization is apparent—anteriorly, particularly in association with the jaws, branchial areas of the head region, and pectoral fins, and posteriorly in association with the pelvic fins.

PECTORAL FIN MUSCLES

Extensor or Abductor

This fan-shaped muscle appears on the dorsal surface of the fin (Figure 17.5).

Origin: Scapular process and adjacent fascia

Insertion: Basal and radial pterygiophores

Action: Extends or abducts (elevates) the pectoral fin

Flexor or Adductor

This fan-shaped muscle appears on the ventral surface of the fin (Figure 17.6).

Origin: Coracoid portion of the scapulocoracoid bar

Insertion: Basal and radial pterygiophores

Action: Flexes or adducts (depresses) the pectoral fin

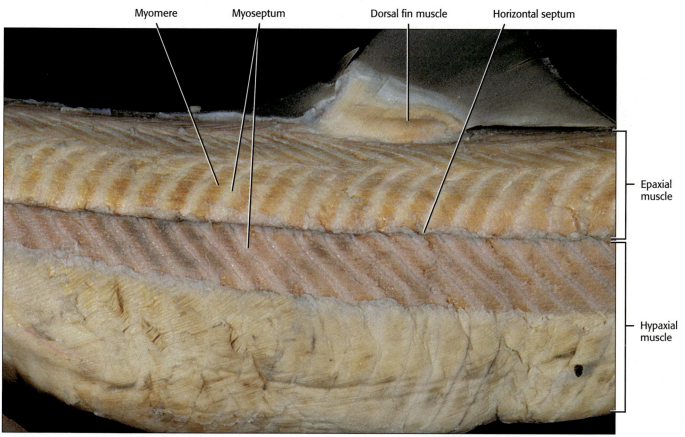

Myomere   Myoseptum   Dorsal fin muscle   Horizontal septum

Epaxial muscle

Hypaxial muscle

FIGURE 17.4 Trunk muscles: Basic design.

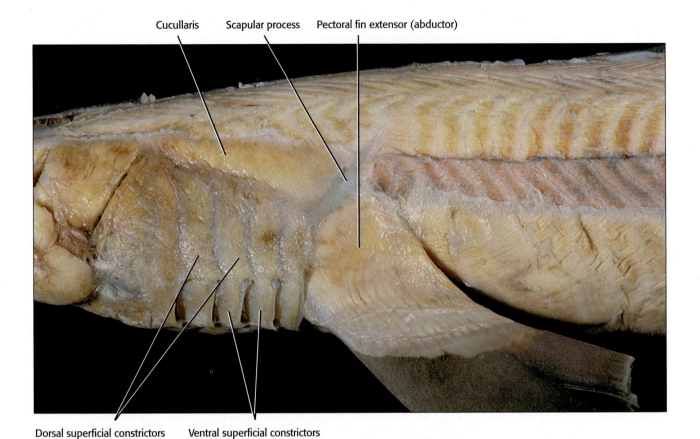

FIGURE 17.5 Branchial, trunk, and pectoral fin muscles.

FIGURE 17.6 Trunk and pectoral fin muscles: Ventral view.

PELVIC FIN MUSCLES

Extensor or Abductor

This muscle appears on the dorsal surface of the fin (Figure 17.7 and Figure 17.8).

Origin: From the trunk myomeres, iliac process of puboischiadic bar and metapterygium

Insertion: Pterygiophores; in males, slips of the muscle extend onto the surface of the clasper and insert there

Action: Extends or abducts (elevates) the pelvic fin; in males, the slips inserted on the clasper move that organ during copulation

Flexor or Adductor

This is a ventral muscle consisting of a proximal and distal portion (Figure 17.9 and Figure 17.10).

Origin: Proximal portion: linea alba, and puboischiadic bar. Distal portion: metapterygium

Insertion: Proximal portion: metapterygium. Distal portion: pterygiophores; in males, slips of the muscle extend onto the clasper and insert there

Action: Flexes or adducts (depresses) the pelvic fin; in males, the slips inserted on the clasper move that organ during copulation

FIGURE 17.7 Pelvic fin muscles: Dorsal view (male).

Pelvic fin flexor (adductor)

FIGURE 17.9 Pelvic fin muscles: Ventral view (male).

Pelvic fin extensor (abductor)

FIGURE 17.8 Pelvic fin muscles: Dorsal view (female).

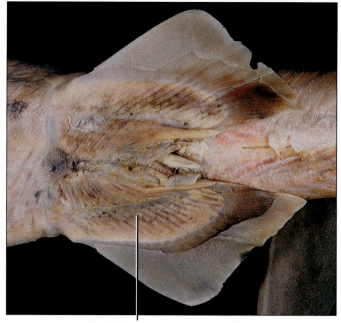

Pelvic fin flexor (adductor)

FIGURE 17.10 Pelvic fin muscles: Ventral view (female).

DORSAL FIN MUSCLE

This is a small muscle at the base of the dorsal fins (Figure 17.4).

Origin: Dorsal trunk myomeres

Insertion: Pterygiophores

Action: Stabilization of the fin

MUSCLES OF THE GILL ARCHES OR THEIR DERIVATIVES

Among vertebrate gnathostomes, typically during embryonic development, a series of seven pairs of gill arches develop. The first pair has been modified into the jaw skeleton, the upper palatoquadrate, and the lower Meckel's cartilage (mandible). The second pair is recognized as the hyoid arch and is separated from the first by the spiracle, a modified gill slit. The next five gill arches constitute the branchial skeleton supporting the gills. Notice that a gill slit precedes each of the branchial gill arches, resulting in five gill slits in dogfish sharks.

In the following discussion we will separate muscles into branchiomeric and hypobranchial muscles. These muscles are considered individually because they derived from two different sources of mesoderm in the embryo. A clue to their different embryonic origin is illustrated by the fact that branchiomeric muscles are innervated by cranial nerves and hypobranchial muscles are innervated by spinal nerves. In our treatment, the muscles of each gill arch will be discussed relative to that gill arch.

BRANCHIOMERIC MUSCLES

Visceral Arch I (Mandibular Arch)

Innervated by Cranial Nerve V (Trigeminal)

Levators

Levator palatoquadrati This is a narrow band-like muscle that lies just anterior to the spiracularis muscle and the spiracle (Figure 17.11).

Origin: Otic capsule

Insertion: Palatoquadrate

Action: Elevates the palatoquadrate (upper jaw)

Dorsal Constrictors

Spiracularis The spiracularis is a narrow, band-like muscle that lies just anterior to the spiracle (Figure 17.11). It adheres closely to the levator palatoquadrati muscle and wall of the spiracle.

Origin: Otic capsule

Insertion: Palatoquadrate cartilage

Action: Elevates the palatoquadrate (upper jaw)

Adductor mandibulae This is a large globular muscle at the angle of the jaw (Figure 17.11 and Figure 17.13).

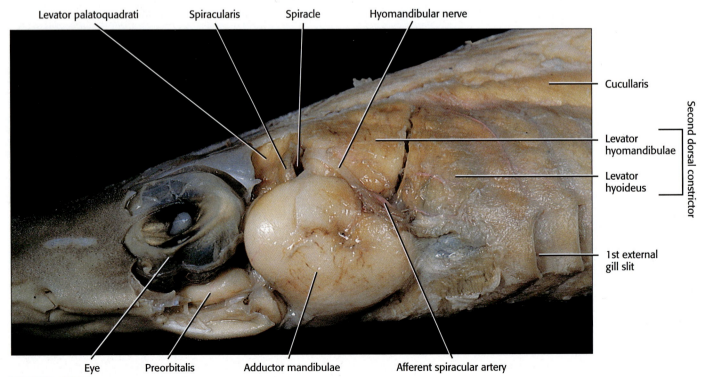

FIGURE 17.11 Muscles of visceral arch I (mandibular).

Origin: Posterior portion of palatoquadrate

Insertion: Meckel's cartilage (mandible)

Action: Elevates Meckel's cartilage (closes jaw)

Preorbitalis This somewhat conical muscle lies deep between the lower portion of the eye orbit and palatoquadrate and close to a portion of the labial cartilage (Figure 17.11). To expose this muscle, the connective tissue and skin have to be removed carefully just ventral to the eye and dorsal to the palatoquadrate. The muscle lies medially and adjacent to the dorsal portion of the labial cartilage.

Origin: Ventral surface of the chondrocranium

Insertion: Joins the adductor mandibulae muscle and with it inserts on Meckel's cartilage

Action: Elevates Meckel's cartilage (closes jaw)

Ventral Constrictors

Intermandibularis This is a broad, thin muscle that covers the ventral surface of the shark posterior to Meckel's cartilage (Figure 17.6).

Origin: Meckel's cartilage

Insertion: Midventral connective tissue raphe

Action: Elevates the floor of the mouth, thereby reducing the volume of the oral cavity

Visceral Arch II (Hyoid Arch)

Innervated by Cranial Nerve VII (Facial)

Dorsal Constrictors

Second Dorsal Constrictor This broad muscle extends from the spiracle to the raphe of the first gill slit. Some anatomists recognize the anterior portion as the levator hyomandibulae muscle, whereas the posterior portion as the levator hyoideus (Figure 17.11).

Origin: Otic capsule

Insertion: Hyomandibular cartilage

Action: Compresses gill pouches

Ventral Constrictors

Interhyoideus This is a thin sheet of muscle dorsal to the intermandibularis muscle (Figure 17.12).

To study this muscle, with a new blade in your scalpel, carefully make a shallow incision (approximately 1/4 to 1/2 mm deep) through the midventral raphe of the intermandibularis muscle and underlying interhyoideus muscle. Loosen the posterior margin from the coracoarcual muscle on one side, and carefully reflect that half laterally. The underlying sheet of muscle is the interhyoideus. Note the difference in the direction of muscle fibers.

Origin: Ceratohyal

Insertion: Midventral raphe

Action: Compresses gill pouches

Visceral Arches III–VII

Muscles of gill arch III are innervated by Cranial Nerve IX (Glossopharyngeal), and muscles of gill arches IV–VII are innervated by Cranial Nerve X (Vagus)

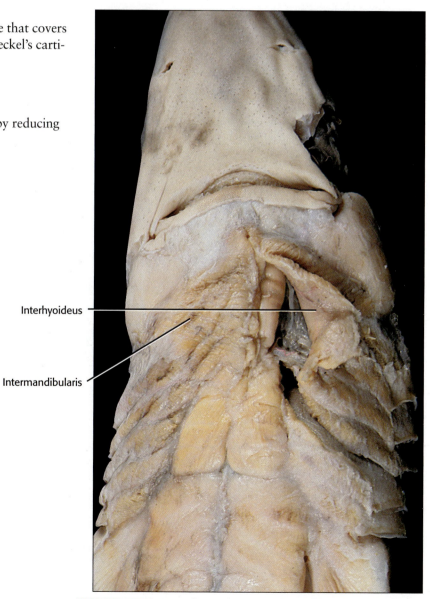

Interhyoideus

Intermandibularis

FIGURE 17.12 Ventral constrictor muscles of visceral arches I and II.

Levators

Cucullaris This is an elongated triangular muscle resulting from the fusion of the levators of the branchial arches (III–VII). It extends from approximately the posterior border of the levator hyomandibulae to the **scapular process** (Figure 17.5, Figure 17.12, and Figure 17.13).

Origin: Fascia of the epaxial muscles

Insertion: Scapular process of the scapulocoracoid bar

Action: Elevates the scapular process

Interarcual Muscles

To gain access to these muscles, separate the ventral border of the cucullaris from the dorsal edge of the underlying gill pouches on one side of your specimen (Figure 17.14 and Figure 17.5). This exposes the anterior cardinal sinus, part of the venous portion of the circulatory system.

You may encounter coagulated blood or sometimes blue latex injected into the venous system. Clean out the sinus and expose the pharyngobranchial cartilages by stripping off the whitish membrane of the sinus. Any nerves that appear as whitish or cream-colored bands should not be removed or cut (Figure 17.15).

With a pair of scissors, make a vertical cut dorsally and ventrally to expose the gill chamber. Push the anterior half forward and carefully peel the gill lamellae and gill membrane from the posterior half of the chamber to produce a specimen resembling Figure 17.15. You now should be able to observe the pharygobranchial, epibranchial, ceratobranchial, and possibly the hypobranchial cartilages, as well as gill rays and associated muscles (Figure 16.11 and Figure 16.12).

Observe the following muscles:

Dorsal Interarcual These are slender band-like muscles extending between adjacent pharyngobranchial cartilages (Figure 17.15).

Origin: Pharyngobranchial cartilage

Insertion: Adjacent posterior pharyngobranchial cartilage

Action: Pulls gill arches anteriorly

Lateral Interarcual These are somewhat broader band-like muscles extending between pharyngobranchial and epibranchial cartilages of the same gill arch (Figure 17.15).

Origin: Pharygobranchial cartilage

Insertion: Epibranchial cartilage of the same arch

Action: Adducts the cartilages, reducing the angle between them

Branchial Adductor These muscles appear at the angle between the epibranchial and ceratobranchial cartilages of the same gill arch (Figure 17.14).

Origin: Epibranchial cartilage

Insertion: Certatobranchial cartilage of the same arch

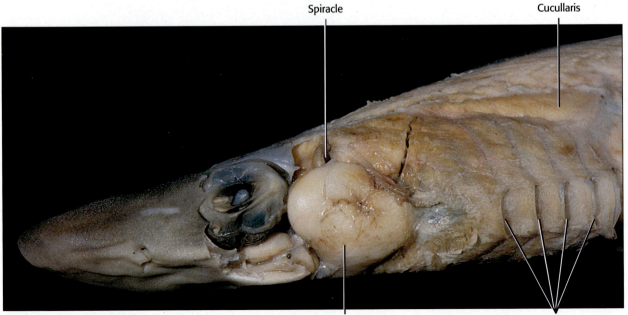

Spiracle   Cucullaris

Adductor mandibulae   External gill slits (1–4)

FIGURE 17.13 Muscles of visceral arch II (hyoid) and visceral arches III–VII (branchial arches).

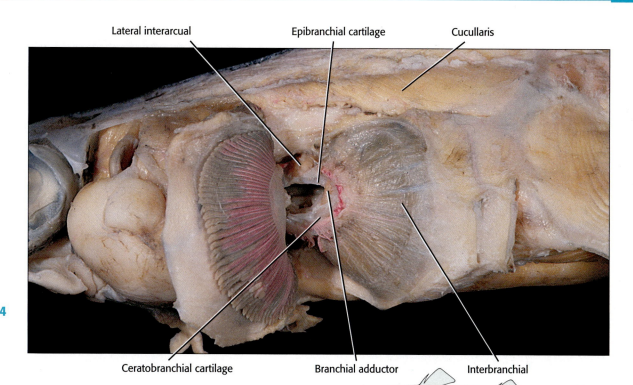

FIGURE 17.14
Interarcual
muscles I.

Lateral interarcual   Epibranchial cartilage   Cucullaris

Ceratobranchial cartilage   Branchial adductor   Interbranchial

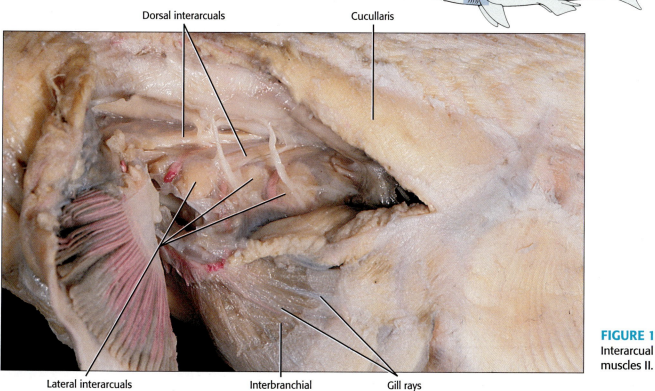

Dorsal interarcuals   Cucullaris

Lateral interarcuals   Interbranchial   Gill rays

FIGURE 17.15
Interarcual
muscles II.

Action: Adducts the cartilages reducing the angle between them

Interbranchial These circularly arranged muscles, actually derivatives of the constrictors, occur between gill rays (Figure 17.14 and Figure 17.15). They are mentioned here

because they occur in conjunction with the levators just discussed.

Origin: Gill ray

Insertion: Gill ray

Action: Adducts the cartilages of the gill arch

The collective effect of the interarcual muscles just described is to compress the gill pouches, causing water to be expelled from them.

Superficial Constrictors

Dorsal and Ventral Constrictors These muscles extend obliquely between adjacent connective tissue raphes associated with each of the gill slits (Figure 17.5 and Figure 17.6).

Origin: Raphe

Insertion: Raphe

Action: Compression of the gill pouches

HYPOBRANCHIAL MUSCLES

Opening the mouth in a dense medium, such as water, is difficult and requires considerable force for an aquatic vertebrate such as a fish. Consequently, it is not too surprising that sharks have several muscles, called hypobranchial muscles, occurring below the respiratory system, which assist in opening the mouth and processing food.

Recall that hypobranchial muscles are innervated by spinal nerves. In your study of comparative anatomy, you will observe that tetrapods, conversely (because they live in air that is considerably less dense than water), exhibit much reduced jaw opening musculature. Derivatives of the hypobranchial muscles of fish ancestors are employed for a variety of different purposes in tetrapods.

The following musculature can be observed in the shark:

Coracomandibularis This is a single strap-like muscle extending from the area of the pectoral girdle to the lower jaw (Figure 17.16). Because this muscle is paired in the embryo but fuses in the adult, sometimes incompletely, it may separate into two parts.

Origin: Fascia of coracoarcual muscle

Insertion: Meckel's cartilage

Action: Depresses lower jaw (opens mouth)

Coracoarcual These paired, somewhat triangular muscles occur posterior to the intermandibularis muscle (Figure 17.16).

Origin: Coracoid portion of the scapulocoracoid bar

Insertion: Coracohyoid muscle

Action: Assists in depressing lower jaw (opens the mouth)

Coracohyoid This is a paired, strap-like muscle lying anterior to the coracoarcual muscle and dorsolateral to the coracomandibularis muscle. To observe the coracohyoid, isolate the coracomandibularis muscle, slip a steel probe underneath the middle of the coracomandibularis muscle,

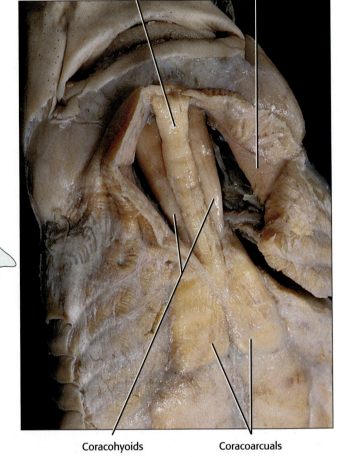

Coracomandibularis   Interhyoideus

Coracohyoids   Coracoarcuals

FIGURE 17.16 Hypobranchial muscles I.

and with a sharp scalpel, cut perpendicularly through the muscle fibers using the probe as a guide. Reflect the cut ends of the coracomandibularis muscle. The grayish pink mass lying anteriorly on the surface of the coracohyoid muscle is the **thyroid gland** (Figure 17.17).

Origin: Coracoarcual muscle

Insertion: Basihyal cartilage

Action: Assists in depressing the lower jaw (opens the mouth)

Coracobranchial This is a deep group of muscles located dorsolaterally to the coracoarcual and coracohyoid muscles. To see these muscles, isolate the coracohyoid, insert a steel probe under this muscle, and with a sharp scalpel, cut across the fibers. Reflect the cut ends of the coracohyoid muscle. Some of these muscles extend dorsal to the coracoarcual muscle (Figure 17.18).

Origin: Coracoid portion of the scapulocoracoid bar and fascia of the dorsal surface of the coracoarcual muscle

Insertion: Hypobranchial cartilages

Action: Depresses the lower jaw (opens the mouth) and stabilizes the pericardial cavity

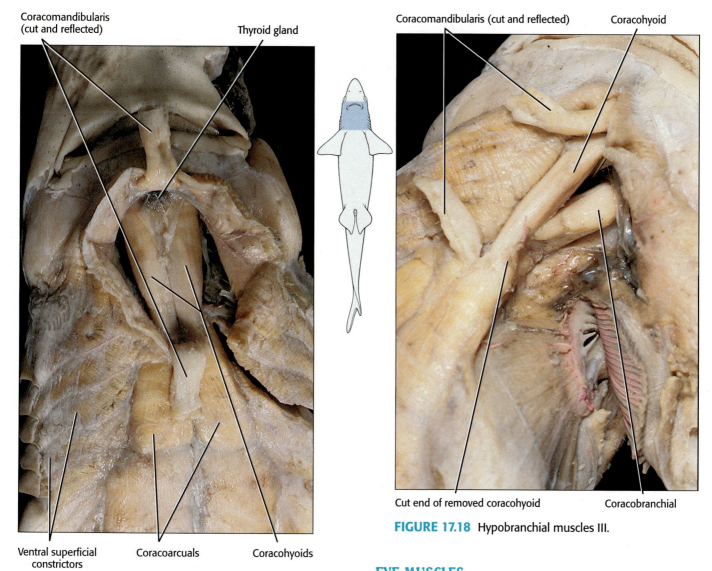

Coracomandibularis (cut and reflected)

Thyroid gland

Ventral superficial constrictors

Coracoarcuals

Coracohyoids

FIGURE 17.17 Hypobranchial muscles II.

Coracomandibularis (cut and reflected)

Coracohyoid

Cut end of removed coracohyoid

Coracobranchial

FIGURE 17.18 Hypobranchial muscles III.

EYE MUSCLES

Although the extrinsic eye muscles, responsible for positioning the eyeball, could be discussed here, they will be dealt with in association with the eye and its anatomy in Chapter 22.

CHAPTER EIGHTEEN

Dogfish Shark — Body Cavities and Mesenteries

An extensive body cavity, the coelom, typical of craniates, is present in the shark. Also typically, the coelom became partitioned by the transverse septum into a pericardial cavity and a pleuroperitoneal cavity. The coelom is lined with mesoderm and the organs projecting into it are also covered with an extension of that mesoderm (see Chapter 7).

OPENING THE SHARK

Injected specimens usually have a slit in the ventral body wall. If the incision is anterior, with the tips of scissors held parallel to the body wall, carefully extend this incision anteriorly to the scapulocoracoid bar (base of the pectoral fin) along a longitudinal line about 1–2 cm from the midline (cut #1). Carefully lift the anterior cut edge and observe the extent of the falciform ligament.

To avoid destruction of the falciform ligament, make a flap as follows. With scissors, make a transverse cut approximately 1.3 cm posterior to the attachment of the ligament to the ventral body wall (cut #2). Cut anteriorly from the edge of the transverse cut to the scapulocoracoid bar (cut #3). Lift the flap and observe the falciform ligament. Make a lateral cut on either side from the anterior end of the flap, following the edge of the scapulocoracoid bar anterior to the pectoral fins (cut #4) (Figure 18.1 and Figure 18.2).

Extend the original longitudinal incision posteriorly to approximately

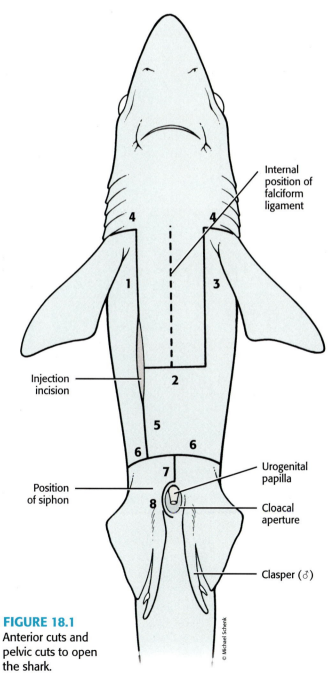

FIGURE 18.1
Anterior cuts and pelvic cuts to open the shark.

© Michael Schenk

153

FIGURE 18.2 Completed anterior cuts.

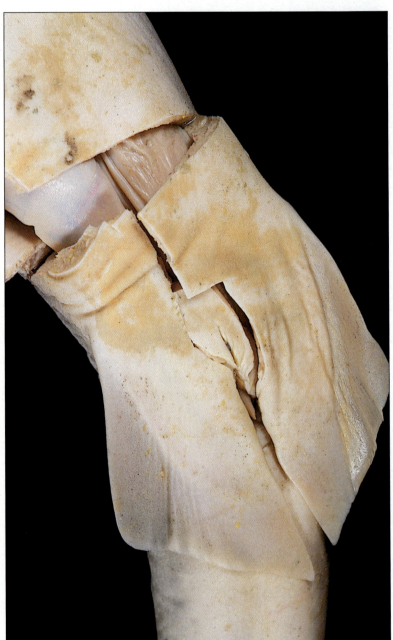

FIGURE 18.3 Completed pelvic cuts.

3.3 cm from the anterior edge of the pelvic fin (cut #5). Make lateral cuts at this point anterior to the pelvic fins (cut #6). Cut posteriorly along the midline from the cut edge toward the cloacal aperture (cut #7) and then cut around the aperture, avoiding the rectum (cut #8) (Figure 18.1 and Figure 18.3). [**Note:** if your specimen is a male, reflect the siphons before making the rectum avoidance cut.] If the injection incision is posterior, complete similar incisions. When you reflect the lateral body wall flaps, the cavity now revealed is the pleuroperitoneal cavity (Figure 18.4).

 While exposing the pericardial cavity, take great care not to damage the transverse septum associated with the dorsal aspect of the scapulo-coracoid bar and the heart with its associated vessels.

With sharp scissors, make a 7.5 cm midventral cut anterior to the scapulocoracoid bar. Carefully remove the

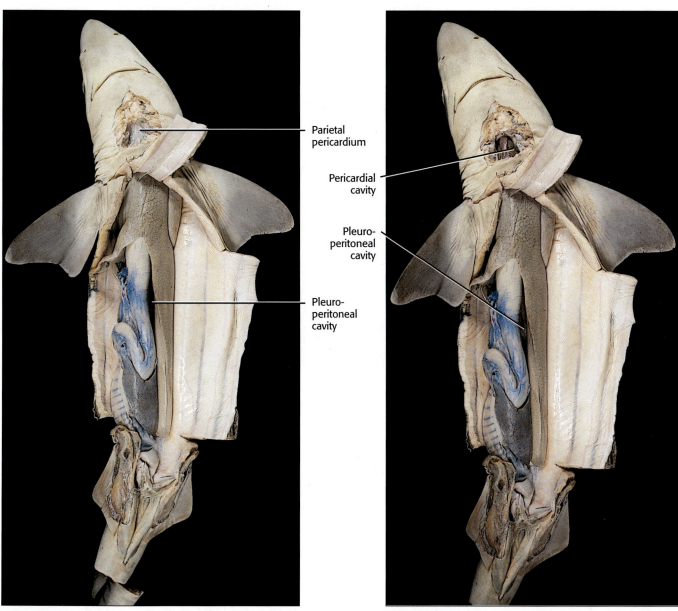

Parietal
pericardium

Pericardial
cavity

Pleuro-
peritoneal
cavity

Pleuro-
peritoneal
cavity

FIGURE 18.4 Pleuroperitoneal cavity with intact parietal pericardium.

FIGURE 18.5 Pleuroperitoneal and pericardial cavities.

hypobranchial musculature overlying the heart by snipping small pieces at a time. You probably will have to remove approximately 1 to 1.5 cm of muscle until you reach a white membrane, the parietal pericardium (Figure 18.4). Lift the pericardium with forceps, and carefully cut through the pericardium, exposing the heart. Now you can observe the extent and subdivision of the body cavity of the shark (Figure 18.5).

MESENTERIES

 Because mesenteries are fragile organs, great care must be exerted in moving the viscera around during the search for them.

Derivatives of the Dorsal Mesentery

To observe the **greater omentum** (mesogaster), extending from the dorsal body wall to the dorsal surface of the esophagus and stomach, carefully pull the stomach away from the body wall. Note the **gastrosplenic ligament** extending between the stomach and the spleen by gently separating the spleen from the posterior surface of the stomach. The **mesentery proper**, supporting almost the entire length of the valvular intestine, can be observed by gently pulling the valvular intestine laterally. Supporting the rectal gland and rectum is the small **mesocolon** (Figure 18.6 and Figure 18.7).

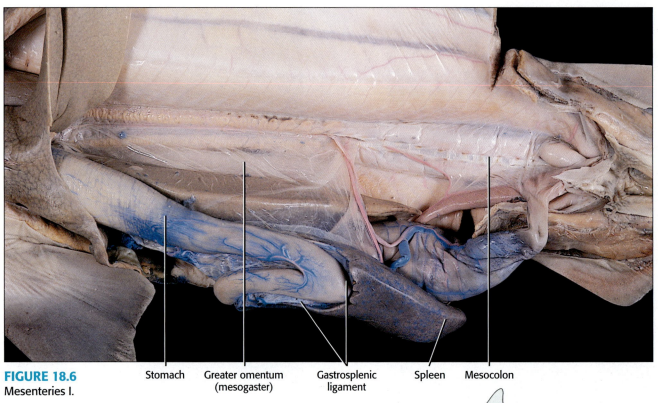

FIGURE 18.6
Mesenteries I.

Stomach　　Greater omentum (mesogaster)　　Gastrosplenic ligament　　Spleen　　Mesocolon

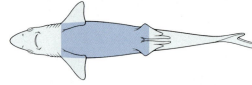

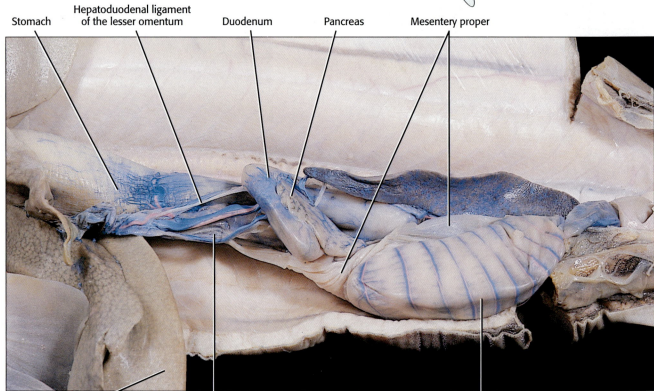

Stomach　　Hepatoduodenal ligament of the lesser omentum　　Duodenum　　Pancreas　　Mesentery proper

Liver　　Gastrohepatic ligament of the lesser omentum　　Valvular intestine

FIGURE 18.7　Mesenteries II.

Derivatives of the Ventral Mesentery

The **falciform ligament** extends from the anterior end of the liver to the midventral body wall. In the female, find the opening of the oviduct, the **ostium tubae,** into the peritoneal cavity (Figure 18.8).

The **lesser omentum** (gastrohepatoduodenal ligament) extends from the liver to the stomach and duodenum of the small intestine. Two regions are recognized, the **hepatoduodenal ligament,** extending from the liver to the duodenum and enclosing the bile duct and blood vessels associated with the digestive organs, and the **gastrohepatic ligament,** extending from the liver to the lesser curvature of the stomach (Figure 18.7).

The following mesenteries are not derivatives of either the dorsal or ventral mesentery but are formed separately during development of the reproductive systems. The **mesorchium,** in the male, suspends the testes from the dorsal body wall (Figure 18.9). In the female, the **mesovarium** suspends the ovaries and the **mesotubarium** suspends the oviduct from the dorsal body wall (Figure 18.10).

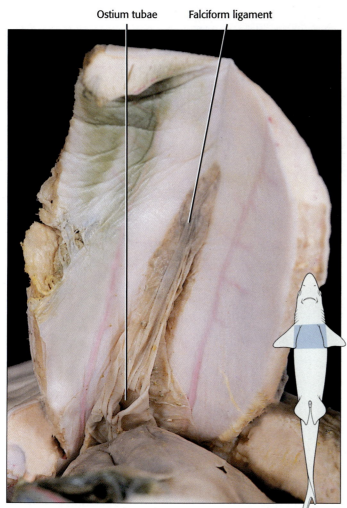

Ostium tubae   Falciform ligament

FIGURE 18.8 Falciform ligament with ostium tubae.

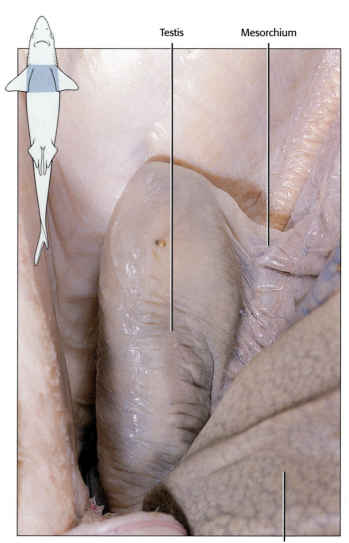

Testis   Mesorchium

Liver

FIGURE 18.9 Reproductive mesenteries: Male.

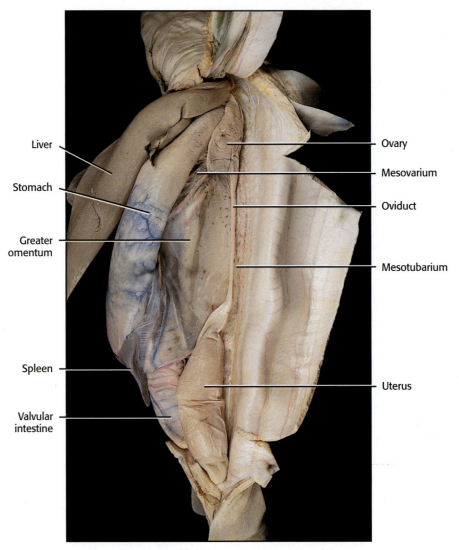

FIGURE 18.10 Reproductive mesenteries: Female.

CHAPTER NINETEEN

Dogfish Shark — Digestive and Respiratory Systems

The digestive and respiratory anatomy of the shark probably illustrates an early evolutionary development in craniates. To obtain an understanding of the evolutionary history of the digestive and respiratory systems, review Chapter 8.

To facilitate the study of the digestive and respiratory systems in the shark, use a sharp and sturdy pair of scissors, and cut through the angle of the mouth on the same side used for dissecting the branchial musculature. Extend the cut through the middle of the gills and gill slits, then through the scapulocoracoid bar, continuing the cut 2 cm farther through the body wall (Figure 19.1). Now make a transverse cut with scissors until reaching the esophagus, then cut horizontally through the esophagus, and continue until reaching the gill slits on the opposite side (Figure 19.2).

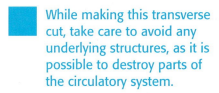 While making this transverse cut, take care to avoid any underlying structures, as it is possible to destroy parts of the circulatory system.

DIGESTIVE SYSTEM

Along the margin of the jaws are one or two rows of functional serrated, triangular **teeth** (Figure 19.2, Figure 19.3, and Figure 19.4). Notice the many rows of replacement teeth along the inner surface of the jaw, which move into place when a functional tooth is lost. Teeth that are replaced continuously are known as polyphyodont. Shark teeth generally are similar in structure and function, so they are known as homodont teeth. In contrast to those of the lamprey, shark teeth are true vertebrate teeth. True teeth are derived from mesodermal, ectodermal, and neural crest tissues. Because the

FIGURE 19.1 Cut to reveal orobranchial region.

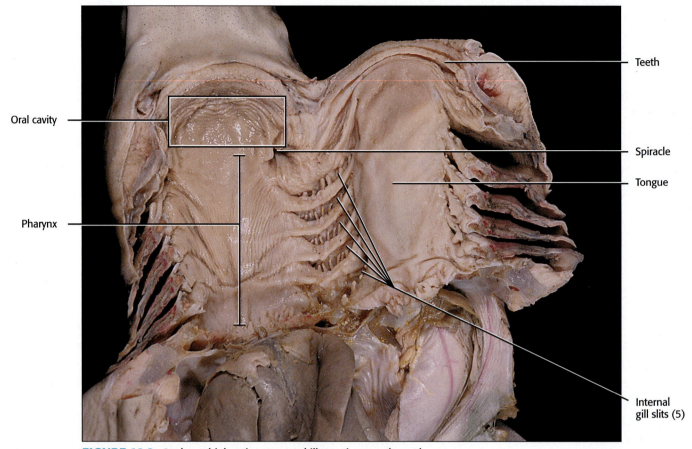

Teeth

Oral cavity

Spiracle

Tongue

Pharynx

Internal
gill slits (5)

FIGURE 19.2   Orobranchial region exposed illustrating esophageal cut.

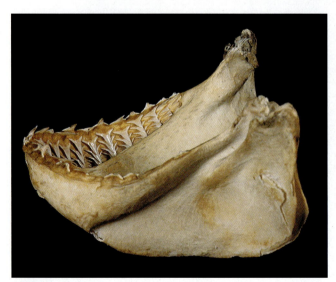

FIGURE 19.3   Shark teeth: Lateral view.

FIGURE 19.4   Shark teeth: Medial view.

development and structure of shark teeth is similar to that of placoid scales, some biologists believe they are modified scales. (See page 29, Figure 4.1.)

The **oral cavity** extends from the lingual surface of the teeth to the level of the spiracle. The slightly raised floor of the oral cavity supported by the hyoid apparatus

is sometimes referred to as the "**tongue.**" No muscles are associated with this structure, so it is not considered to be a true tongue, but it does contribute to the base of the true tongue in tetrapods (Figure 19.2 and Figure 19.5). Notice the "bumps" on the tongue, which probably are taste buds. If you examine the lining of the oral cavity and pharynx, you may observe other taste buds.

The **pharynx** is the space, common to the digestive and respiratory systems, that extends from the spiracle to the opening of the esophagus. In addition to the **spiracle**, the

to "overfeed" by permitting the stomach to expand, are characteristic of many vertebrates. Not all vertebrates can drive up to a fast-food joint whenever they are hungry! Some of you may find the shark's last meal remaining in its stomach, and you also will notice that rugae are not always readily visible (Figure 19.9).

The pyloric sphincter controls the passage of food from the stomach into the very short proximal portion of the intestine, the **duodenum**. Food then moves into the **valvular intestine**, so-called because it contains a **spiral valve**. The intestine is the major site of digestion. (Figure 19.10). If we compare the ratio of small intestine length to body length among a number of vertebrates (*e.g.*, cat, mudpuppy, and shark), we will find that the shark has the shortest intestinal/body length ratio, as the spiral valve greatly increases the surface area and precludes a long intestine. The remnants

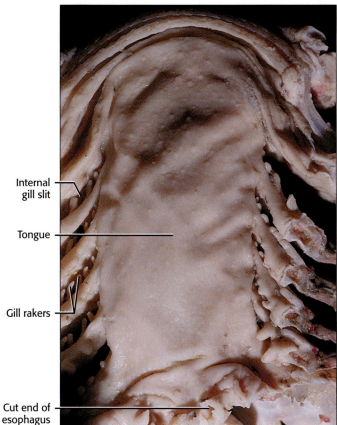

Internal gill slit

Tongue

Gill rakers

Cut end of esophagus

FIGURE 19.5 Tongue.

pharynx is pierced by five **internal gill slits** (Figure 19.2).

Externally, the **esophagus** and the **stomach** are difficult to differentiate. The stomach is elongated and J-shaped, with the curved portion of the J facing left. The elongated major portion, the **body**, separates the short upper portion, the **cardiac**, from the lower, short, curved **pyloric** region. The thick **pyloric sphincter** (pylorus) marks the caudal end of the stomach, separating it from the small intestine (Figure 19.8). The outer convex surface, the **greater curvature**, and the inner concave surface, the **lesser curvature**, are major landmarks of the stomach (Figure 19.6).

■ With a pair of scissors, make a 7.5 cm midventral incision in the esophagus, beginning with the cut edge.

Papillae line the lumen of the esophagus, and longitudinal folds, **rugae**, are characteristic of the stomach lining (Figure 19.7 and Figure 19.8). Perhaps papillae assist in swallowing. Rugae, which allow animals

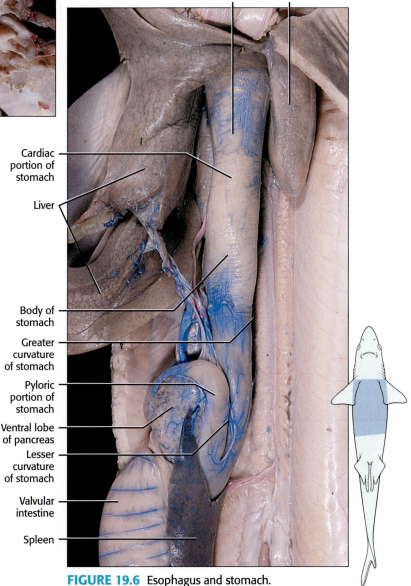

Esophagus   Testis

Cardiac portion of stomach

Liver

Body of stomach

Greater curvature of stomach

Pyloric portion of stomach

Ventral lobe of pancreas

Lesser curvature of stomach

Valvular intestine

Spleen

FIGURE 19.6 Esophagus and stomach.

of digestion then move into the distal portion of the intestine, the **rectum**, terminating in an **anus** opening into the **cloaca** (Figure 19.10).

The cloaca is separated by a shelf of tissue into a ventral **coprodeum** into which the anus opens, and a dorsal **urodeum** into which the urinary and reproductive systems open (Figure 19.11 and Figure 19.12). Notice the fingerlike **rectal gland** that opens into the rectum (Figure 19.13). The function of the rectal gland, in spite of its position, is to secrete a highly concentrated sodium chloride (salt) solution into the rectum, thereby aiding in osmoregulation.

Papillae of esophagus

Rugae of stomach

FIGURE 19.7 Internal anatomy of the esophagus and stomach.

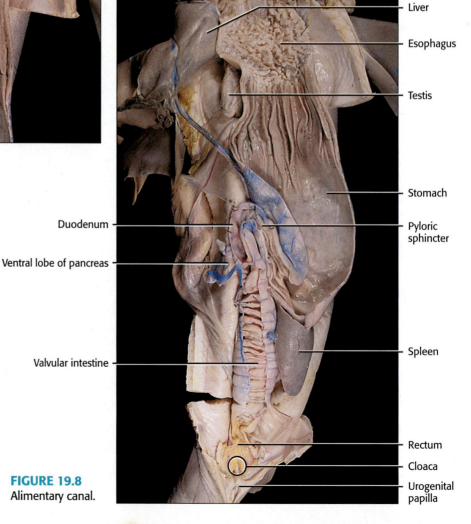

Oral cavity

Tongue

Pharynx

Liver

Esophagus

Testis

Stomach

Pyloric sphincter

Duodenum

Ventral lobe of pancreas

Valvular intestine

Spleen

Rectum

Cloaca

Urogenital papilla

FIGURE 19.8 Alimentary canal.

To appreciate the construction of the spiral valve, make a 5 cm longitudinal incision through the valvular intestinal wall, avoiding the major blood vessels associated with this organ. Carefully separate the cut edges and observe the valve. Notice that the spiral valve is absent from the duodenum and rectum (Figure 19.10).

Accessory Digestive Organs

Two major accessory digestive organs, the liver and the pancreas, are derivatives of the primitive gut. The liver and pancreas have ducts that empty into the duodenum.

In most sharks the liver is very large, making up approximately 25% of the total body weight. It is advantageous for an aquatic vertebrate, such as a fish to reduce its

FIGURE 19.9 Shark's last meal.

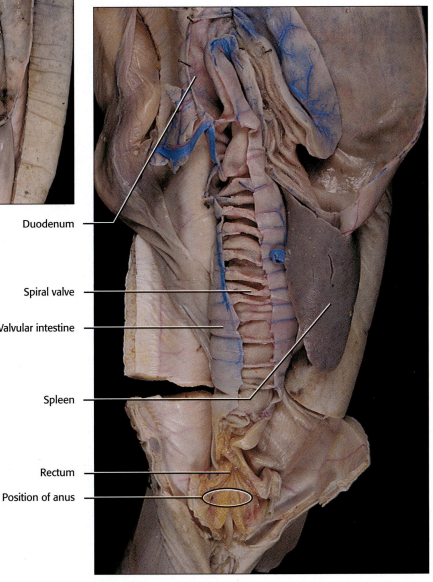

Duodenum

Spiral valve

Valvular intestine

Spleen

Rectum

Position of anus

FIGURE 19.10 Valvular intestine: Spiral valve.

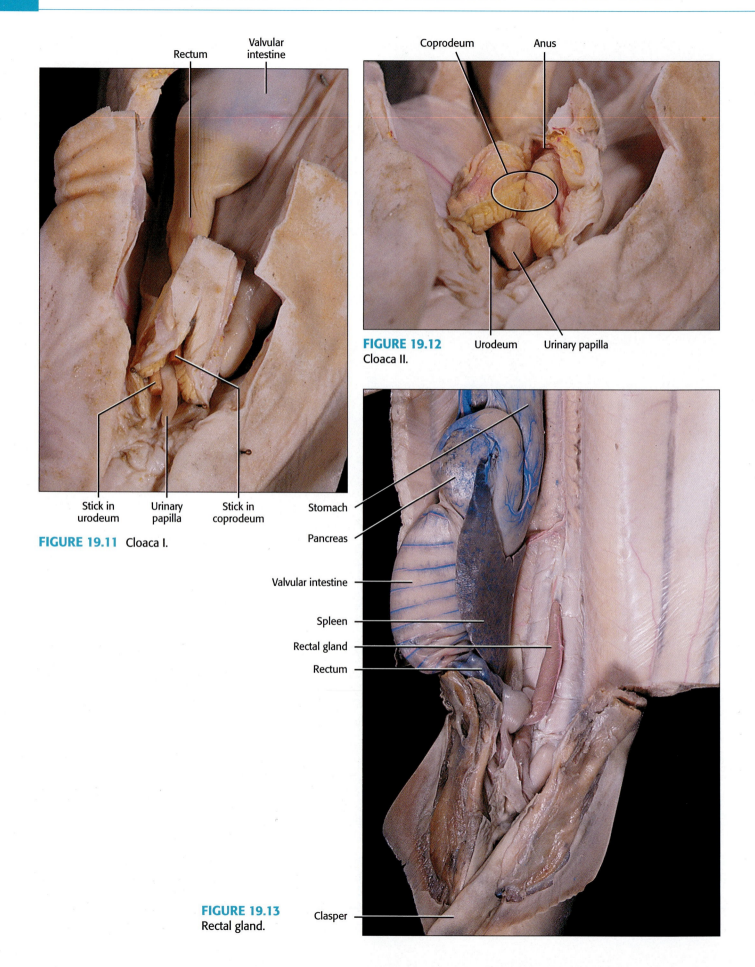

Rectum

Valvular intestine

Stick in urodeum

Urinary papilla

Stick in coprodeum

FIGURE 19.11 Cloaca I.

Coprodeum

Anus

Urodeum

Urinary papilla

FIGURE 19.12
Cloaca II.

Stomach

Pancreas

Valvular intestine

Spleen

Rectal gland

Rectum

FIGURE 19.13
Rectal gland.

Clasper

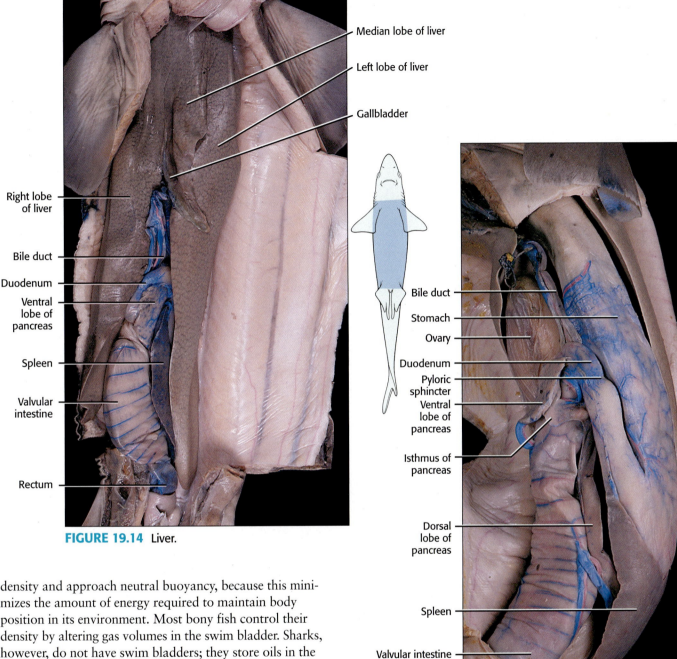

Median lobe of liver

Left lobe of liver

Gallbladder

Right lobe
of liver

Bile duct

Duodenum

Ventral
lobe of
pancreas

Spleen

Valvular
intestine

Rectum

FIGURE 19.14 Liver.

Bile duct

Stomach

Ovary

Duodenum

Pyloric
sphincter

Ventral
lobe of
pancreas

Isthmus of
pancreas

Dorsal
lobe of
pancreas

Spleen

Valvular intestine

FIGURE 19.15 Pancreas.

density and approach neutral buoyancy, because this mini-
mizes the amount of energy required to maintain body
position in its environment. Most bony fish control their
density by altering gas volumes in the swim bladder. Sharks,
however, do not have swim bladders; they store oils in the
liver to reduce their body density.

Identify the large **left, right,** and much shorter **median
lobes** of the **liver.** Notice the elongate **gallbladder** along
the right edge of the median lobe of the liver. Bile is secreted
by the liver, and conveyed by hepatic ducts (do not attempt
to locate these) in the liver to the gallbladder, where it is
stored. Follow the grayish-tan, flattened **bile duct** leading
from the gallbladder to the duodenum within the hepatoduodenal ligament (Figure 19.14).

The **pancreas** consists of an obvious **ventral lobe,** closely
applied to the ventral surface of the duodenum, and a **dorsal
lobe,** located between the dorsal surface of the duodenum
and the greater curvature of the stomach (see page 60). A
narrow **isthmus** connects the two lobes (Figure 19.15). A

single whitish **pancreatic duct** drains the pancreas from the
caudal end of the ventral lobe, emptying into the duodenum. Carefully pick away pancreatic tissue from the caudal
end of the ventral lobe to expose the small, thin duct.

The large, gray organ suspended by the gastrosplenic
ligament from the greater curvature of the stomach is the
spleen (Figure 19.10). The spleen is not a digestive organ
but, rather, a large mass of lymphoid tissue, identified
here because it is a prominent organ among the viscera. In
vertebrates, its major functions include production, storage,

and destruction of erythrocytes. Another important function is the destruction of potentially harmful invaders (*e.g.*, bacteria, viruses) similar to other lymph organs.

RESPIRATORY SYSTEM

Extracting oxygen from an aquatic medium is demanding and difficult. Compared to air, water is dense and viscous, and it contains a small volume of oxygen per unit volume of respiratory medium. For these reasons, large volumes of water have to be moved past large respiratory surface areas. A wide variety of anatomical solutions have evolved in fishes—among them, both internal and external gills, skin, and linings of various parts of the alimentary canal. The vast majority of fish, however, utilize a gill respiratory mechanism. All vertebrates demonstrate an intimate relationship between the respiratory portions of the circulatory system and the respiratory organs themselves, but you will be able to appreciate this close relationship especially well in the shark.

To maximize the oxygen saturation of the respiratory blood in most fish, the flow of water must be continuous and unidirectional. To produce this type of flow in a fish, such as a shark, the orobranchial (oral and pharyngeal) volume is increased by muscular contraction, thereby reducing the water pressure in the pharynx and allowing environmental water to flow passively through the mouth and spiracle into the oral cavity and pharynx because the environmental water pressure is greater than the pharyngeal pressure.

During this phase, the mouth and spiracle are open while the interbranchial septa (gill slits) are shut, preventing the escape of incoming water. The volume of the parabranchial cavity (the cavity between the gills and the inner body wall) is increased, reducing the pressure in this cavity and causing water to be sucked through the internal gill slits into this space from the orobranchial chamber that now is being compressed by muscular contraction. This raises the pressure and forces the water into the low-pressure parabranchial chamber.

As water passes from the orobranchial chamber to the parabranchial chamber, gases are exchanged across the gill capillary/gill filament membrane interface. With the compression of the parabranchial chamber while the mouth and spiracle are closed, water is expelled through the open external gill slits. This sequence is repeated continually, maintaining a continuous and unidirectional water flow.

Internal Gill Slits

Observe the first opening into the pharynx, the **spiracle**, a modified internal gill slit. Following it is a row of five typical **internal gill slits**, each leading into a **gill chamber**. The part of the gill chamber between the gills and the inner body wall is the **parabranchial chamber**. A series of obvious **gill rakers** projecting from the inner surface of the **gill arch** prevents food particles from clogging the gill chamber (Figure 19.16).

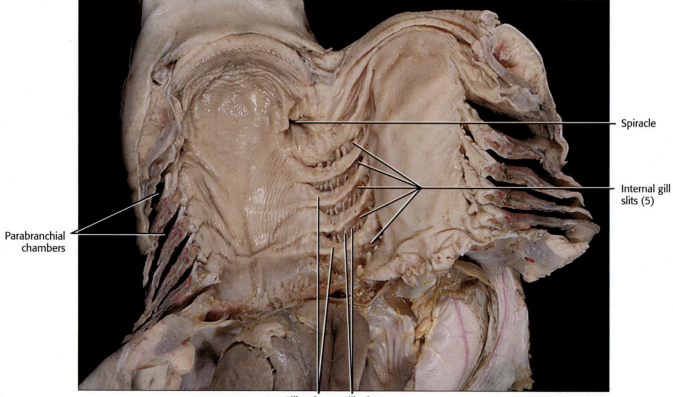

Spiracle

Internal gill slits (5)

Parabranchial chambers

FIGURE 19.16 Internal respiratory anatomy.

Gill arches   Gill rakers

Gills

Shark gills are known as septal gills because they possess an elongate **interbrachial septum** extending from the gill arch to the outer body wall, where they are attached, alternating with the gill slits. The septum is supported by cartilaginous **gill rays**. Attached to the surface of the septum are the obvious ridged structures, the **primary gill filaments** (Figure 19.17). Thin, folded **secondary gill filaments** are densely packed in a parallel series perpendicular to the primary gill filaments. With the exception of the spiracle and the sixth gill, primary filaments occur on both sides of the interbranchial septum of each gill arch; they are called **holobranchs** (Figure 19.18).

The gills associated with the spiracle and the sixth gill have primary filaments on only one side; they are known as **hemibranchs** (Figure 19.18 and Figure 19.19). Further,

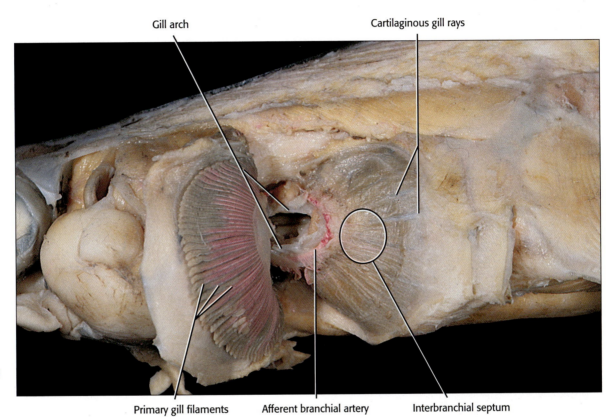

Gill arch

Cartilaginous gill rays

FIGURE 19.17
Gill anatomy.

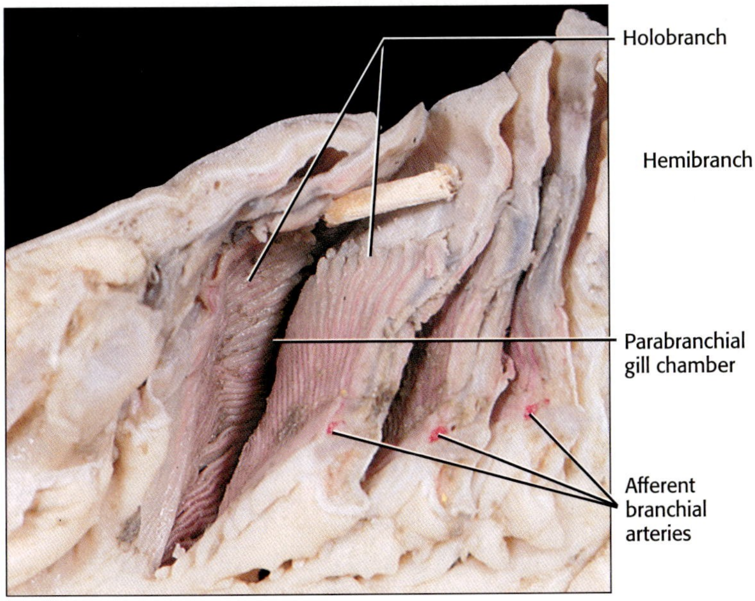

Primary gill filaments

Afferent branchial artery

Interbranchial septum

Holobranch

Hemibranch

Parabranchial gill chamber

Afferent branchial arteries

FIGURE 19.18 Holobranch (right side).

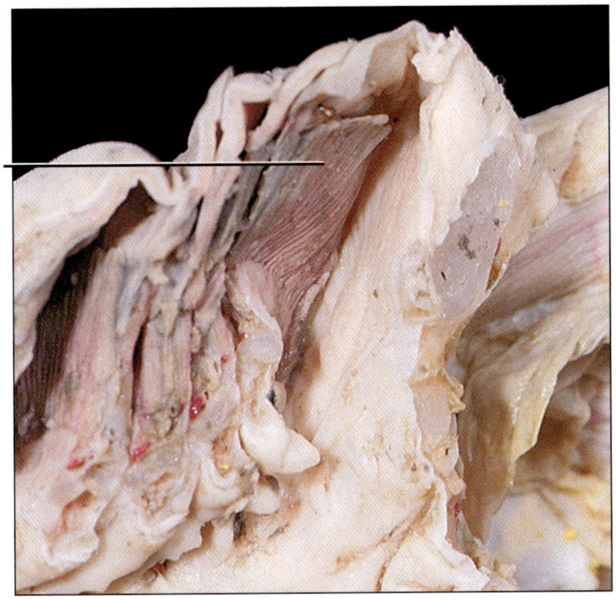

FIGURE 19.19 Hemibranch (right side).

the spiracular hemibranch has a peculiar arterial blood supply in that it receives oxygenated blood and therefore is known as a **pseudobranch**.

Observe the red vessel, the **afferent branchial artery**, lying on the surface of the interbranchial septum (Figure 19.17 and Figure 19.18). This vessel branches into numerous capillaries in the secondary lamellae, where gaseous exchange occurs, with the blood now flowing away from the secondary lamellae through efferent branchial arteries to the dorsal aorta. In this gas-exchange process, the respiratory stream of water flows from the orobranchial side of the gills to the parabranchial side of the gills and the blood flows in the secondary gill filaments from the parabranchial side to the orobranchial side. This anatomical arrangement constitutes a countercurrent flow mechanism and is essential in permitting maximal oxygen saturation of the blood.

CHAPTER TWENTY

Dogfish Shark — Urogenital System

The urogenital system consists of the urinary system and the reproductive system.

URINARY SYSTEM

Kidneys and Ducts

In most vertebrates the kidneys are glomerular and retroperitoneal. The shark is no exception. An organ is retroperitoneal when it resides behind the parietal peritoneum. To observe the kidneys, make a longitudinal slit lateral to the kidney with your scissors through the parietal peritoneum covering the ventral surface. If your specimen is a male, be extremely careful not to destroy the cream-colored, highly convoluted duct lying on the surface of the kidney. Cautiously remove the peritoneum covering the kidney.

In males and females alike, the opisthonephric **kidneys** are elongate, brownish, straplike organs lying against the dorsal body wall on either side of the dorsal aorta and extending almost the entire length of the pleuroperitoneal cavity (Figure 20.1). In the male, the cranial end of the kidney, **Leydig's gland** (Figure 20.4), has no urinary function but contributes to the reproductive fluid. In the female, the anterior end of the kidney is degenerate and nonfunctional. In both sexes, the caudal portion is urinary in function.

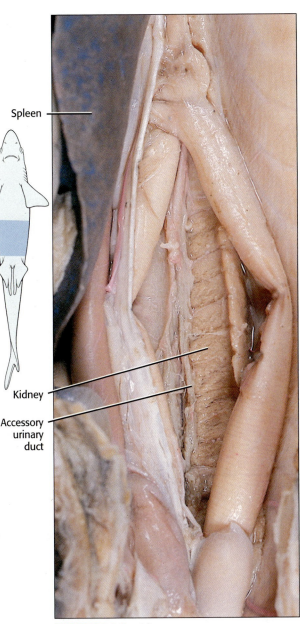

Spleen

Kidney

Accessory urinary duct

FIGURE 20.1 Kidney.

In the embryo of both sexes, the kidney is drained by the **archinephric duct.** In the mature female, the caudal portion of the kidney continues to be drained by the archinephric duct, which is difficult to find, and empties into the cloaca through the **urinary papilla.** In the mature male, however, the archinephric duct is modified into a reproductive duct (discussed later in the chapter), and has no urinary function. Urine is drained from the caudal end of the kidney in males through an **accessory urinary duct** into the cloaca through the **urogenital papilla** (Figure 20.1).

REPRODUCTIVE SYSTEM

Female

The primary sex organs are the **ovaries,** dorsal to the liver. They are suspended from the cranial dorsal body wall by a supporting mesentery, the **mesovarium.** If the shark is mature, the surface of the ovary will appear somewhat like a bag of marbles, which in reality consist of ova in various stages of development (Figure 20.2). In immature females, the reproductive system is not well developed. The ovary is flattened and small, and the oviducts appear as slender tubes with no supporting mesenteries. Observe the reproductive organs in a mature specimen.

Common to all female vertebrates is a system of ducts that enable the ova or oocytes to be transported from the peritoneal cavity to the outside. Among mature sharks, the **oviducts** extend from a common opening, the **ostium tubae,** located in the midventral portion of the **falciform ligament** (Figure 20.3). The oviduct curves dorsally around the liver, continues posteriorly along the dorsal body wall, enlarging because of the presence of a **shell** or **nidamental gland** at about the level of the ovary, then proceeds caudally to a second enlargement, the **uterus,** which terminates in an opening into the **cloaca** (Figure 20.2 and Figure 20.3).

Ovulation results from the rupture of the wall of the ovary, releasing the mature ovum into the peritoneal cavity. The heavily yolked eggs enter the oviduct through the ostium tubae. Fertilization occurs in the proximal portion of the oviduct anterior to the shell gland. In many species of sharks, such as the dogfish, that give birth to living young— termed viviparous—as the eggs pass through the shell gland, a thin, membranous shell is secreted around them. These embryos are retained within the reproductive tract, sometimes for many months, where they hatch and at times become associated in some fashion with the inner lining of the female uterus, where they may gain some nutrients. Other sharks are oviparous—they deposit eggs encased in a heavy protective shell in the environment.

Male

This description is applicable to mature males. The primary sex organs are the **testes,** occurring dorsal to the liver, suspended by a supporting mesentery, the **mesorchium.** Sperm

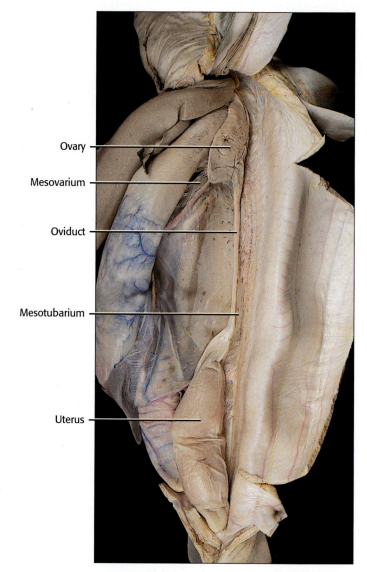

Ovary

Mesovarium

Oviduct

Mesotubarium

Uterus

FIGURE 20.2 Female reproductive system I.

are conveyed from the testes through minute tubules, the efferent ductules, in the **mesorchium,** into the **epididymis,** modified kidney tubules located in the cranial portion of the kidney, referred to as Leydig's gland, no longer urinary in function but secreting a fluid similar to the seminal fluid produced by amniotes. In addition, cells in the Leydig's gland occurring between the seminiferous tubules produce testosterone.

 To better view the ducts, carefully remove the parietal peritoneum from the vas deferens and seminal vesicles. Because the sperm sacs are more delicate, we do not recommend removing the peritoneum from them.

From the epididymis, sperm move into the highly coiled **ductus deferens** (vas deferens), the modified archinephric

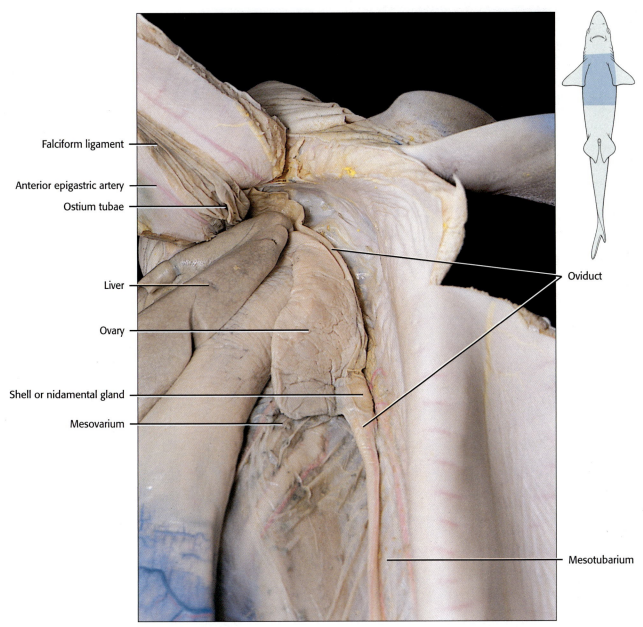

Falciform ligament

Anterior epigastric artery

Ostium tubae

Liver

Ovary

Shell or nidamental gland

Mesovarium

Oviduct

Mesotubarium

FIGURE 20.3 Female reproductive system II.

duct, lying on the ventral surface of the kidney and extending from the level of the testis caudally, where it enlarges into straight **seminal vesicles** that terminate in a bilobed **sperm sac** opening into the base of the **urogenital papilla** (Figure 20.4 and Figure 20.5). In immature males, reproductive structures are difficult to identify. This is why we suggest that you study a mature male).

In vertebrates, when shelled eggs or living young are produced, internal fertilization is necessary. To introduce sperm into the reproductive system, copulatory organs often have evolved to facilitate this behavior. In sharks, these organs are called **claspers** and are the modified pelvic fins of males. The dorsal surface of the clasper is grooved, permitting the transfer of sperm. During this process, only

one clasper is inserted into the female and is held in place by a **spine** on the distal, lateral surface of the clasper (Figures 20.6A and 20.6B).

Recall that you should have exposed these structures already when skinning the male pelvic fin for muscle dissection. Water is pumped into the **siphons**, sac-like organs overlying the ventral surface of the pelvic fins, through muscular movement of the fins. Sperm and accessory secretions of Leydig's gland are ejaculated from the urogenital papilla into the groove and are flushed with the water and secretions of the siphon into the female's reproductive tract. The sperm then swim cranially up the female reproductive tract, where fertilization takes place.

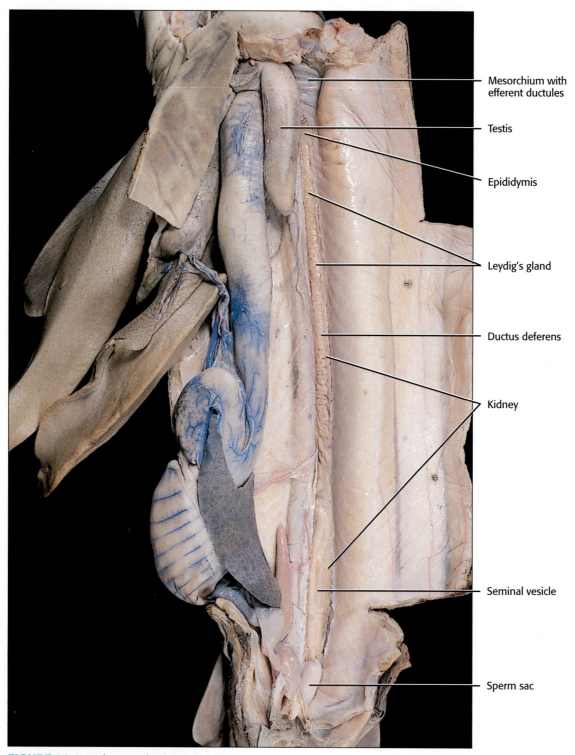

Mesorchium with
efferent ductules

Testis

Epididymis

Leydig's gland

Ductus deferens

Kidney

Seminal vesicle

Sperm sac

FIGURE 20.4 Male reproductive system I.

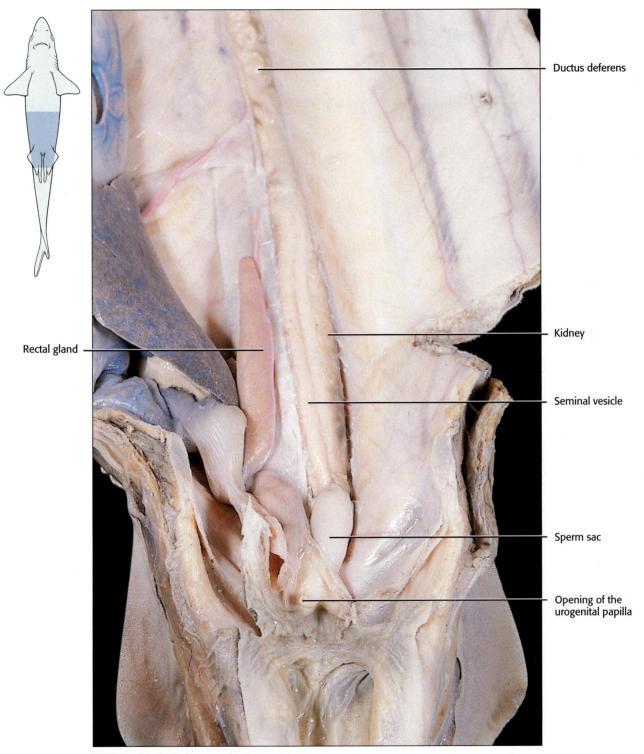

Ductus deferens

Kidney

Seminal vesicle

Rectal gland

Sperm sac

Opening of the
urogenital papilla

FIGURE 20.5 Male reproductive system II.

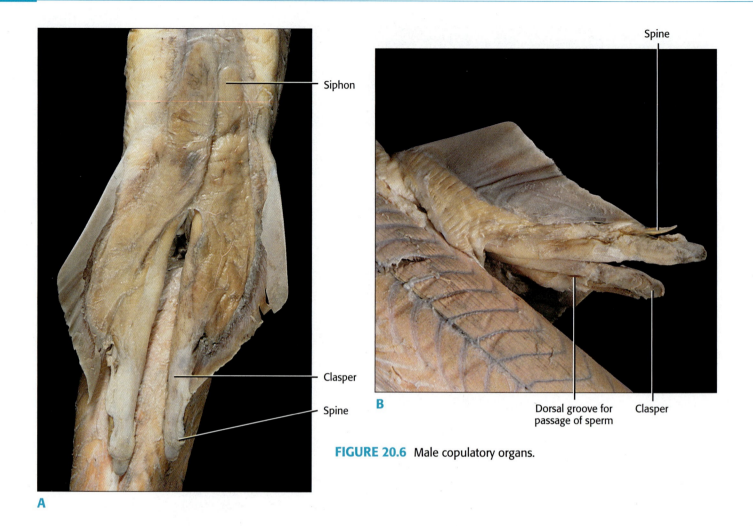

Siphon

Clasper

Spine

A

Spine

B

Dorsal groove for
passage of sperm

Clasper

FIGURE 20.6 Male copulatory organs.

CHAPTER TWENTY-ONE

Dogfish Shark — Circulatory and Lymphatic Systems

Within the circulatory system of vertebrates, it is possible to encounter branching diversity. Your specimen may exhibit a slightly different pattern from those of your classmates. For example, in most sharks, the gastrosplenic and posterior intestinal arteries arise independently from the dorsal aorta, but they may branch from a common stem from the dorsal aorta. **Therefore, it is important to read the text and to examine the specimens of your classmates.**

HEART AND ARTERIES

Heart and Afferent Branchial Arteries

During your examination of the body cavities in the shark, the heart was exposed. At this point you must exercise great care in further exposure of the heart and associated arteries. Carefully trim and pick away enough surrounding tissue with forceps to identify and study the heart, ventral aorta, and distribution of the afferent branchial arteries.

The anatomy of the heart of the shark consists of the following, beginning with the anterior end: **conus arteriosus, ventricle, atrium,** and **sinus venosus,** now arranged in an S-shaped configuration, in contrast to the ancestral condition (Figure 21.1 and Figure 21.2).

In the shark, venous blood, which is oxygen-poor, flows into the **sinus venosus,** a dorsally located, triangular, thin-walled chamber. Blood is pumped through the **sinoatrial valve** into the more muscular **atrium,** which appears as a darkly pigmented, pouch-like chamber that projects on either side of the ventral muscular ventricle. From the atrium, blood is pumped through the **atrioventricular valve** into the **ventricle.** The ventricle pumps blood into the tube-like **conus arteriosus,** containing several sets of

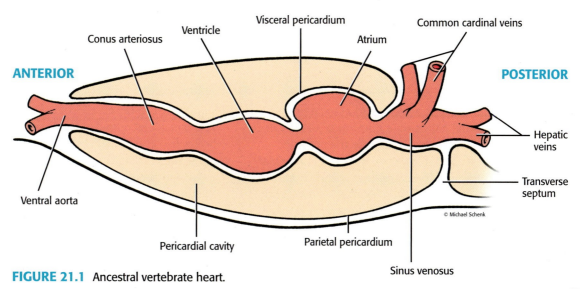

FIGURE 21.1 Ancestral vertebrate heart.

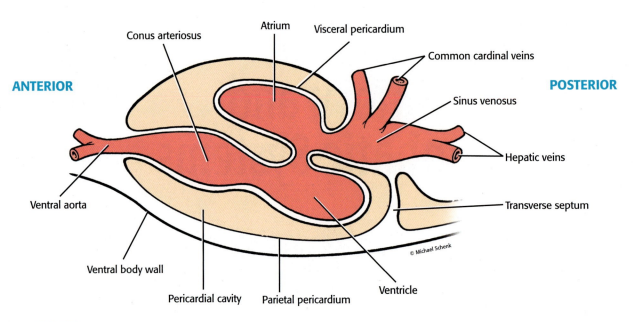

FIGURE 21.2 Shark heart: S-shaped configuration.

semilunar valves, arranged in rows (Figure 21.3, Figure 21.4, Figure 21.5, Figure 21.6, and Figure 21.7). In general, heart valves prevent the backwash of blood, thereby maintaining a unidirectional blood flow.

Blood continues from the conus into the **ventral aorta**, a whitish tube from which **five pairs of afferent branchial arteries** branch (Figure 21.8). The fifth and fourth pairs arise almost at the base of the aorta and pass to the posterior gills. The third pair originates about 8mm anterior to these and passes to the gills. Approximately 1.25 cm farther, the ventral aorta bifurcates and then gives rise to the second and first pairs of arteries, likewise passing to the gills.

 Trace the afferent branchial arteries to their respective gill lamellae.

Efferent Branchial Arteries, Dorsal Aorta, and Its Branches

 With scissors, make a slit through the mucous lining of the roof of the oral cavity and pharynx and carefully remove the lining and associated connective tissue, exposing a number of vessels injected with red latex. To expose the collector loops, carefully cut the cartilage of the branchial arches, watching to ensure that blood vessels are not in the way of the cuts. Work slowly and cautiously to achieve a good dissection. Several hours may be required.

Blood becomes oxygen-rich in the capillaries of the gills and is collected by four complete **collector loops** draining

the first four holobranchs, and an incomplete collector loop draining the last gill, a hemibranch. A complete collector loop, surrounding an internal gill slit, consists of an anterior **pretrematic branch** connected to a posterior **posttrematic branch** of an efferent branchial artery. Notice the short branches connecting the pretrematic and posttrematic arterial branches. The last collector loop consists only of the pretrematic branch. Blood from the collector loops flows through four **efferent branchial arteries**, and their convergence forms the **dorsal aorta** (Figure 21.9 and Figure 21.10). Anteriorly, the dorsal aorta continues as a pair of vessels.

From the dorsal end of the pretrematic branch of the first collector loop originates a pair of **hyoidean arteries**, which join with the paired dorsal aortae, forming the **internal carotid arteries** that unite anteriorly, forming a **single internal carotid artery**. The internal carotid artery then passes through the carotid foramen of the chondrocranium to supply the brain.

Prior to the union of the internal carotid arteries, a pair of lateral **stapedial arteries** originates and passes deeply to supply the eye and rostral area. In sharks this artery passes near the hyomandibular cartilage, which in tetrapod vertebrates becomes the stapes of the middle ear—hence, the name stapedial artery (Figure 21.11 and Figure 21.12).

From about the middle of the pretrematic branch of the first collector loop arises the **afferent spiracular artery**, which transports blood to the spiracle. From the spiracle, the **efferent spiracular artery** continues anteriorly to join the single carotid artery within the chondrocranium. From the ventral end of the pretrematic branch of the first collector loop originates the inconspicuous **external carotid artery**,

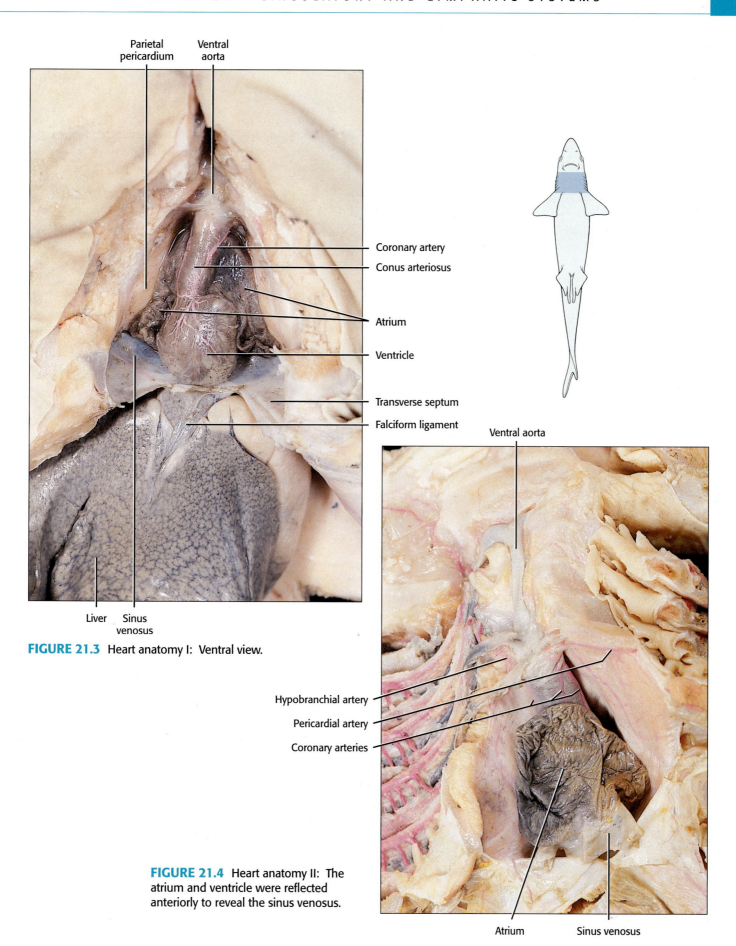

Parietal pericardium   Ventral aorta

Coronary artery

Conus arteriosus

Atrium

Ventricle

Transverse septum

Falciform ligament

Liver   Sinus venosus

FIGURE 21.3 Heart anatomy I: Ventral view.

Ventral aorta

Hypobranchial artery

Pericardial artery

Coronary arteries

FIGURE 21.4 Heart anatomy II: The atrium and ventricle were reflected anteriorly to reveal the sinus venosus.

Atrium   Sinus venosus

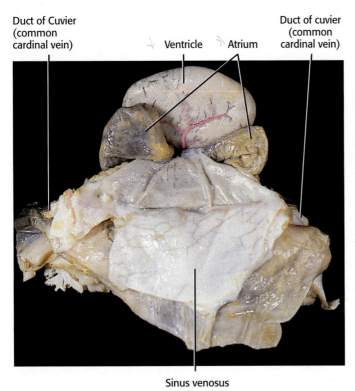

Duct of Cuvier (common cardinal vein)   Ventricle   Atrium   Duct of cuvier (common cardinal vein)

Sinus venosus

FIGURE 21.5 Heart anatomy III: As seen from a posterior view.

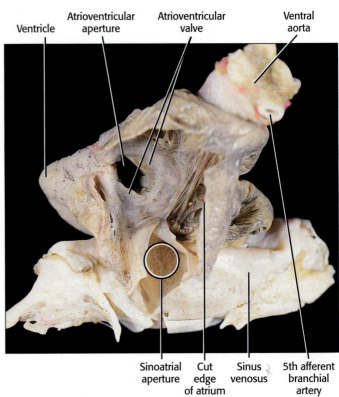

Ventricle   Atrioventricular aperture   Atrioventricular valve   Ventral aorta

Sinoatrial aperture   Cut edge of atrium   Sinus venosus   5th afferent branchial artery

FIGURE 21.6 Heart anatomy IV: Dorsal view. A cut was made through the atrium to reveal the atrioventricular valve.

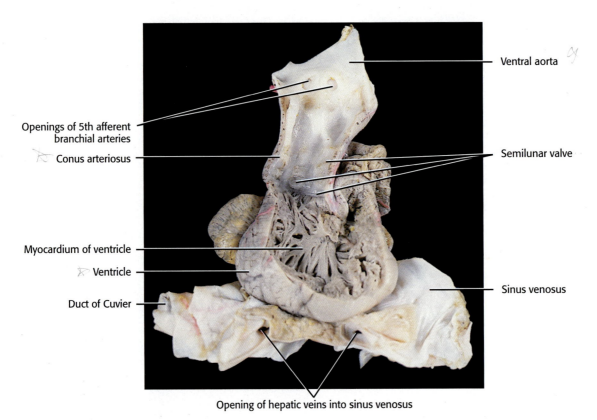

Ventral aorta

Openings of 5th afferent branchial arteries

Conus arteriosus

Semilunar valve

Myocardium of ventricle

Ventricle

Sinus venosus

Duct of Cuvier

Opening of hepatic veins into sinus venosus

FIGURE 21.7 Heart anatomy V: Ventral view. A frontal cut was made through the ventricle and a ventral slit was made through the conus arteriosus and base of the ventral aorta to reveal the myocardium of the ventricle and valves of the conus arteriosus, respectively.

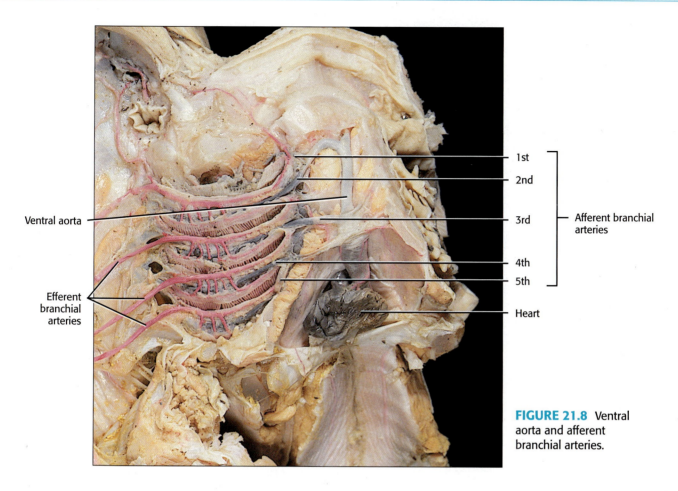

Ventral aorta

Efferent
branchial
arteries

1st
2nd
3rd
4th
5th

Afferent branchial
arteries

Heart

FIGURE 21.8 Ventral aorta and afferent branchial arteries.

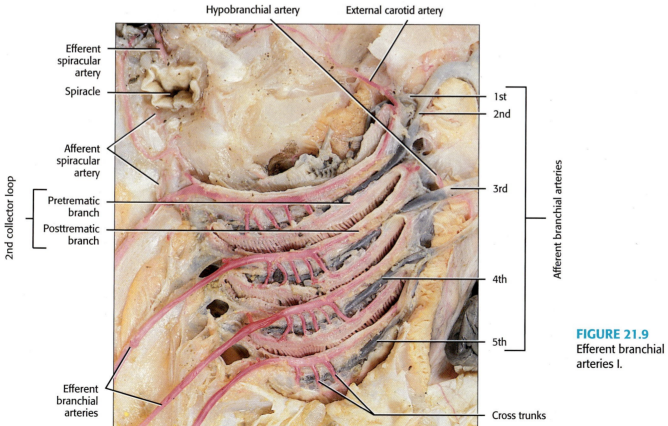

Hypobranchial artery

External carotid artery

Efferent
spiracular
artery

Spiracle

Afferent
spiracular
artery

2nd collector loop

Pretrematic
branch

Posttrematic
branch

Efferent
branchial
arteries

1st
2nd

3rd

4th

5th

Afferent branchial arteries

Cross trunks

FIGURE 21.9 Efferent branchial arteries I.

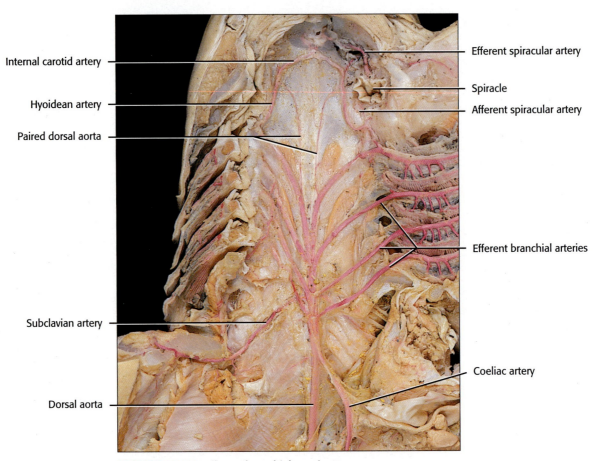

Internal carotid artery

Hyoidean artery

Paired dorsal aorta

Subclavian artery

Dorsal aorta

Efferent spiracular artery

Spiracle

Afferent spiracular artery

Efferent branchial arteries

Coeliac artery

FIGURE 21.10 Efferent branchial arteries II.

which supplies blood to the mandible (Figure 21.9, Figure 21.11, and Figure 21.12).

The **pharyngoesophageal artery** originates from the second efferent branchial artery to supply blood to the esophagus and pharynx. A **hypobranchial artery** arises from the second collector loop, sometimes with branches from other collector loops, terminating as **pericardial arteries**, which transport blood to the pericardium, and **coronary arteries**, which supply blood to the heart (Figure 21.3, Figure 21.4, Figure 21.11, Figure 21.12, and Figure 21.14).

Let us now consider the architecture of vessels associated with the single dorsal aorta. Serially paired **intersegmental arteries** arise along the length of the dorsal aorta, carrying blood to adjacent tissues (Figure 21.13). A pair of lateral **subclavian arteries**, embryologically derived from a pair of these segmental arteries, arises from the dorsal aorta just anterior to the fourth efferent branchial arteries. As is common in the anatomy of arteries, divergent branching occurs in the subclavian artery. The first branch is the **lateral artery**, roughly following the lateral line and supplying muscles in this region. The second is the **ventrolateral artery**, supplying neighboring muscles of the ventrolateral body wall. The third is the **anterior epigastric artery**, located approximately 1.25 cm from the midventral line (Figure 20.3 and Figure

21.11). It probably will appear on the under surface of the flap cut to preserve the falciform ligament. At almost the exact point where the ventrolateral artery arises, the subclavian continues deep as the **brachial artery**, supplying the tissues of the pectoral fin and the surrounding area (Figure 21.11 and Figure 21.14).

In vertebrates, the arterial blood supply to the abdominal viscera generally follows a common plan of three major trunks: a coeliac, an anterior mesenteric, and a posterior mesenteric artery. In sharks, the pattern usually consists of four main arteries. Two of these, however, may arise from a common stem, giving the impression that there are three. The most cranial, the **coeliac artery**, arises just after the dorsal aorta enters the pleuroperitoneal cavity (Figure 21.11). Almost immediately, the **ovarian artery** in the female (Figure 21.15) and the **testicular artery** in the male (Figure 21.16), transporting blood to the gonads, arise from the coeliac.

Additional small branches to the esophagus and cardiac portion of the stomach branch from the coeliac artery. The coelic then continues as an unbranched artery, entering the gastrohepatic ligament, where it gives rise to two vessels, the **gastrohepatic** and the **pancreaticomesenteric arteries** (Figure 21.17).

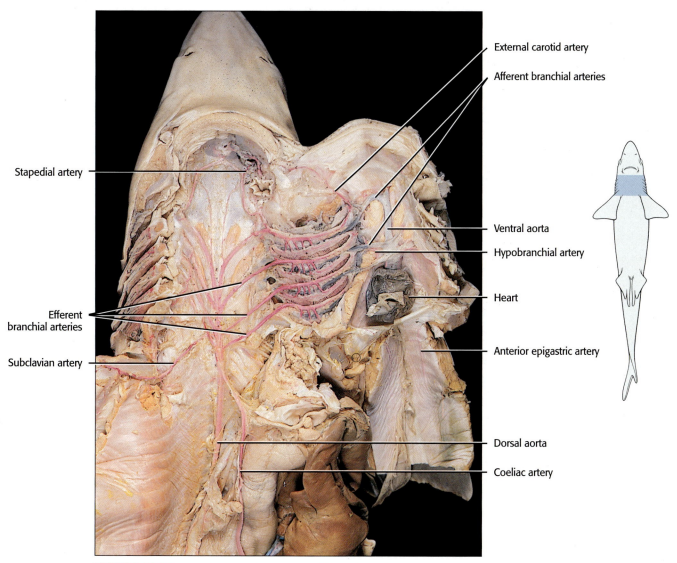

FIGURE 21.11 Branchial circulation (mandible reflected).

The gastrohepatic artery gives rise to the **gastric** and the **hepatic arteries**. The hepatic artery turns cranially, coursing alongside the common bile duct, and enters the liver. The **gastric artery** divides into a **dorsal** and a **ventral gastric artery**, supplying the dorsal and ventral portions of the body of the stomach, respectively. The **pancreatico-mesenteric artery** gives rise to small branches to the pyloric stomach and ventral lobe of the pancreas, a **duodenal artery** to the duodenum, and continues as the large **anterior intestinal artery** situated along the right side of the valvular intestine, which gives off ring-like branches to the valve (Figure 21.18). The **intraintestinal artery** arises from the pancreaticomesenteric artery and supplies the internal anatomy of the intestine (Figure 21.19).

From the dorsal aorta a large **anterior mesenteric artery** courses to the left side of the valvular intestine, giving off ringlike branches to the spiral valve, which anastomose

with those of the anterior intestinal artery (Figure 21.20). Very close, and posterior to, but sometimes arising from a common stem with the anterior mesenteric artery is the **gastrosplenic (or lienogastric) artery**. It transports blood to the spleen, stomach, and dorsal lobe of the pancreas. The smaller **posterior mesenteric artery** arises from the dorsal aorta and supplies blood to the rectal gland (Figure 21.20).

Two **iliac arteries** originate from the dorsal aorta and pass laterally, giving off small cloacal branches, and continue as the **femoral arteries** into the pelvic fins. Observe the **posterior epigastric arteries** that course along the lateral portions of the body wall to anastomose with the anterior epigastric arteries discussed previously (Figure 21.21). The dorsal aorta continues into the tail as the **caudal artery**, lying in the hemal arch (Figure 21.22).

During your careful dissection of the arteries of the pleuroperitoneal cavity, you probably have noticed numerous

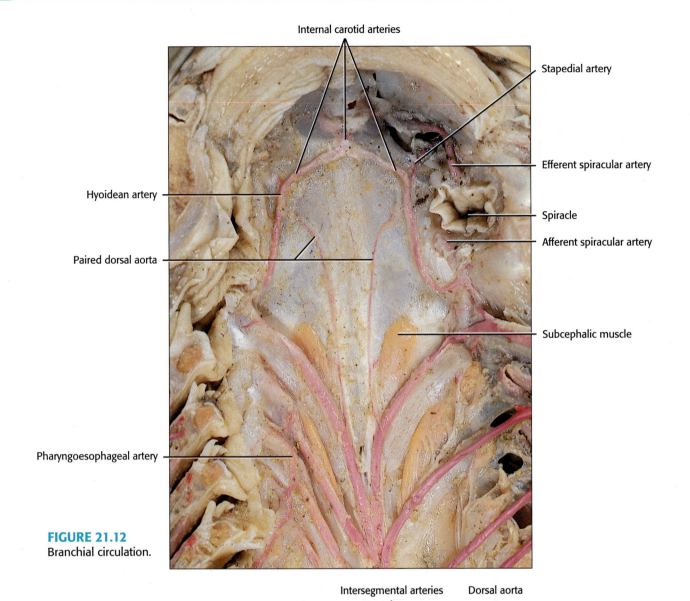

Internal carotid arteries

Stapedial artery

Hyoidean artery

Efferent spiracular artery

Spiracle

Paired dorsal aorta

Afferent spiracular artery

Subcephalic muscle

Pharyngoesophageal artery

FIGURE 21.12
Branchial circulation.

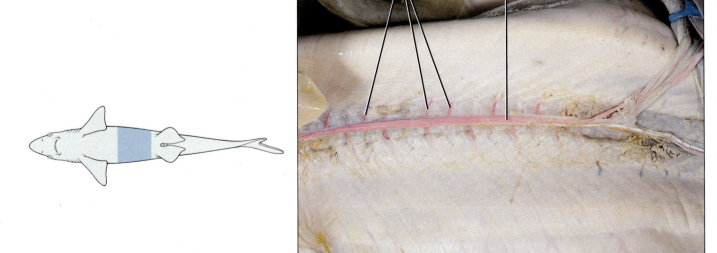

Intersegmental arteries   Dorsal aorta

FIGURE 21.13 Dorsal aorta with segmental arteries.

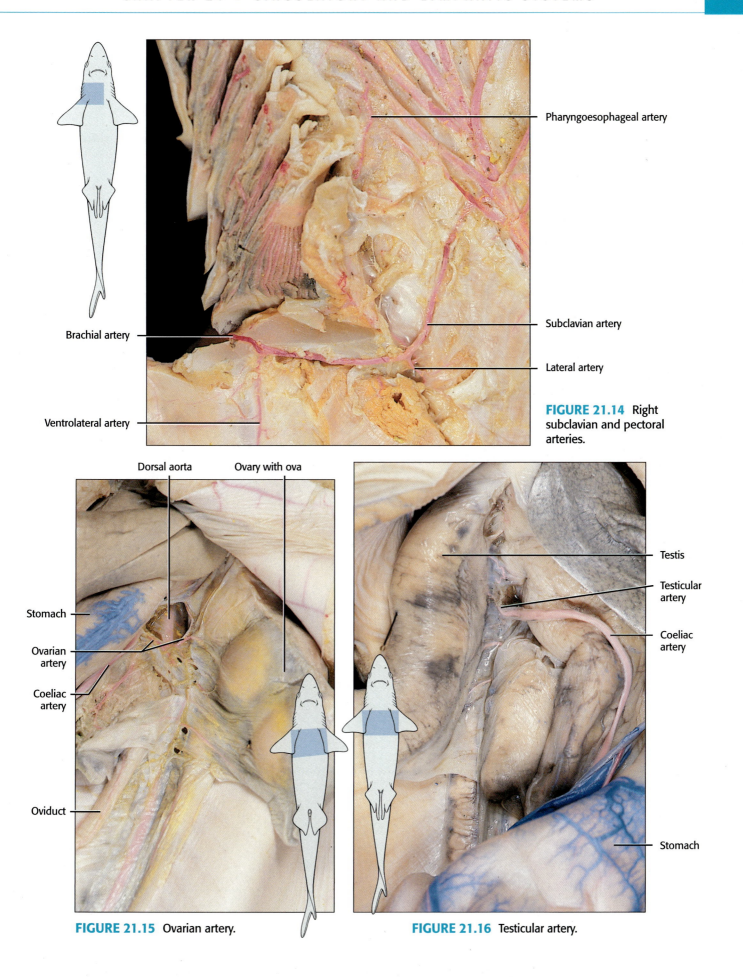

Pharyngoesophageal artery

Subclavian artery

Lateral artery

Brachial artery

Ventrolateral artery

FIGURE 21.14 Right subclavian and pectoral arteries.

Dorsal aorta

Ovary with ova

Stomach

Ovarian artery

Coeliac artery

Oviduct

FIGURE 21.15 Ovarian artery.

Testis

Testicular artery

Coeliac artery

Stomach

FIGURE 21.16 Testicular artery.

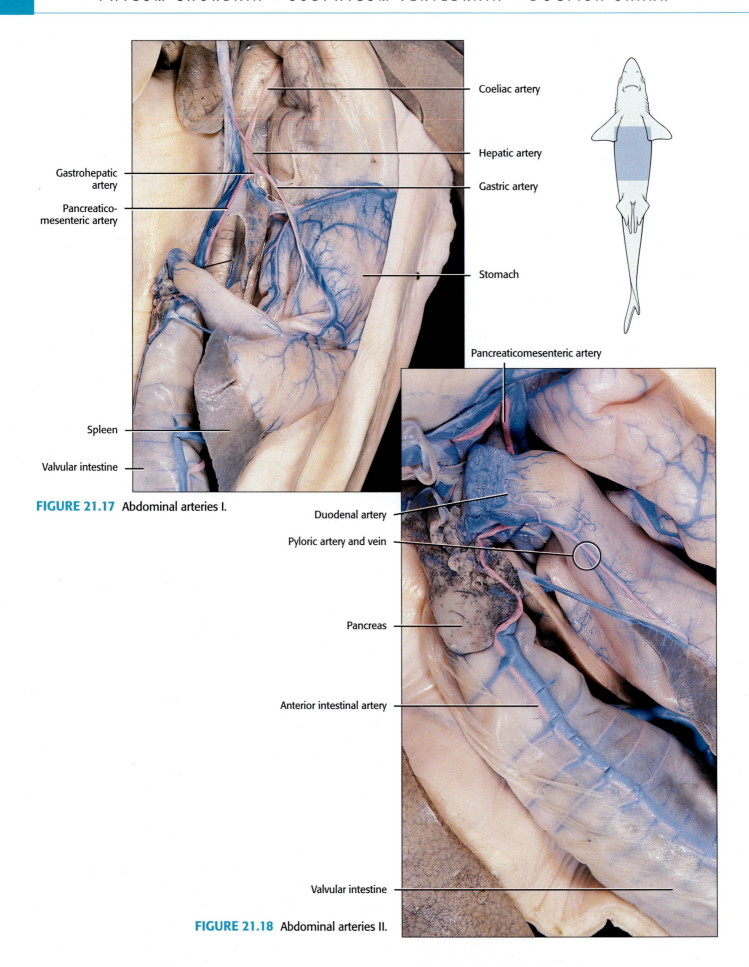

Coeliac artery

Hepatic artery

Gastrohepatic artery

Gastric artery

Pancreatico-mesenteric artery

Stomach

Spleen

Valvular intestine

FIGURE 21.17 Abdominal arteries I.

Pancreaticomesenteric artery

Duodenal artery

Pyloric artery and vein

Pancreas

Anterior intestinal artery

Valvular intestine

FIGURE 21.18 Abdominal arteries II.

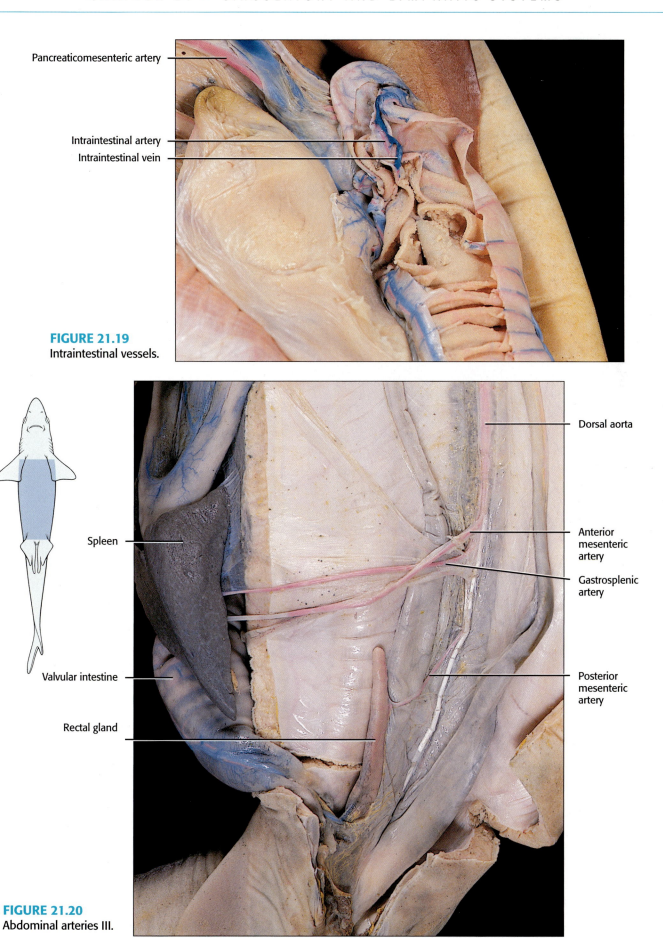

Pancreaticomesenteric artery

Intraintestinal artery
Intraintestinal vein

FIGURE 21.19
Intraintestinal vessels.

Dorsal aorta

Spleen

Anterior mesenteric artery

Gastrosplenic artery

Valvular intestine

Posterior mesenteric artery

Rectal gland

FIGURE 21.20
Abdominal arteries III.

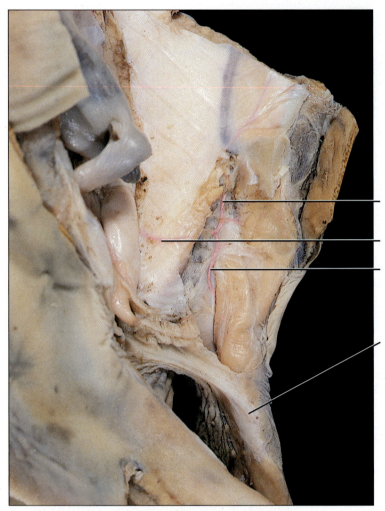

FIGURE 21.21 Arteries of the left pelvic region.

Posterior epigastric artery

Iliac artery

Femoral artery

Left pelvic fin

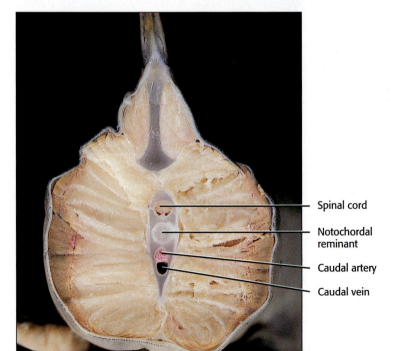

Spinal cord

Notochordal reminant

Caudal artery

Caudal vein

FIGURE 21.22 Caudal vessels.

serially arranged segmental arteries supplying the tissues of their respective segments. Those modified segmental arteries associated with the kidneys are the **renal arteries**.

VENOUS SYSTEM

Portal Systems

Veins in which blood is transported from capillaries in tissues of one or more organs to capillaries in tissues of the same or different organs, before reaching the heart, form a portal system. (See page 78.)

Renal Portal System

In the renal portal system of the shark, the blood originates in the capillaries of the caudal tissues and terminates in the capillaries within the kidneys. The tail is drained by the **caudal vein**, lying ventral to the artery in the hemal arch. At the anus, the caudal vein bifurcates into a pair of **renal portal veins**, passing lateral and dorsal to the kidney and giving off afferent renal veins that carry caudal venous blood (Figure 21.23). Capillaries connect these with the efferent renal veins draining the kidneys into the posterior cardinal veins.

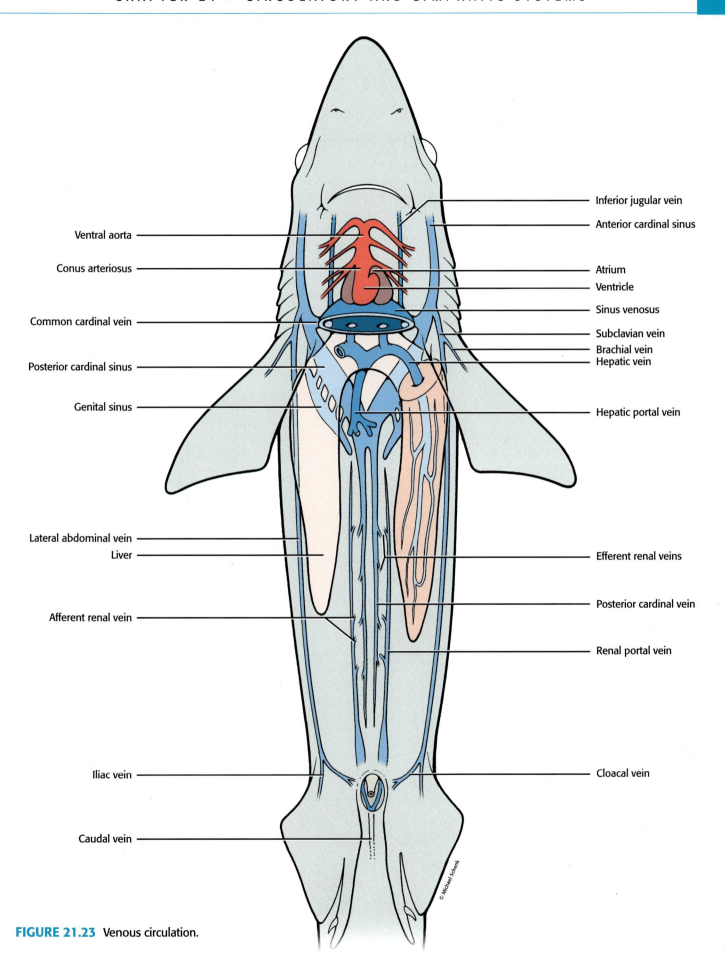

Inferior jugular vein

Anterior cardinal sinus

Ventral aorta

Conus arteriosus

Atrium

Ventricle

Sinus venosus

Common cardinal vein

Subclavian vein
Brachial vein
Hepatic vein

Posterior cardinal sinus

Genital sinus

Hepatic portal vein

Lateral abdominal vein

Liver

Efferent renal veins

Afferent renal vein

Posterior cardinal vein

Renal portal vein

Iliac vein

Cloacal vein

Caudal vein

© Michael Schenk

FIGURE 21.23 Venous circulation.

Hepatic Portal System

In the hepatic portal system of all vertebrates, venous blood from the capillaries of the abdominal viscera is transported to the liver "capillaries." By way of this circuit, the liver has access to all chemical substances transported from the viscera. Because the liver is the site of numerous metabolic activities, as well as the body's detoxification center and storage depot, the hepatic portal circulation is appropriate to permit the liver to buffer internal body physiology. For example, large amounts of glucose or toxins from digestive organs released into
the general circulation could severely disrupt homeostasis.

Three major tributaries—the gastric, the lieno-mesenteric, and the pancreaticomesenteric veins—join to form the **hepatic portal vein** (Figure 21.24 and Figure 21.25). The **gastric vein**, formed by the **dorsal** and **ventral gastric veins**, drains the stomach. The **lienomesenteric vein**, lying along the dorsal surface of the dorsal lobe of the pancreas, is formed by the coalescence of the **posterior intestinal vein** from the left side of the valvular intestine and the **posterior lienogastric vein** draining the spleen and the stomach. Small **pancreatic veins** from the dorsal lobe of the pancreas join the lienomesenteric vein as it passes through the pancreas (Figure 21.24 and Figure 21.25).

The **pancreaticomesenteric vein** is formed by the **anterior intestinal vein**, draining the right side of the valvular intestine, the **anterior lienogastric vein**, draining the spleen and portions of the pylorus of the stomach, a small **pyloric vein** further draining the pylorus, and the **intraintestinal vein**, draining the internal anatomy of the intestine. A small **choledochal vein** from the gallbladder running along the common bile duct enters the hepatic portal vein. The hepatic portal vein enters the liver and subdivides into a number of smaller branches that open into the hepatic sinuses. Blood is drained from the hepatic sinuses through a pair of hepatic veins into the sinus venosus (Figure 21.24 and Figure 21.25).

Systemic Veins

Because of the unusual branchial circulation of fish, in which blood is transported totally within arteries, no veins are found in the gill region. The venous portion of the circulatory system is a

combination of formed veins associated with larger ill-defined spaces called sinuses, and the blood carried here empties into the sinus venosus of the heart. A general rule for vertebrate circulation is that arteries and veins associated with organs or organ systems often run close to one another and share common names (e.g., subclavian artery and subclavian vein).

To enhance your understanding of blood circulation and because the venous system carries oxygen-poor blood, we will begin the discussion of the systemic veins from the distal areas of the body toward the heart. The **caudal vein** drains the tail area (Figure 21.22). At the level of the anus, it bifurcates into the **renal portal veins**, sending this blood through the kidney. After passing through the kidney, the blood flows into the **posterior cardinal veins**, lateral to the dorsal aorta, which widen into the **posterior cardinal sinuses**, the major systemic veins of the trunk.

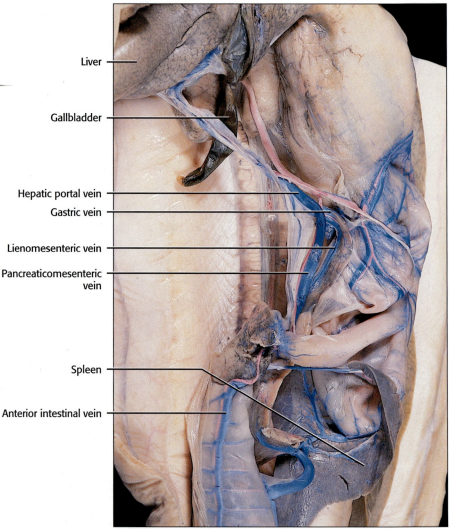

FIGURE 21.24 Hepatic portal circulation I.

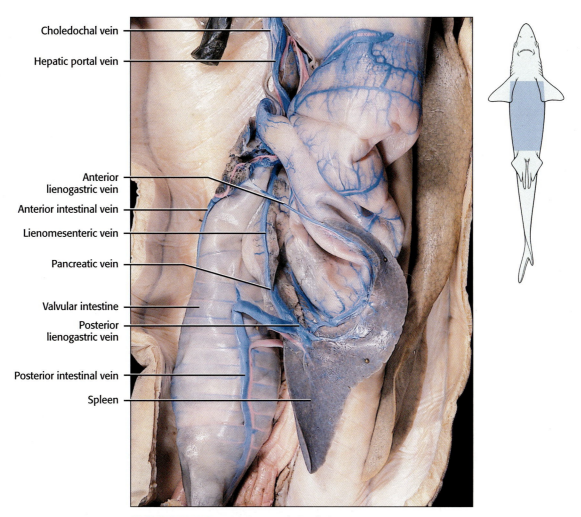

Choledochal vein

Hepatic portal vein

Anterior lienogastric vein

Anterior intestinal vein

Lienomesenteric vein

Pancreatic vein

Valvular intestine

Posterior lienogastric vein

Posterior intestinal vein

Spleen

FIGURE 21.25 Hepatic portal circulation II.

 Follow the posterior cardinal veins anteriorly until they expand into a common space that diverges into a pair of sinuses.

Associated with the gonads are **genital sinuses,** draining the ovaries or testes, which empty into the posterior cardinal sinuses. The posterior cardinal sinuses empty into the **common cardinal veins,** or **ducts of Cuvier,** which in turn empty into the sinus venosus of the heart (Figure 21.23 and Figure 21.26).

A second posterior drainage route (Figure 21.23) consists of the **femoral veins,** draining the pelvic fins, continuing as the **iliac veins** and joined by the **cloacal veins,** draining the cloacal area to form the **lateral abdominal veins,** coursing anteriorly along the lateral portion of the wall of the pleuroperitoneal cavity. Serial **intersegmental veins** draining blood from adjacent trunk areas empty into the lateral abdominal veins. The **brachial veins** draining the pectoral fin join with lateral abdominal veins to form the

subclavian veins, which empty into the common cardinal veins. **Subscapular veins,** draining the "shoulder" region, join the proximal end of the brachial veins.

Drainage of the head region (Figure 21.23) of the shark is accomplished by numerous sinuses emptying into the much larger and more easily identifiable **anterior cardinal sinuses.** During dissection of the branchial muscles, the large space encountered dorsal to the gill arches was the anterior cardinal sinus. Small **inferior jugular veins,** draining the ventral portion of the pharynx, empty into the common cardinal veins.

LYMPHATIC SYSTEM

The lymphatic system is thought to be absent in the chondrichthyes. In the shark, however, a number of sinusoids drain extracirculatory fluids into the veins and function in a fashion analogous to the lymphatic system in other vertebrates.

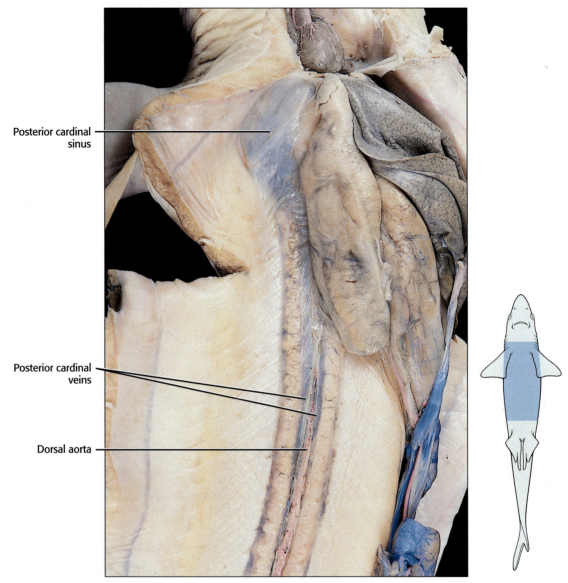

Posterior cardinal sinus

Posterior cardinal veins

Dorsal aorta

FIGURE 21.26 Posterior cardinal veins and sinus.

C H A P T E R   T W E N T Y - T W O

Dogfish Shark — Nervous System and Sense Organs

22

The central nervous system consists of the hollow dorsal nerve cord, the brain, the cranial and spinal nerves, and nerves of the autonomic nervous system. We include sense organs in this chapter because sensory information is transmitted from the sense organs to the central nervous system.

BRAIN AND NERVES

The Brain: Dorsal View

Because the brain is housed within the chondrocranium, it has to be exposed along with the cranial nerves, the olfactory apparatus, the ampullae of Lorenzini, the eyeball, and the inner ear. The chondrocranium was exposed during the skinning process. Perhaps your instructor will have additional comments regarding the delicate dissection of the brain, cranial nerves and semicircular canals.

 During the following dissection, chips of cartilage likely will come loose suddenly and may fly into your eyes. The use of goggles is recommended during this procedure.

With a sharp scalpel held at an acute angle, carefully shave and chip away the cartilage of the dorsal surface of the chondrocranium, being particularly watchful for the presence of the semicircular ducts in the vicinity of the otic capsule. The first hint of these ducts usually is indicated by a difference between the opacity of the canals and the surrounding cartilage, enabling you to see the ducts below the cartilage.

Be particularly cautious in removing the cartilage over the roof of the brain, especially the diencephalon, from which the extremely delicate, thread-like epiphysis or pineal gland projects (see Figure 23.2, page 208).

In the adult shark brain, like other vertebrate brains including the lamprey, five regions—telencephalon, the diencephalon, the mesencephalon, the metencephalon, and the myelencephalon—can be identified. To identify regions, structures, nerves, and other important areas discussed in this chapter, see Figure 22.1–Figure 22.7.

The **telencephalon** consists of the **olfactory bulbs, olfactory tracts,** and the **cerebral hemispheres** (Figure 22.1, Figure 22.2, and Figure 22.10). Most of the telencephalon in "primitive" vertebrates is devoted to olfaction. Therefore, these brains often are spoken of as "smelling brains."

Anteriorly, the olfactory bulb abuts the **olfactory sac,** lined with an olfactory epithelium containing sensory neurons, which are responsible for detecting chemicals dissolved in the environment. These sensory neurons enter the bulbs as the **olfactory nerve** (cranial nerve I) (Figure 22.6), where they synapse with a second set of neurons that travel within the **olfactory tracts,** terminating in the ventrolateral portion of the cerebral hemispheres, where interpretation of the chemicals is accomplished or smelled. Notice that the cerebral hemispheres appear as slightly bulged areas and are not the obvious convoluted area seen in mammals, for example.

The **diencephalon** occurs just posterior to the telencephalon. The diencephalon consists of a thin roof or the **epithalamus,** the lateral walls or the **thalamus,** and the floor or the **hypothalamus** (Figure

22.2 and Figure 22.8). From the **habenular region**, a thickened area of the epithalamus, projects the **epiphysis** (pineal gland), a fine, stalklike dorsal projection. Dorsally, the diencephalon appears as a sunken area covered by a membrane, the **tela choroidea**, which consists of the ependyma, the inner epithelial lining of the brain and spinal cord, along with the outer menninx. Projecting from the inner surface of the tela choroidea is a vascular choroid plexus. Associated with the diencephalon is the optic nerve (cranial nerve II), carrying sensory fibers that transmit visual impulses originating in the retina of the eye. These fibers cross over in the optic chiasma and enter the brain on the opposite side of their origin.

Just posterior to the diencephalon are the prominent, dorsal **optic lobes** of the **tectum** of the **mesencephalon**, whose floor is the **tegmentum**. The **oculomotor nerve** (cranial nerve III) is associated with the mesencephalon and carries primarily motor impulses to four of the six extrinsic eye muscles. A small **trochlear nerve** (cranial nerve IV) emerges dorsally from the roof of the mesencephalon and innervates another of the six extrinsic eye muscles (Figure 22.4, Figure 22.6, Figure 22.7, and Figure 22.17).

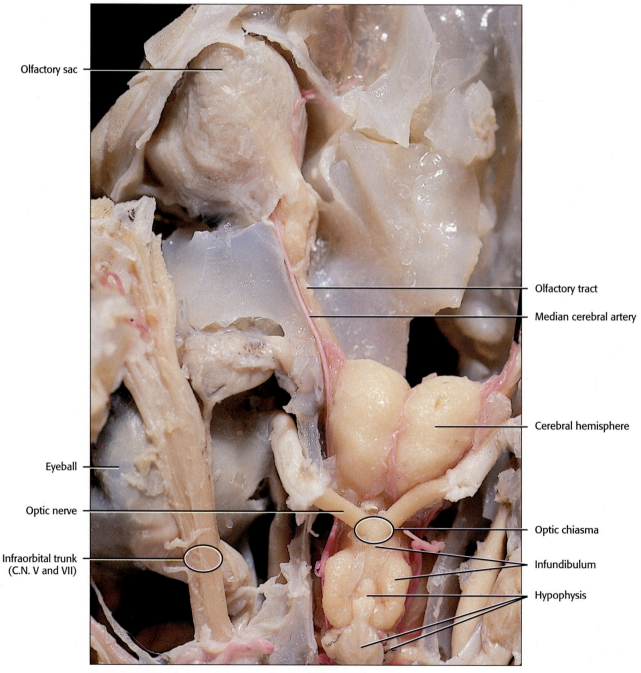

Olfactory sac

Olfactory tract

Median cerebral artery

Cerebral hemisphere

Eyeball

Optic nerve

Optic chiasma

Infraorbital trunk
(C.N. V and VII)

Infundibulum

Hypophysis

FIGURE 22.1 Brain and cranial nerves I: Ventral view.

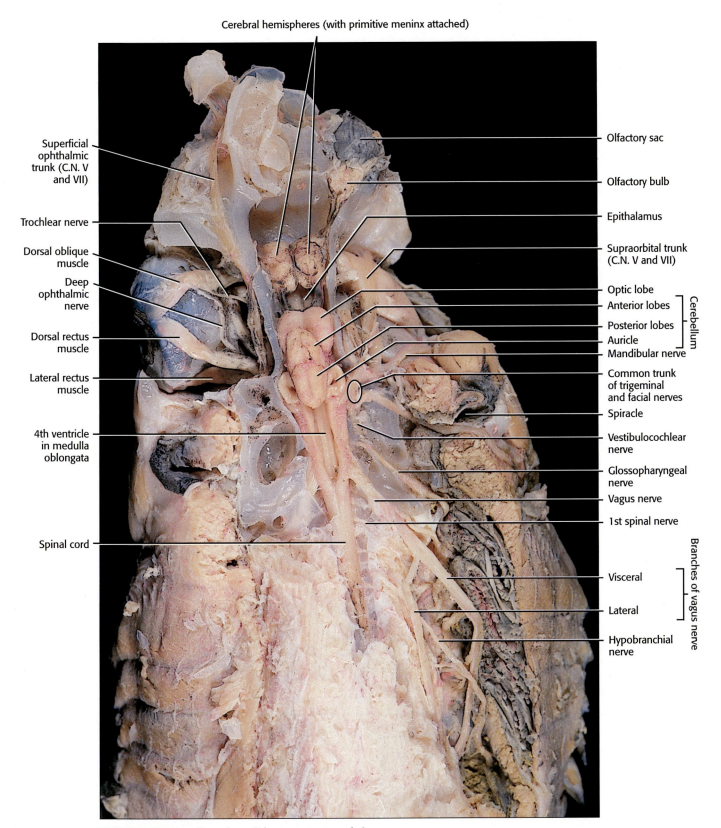

Cerebral hemispheres (with primitive meninx attached)

Superficial ophthalmic trunk (C.N. V and VII)

Trochlear nerve

Dorsal oblique muscle

Deep ophthalmic nerve

Dorsal rectus muscle

Lateral rectus muscle

4th ventricle in medulla oblongata

Spinal cord

Olfactory sac

Olfactory bulb

Epithalamus

Supraorbital trunk (C.N. V and VII)

Optic lobe

Anterior lobes

Posterior lobes

Auricle

Mandibular nerve

Common trunk of trigeminal and facial nerves

Spiracle

Vestibulocochlear nerve

Glossopharyngeal nerve

Vagus nerve

1st spinal nerve

Visceral

Lateral

Hypobranchial nerve

Cerebellum

Branches of vagus nerve

FIGURE 22.2 Brain and cranial nerves II: Dorsal view.

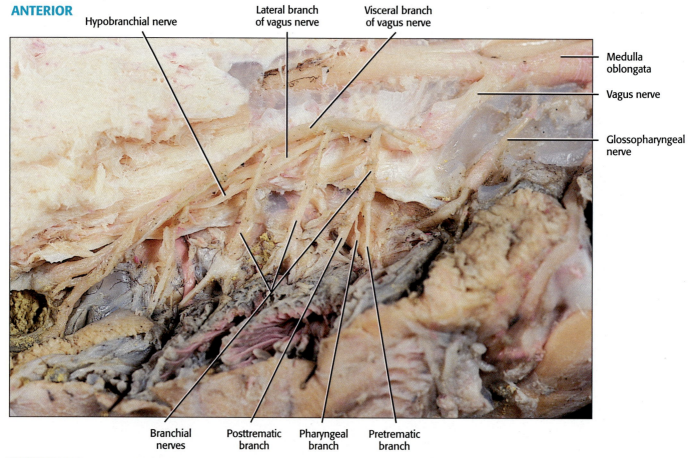

ANTERIOR

Hypobranchial nerve

Lateral branch of vagus nerve

Visceral branch of vagus nerve

Medulla oblongata

Vagus nerve

Glossopharyngeal nerve

Branchial nerves

Posttrematic branch

Pharyngeal branch

Pretrematic branch

FIGURE 22.3 Brain and cranial nerves III: Lateral view.

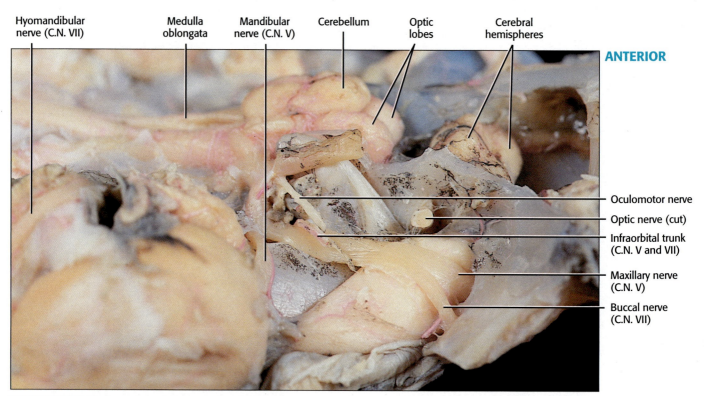

Hyomandibular nerve (C.N. VII)

Medulla oblongata

Mandibular nerve (C.N. V)

Cerebellum

Optic lobes

Cerebral hemispheres

ANTERIOR

Oculomotor nerve

Optic nerve (cut)

Infraorbital trunk (C.N. V and VII)

Maxillary nerve (C.N. V)

Buccal nerve (C.N. VII)

FIGURE 22.4 Brain and cranial nerves IV: Lateral view.

The **cerebellum** and anterior portion of the **medulla oblongata** constitute the metencephalon. A longitudinal and a transverse groove divide the cerebellum into four parts. The anterior pair overhangs the optic lobes, and the posterior pair overhangs the anterior portion of the medulla oblongata. The **auricles**, the paired earlike appendages of the cerebellum, are located on either side of the posterior lobes of the cerebellum. The **medulla oblongata**, Y-shaped and continuous with the spinal cord, is the major part of the myelencephalon (Figure 22.2, Figure 22.3, Figure 22.4, Figure 22.5, Figure 22.6, and Figure 22.7).

Cranial nerves V, VI, VII, VIII, IX, and X are serially associated with the medulla oblongata, beginning at the anterior end (Figure 22.6 and Figure 22.7). In the shark, the **trigeminal** (cranial nerve V) and **facial** (cranial nerve VII) are intimately associated and carry motor and sensory impulses to and from the rostral, oral, and facial regions of the head. Lateral line impulses are carried by the facial. The **abducens** (cranial nerve VI) innervates the last of the extrinsic eye muscles. The **vestibulocochlear** (cranial nerve VIII) carries impulses related to equilibrium and hearing. The **glossopharyngeal** (cranial nerve IX) carries sensory and motor impulses associated with muscles, gill filaments of the first gill pouch, and the pharyngeal area. The **vagus** (cranial nerve X) carries sensory impulses from and motor impulses to the remaining four gill pouches, the posterior pharyngeal area, the lateral line system, the heart, and abdominal viscera.

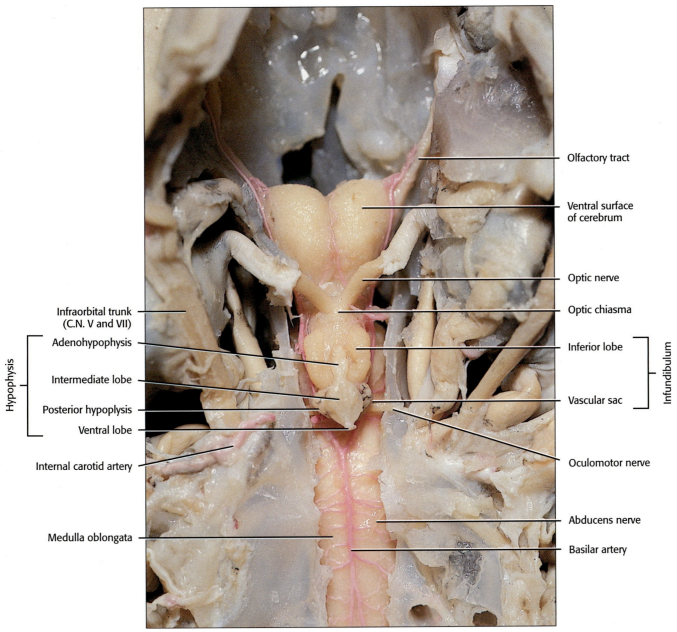

FIGURE 22.5 Brain and cranial nerves V: Ventral view.

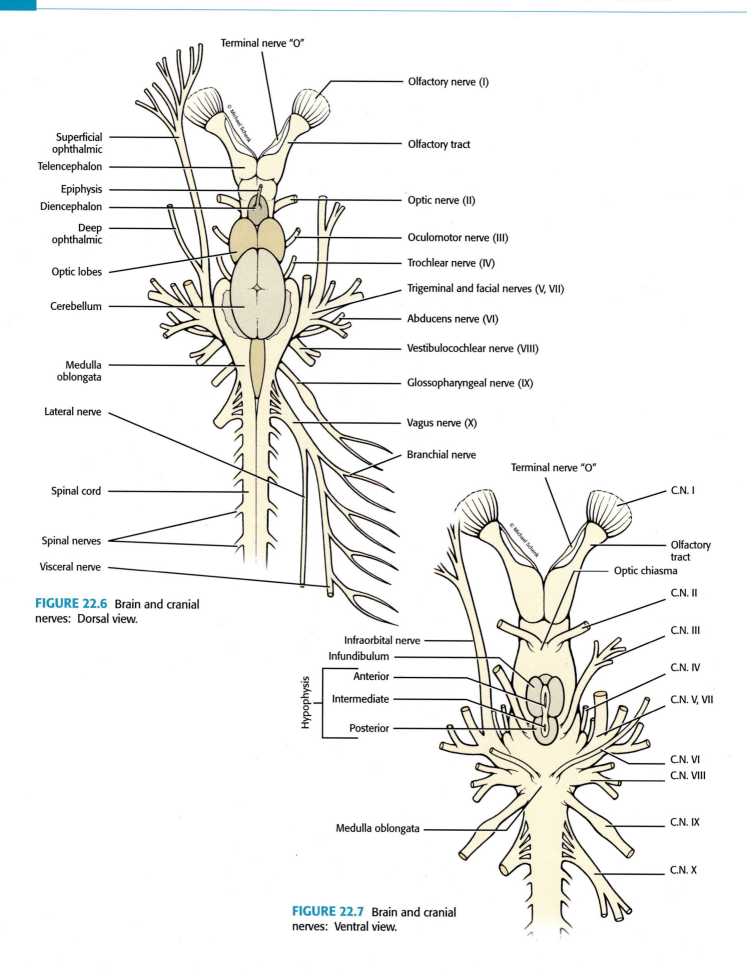

Terminal nerve "O"

Olfactory nerve (I)

Superficial ophthalmic

Olfactory tract

Telencephalon

Epiphysis

Optic nerve (II)

Diencephalon

Deep ophthalmic

Oculomotor nerve (III)

Trochlear nerve (IV)

Optic lobes

Trigeminal and facial nerves (V, VII)

Cerebellum

Abducens nerve (VI)

Vestibulocochlear nerve (VIII)

Medulla oblongata

Glossopharyngeal nerve (IX)

Lateral nerve

Vagus nerve (X)

Branchial nerve

Spinal cord

Spinal nerves

Visceral nerve

FIGURE 22.6 Brain and cranial nerves: Dorsal view.

Terminal nerve "O"

C.N. I

Olfactory tract

Optic chiasma

Infraorbital nerve

Infundibulum

C.N. II

Hypophysis

Anterior

C.N. III

Intermediate

C.N. IV

Posterior

C.N. V, VII

C.N. VI

C.N. VIII

Medulla oblongata

C.N. IX

C.N. X

FIGURE 22.7 Brain and cranial nerves: Ventral view.

THE BRAIN: VENTRAL VIEW

 To observe the brain in ventral view, the mandible is reflected laterally so the roof of the oral cavity is exposed.

Cut through the oral membrane of the roof and remove it, exposing the ventral surface of the chondrocranium. Carefully shave the cartilage, beginning at the anterior end, first exposing the olfactory bulbs and tracts and the optic chiasma. Proceed posterior to the chiasma, being careful in the obvious bulged area (sella turcica), as the hypophsis and infundibulum are attached firmly to the area by connective tissue. Exposing the brain posterior to the sella is advisable so the cartilage around the sella can be carefully shaved and chipped away from the hypophysis and infundibulum.

The **infundibulum**, projecting from the hypothatlamus, connects the hypothalamus to the **hypophysis** (pituitary gland). The **optic nerves** (cranial nerve II), carry visual impulses from the retina of the eyes to the **optic chiasma**, the point at which the neurons in the nerves cross to opposite sides and continue as the optic tracts, which travel through the diencephalon into the roof of the optic lobes or the optic tectum (Figure 22.5 and Figure 22.7).

Ventricles of the Brain

Because the brain in vertebrates evolved as a specialization of the anterior end of the dorsal hollow nerve cord, it should not be surprising to learn that the brain possesses cavities or ventricles. Vertebrates have four ventricles, continuous with each other and the spinal cord. **Ventricles I** and **II** are located within the left and right cerebral hemispheres, respectively. A narrow canal, the foramen of Monro, occurring in the diencephalon, connects the lateral ventricles (I and II) to **ventricle III**. The narrow **aqueduct of Sylvius** connects ventricle III to **ventricle IV** in the medulla oblongata (Figure 22.8).

In anamniotes (fish and amphibians), other portions of the brain contain cavities as well. A pair of cavities known as the **optic ventricles** occur in the optic lobes of the mesencephalon. These cavities, in turn, communicate posteriorly with a **cerebellar cavity** and anteriorly with the third ventricle. The aqueduct, in passing from the third ventricle to the fourth ventricle, also communicates with the cavities of the mesencephalon and the cerebellum.

Cerebrospinal Fluid Circulation and Meninges

Like the lamprey and most other fishes, the surfaces of the brain and spinal cord of the shark are covered by a delicate protective covering of tissue, the **primitive meninx**, which appears as a spidery, often blackish, surface layer.

The tela choroidea, consisting of thin, highly vascularized tissues, forms the roof of the ventricles. The choroid plexus, or vascularized portion of the tela choroidea, secretes a thin cerebrospinal fluid that circulates primarily within the cavities of the brain and spinal cord. Only a small proportion of the cerebrospinal fluid circulates within the space

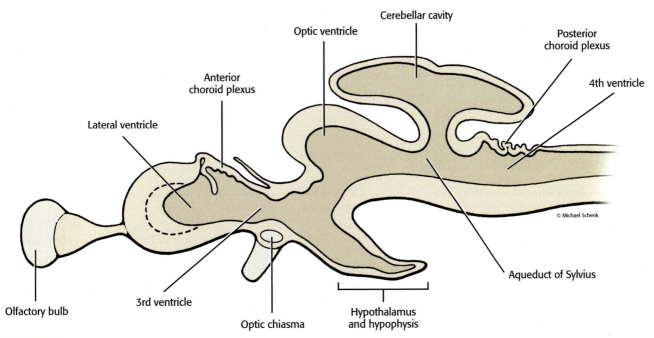

FIGURE 22.8 Ventricles of the brain.

surrounding the central nervous system. This space occurs between the brain and spinal cord and their covering, the meninx or meninges.

The Occipital Nerves, Hypobranchial Nerve, Spinal Nerves, and Spinal Cord

Just posterior to the **vagus nerve,** observe two or three occipital nerves arising from the posterolateral aspect of the medulla oblongata. These join with the first few spinal nerves to form a **hypobranchial nerve,** which crosses over the visceral branch of the vagus nerve and turns anteriorly just posterior to the last gill slit to supply the hypobranchial musculature (Figure 22.3). Amniotes have two additional cranial nerves—the spinal accessory (XI), derived from a portion of the vagus, and the hypoglossal (XII), a homolog of the hypobranchial nerve—bringing the total to 12 cranial nerves.

TABLE OF CRANIAL NERVES

Name	Number	Neuron Type Sensory	Neuron Type Motor	Distribution
Terminal**	O	X		Neurosensory cells of nasal epithelium
Olfactory	I	X		Neurosensory cells of nasal epithelium
Optic	II	X		Sensory fibers of retina
Oculomotor	III	†	X	Innervates the following extrinsic eye muscles: dorsal rectus, ventral rectus, medial rectus and ventral oblique muscles. Innervates intrinsic ciliary muscles.
Trochlear	IV	†	X	Innervates an extrinsic eye muscle, the dorsal oblique muscle.
Trigeminal	V	X	X	This nerve consists of four branches, all associated with the facial and jaw regions. Ophthalmic branches Deep: innervates the skin in region of eye and rostrum and eyeball Superficial: skin of the dorsal and lateral head region Maxillary branch: innervates skin of upper jaw and ventral part of the rostrum Mandibular branch: innervates skin and muscles of lower jaw
Abducens	VI	†	X	Innervates an extrinsic eye muscle, the lateral rectus
Facial	VII	X	X	This nerve consists of four branches, all associated with the facial and jaw regions. Superficial Ophthalmic: lateral line system of dorsal and lateral portion of the head Buccal: Associates with the maxillary branch of the Trigeminal Nerve (V) as the *infraorbital nerve* and carries sensory impulses from the lateral line and ampullae of Lorenzini of the upper jaw and ventral portion of the rostrum Hyomandibular: innervates hyoid muscles, carries impulses from the lateral line system of lower jaw, oral lining and tastebuds Palatine: carries impulses from the lining of the mouth and taste buds
Vestibulocochlear	VIII	X		Carries impulses from the saccule, utricle and semicircular canals of the inner ear
Glossopharyngeal	IX	X	X	There are four branches of this nerve. Dorsal: carries sensory impulses from the lateral line and skin Pretrematic and Postrematic, associated with structures of the first gill pouch: carries sensory and sensory and motor impulses, respectively Pharyngeal: carries sensory impulses from the pharynx
Vagus	X	X	X	There are two major branches of this nerve. Visceral: carries sensory impulses from and motor impulses to the remaining four gill pouches via Pretrematic, Postrematic and Pharyngeal branches, heart and abdominal viscera Lateral: carries sensory impulses from the lateral line system of the trunk and tail

** After the identification of cranial nerves I–X, a small inconspicuous nerve, the terminal nerve (cranial nerve O), lying medial to the olfactory tract, was discovered. It appears to carry impulses from the olfactory sac to the cerebrum. Its function is unknown but has been associated with chemosensory functions.

† Although cranial nerves III, IV, and VI are considered motor nerves, they carry proprioceptive fibers, which are sensory and transmit information concerning muscle status.

Note: Some anatomists consider lateral line nerves to be cranial nerves distinct from the facial (VII) and vagus (X) nerves.

Carefully chip away the cartilage of the neural arches of the vertebrae to expose the spinal cord, which is continuous with the medulla oblongata. Serially arranged spinal nerves continue to arise from the spinal cord, innervating the corresponding posterior segments. These are buried deep within the body musculature.

Cervicobrachial and Lumbosacral Plexus

The ventral roots of the spinal nerves in the region of the paired appendages are somewhat united to form plexi, loose networks of interconnected nerves. The muscles of the pectoral fin are innervated by the cervicobrachial plexus, and the muscles of the pelvic fin are innervated by the lumbosacral plexus. These are difficult to dissect.

AUTONOMIC NERVOUS SYSTEM

The autonomic nervous system of the shark, which is not subdivided into sympathetic and parasympathetic divisions, consists of a double chain of ganglia that courses along the dorsal aspect of the body wall. This system is difficult to dissect.

SENSE ORGANS

The sense organs discussed next are the olfactory apparatus, the lateral line system, the auditory and equilibrium apparatus, and the eye and extrinsic eye muscles.

Olfactory Apparatus

The sense of smell in sharks is well developed and is a major tool employed to locate food as well as engage in social behavior. Water currents enter through the lateral **incurrent opening** of the nostril, pass over the olfactory epithelium of the olfactory sac which is folded into a complex series of **olfactory lamellae,** and exit through the medial **excurrent opening** of the nostril (Figure 22.9). Olfactory receptors reside in the epithelium. During your dissection of the brain, you probably will have exposed this apparatus, consisting of the **olfactory sac, olfactory bulb,** and **olfactory tract,** connected to the cerebral hemispheres (Figure 22.10).

Lateral Line System

The lateral line system is an ancient sensory system that already was present in the first recognizable vertebrates, the "ostracoderms." It apparently evolved to permit aquatic vertebrates to detect water movement, allowing them to orient themselves within a dynamic environment. A wide variety of information is available through this seemingly simple system, such as whether water flow is from head to tail or vice versa, direction of the disturbance in the water, size of the object making the disturbance, whether the object is approaching, receding or stationary, and so on. A lateral line system is found among fishes and larval amphibians and a few aquatic adult amphibians, but not among amniotes.

Externally, the **lateral line,** appearing as a series of small openings, extends from just posterior to the eye to the tail (Figure 22.11). The anterior portion of the lateral line system, located on the dorsal and ventral regions of the snout and jaws has evolved in sharks into a highly specialized sensory mechanism, the **ampullae of Lorenzini** (Figure 22.12). The function of these organs is to detect minute electrical currents such as those produced by living muscle. Therefore, the shark can detect even prey that are camouflaged by sand and are visually invisible.

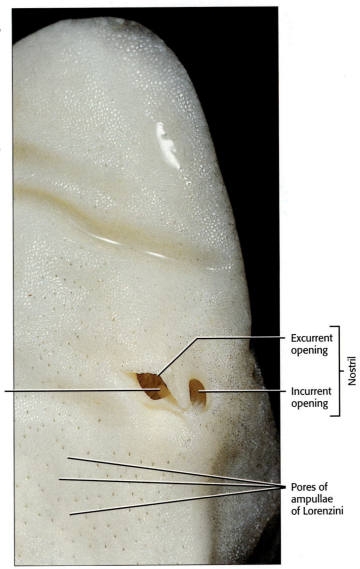

Olfactory lamellae

Excurrent opening

Incurrent opening

Nostril

Pores of ampullae of Lorenzini

FIGURE 22.9 Nostrils.

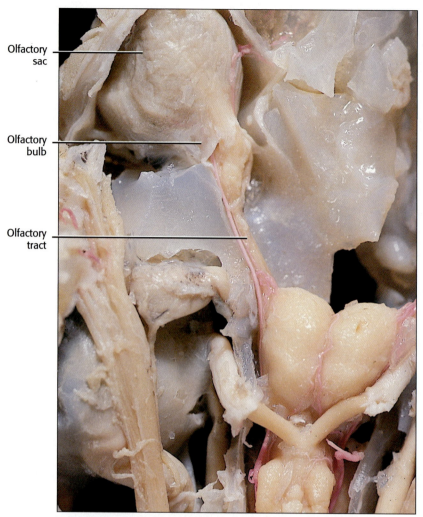

Olfactory sac

Olfactory bulb

Olfactory tract

FIGURE 22.10 Olfactory apparatus: Ventral view.

Auditory and Equilibrium Apparatus

The auditory and equilibrium apparatus in vertebrates is thought to have evolved from an anterior portion of the lateral line system. In sharks, this mechanism operates primarily as an equilibrium apparatus. Whether sharks can "hear" is not clear. During the dissection of the brain and cranial nerves, this apparatus may have been exposed.

To expose this complex, you must carefully shave thin layers of the cartilage of the otic capsule of the chondrocranium, constantly watching for the semicircular ducts, lying in a space shaped similar to that of the apparatus. The first hint will be a clear tube-like space or canal of the cartilaginous inner ear labyrinth, containing the semicircular ducts, and appearing just below the surface of the surrounding opaque cartilage of the chondrocranium (Figure 22.13).

Continue to cautiously remove thin slivers of cartilage dorsally and laterally, exposing the three semicircular ducts—the anterior, the posterior and the lateral.

Lateral line pores (small white dots)

FIGURE 22.11
Lateral line.

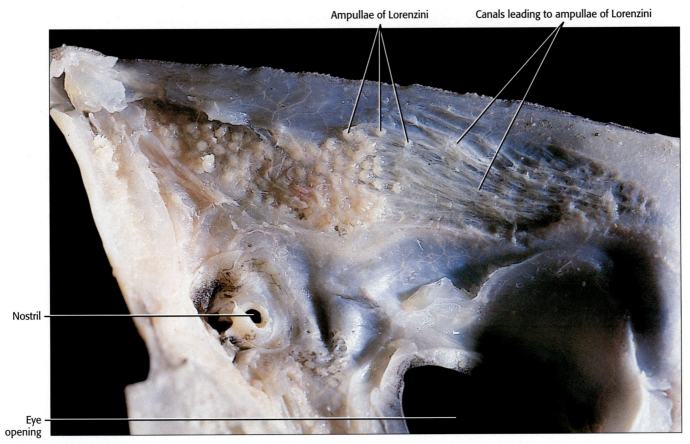

FIGURE 22.12 Ampullae of Lorenzini on undersurface of skin.

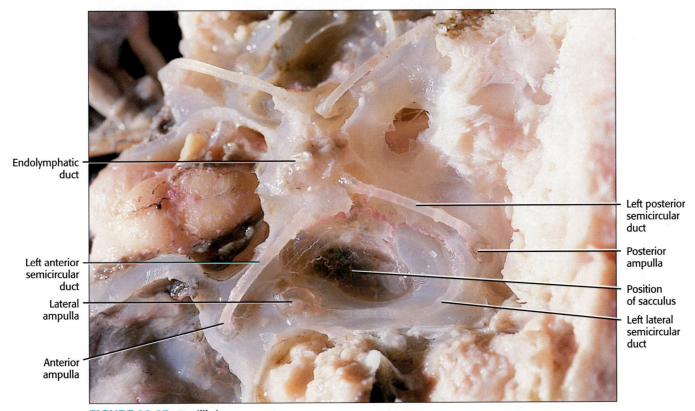

FIGURE 22.13 Equilibrium apparatus.

Recall that the lamprey has two semicircular ducts, but the hagfish only one. However, all gnathostomes possess three, oriented similar to those of the shark. These ducts are tubelike and are semicircular in shape. A saclike swelling, the **ampulla**, occurs at the ventral end of each of the semicircular ducts at their point of attachment to the utriculus. The anterior and posterior semicircular ducts are primarily vertical, whereas the lateral semicircular duct is horizontal in position. The rest of the complex consists of an **anterior** and a **posterior utriculus** associated with the anterior and lateral semicircular ducts and the posterior semicircular duct, respectively. A baglike **sacculus** is attached to the anterior utriculus (Figure 22.13).

A posterior bulge of the wall of the sacculus, the lagena, is the forerunner of the later cochlea duct, associated with "hearing" in mammals (Figure 22.13). Leading from the sacculus is the **endolymphatic duct**, which opens to the outside through the **endolymphatic pore** (Figure 22.14 and Figure 22.15). The cartilaginous labyrinth is filled with a fluid known as perilymph, and the semicircular ducts,

utriculus, and sacculus are filled with a fluid known as endolymph. Although we have discussed this complex as individual parts, we stress that the entire apparatus is interconnected, and if one of the parts is nicked, the entire apparatus will collapse, as the endolymph will escape.

The fibrous tufts of the vestibulocochlear nerve receive impulses from cristae in the ampullae, maculae in the anterior and posterior utriculus, the lagena, and the sacculus. The cristae and maculae consist of groups of cells that transmit sensory information to the shark's brain. Movement of the endolymph in the utriculus, lagena, and sacculus convey information of linear acceleration and position of the head, and movement of endolymph in the semicircular ducts conveys turning movements of the body.

Eye and Extrinsic Eye Muscles

The anatomy of vertebrate eyes is highly similar, with some differences in the physiology of the eye. The visual sense becomes important to the shark during close encounters in its daily life, feeding, schooling, mating, and the like.

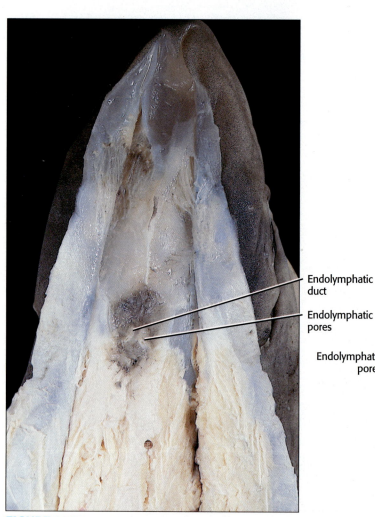

Endolymphatic duct

Endolymphatic pores

Endolymphatic pores

FIGURE 22.14 Endolymphatic ducts and pores.

FIGURE 22.15 Endolymphatic pores.

External Structures of the Eye

 Your instructor may direct some students in the class to dissect the shark's eye to study the extrinsic eye muscles, optic nerve, and other structures, and a second group may dissect a shark's eye to observe the internal anatomy. Obviously, both groups must study both preparations.

To remove the eyeball, cut each extrinsic eye muscle approximately 1.25 cm from its insertion on the eyeball. Sever the optic nerve and remove the eyeball from the orbit, keeping its correct orientation with respect to right and left. To facilitate study of the eye muscles, carefully remove enough white connective tissue, avoiding removal of any pink, tan, or gray structures that may be muscles of the eyeball. The optic nerve, which is white and is located on the back and near the center of the eyeball, must be conserved. As you remove the eyeball, you will notice that the shark's eye is supported by a cartilaginous pedicel. It is shaped like a small golf tee and is not found in most vertebrates (Figure 22.16).

All vertebrates possess six extrinsic eye muscles—a ventral oblique, a dorsal oblique, an anterior rectus, a posterior rectus, a dorsal rectus, and a ventral rectus. To identify these muscles correctly, it is imperative to determine whether the eyeball is left or right. A key to this determination is to identify the oblique muscles. The obliques originate and insert anteriorly on the eyeball. The **ventral oblique** inserts anteroventrally on the eyeball and near the insertion of the **ventral rectus muscle**. Identify the **dorsal oblique muscle**, whose insertion is anterodorsal and near the insertion of the **dorsal rectus**. The insertion of the **anterior rectus muscle** appears anteriorly between the insertion of the obliques and opposite the insertion of the **posterior rectus muscle** posteriorly (Figure 22.17 and Figure 22.18).

Sharks have immovable **eyelids**. Therefore, muscles found in some other vertebrates with movable eyelids are not found in the shark (Figure 22.19).

Eyeball moisture for most terrestrial vertebrates affords a well-lubricated surface to assure that the surface remains clean and can function as part of the lens system. To provide this moisture, glands are present. Because sharks are aquatic vertebrates, they do not have glands to bathe the eye.

Internal Structures of the Eye

To prepare a specimen that is superior for the study of internal structures, insert the tip of a sharp scalpel into the eyeball and make a small slit approximately 0.5 cm from the posterior edge of the cornea, and with scissors continue to cut around the circumference of the eyeball to

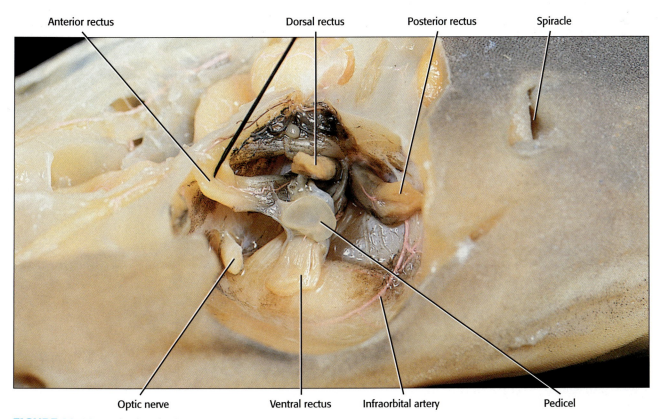

Anterior rectus   Dorsal rectus   Posterior rectus   Spiracle

Optic nerve   Ventral rectus   Infraorbital artery   Pedicel

FIGURE 22.16 Pedicel of the left eye.

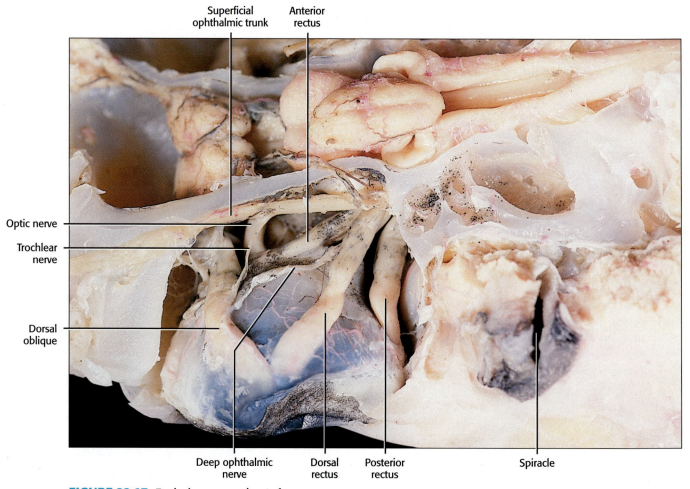

Superficial ophthalmic trunk

Anterior rectus

Optic nerve

Trochlear nerve

Dorsal oblique

Deep ophthalmic nerve

Dorsal rectus

Posterior rectus

Spiracle

FIGURE 22.17 Extrinsic eye muscles: Left eye.

separate it into anterior and posterior parts (Figure 22.20).

A gelatinous substance, the **vitreous humor**, occurs between the lens and the retina. It functions in holding the retina and lens in place and serves as a refractive medium as part of the lens system of the eye. Between the cornea and the ciliary body is the **aqueous humor**. It keeps the structures of the eye in position and serves as part of the lens system of the eye. Further, it supplies the structures in this area with nutrients and oxygen while removing metabolites.

Three distinct concentric tunics or layers form the eyeball. In the posterior end of your preparation, identify the three layers (Figure 22.19 and Figure 22.20).

1. The outer layer is a tough white, fibrous coat, the opaque **sclera**. In the shark, the sclera is primarily cartilaginous. Over the anterior surface of the eyeball, the sclera is modified as a transparent window called the **cornea**.

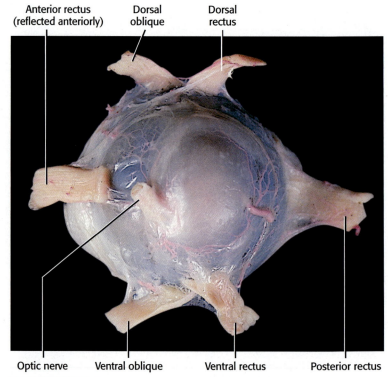

Anterior rectus (reflected anteriorly)

Dorsal oblique

Dorsal rectus

Optic nerve

Ventral oblique

Ventral rectus

Posterior rectus

FIGURE 22.18 Extrinsic eye muscles II: Right eye.

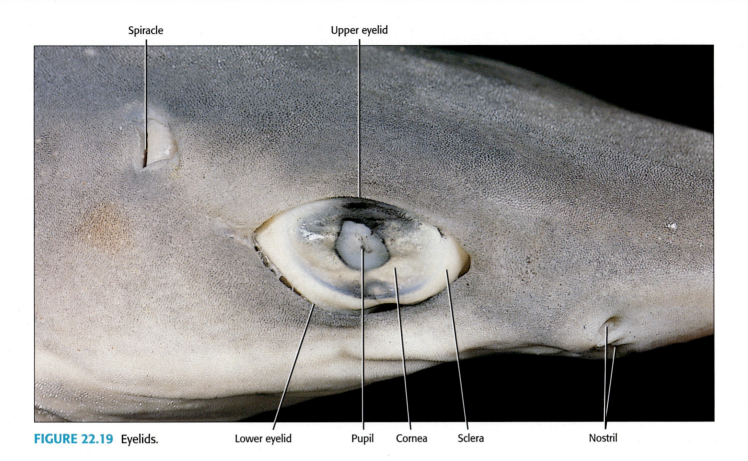

FIGURE 22.19 Eyelids.

Spiracle

Upper eyelid

Lower eyelid Pupil Cornea Sclera Nostril

Ciliary body Iris Pupil Retina (inner layer — notice fold) Choroid (beneath retina)

Sclera Lens Optic disk

FIGURE 22.20 Internal eye anatomy.

2. The middle layer, the **choroid**, is heavily pigmented and appears black.

3. The **retina,** the innermost tunic, consists of an outer pigmented layer abutting the choroid and an inner neural layer abutting the vitreous humor.

The neural layer will appear creamy, folded, and displaced from its normal position as a result of preservation, whereas the pigmented layer typically adheres to the choroid. Concentrations of reflective crystals in the choroid characterize a **tapetum lucidum,** common in vertebrates that are active in subdued light, including the shark.

Have you noticed that cat eyes, when suddenly illuminated by incidental light in the dark, appear yellow or green? This is light reflecting from the surface of the tapetum lucidum. When the sheep eye is dissected, the tapetum will be obvious, but in the shark this reflective area may be masked by adjacent pigments, depending upon whether the eye is light-adapted or dark-adapted. Therefore, if light-adapted, it will not be obvious.

Look closely into the posterior half of the eyeball and observe a small light dot where the retina is attached to the sclera. This point of attachment is the **optic disk**. The optic disk is the area known as the "blind spot," as it has no photoreceptors, rods, or cones. The blind spot is the site through which the neurons of the nervous layer of the retina leave the eyeball and enter the optic nerve. Observe that the optic nerve is located exactly opposite the optic disc (Figure 22.20).

Anteriorly, the choroid is modified as the **iris** and the **ciliary body**. An opening in the iris, the **pupil**, permits passage of light. Delicate **zonule fibers** of the suspensory ligament extend from the ciliary body to the lens.

Among terrestrial vertebrates, the cornea strongly refracts (bends) light rays as they pass through it. Therefore, it is the main focusing element of the lens system of the eye. The lens does the fine-focusing. Because the cornea and water have approximately the same refractive index (focusing power), the cornea causes no refraction of light rays in fishes. The lens in fishes is the major focusing structure and, consequently, is spherical, in contrast to the lenses of terrestrial vertebrates, which are much less convex.

 Remember that the more convex a lens is, the more strongly it refracts light rays. In life, the lens is transparent, but in preserved sharks the lens may appear opaque.

CHAPTER TWENTY-THREE

Dogfish Shark — Endocrine System

A lthough the endocrine organs and tissues are known collectively as a system, they are generally not contiguous. The functional units of the endocrine system are called glands and they are ductless.

Secretions of these glands, called hormones, act as regulatory chemicals, often affecting the functions of target tissues or other glands in the system. In contrast to the excretion of exocrine or ducted glands, hormones are transported by the circulatory system to their target tissues, where they bind to specific receptor sites of the target tissues causing their characteristic activities.

HYPOPHYSIS OR PITUITARY

This significant endocrine gland is located on the ventral side of the brain. Anatomically, it consists of an anterior **adenohypophysis**, an **intermediate lobe**, a **ventral lobe**, and the **posterior hypophysis**. The embryology of this essential gland is interesting in that the neurohypophysis, or posterior hypophysis, develops from an evagination of the diencephalons, whereas the adenohypophysis and intermediate lobe are derived from an evagination of the rostral gut. The close proximity of these lobes results in what appears as a relatively compact mass. The hypophysis is an organ that produces some hormones of its own and also stores other hormones produced by cells in the hypothalamus. These hormones affect a myriad of organ activities.

In the shark, the hypophysis is attached to the **infundibulum**, extending from the hypothalamus of the diencephalon. The infundibulum consists of a pair of anterior **inferior lobes** and a pair of thin-walled **vascular sacs** (Figure 23.1).

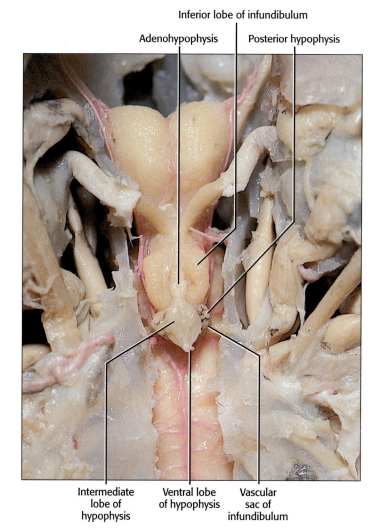

Inferior lobe of infundibulum
Adenohypophysis
Posterior hypophysis
Intermediate lobe of hypophysis
Ventral lobe of hypophysis
Vascular sac of infundibulum

FIGURE 23.1 Hypophysis.

Pineal Gland or Epiphysis

The primitive pineal gland appears as a slender stalk of tissue attached to the roof of the diencephalon. Its function is not well understood but probably is light-sensitive and may play a role in pigmentation and reproduction (Figure 23.2).

 Take care not to destroy this organ in the process of exposing the brain.

Thyroid Gland

In the shark, the thyroid gland appears as a grayish pink mass of tissue lying between the coracomandibularis and coracohyoid muscles (Figure 23.3). The function of this gland is general tissue metabolism.

Thymus Gland

Masses of tissue located deep above most of the gill slits may be thymus tissue. Like other thymus tissue of tetrapods, the thymus plays a role in establishing the immune system.

Adrenal Gland

In the shark, the adrenal gland consists of chromaffin bodies (medulla) and an interrenal gland (cortex). Contrary to what may be observed in mammals, which have a well-formed gland, in the shark these two tissues consist of separate isolated masses between the kidneys.

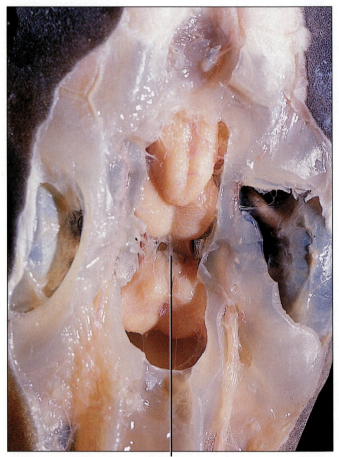

Pineal gland

FIGURE 23.2 Pineal gland.

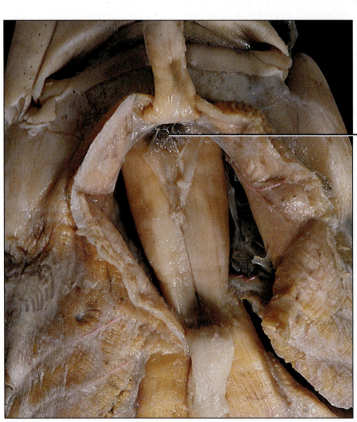

Thyroid gland

FIGURE 23.3 Thyroid gland.

The adrenal medulla tissue secretes epinephrine and norepinephrine, intensifying and prolonging the characteristic "fight or flight" syndrome of the sympathetic nervous system, whereas the adrenal cortex secretes adrenocorticoids, which affect metabolic activities. These tissues are difficult to find.

Pancreas

In the shark, the pancreas consists of two distinct lobes connected by an isthmus—the ventral lobe, closely applied to the ventral surface of the duodenum, and a dorsal lobe located between the dorsal surface of the duodenum and the greater curvature of the stomach (Figure 23.4). The major mass of tissue in this gland is exocrine and is concerned with production of digestive enzymes and buffering fluid, which are released into the small intestine. Clusters of endocrine tissue secrete hormones associated with carbohydrate metabolism.

Ovaries

The ovaries, the primary female sex organs, are best observed in mature specimens. They are located dorsal to the liver (Figure 23.5). In addition to oocytes, the ovaries produce hormones that stimulate and maintain sexual maturity and promote reproductive behavior.

Testes

The testes, the primary male sex organs, are best observed in mature specimens. They are located dorsal to the liver (Figure 23.6). In addition to sperm, the testes produce hormones that stimulate and maintain sexual maturity and promote reproductive behavior.

Other Endocrine Tissues

Tissues in other organs (e.g., the kidneys and small intestine) produce hormones that affect local physiologies in those organs or related organs in the same system.

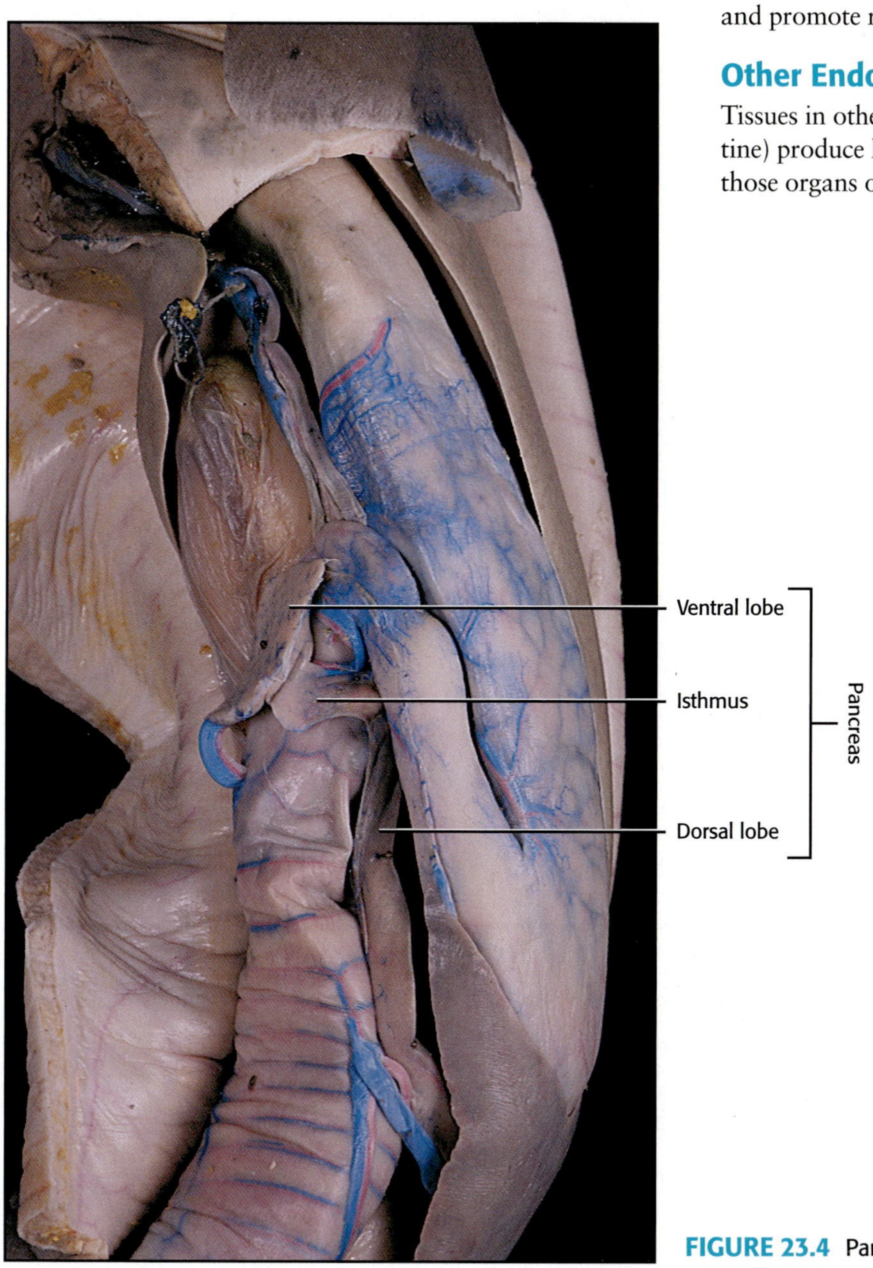

— Ventral lobe
— Isthmus
— Dorsal lobe
Pancreas

FIGURE 23.4 Pancreas.

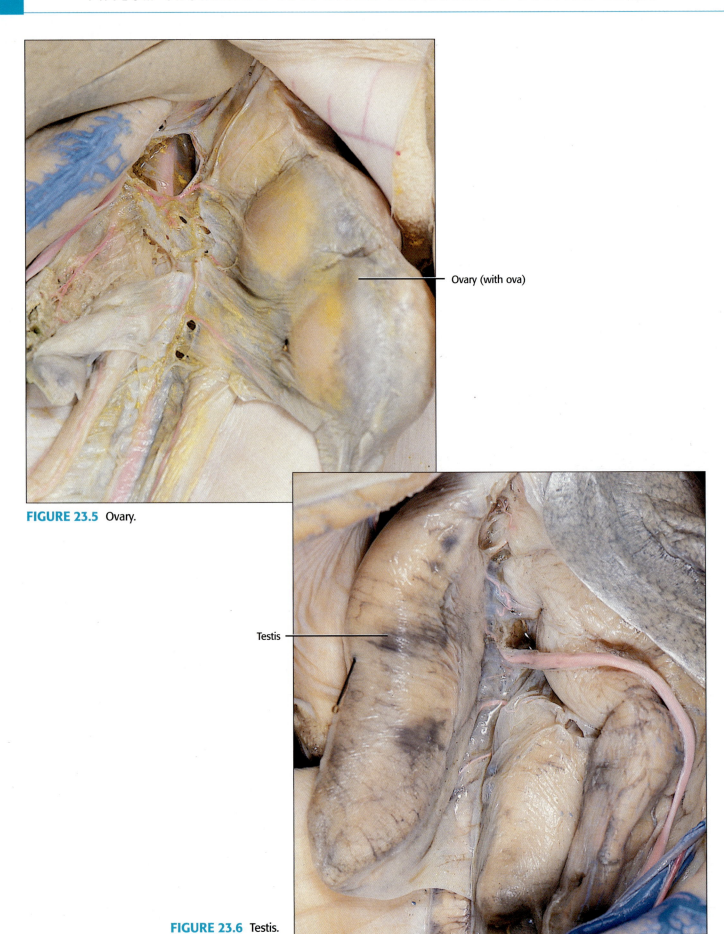

Ovary (with ova)

FIGURE 23.5 Ovary.

Testis

FIGURE 23.6 Testis.

Phylum Chordata / Subphylum Vertebrata

Mudpuppy
(*Necturus*)

The first tetrapods were tailed amphibians. Salamanders, members of the class Amphibia, order Caudata, are thought to represent the least modified members of this oldest group of tetrapods. Salamanders are tailed amphibians with unspecialized appendages held and used in what appears to be the primitive mode. Watching a salamander move, one will notice that the body moves in a side-to-side motion that is essentially characteristic of fishes swimming. The major difference when this movement occurs on land is that the appendages are used to brace the body and prevent it from being propelled sideways willy-nilly.

Necturus is a permanently aquatic member of the amphibian order Caudata. Like several other species of salamanders, it is a larval vertebrate that has become capable of reproduction. In salamanders, delayed maturation of the body, in contrast to early development of the reproductive system, seems to be the reason for their conditon (Gould, 1977). Sometimes, during the course of evolution, a shift in developmental timing occurs, so that some organs or systems develop at a rate different from that of other systems. Such a shift in timing is called heterochrony and is seen in several salamanders. This phenomenon is known as neoteny, a form of paedomorphosis. In the mudpuppy, we find larval characteristics such as external gills, gill slits, no eyelids, no external ear, and a lateral line system. Some internal anatomy also belies their larval condition. Because *Necturus* combines interesting larval and adult characteristics and is one of the largest readily available salamanders, it makes a good specimen for dissection.

Mudpuppy — External Anatomy and Integumentary System

24

EXTERNAL ANATOMY

The body of the mudpuppy consists of a **head**, delimited posteriorly by the **gular fold**, a **trunk**, extending to the posterior border of the **cloacal opening** and a **tail** extending from the posterior border of the cloaca to the tip. Observe that the laterally flattened tail is specialized as a propulsive organ. While swimming by lateral movement of the tail, the legs are held adpressed to the sides of the body. Notice that the legs are short and the number of **digits** has been reduced to four on each appendage (Figure 24.1 and Figure 24.2).

The head is flattened dorsoventrally with a terminal **mouth** bounded by **lips** (Figure 24.3). Perhaps the head shape is related to its feeding habits, which include capturing small fish, crayfish, and mollusks, or it may generate lift during swimming. Its small, laterally placed **eyes** are lidless. Notice the small **external naris** just above the margin of the upper lip (Figure 24.4).

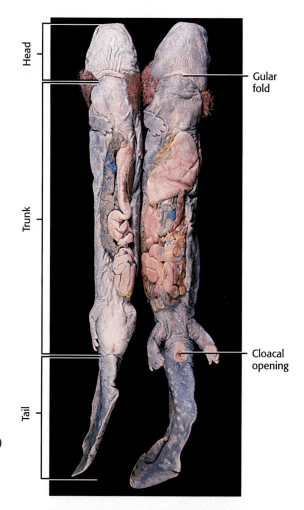

FIGURE 24.1 *Necturus* external features: Male (left) and female (right).

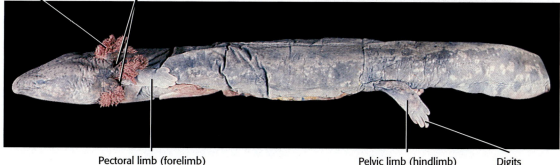

FIGURE 24.2 External features.

213

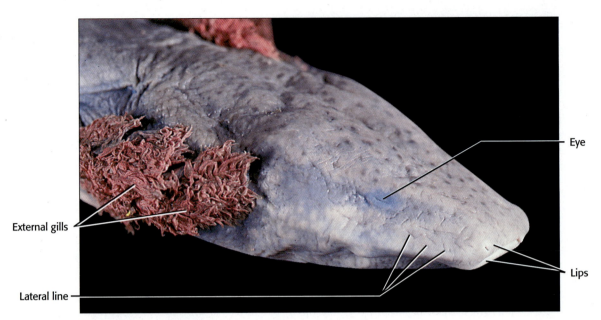

FIGURE 24.3 External features: Head region, dorsal view.

Perhaps one of the most noticeable characteristics consists of the three pairs of lateral **external gills**, whose bases are separated by two pairs of **gill slits** (Figure 24.2). A **lateral line system** appears as muted indentations in the vicinity of the head and trunk (Figure 24.4).

INTEGUMENTARY SYSTEM

As in all vertebrates, the skin consists of two layers—the outer multilayered (stratified) **epidermis** lying on top of a **dermis** and separated from it by a noncellular **basement membrane**. The deepest layer of the epidermis, consisting of columnar cells, the **stratum germinativum**, gives rise to cell layers that are pushed toward the surface to replace cells lost through normal wear and tear and, in amphibians, shed periodically in sheets. Because the terrestrial environment is arid, causing evaporation, we expect that some mechanism might have evolved in amphibians to prevent excessive cutaneous water loss. Keratinization of outer layers of the epidermis did evolve, producing a **stratum corneum**, especially in adult amphibians. This mechanism, however, apparently does not ensure that water loss will not occur across the skin (Figure 24.5 and Figure 24.6).

Furthermore, the skin of amphibians is highly vascularized and functions as a respiratory organ. Because the skin of amphibians serves as a respiratory organ, it must be kept constantly moist. General **mucous glands**, producing copious amounts of mucus to maintain a moist respiratory surface, and **poison glands**, modified mucous glands responsible for secretion of some of the most virulent known toxins, are both constituents of the skin. These cells are derived from the epidermis but come to lie in the dermis. Perhaps the cutaneous respiratory function of amphibian skin precluded the evolution of antidehydration mechanisms (Figure 24.5 and Figure 24.6).

FIGURE 24.4 Lateral line.

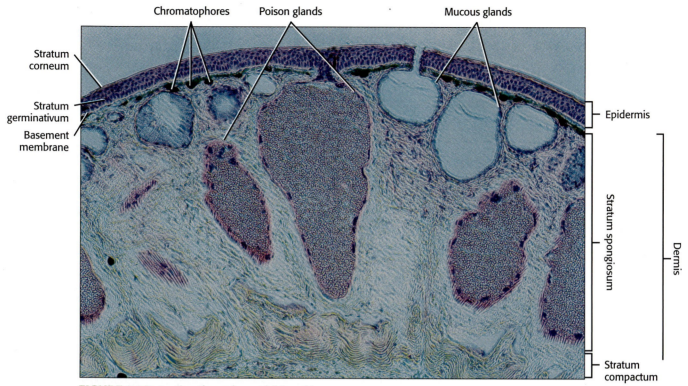

Chromatophores   Poison glands   Mucous glands

Stratum corneum

Stratum germinativum

Basement membrane

Epidermis

Stratum spongiosum

Dermis

Stratum compactum

FIGURE 24.5 Section through amphibian skin.

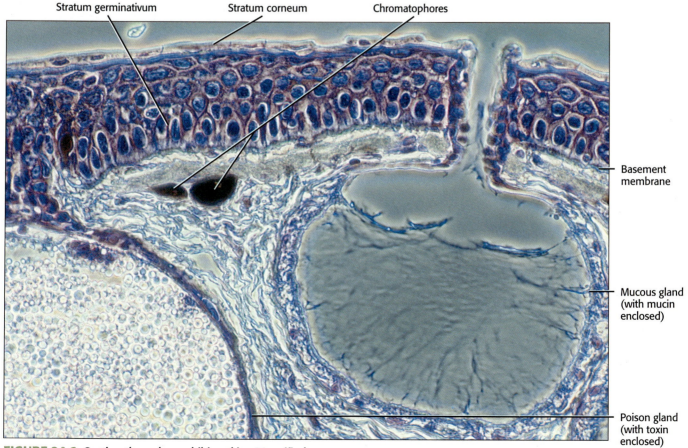

Stratum germinativum   Stratum corneum   Chromatophores

Basement membrane

Mucous gland (with mucin enclosed)

Poison gland (with toxin enclosed)

FIGURE 24.6 Section through amphibian skin: Magnified.

The dermis, also layered, includes an outer **stratum spongiosum**, consisting largely of connective tissue, and an inner **stratum compactum**, more heavily infiltrated with collagen fibers. Various cellular inclusions, blood vessels, nerves, and so on, are found in the dermis. Commonly, **chromatophores** and reflective crystals are present in the dermis, giving amphibians their specific color and skin patterns (Figure 24.5, Figure 24.6, and Figure 24.7).

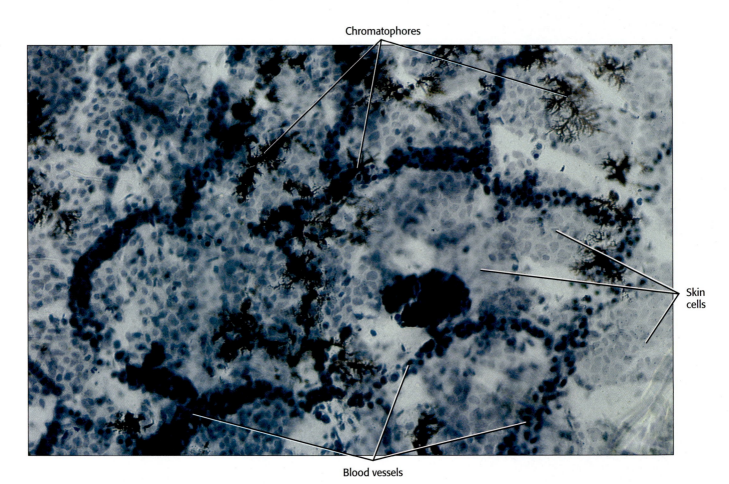

FIGURE 24.7 Skin with chromatophores and blood vessels.

CHAPTER TWENTY-FIVE

Mudpuppy — Skeletal System

25

I n contrast to the cartilaginous skeleton of the shark, the skeleton of *Necturus* consists of bone and cartilage. Like the dogfish shark, however, its skeleton can be subdivided into an axial and appendicular division. Contained within the axial division are the skull, the mandible, the hyoid apparatus, the vertebral column, the sternum, and ribs. Within the appendicular division are the pectoral girdle and forelimb and the pelvic girdle and hindlimb (Figure 25.1 and Figure 25.2).

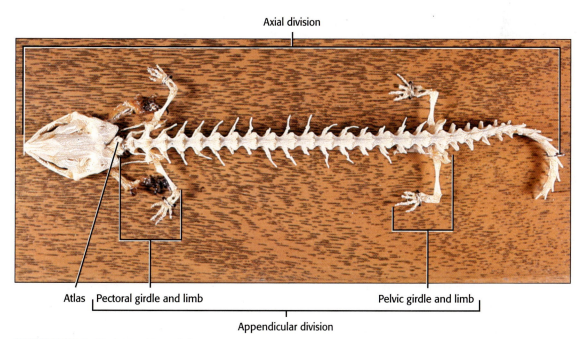

Axial division

Atlas | Pectoral girdle and limb

Pelvic girdle and limb

Appendicular division

FIGURE 25.1 Skeleton: Dorsal view.

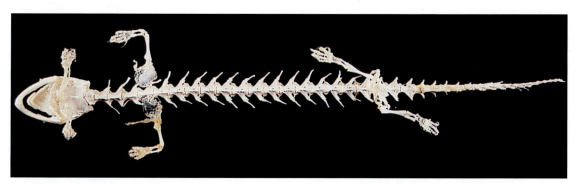

FIGURE 25.2 Skeleton: Ventral view.

AXIAL DIVISION
Skull
Chondrocranium

The skull of amphibians is considered highly specialized among tetrapods. Although a number of larval characteristics make neotenic amphibians such as *Necturus* poor representatives of the ancestral adult condition, the study of *Necturus* is valuable to obtain an overall appreciation of a tetrapod skull.

The skull in *Necturus,* as in most bony vertebrates, consists of an embryonic cartilaginous chondrocranium that has been overlain by bone, largely obscuring the chondrocranium. Bones derived from the splanchnocranium also contribute to the skull.

The **chondrocranium**, or braincase, supporting the brain and associated sense organs develops from a series of cartilages and ossified sites. The **trabeculae**, a pair of elongate parallel bars, fuse anteriorly, forming the horizontal **ethmoid plate**, leaving a pair of short **trabecular horns** projecting anteriorly from the ethmoid plate. As the trabeculae converge to form the ethmoid plate, a loosely articulated pair of **antorbital cartilages** project laterally (see Figure 25.3).

Posteriorly, paired **parachordal bars** extend from the trabeculae, forming plates associated laterally with the **quadrate cartilage and bone** (portions of the mandibular arch, visceral arch I of the splanchnocranium) and medially with the **otic capsules** surrounding the inner ear. The otic capsules are united by means of an occipital series consisting of a ventral **basioccipital** and paired **exoccipital bones**,

bearing the **occipital condyles** for articulation with the atlas (Figure 25.11).

The otic capsules are united dorsally by means of a **synotic tectum** (roof uniting the otic capsules). An anterior **prootic** and posterior **opisthotic bone** are associated with the otic capsule. The **fenestra ovalis**, or **oval window**, pierces the cartilage between the prootic and opisthotic bone (Figure 25.3).

The columella, a middle ear bone peculiar to the amphibians, consists of the stapes derived from the hyomandibula of the hyoid arch (visceral arch II of the splanchnocranium) and the unique amphibian operculum derived from the otic capsule. It extends between the tympanic membrane (eardrum) and the fenestra ovalis. In other tetrapods, the stapes develops mainly from the hyomandibula. It is interesting that the hyomandibula begins as a jaw "prop" in fish, but in tetrapods becomes a sound transmission organ (see Figure 5.4).

Dermatocranium

The bones of the dermatocranium, of dermal origin, roof the skull and cover the palate, forming **sutures** (immovable joints) between them.

> Dorsally, beginning at the anterior end, study the following bones of the roof of the skull.

The paired, tooth-bearing, V-shaped **premaxillae** form the snout tip. The external nares are associated with these

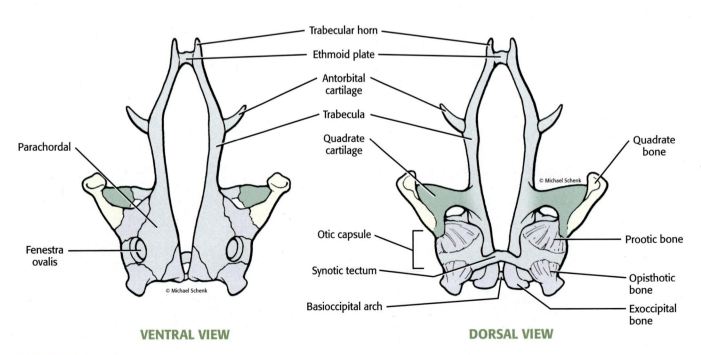

VENTRAL VIEW   **DORSAL VIEW**

FIGURE 25.3 Chondrocranium of *Necturus.*

bones. The paired maxillae, important tooth-bearing bones in most other tetrapods, are absent in *Necturus*. The paired elongate **frontals** extend posteriorly, articulating with the paired **parietals**. The eye orbit lies ventral to the anterior processes of the parietal. Note that the medial arms of the premaxillae overlie the frontals. Lateral to the parietal is a slender **squamosal** articulating with the **quadrate** and the **otic capsule** (Figure 25.4 and Figure 25.5).

On the ventral aspect of the skull, study the following dermal bones of the palate.

The bulk of the palate is the medial **parasphenoid**, extending from the occipital condyles to the tooth-bearing anteriolateral **vomers** posterior to the premaxillae. Note the cartilaginous **ethmoid plate** at the anterior end of the

parasphenoid and between the vomers. Articulating with the vomers are the tooth-bearing **pterygoids**, which also articulate posteriorly with the quadrates (Figure 25.6). The internal nares enter the oral cavity in the vicinity of the pterygoid–vomer suture and are best identified in a preserved animal.

Splanchnocranium

Mandible   The mandible or Meckel's cartilage of the first visceral arch of *Necturus* is covered by three dermal bones. Most of the lateral surface of the mandible is covered by the tooth-bearing **dentary**. The major portion of the medial surface is covered by the **angular**. A smaller tooth-bearing **splenial** is located dorsomedially. The posterior articulating end of the mandible (Meckel's cartilage) remains cartilaginous in *Necturus* but in most tetrapods becomes ossified as the articular, establishing the typical nonmammalian tetrapod jaw articulation between the quadrate of the upper jaw and the articular of the lower jaw. The teeth of *Necturus* are homodont, or all of the same type, occurring as a double row in the upper jaw and a single row in the mandible (Figure 25.7, Figure 25.8, and Figure 25.9).

Hyoid Apparatus   This complex consists of portions of visceral arches II–V. Portions of the hyoid arch (II), paired hypohyals and ceratohyals, make up the major part of this apparatus which supports the tongue (see Figure 5.3A). A basibranchial ties the halves of the hyoid arch together and extends to the halves of the third visceral arch, where a second basibranchial projects posteriorly. Parts of visceral arches III–V, consisting of ceratobranchial and epibranchial cartilages, articulate with the hyoid arch and support the

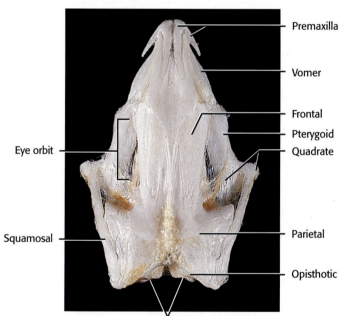

FIGURE 25.4 Skull: Dorsal view.

FIGURE 25.5 Skull: Lateral view.

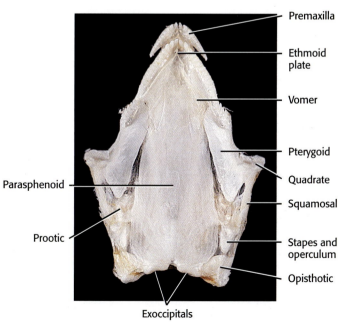

FIGURE 25.6 Skull: Ventral view.

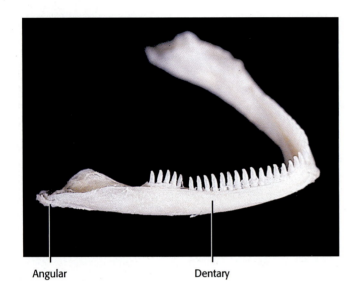

FIGURE 25.7 Mandible: Lateral view.

Angular   Dentary

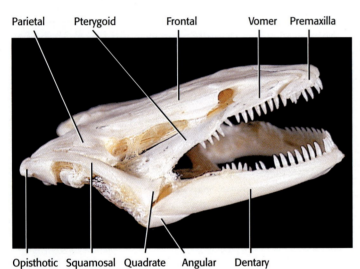

Parietal   Pterygoid   Frontal   Vomer   Premaxilla

Opisthotic Squamosal Quadrate   Angular   Dentary

FIGURE 25.9 Articulated mandible and skull.

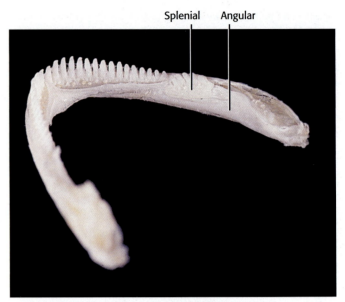

Splenial   Angular

FIGURE 25.8 Mandible: Medial view.

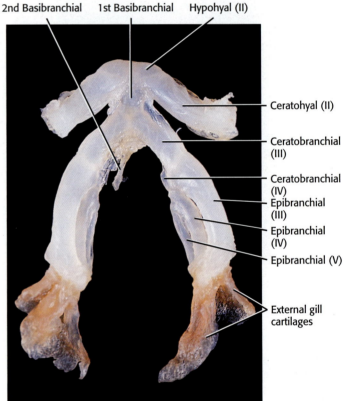

2nd Basibranchial   1st Basibranchial   Hypohyal (II)

Ceratohyal (II)

Ceratobranchial (III)

Ceratobranchial (IV)
Epibranchial (III)
Epibranchial (IV)

Epibranchial (V)

External gill cartilages

FIGURE 25.10 Hyoid apparatus.

three pairs of external gills (Figure 25.10). Two external gill slits occur, one between elements of visceral arches III and IV and the second between visceral arches IV and V.

Vertebral Column

Similar to fishes, the vertebral column of early amphibians continued to consist primarily of trunk and caudal vertebrae. Limited regional differentiation occurred among amphibians. In *Necturus*, the evolution of a single cervical vertebra, the atlas, permitted nodding movements of the head, and a single sacral vertebra linked the pelvic girdle to the vertebral column. More than likely, separation of the head from the pectoral girdle and the evolution of limbs played a pivotal role in vertebral regional differentiation.

Terrestrial locomotion demands a weight-bearing vertebral column, as the body is no longer buoyed by water. In a column with independent vertebrae, the uneven stress of

body weight pulling on it would cause deformation of the vertebral column. Therefore, vertebrae became mutually supportive, with areas of articulation evolving between them. Locomotion among primitive tetrapods continued to be a "swimming" movement on land, but now the limbs helped to stabilize the body somewhat and reduced the lateral body transfer so common among fish. Vertical flexion and extension also are part of terrestrial locomotion. These became more important and obvious among some reptiles and mammals.

Vertebral Regions

 Obtain a prepared skeleton of *Necturus* and study the vertebral column and ribs.

Cervical

A single **cervical vertebra**, the **atlas**, which articulates with the two **occipital condyles** of the skull, allowed the

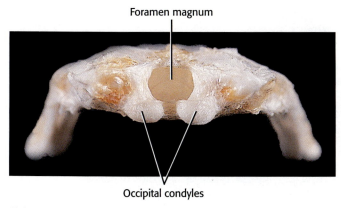

Foramen magnum

Occipital condyles

FIGURE 25.11 Skull: Posterior view.

development of a neck region with head movement independent of the vertebral column (Figure 25.1 and Figure 25.11).

Trunk

Following the atlas are approximately 17 **trunk vertebrae**. Each consists of a biconcave or amphicoelous **centrum** with persistent notochordal tissue occurring between adjacent vertebrae. The spinal cord lies in the vertebral canal formed dorsally by the **neural arches** with the centrum. A **neural spine** projects posteriorly from the arch. At the anterior end of the neural arch is a pair of **prezygapophyses**, and a pair of **postzygapophyses** projects from the posterior end of the neural arch. The prezygapophyses of a vertebra overlap with the postzygapophyses of an anterior, adjacent vertebra. This articulation stabilizes and reduces vertical bending of the vertebral column. A caudally projecting **tranverse process** consists of a dorsal **diapophysis** arising from the neural arch and a ventral **parapophysis** arising from the centrum. Ribs articulate with these processes. Thin wedges of cartilage are found between all adjacent vertebrae (Figure 25.1, Figure 25.12, Figure 25.13, and Figure 25.14).

Sacral

A single **sacral vertebra** with its rib modified for articulation with the pelvic girdle generally occurs in the 19th position (Figure Figure 25.1 and Figure 25.15).

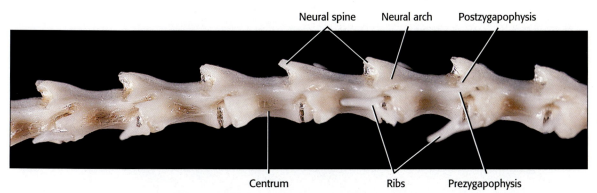

Neural spine   Neural arch   Postzygapophysis

Centrum   Ribs   Prezygapophysis

FIGURE 25.12 Trunk vertebrae: The anterior end is on the right.

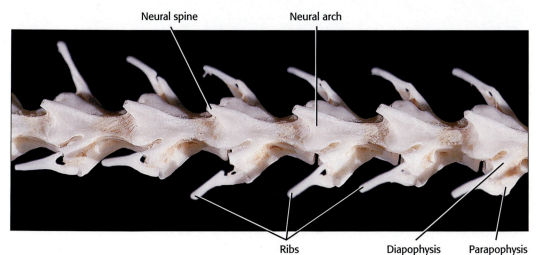

Neural spine   Neural arch

Ribs   Diapophysis   Parapophysis

FIGURE 25.13 Trunk vertebrae: The anterior end is on the right.

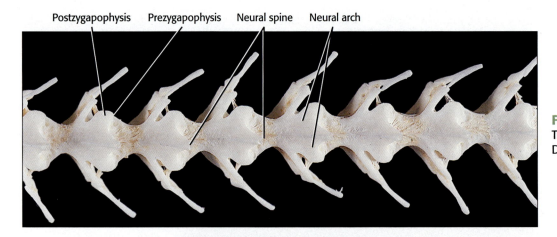

Postzygapophysis   Prezygapophysis   Neural spine   Neural arch

FIGURE 25.14
Trunk vertebrae:
Dorsal view.

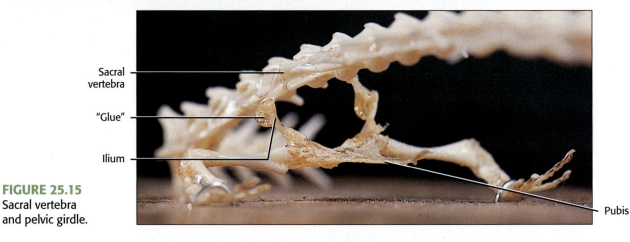

Sacral
vertebra

"Glue"

Ilium

Pubis

FIGURE 25.15
Sacral vertebra
and pelvic girdle.

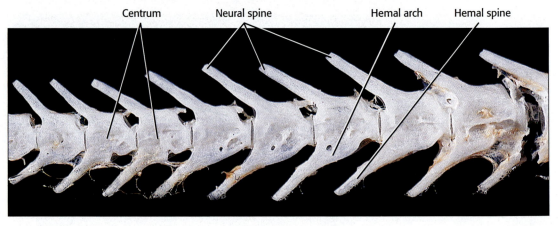

Centrum   Neural spine   Hemal arch   Hemal spine

FIGURE 25.16 Caudal
vertebrae. The anterior
end is on the right.

Caudal

A variable number of **caudal vertebrae** follow the sacral region. The anatomy of these vertebrae is similar to those of the trunk region. The neural spines are somewhat more prominent, the transverse processes dwindle in importance toward the tail tip, and none of the caudal vertebrae has ribs. A notable exception is the **hemal arch** with a spine attached to the ventral surface of the centrum. The caudal artery and vein pass through the arch (Figure 25.1 and Figure 25.16).

Ribs

The tetrapod two-headed or **bicipital ribs** articulate with the transverse processes of the trunk vertebrae. The **tuberculum** articulates with the diapophysis, and the **capitulum** with the parapophysis (Figure 25.13 and Figure 25.17).

Sternum

Although other salamanders have a small cartilaginous sternum, *Necturus* does not. Some small pieces of cartilage embedded in the myosepta of the thoracic musculature may represent its sternum.

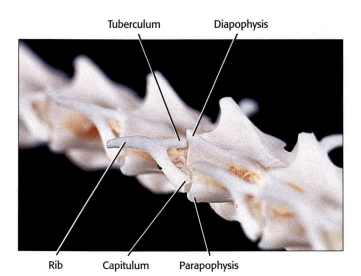

Tuberculum   Diapophysis

Rib   Capitulum   Parapophysis

FIGURE 25.17 Trunk vertebrae with ribs.

APPENDICULAR DIVISION
Pectoral Girdle and Forelimb

Pectoral Girdle

In contrast to fish in which the pectoral girdle is firmly articulated with the skull, in tetrapods this connection has been lost and the pectoral girdle now rests in a muscular sling. This change freed the skull for independent movement and also provided greater freedom of movement associated with the anterior appendages.

In the girdle of the ancestors of tetrapods, bones derived from the dermal armor lie over the endochondral elements. A reduction of the dermal portion of the pectoral girdle of tetrapods resulted in an emphasis of the endochondral portion. An exception is the dermal clavicle, which does not occur in many amphibians but tends to be a prominent bone in most other tetrapods.

Broad plates of cartilage, the paired **coracoids**, project medially to overlap in the midline. A coracoid foramen, permitting the passage of blood vessels and nerves, pierces the coracoid plate. The **procoracoid**, a lateral process of the coracoids, extends anteriorly. An ossified **scapula**, with a distal **suprascapular cartilage**, projects dorsally from the rim of the **glenoid fossa**, with which the humerus articulates (Figure 25.18).

Forelimb

The posture in caudate amphibians most closely approaches the ancestral condition in tetrapods. Both the forelimbs and hindlimbs are extended out to the sides of the animal, with the upper segments parallel to the substratum and at right angles to the lower segments, with the hand and foot flexed at

the wrist and ankle, resulting in the hand and foot resting flat on the substratum. This posture results in the weight of the body being almost entirely supported by the upper segments and requires the broad expanses of coracoid surface area to serve as origin points for leg bracing and stabilizing muscles.

The forelimb consists of the **brachium** or upper arm, the **antebrachium** or forearm, and the **manus** or wrist and hand. A single bone, the **humerus**, is found in the brachium. The joint between the brachium and the antebrachium is the **cubitus** or elbow, forming the proximal flexure. Two bones, the medial **radius** and the lateral **ulna**, are found in the antebrachium. The proximal portion of the manus, the **carpus** or wrist, consists of six small bones in three rows and articulates with the forearm. It is the distal point of flexure in the forelimb. The distal portion of the manus terminates in four **digits**, the pollex or thumb probably having been lost. The intermediate series of bones of the manus are the **metacarpals**, one per digit. Two distal bones or **phalanges** are found in digits 2, 3, and 5, and three are found in digit 4 (Figure 25.18).

Pelvic Girdle and Hindlimb

Pelvic Girdle

Ancestrally, the pelvic girdle in tetrapods consisted of paired elements—an anteriorly projecting pubis, and a posteriorly

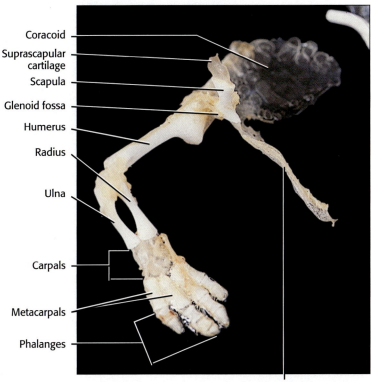

Coracoid

Suprascapular cartilage

Scapula

Glenoid fossa

Humerus

Radius

Ulna

Carpals

Metacarpals

Phalanges

Procoracoid

FIGURE 25.18 Right pectoral girdle and forelimb: Dorsal view.

projecting ischium and dorsal ilium. At the site of articulation of the three bones, is the **acetabulum,** a depression to accommodate the articulation of the head of the femur.

Unlike the pectoral girdle in *Necturus,* the paired anterior cartilaginous **pubes** and bony posterior **ischia** are united medioventrally in a **symphysis** to form a **puboischiadic plate.** The **obturator foramen** pierces the pubis to permit the passage of a nerve by the same name. Extending from this plate are the dorsal **ilia,** which articulate with the modified rib of the sacral vertebra, thus forming a stable connection between the pelvis and the vertebral column. The **pelvic canal,** a space enclosed by the pelvic girdle, permits organs of the digestive and urogenital systems to gain access to the cloaca (Figure 25.15 and Figure 25.19).

Hindlimb

The hindlimb and forelimb have the same basic structural plan. The hindlimb consists of the **thigh** or upper leg, the **shank** or lower leg, and the **pes** or foot. A single bone, the **femur,** occurs in the thigh. A joint, the **genu** or knee, occurs between the thigh and the shank. A medial **tibia** and a lateral **fibula** occur in the shank. The proximal portion of the **pes,** the **tarsus** or ankle, consisting of six small bones in three rows, articulates with the shank. The distal portion of the pes terminates in four **digits,** the hallux or great toe probably having been lost. The intermediate series of bones of the pes consists of one **metatarsal** per digit. Two terminal **phalanges** or distal bones are found in digits 2, 3, and 5, and three occur in digit 4 (Figure 25.19).

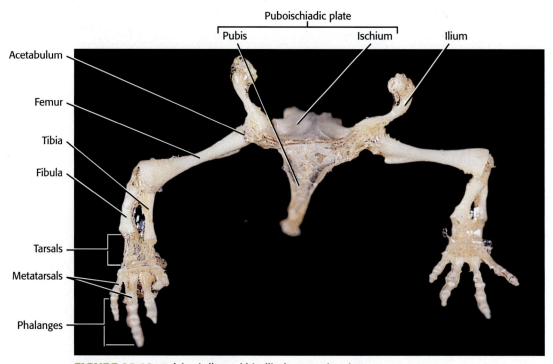

FIGURE 25.19 Pelvic girdle and hindlimb: Anterior view.

CHAPTER TWENTY-SIX

Mudpuppy — Muscular System

S ome of the most fascinating aspects of comparative anatomy are the homologies that can be deduced from the anatomy of previous animals studied. This is particularly evident in the study of muscles. You might wonder how comparative anatomists can state with some confidence that a muscle in *Necturus* is homologous with a shark muscle. An anatomical "rule" for muscles and their nerve innervation is that the nerve generally retains its fidelity to the muscle, no matter where the muscle may migrate or how it may differentiate. For example, in mammals, the musculature of the diaphragm is innervated by the phrenic nerve originating in the cervical region.

Why should a muscle mass situated between the pleural and abdominal cavities of the midbody be innervated by a cervical or neck nerve? The answer seems to be that the diaphragm muscles in the embryo are derived from muscles in the neck region that migrate posteriorly to join with the transverse septum to form the muscular diaphragm, and the phrenic nerve that innervates them migrates right along with the muscle. Comparative embryology of muscles and similarities in the origin, insertion, and action of muscles offer strong supporting evidence for these homologies as well.

SKINNING

Notice the sizable incision in the belly where latex was injected into the circulatory system. As a consequence, the viscera protrude through the cut. With a pair of scissors, make a middorsal slit in the skin from the snout to about the base of the tail. Hold the points of the scissors away from underlying tissues.

 Be cautious when skinning along the incision so adjacent muscles, viscera, mesenteries, and blood vessels remain intact.

You will immediately determine that the mudpuppy is easier to skin than the shark, but you still will have to exercise your usual common sense. Areas where the skin tends to adhere more tightly include the eye, the gular fold, the cloaca, the branchial levators, the muscles of the gills, and the gills themselves. Do not attempt to skin the gills, as they are easily pulled off. Skin the rest of the entire animal. To prepare a superior study specimen requires about 1.5 hours.

Amphibian muscle is moist, gelatinous, and fragile. You must take great care in the removal of surrounding connective tissue and during the separation of individual muscles. [**Note:** To keep the specimen moist and free of decay, we recommend that it be wetted down with a liquid preservative during the entire study.]

 We strongly urge that the cleaning process and dissection of muscles be completed using a dissecting microscope, if available.

AXIAL MUSCLES

Similar to fish, the axial musculature consists of serially arranged **myomeres** separated by **myosepta,** which are further subdivided dorsally and ventrally as well as laterally by distinct sheets of connective tissue. For example, the **linea alba** separates the left and right trunk musculature midventrally, and the **horizontal septum** separates the epaxial and hypaxial musculature. In *Necturus,* the position of the horizontal septum lies more dorsal than in the shark (Figure 26.1).

Epaxial Muscles

Dorsalis Trunci

The epaxial musculature of the shark is homologous with the single **dorsalis trunci** muscle extending from the head just posterior to the mandibular adductor muscles to the tail tip, where it narrows considerably and lies laterally because of the shape of the tail (Figure 26.2). Deeper slips of this muscle—the interspinalis and the intertransversarius—extend between adjacent vertebrae.

Origin: Transverse process of a vertebra and associated myosepta

Insertion: Transverse process of an adjacent vertebra and associated myosepta

Action: Flexes body laterally and dorsally during dorsoventral bending of the vertebral column

Hypaxial Muscles

In contrast to the epaxial musculature, the single layer of hypaxial muscles of the abdominal region of the shark is homologous with the following four individual muscles in *Necturus.* In addition to these hypaxial derivatives, the subvertebralis muscles extend between adjacent vertebrae, ventral to the transverse processes, and bring about ventral bending during dorsoventral bending of the vertebral column.

Rectus Abdominis

The **rectus abdominus,** a narrow band of muscle on either side of the linea alba, extends from the anterior pubis to the hypobranchial region (Figure 26.3 and Figure 26.4).

Origin: Anterior edge of the pubis

Insertion: Rectus cervicis

Action: Flexes the trunk ventrally

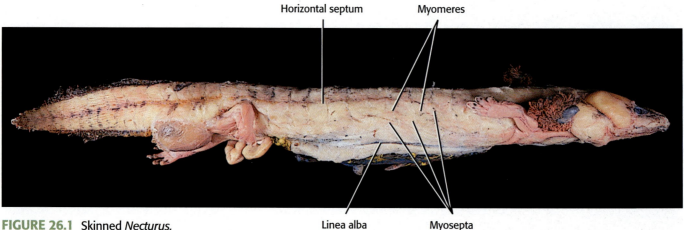

Horizontal septum   Myomeres

Linea alba   Myosepta

FIGURE 26.1 Skinned *Necturus.*

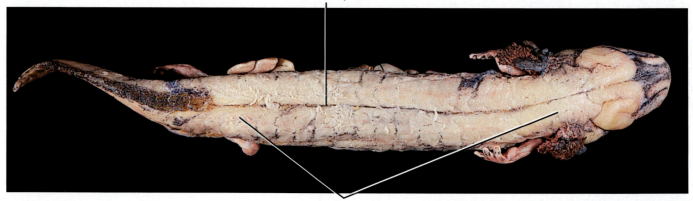

Middorsal septum

Dorsalis trunci

FIGURE 26.2 Dorsal muscles.

Notice that the following abdominal muscles are arranged in three layers: the outer external oblique, the middle internal oblique, and the innermost transversus abdominis.

External Oblique

The **external oblique** is a sheet of muscle whose fibers extend obliquely and tilt downward posteriorly (Figure 26.4).

Origin: Transverse processes of vertebrae

Insertion: Connective tissue of rectus abdominis

Action: Compresses the viscera and aids in locomotion

Internal Oblique

The **internal oblique** is a sheet of muscle medial to the external oblique whose fibers extend obliquely and tilt downward anteriorly (Figure 26.4)

Origin: Transverse processes of vertebrae

Insertion: Connective tissue of rectus abdominis

Action: Compresses the viscera and aids in locomotion

Transversus Abdominis

The **transversus abdominis** is a sheet of muscle medial to the internal oblique and abutting the parietal peritoneum, and whose fibers extend transversely (Figure 26.4).

Origin: Transverse processes of vertebrae

Insertion: Connective tissue of rectus abdominis

Action: Compresses the viscera and aids in locomotion

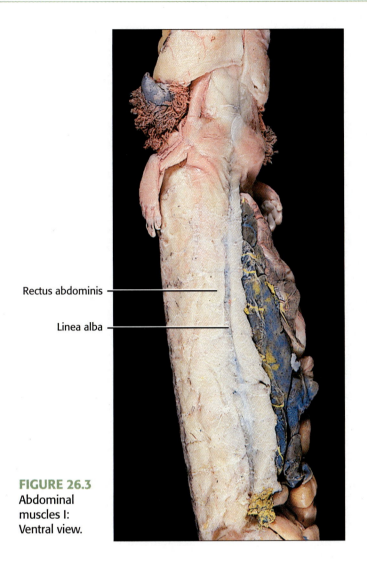

Rectus abdominis

Linea alba

FIGURE 26.3
Abdominal muscles I: Ventral view.

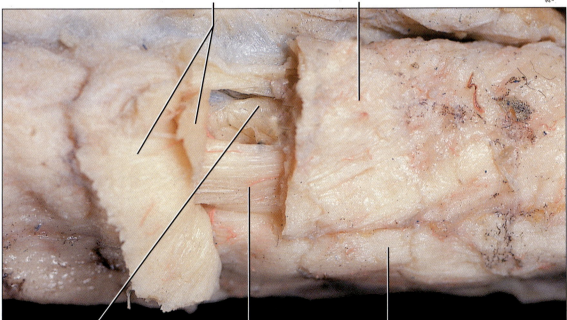

External oblique (reflected)　　　External oblique

Transversus abdominis　　　Internal oblique　　　Rectus abdominis

FIGURE 26.4
Abdominal muscles II: Lateral view.

DORSAL AND LATERAL MUSCLES OF THE HEAD

The musculature of visceral arch I in the shark is homologous with two adductor muscles, mandibulae anterior and externus.

Adductor Mandibulae Anterior

The **adductor mandibulae anterior** is a thick muscle on either side of the middorsal line of the head, extending from the eye to the back of the head (Figure 26.5).

Origin: Frontal and parietal bones

Insertion: Dentary

Action: Elevates the lower jaw (closes the mouth)

Adductor Mandibulae Externus

The **adductor mandibulae externus** is a broad, thick muscle lateral to the anterior mandibular levator muscle, wrapping around the lateral surface of the head (Figure 26.5).

Origin: Parietal

Insertion: Dentary

Action: Elevates the lower jaw (closes the mouth)

Depressor Mandibulae

The shark hyoid levator (visceral arch II) is homologous with the **depressor mandibulae**, a slender band lying between the **branchiohyoideus** and the adductor mandibulae externus. To locate the depressor, lift and push the adductor mandibulae externus forward (Figure 26.6).

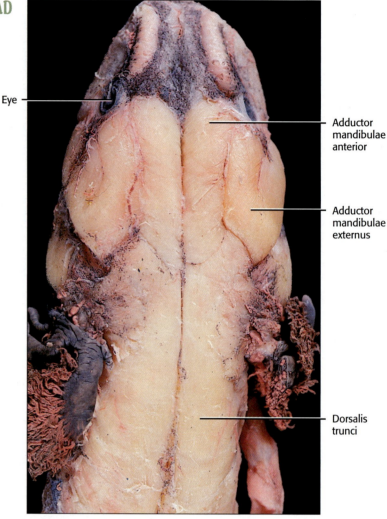

Eye

Adductor mandibulae anterior

Adductor mandibulae externus

Dorsalis trunci

FIGURE 26.5 Muscles of the head.

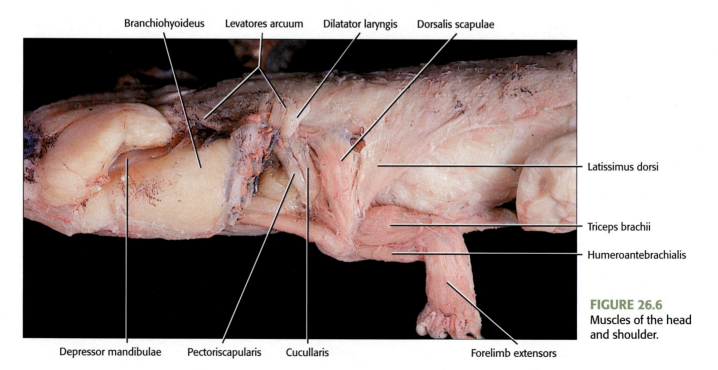

Branchiohyoideus   Levatores arcuum   Dilatator laryngis   Dorsalis scapulae

Latissimus dorsi

Triceps brachii

Humeroantebrachialis

Depressor mandibulae   Pectoriscapularis   Cucullaris   Forelimb extensors

FIGURE 26.6 Muscles of the head and shoulder.

Origin: Squamosal and parietal bones

Insertion: Angular

Action: Depresses lower jaw (opens the mouth)

Compare the mass of the "jaw-opening" and "jaw-closing" muscles. Notice that the adductors are much more heavily developed than the depressor. This is typical among terrestrial vertebrates, as the weight of the lower jaw is almost sufficient to cause the lower jaw to open with no help from muscles.

Branchiohyoideus

Superficial constrictor muscles of visceral arch III in the shark are homologous with the **branchiohyoideus,** a stout muscle lying lateral and posterior to the angle of the jaw (Figure 26.6).

Origin: Ceratobranchial of visceral arch III

Insertion: Ceratohyal (visceral arch II)

Action: Retracts the hyoid apparatus

Levatores Arcuum

The **levatores arcuum** comprise a fan-shaped group of muscles occurring dorsal to the gills, homologous with part of the cucullaris of the shark (Figure 26.6).

Origin: Fascia of the dorsalis trunci

Insertion: Epibranchial cartilages

Action: Gill elevators

Dilatator Laryngis

The constrictor muscles of the posthyoid visceral arches (III–VII) of the shark are homologous with a number of muscles, among which is the **dilatator laryngis,** a narrow muscle occurring posterior to the **levatores arcuum** (Figure 26.6).

Origin: Fascia of the dorsalis trunci

Insertion: Cartilages of the larynx

Action: Expands the larynx

VENTRAL MUSCLES OF THE THROAT

Intermandibularis

The **intermandibularis** is a thin sheet of muscle extending between the halves of the mandible and connected medially in a raphe. It is homologous with the intermandibularis muscle in the shark (Figure 26.7).

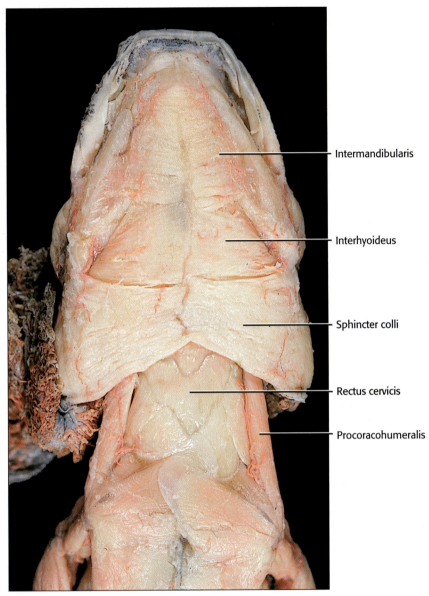

FIGURE 26.7 Superficial muscles of the throat.

Labels: Intermandibularis — Interhyoideus — Sphincter colli — Rectus cervicis — Procoracohumeralis

Origin: Mandible

Insertion: Median raphe

Action: Elevates throat

Interhyoideus

The **interhyoideus** is a thin sheet of muscle lying posterior to the **intermandibularis** and fusing with its posterior edge. It is homologous with the ventral hyoid constrictor of the shark (Figure 26.7).

Origin: Fascia of the depressor mandibulae and the branchiohyoideus

Insertion: Median raphe

Action: Elevates throat

Sphincter Colli

The posterior portion of the **interhyoideus**, known as the **sphincter colli**, is homologous with the posterior portion of interhyoideus of the shark (Figure 26.7).

Origin: Same as the interhyoideus

Insertion: Same as the interhyoideus

Action: Same as the interhyoideus

 To observe the following three muscles, using a pair of sharp scissors and beginning approximately 3mm to one side of the median raphe, cut from the posterior edge of the interhyoideus to the mandible. Be careful to keep the scissor tips pointed up to avoid cutting underlying muscles.

Geniohyoideus

The **geniohyoideus** is a slender band of muscle lying dorsal to the **intermandibularis** and **interhyoideus** muscles. It is homologous with part of the coracomandibular of the shark (Figure 26.8).

Origin: Mandibular symphysis

Insertion: Second basibranchial cartilage

Action: Depress lower jaw and pulls hyoid apparatus anteriorly

Genioglossus

The **genioglossus** is a smaller and narrower band of muscle lying dorsolateral to the **geniohyoid**. It, too, is homologous with a portion of the shark coracomandibular (Figure 26.8).

Origin: On the mandible dorsal to the geniohyoid

Insertion: Tongue

Action: Moves tongue

Rectus Cervicis

The coracohyoid and coracoarcual muscles of the shark are homologous with the **rectus cervicis**, which extends from the supracoracoideus muscle to the base of the geniohyoid. Notice the prominent segmentation (myosepta) (Figure 26.8).

Origin: Rectus abdominis

Insertion: Branchial cartilages

Action: Retracts hyoid and gill apparatus; depresses the head

Portions of the procoracohumeralis, omoarcual, and rectus cervicis have to be removed to expose the following three muscles. Exercise care in this region, as afferent branchial arteries and the pericardial cavity lie beneath the rectus cervicis.

All three of the following muscles are homologous with shark interbranchial muscles.

Subarcuales

The **subarcuales** are composed of a series of short, straplike muscles extending between adjacent visceral arches (Figure 26.9).

Origin: Hyoid arch, visceral arch III, visceral arch IV

Insertion: Visceral arch III, visceral arch IV, and visceral arch V, respectively

Action: Adduction of gills

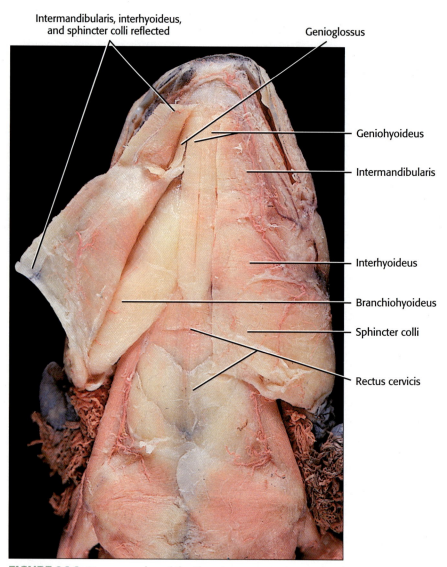

Intermandibularis, interhyoideus, and sphincter colli reflected

Genioglossus

Geniohyoideus

Intermandibularis

Interhyoideus

Branchiohyoideus

Sphincter colli

Rectus cervicis

FIGURE 26.8 Deep muscles of the throat.

Transversi Ventrales

The **transversi ventrales** make up a series of deep muscles whose fibers are oriented transversely (Figure 26.9).

Origin: Rectus cervicis

Insertion: Cartilage of visceral arch V

Action: Pulls gills ventromedially

Depressores Arcuum

The **depressores arcuum** are a series of small muscles located lateral to the **subarcuals** (Figure 26.9).

Origin: From visceral arches III–V

Insertion: Base of the gills

Action: Depresses the gills

VENTRAL PECTORAL MUSCLES

Omoarcual

The **omoarcual,** a small, strap-like muscle, is located medial to the **procoracohumeralis** and is considered to be homologous with the rectus cervicis of the shark (Figure 26.10).

Origin: Rectus cervicis

Insertion: Junction of procoracoid process with coracoid plate

Action: Draws shoulder anteriorly

Procoracohumeralis

The **procoracohumeralis,** a long, strap-like muscle, is located lateral to the **omoarcual** (Figure 26.10). It is homologous with a portion of the pectoral abductor of the shark.

Origin: Lateral surface of the procoracoid

Insertion: Humerus

Action: Pulls humerus anteriorly

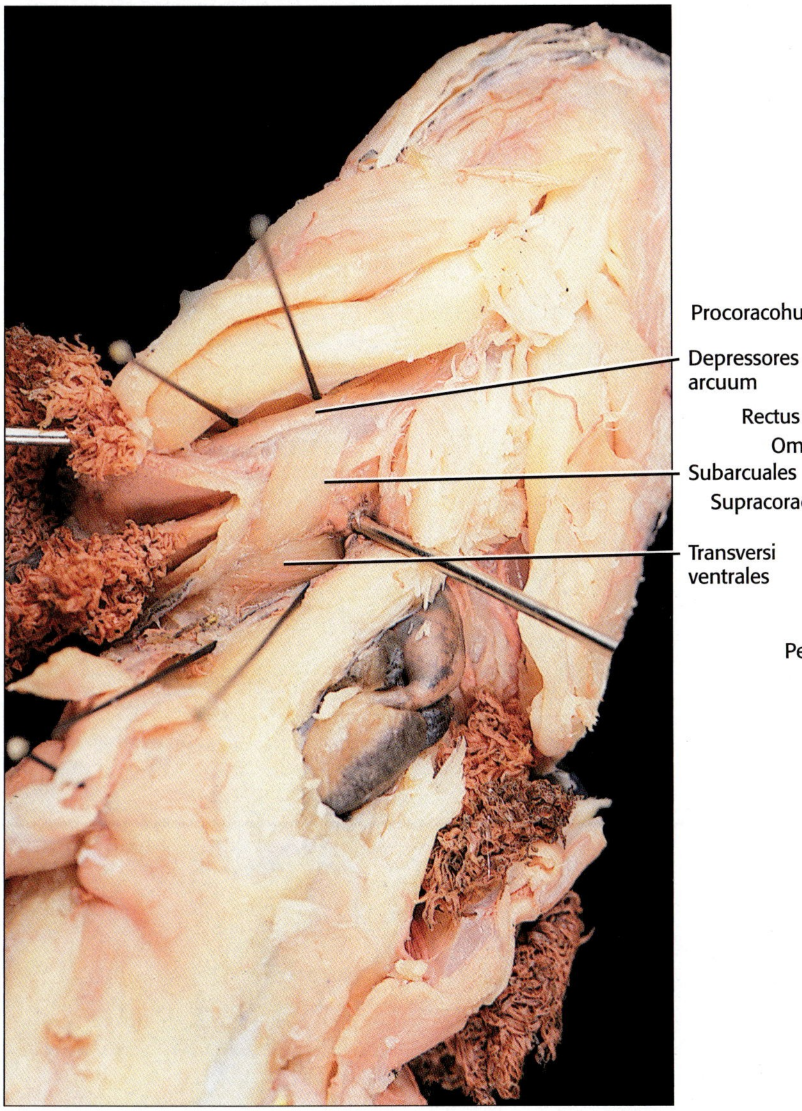

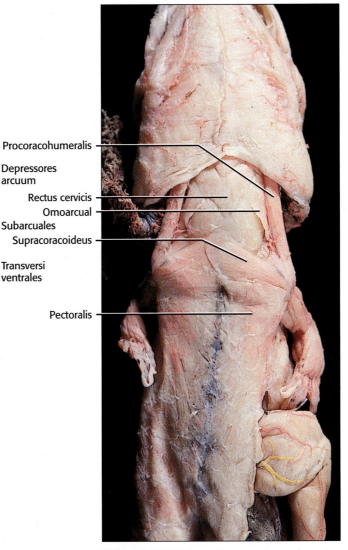

FIGURE 26.9 Deep muscles of the visceral arches (ventral view).

FIGURE 26.10 Ventral pectoral muscles.

Supracoracoideus

The **supracoracoideus**, a fan-shaped muscle, occurs anterior and somewhat dorsal to the anterior portion of the **pectoralis** muscle (Figure 26.10). It is homologous with a portion of the pectoral adductor of the shark.

Origin: Coracoid cartilage

Insertion: Humerus

Action: Adducts humerus

Pectoralis

The **pectoralis** is a large fan-shaped muscle that lies partially over the posterior surface of the **supracoracoideus** (Figure 26.10). It is homologous with a portion of the pectoral adductor of the shark.

Origin: Linea alba

Insertion: Humerus

Action: Adducts humerus

DORSAL/LATERAL/MEDIAL PECTORAL MUSCLES

Pectoriscapularis

The most anterior of a series of four lateral muscles, the **pectoriscapularis** is homologous with the coracoarcual muscles of the shark (Figure 26.6).

Origin: Epibranchial cartilage

Insertion: Scapula

Action: Pulls scapula anteriorly

Cucullaris

The cucullaris, a strap-like muscle is the next in the series. It is homologous with a portion of the cucullaris of the shark (Figure 26.6).

Origin: Fascia of the dorsalis trunci

Insertion: Scapula near shoulder joint

Action: Pulls scapula anteriodorsally

Dorsalis Scapulae

The **dorsalis scapulae**, a somewhat broader, strap-like muscle occurs next (Figure 26.6). It is homologous with a portion of the shark pectoral abductor.

Origin: Suprascapular cartilage

Insertion: Humerus

Action: Pulls humerus anteriorly

Latissimus Dorsi

The last and largest of the series is the triangular **latissimus dorsi**. It is homologous with a portion of the shark pectoral abductor muscle (Figure 26.6).

Origin: Fascia of dorsalis trunci

Insertion: Humerus

Action: Retracts pectoral limb

 To locate the following two medial muscles, cut through the belly of the latissimus dorsi muscle and reflect the ends. Carefully detach the origin of the dorsalis scapulae muscle and gently pull the scapula laterally.

Two muscles that are homologous with a portion of the hypaxial musculature of the shark will be exposed.

Thoraciscapularis

The bandlike **thoraciscapularis** extends from the hypaxial muscles to the **scapula** (Figure 26.11).

Origin: Hypaxial musculature of the trunk

Insertion: Dorsomedial portion of the suprascapula

Action: Retracts and depresses the scapula

Levator Scapulae

A slender bandlike muscle that extends from the skull to the scapula (Figure 26.11), the **levator scapulae** in terrestrial salamanders is known as the opercularis and functions as a sound transmitter.

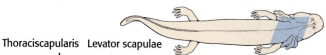

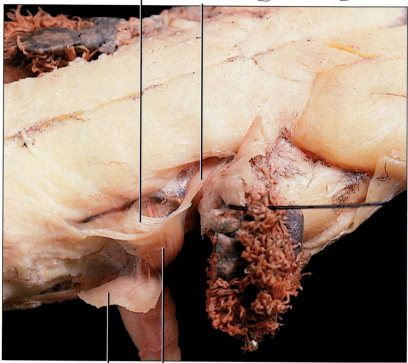

Thoraciscapularis   Levator scapulae

Latissimus dorsi reflected   Dorsalis scapulae reflected

FIGURE 26.11 Medial shoulder muscles.

Origin: Posterior region of the skull

Insertion: Anterior border of the suprascapula

Action: Pulls scapula anteriorly

Subcoracoscapularis

A small muscle found in the shoulder, the **subcoracoscapularis**, is difficult to find and is homologous with a portion of the pectoral musculature of the shark. It is mentioned here because it is homologous with the subscapularis of mammals.

MUSCLES OF THE FORELIMB

Triceps Brachii

The **triceps brachii** is on the dorsal surface of the brachium and is homologous with a portion of the shark fin abductor (Figure 26.6).

Origin: Three heads from the coracoid, scapula, and humerus

Insertion: Ulna

Action: Extension of the forearm (antebrachium)

Coracobrachialis

The **coracobrachialis** muscle is located on the ventromedial surface of the brachium and is homologous with a portion of the adductor of the shark fin (Figure 26.12).

Origin: Coracoid process

Insertion: Humerus

Action: Adducts the forelimb

Humeroantebrachialis

The **humeroantebrachialis**, a ventrolateral muscle, lies next to the **coracobrachialis** and is homologous with

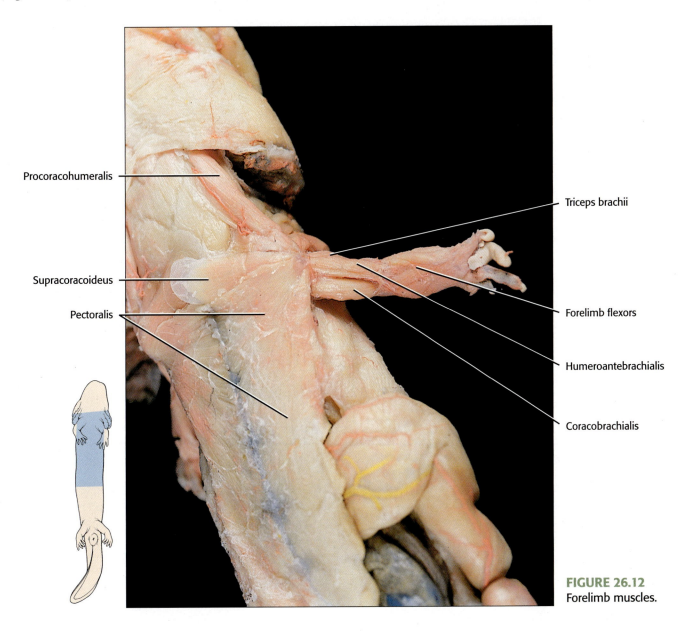

Procoracohumeralis

Supracoracoideus

Pectoralis

Triceps brachii

Forelimb flexors

Humeroantebrachialis

Coracobrachialis

FIGURE 26.12
Forelimb muscles.

a portion of the shark fin adductor (Figure 26.6 and Figure 26.12).

Origin: Humerus

Insertion: Radius

Action: Flexes the forearm (antebrachium)

Flexors and Extensors of the Forearm

The **forelimb flexors** consist of a group of muscles located on the ventral surface of the forearm. They are homologous with a portion of the shark pectoral fin adductors and flex the antebrachium and manus (Figure 26.12). The **forelimb extensors** consist of a group of muscles on the dorsal surface of the forearm. They are homologous with a portion of the shark pectoral fin abductors and extend the antebrachium and the manus (Figure 26.6).

MUSCLES OF THE PELVIC GIRDLE AND HINDLIMB

A large muscle mass, divisible into two muscles, the anterior puboischiofemoralis externus and the posterior puboischiotibialis, lies over the ventral surface of the pelvic girdle.

Puboischiofemoralis Externus

A large muscle, the **puboischiofemoralis externus** is homologous with a portion of the shark pelvic fin adductors (Figure 26.13 and Figure 26.14).

Origin: Ventral surface of the puboischiadic plate

Insertion: Femur

Action: Adducts the hindlimb

Puboischiotibialis

The **puboischiotibialis** is another muscle that is homologous with a portion of the shark pelvic fin adductors (Figure 26.13 and Figure 26.14).

Origin: Ventral surface of the puboischiadic plate

Insertion: Tibia

Action: Adducts the hindlimb

Ischiofemoralis

The **ischiofemoralis** is a small, triangular muscle occurring dorsal to the **puboischiotibialis** and posterior to the **puboischiofemoralis externus** (Figure 26.13 and Figure 26.14). It is homologous with a portion of the shark pelvic fin adductors.

Origin: Ischium

Insertion: Femur

Action: Retracts femur

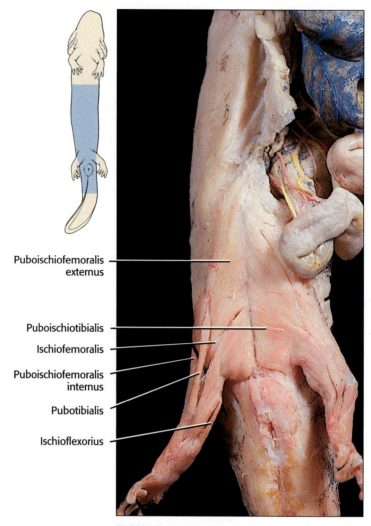

Puboischiofemoralis externus

Puboischiotibialis

Ischiofemoralis

Puboischiofemoralis internus

Pubotibialis

Ischioflexorius

FIGURE 26.13 Pelvic and hindlimb muscles I.

Puboischiofemoralis Internus

The **puboischiofemoralis internus**, a narrow, strap-like muscle, lies along the anterior border of the thigh. It is homologous with a portion of the shark pelvic fin abductors (Figure 26.13 and Figure 26.14). Notice that this muscle can be viewed from both the ventral and the dorsal aspect of the hindlimb.

Origin: Dorsal surface of puboischiadic plate

Insertion: Femur

Action: Pulls hindlimb anteriorly

Pubotibialis

The **pubotibialis**, a narrow, distinctive muscle lies just medial to the **puboischiofemoralis internus** and is homologous with a portion of the shark pelvic fin adductors (Figure 26.13 and Figure 26.14).

Origin: Pubis

Insertion: Tibia

Action: Adducts the hindlimb

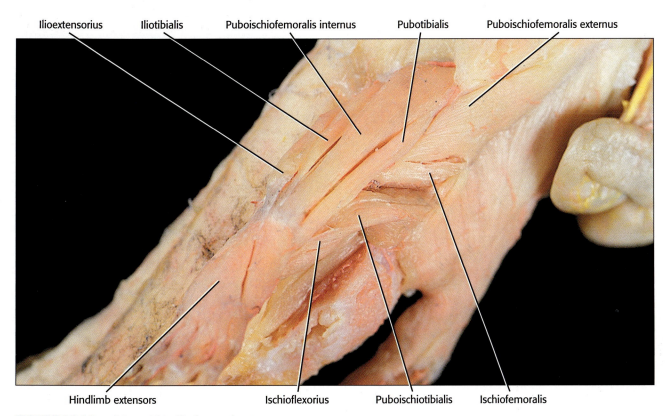

Ilioextensorius   Iliotibialis   Puboischiofemoralis internus   Pubotibialis   Puboischiofemoralis externus

Hindlimb extensors   Ischioflexorius   Puboischiotibialis   Ischiofemoralis

FIGURE 26.14 Pelvic and hindlimb muscles II.

Ischioflexorius

The **ischioflexorius** is located along the ventromedial border of the thigh and adheres closely to the **puboischiotibialis** (Figure 26.13 and Figure 26.14). Ischioflexorius is homologous with a portion of the pelvic fin adductors of the shark.

 As suggested before, a dissecting microscope will greatly facilitate the separation of these two muscles.

Origin: Ischium

Insertion: Fascia of the shank

Action: Flexes shank and foot

 To observe the following three small muscles, carefully remove half of the cloacal tissue.

All of these are homologous with the shark pelvic fin adductors.

Ischiocaudalis

The slender and most medial muscle, the **ischiocaudalis** lies along the rim of the cloaca (Figure 26.15).

Origin: Ischium

Insertion: Caudal vertebrae

Action: Flexes tail

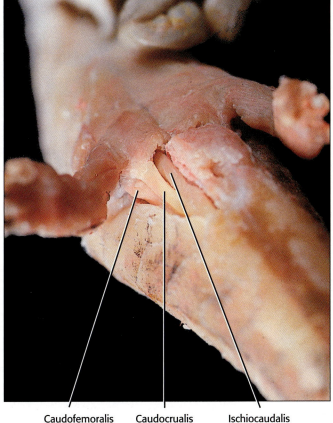

Caudofemoralis   Caudocrualis   Ischiocaudalis

FIGURE 26.15 Pelvic muscles.

Caudocrualis (or Caudopuboischiotibialis)

The second straplike muscle, the **caudocrualis**, occurs lateral to the **ischiocaudalis** (Figure 26.15 and Figure 26.16).

Origin: Caudal vertebrae

Insertion: Posterior surface of the puboischiotibialis muscle
Action: Flexes tail

Caudofemoralis

A straplike muscle, the **caudofemoralis** lies dorsal to the **caudocrualis** (Figure 26.15 and Figure 26.16).

Origin: Caudal vertebrae

Insertion: Femur

Action: Retraction of the thigh

Iliotibialis

The **iliotibialis** is a slender muscle occurring dorsolateral to the **puboischiofemoralis internus**. It is homologous with a portion of the shark pelvic fin abductors (Figure 26.14 and Figure 26.16).

Origin: Ilium

Insertion: Tibia

Action: Abducts the hindlimb

Ilioextensorius

The **ilioextensorius** muscle lies just posterior to the **iliotibialis** and has a tendency to adhere to it. It is homologous with a portion of the shark pelvic fin abductors (Figure 26.14 and Figure 26.16).

Origin: Ilium

Insertion: Tibia

Action: Extends the shank

Iliofibularis

A slender muscle, the **iliofibularis** lies just posterior to the ilioextensorius and is homologous with a portion of the shark pelvic fin abductors (Figure 26.16).

Origin: Ilium

Insertion: Fibula

Action: Abducts the hindlimb

FLEXORS AND EXTENSORS OF THE HINDLIMB

The **hindlimb flexors** consist of a group of muscles on the ventral surface of the hind limb. They are homologous with a portion of the shark pelvic fin adductors and flex the shank and pes (Figure 26.16). The **hindlimb extensors** consist of a group of muscles on the dorsal surface of the hindlimb, are homologous with a portion of the shark pelvic fin abductors, and extend the shank and pes (Figure 26.14).

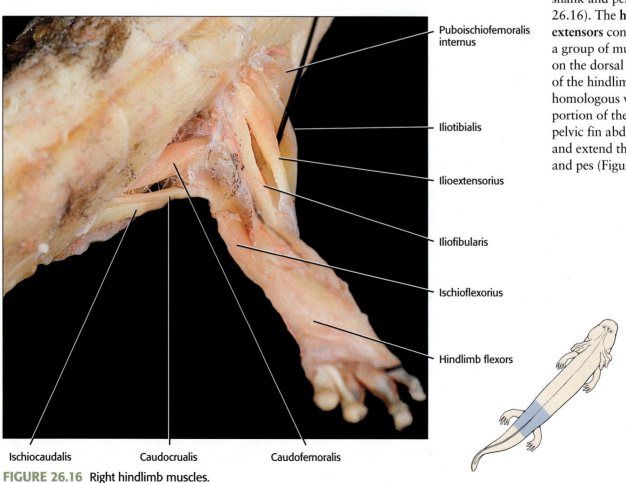

Puboischiofemoralis internus

Iliotibialis

Ilioextensorius

Iliofibularis

Ischioflexorius

Hindlimb flexors

Ischiocaudalis   Caudocrualis   Caudofemoralis

FIGURE 26.16 Right hindlimb muscles.

Mudpuppy — Body Cavities and Mesenteries

BODY CAVITIES

The body cavities in *Necturus* remain essentially identical to those in the lamprey and the shark—an anterior **pericardial cavity** separated by a **transverse septum** from the much larger, posterior **pleuroperitoneal cavity** (Figure 27.1). The linings of the cavities and suspensory membranes also are highly similar. (See Chapter 7.)

OPENING *NECTURUS*

First you will determine the position of the falciform ligament extending between the liver and the midventral body wall. In doing so, you must exercise extreme caution, as all of the membranes in Necturus are small and delicate.

If the slit made in the ventral body wall for injection of the circulatory system is well lateral to the midline, continue to cut anteriorly with scissors to about the posterior border of the rectus cervicis, the approximate level of the position of the transverse septum. Likewise, extend the cut posteriorly to the anterior edge of the cloaca.

Then cut transversely across the ventral body wall just below the level of the posterior border of the falciform ligament to about 1 cm from the midline on the opposite side, and extend the cut anteriorly to the same level as the initial cut. You now have created a flap isolating the falciform ligament (Figure 27.2).Be careful to avoid cutting any blood vessels associated with this membrane.

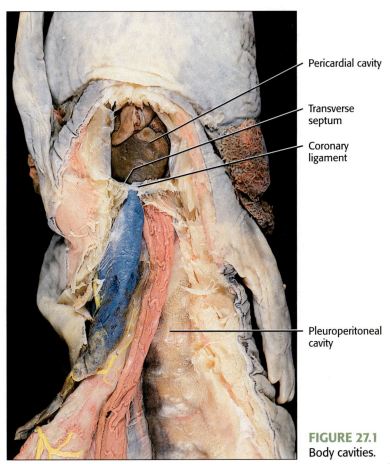

Pericardial cavity

Transverse septum

Coronary ligament

Pleuroperitoneal cavity

FIGURE 27.1
Body cavities.

Look for the urinary bladder at the posterior end of the pleuroperitoneal cavity. It may be tucked under the intestine. Find and identify the median ligament of the bladder, anchoring it to the ventral body wall. Construct a similar flap to isolate this ligament (Figure 27.3).

If the slit is closer to the midline, carefully find the falciform ligament by lifting the usually protuberant liver from the cut edge of the body wall. Extend the cut anteriorly and posteriorly similar to the previous directions, with the exception of veering the cuts farther away from the midline. The posterior flap may be easier to create than the anterior flap.

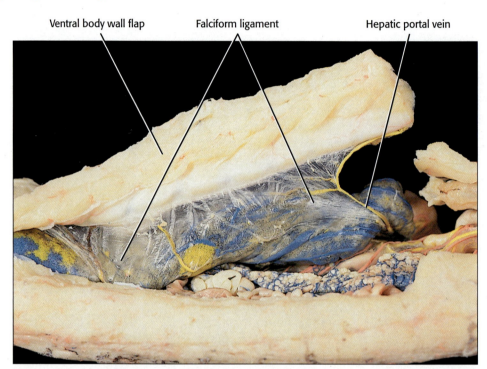

Ventral body wall flap   Falciform ligament   Hepatic portal vein

FIGURE 27.2
Falciform ligament as seen from left side.

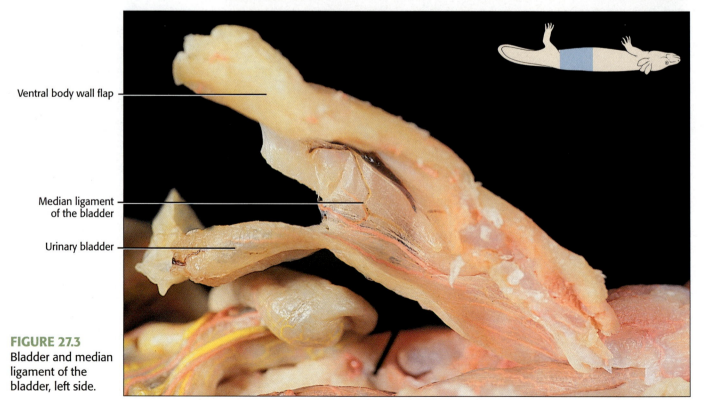

Ventral body wall flap

Median ligament of the bladder

Urinary bladder

FIGURE 27.3
Bladder and median ligament of the bladder, left side.

MESENTERIES

 Remember that these structures are fragile and easily torn, so remain cautious during the search for them.

Derivatives of the Dorsal Mesentery

The **dorsal mesentery** in *Necturus* remains fairly continuous with regional specialization similar to the shark (Figure 27.4). The **mesogaster** (**greater omentum**) extends from the dorsal body wall to the dorsal wall of the stomach. The **gastrosplenic ligament**, actually a portion of the greater omentum, extends between the stomach and the spleen. An extensive **mesentery proper**, containing a multitude of blood vessels, supports the small intestine, and a less extensive **mesocolon** supports the large intestine (Figure 27.5 and Figure 27.6). A **pulmonary ligament** suspends each of the lungs (Figure 27.7). In addition, the right lung has a second

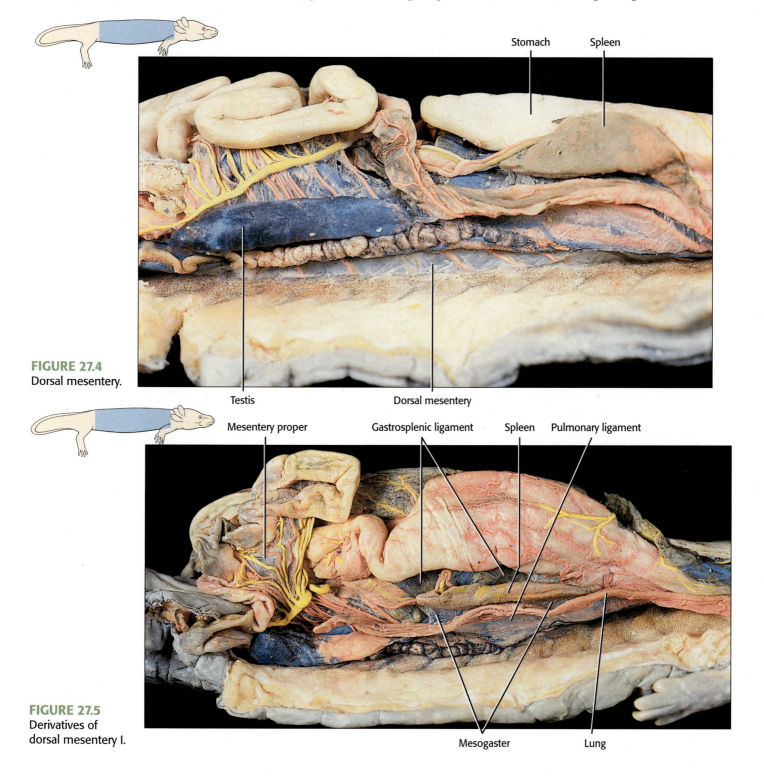

FIGURE 27.4
Dorsal mesentery.

FIGURE 27.5
Derivatives of dorsal mesentery I.

POSTERIOR　　　　　　　　　　　Large intestine　　　Small intestine　　　**ANTERIOR**

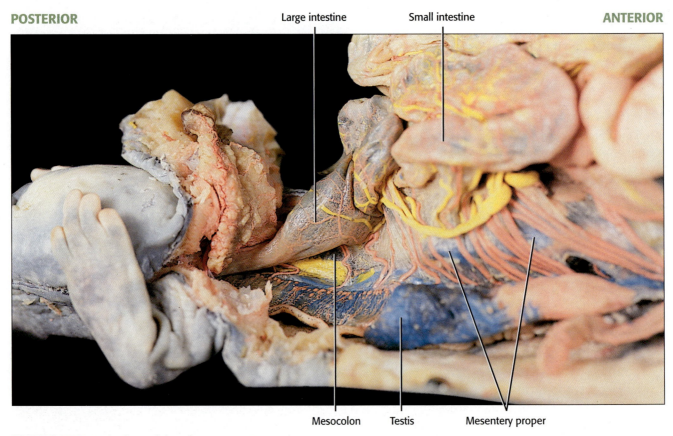

Mesocolon　　Testis　　Mesentery proper

FIGURE 27.6 Derivatives of dorsal mesentery II.

POSTERIOR　　　　　　　　　　　　　Stomach　　　　　Lung　**ANTERIOR**

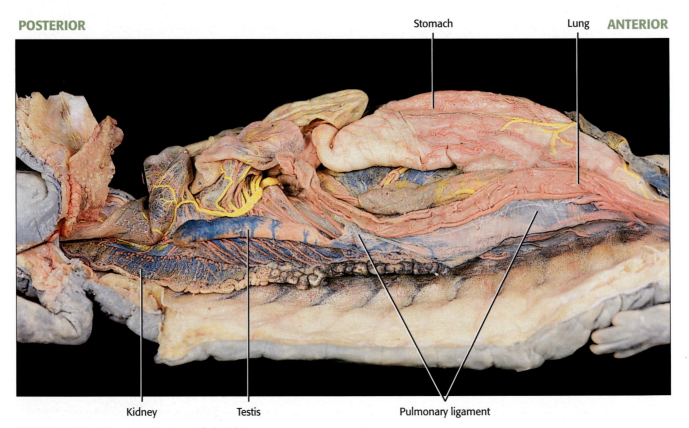

Kidney　　　Testis　　　　　Pulmonary ligament

FIGURE 27.7 Pulmonary ligament (left side).

suspensory membrane, the **hepatocavopulmonary ligament**, from the liver to the right lung. Along the posterior edge of this ligament, the **posterior vena cava** travels toward the liver (Figure 27.8).

Derivatives of the Ventral Mesentery

An extensive **falciform ligament** extends from the midventral body wall to the ventral surface of the liver (Figure 27.1). An anterior extension of the falciform ligament, the **coronary ligament**, anchors the liver to the transverse septum (Figure 27.3). The small **gastrohepatic ligament** extending between the anterior portion of the stomach and the liver can be observed best on the left side. A small **hepatoduodenal ligament** extends between the posterior portion of the liver and the duodenum. This membrane enfolds part of the pancreas and is difficult to isolate (Figure 27.9). Observe the **median ligament of the bladder** anchoring the urinary bladder to the ventral body wall (Figure 27.2).

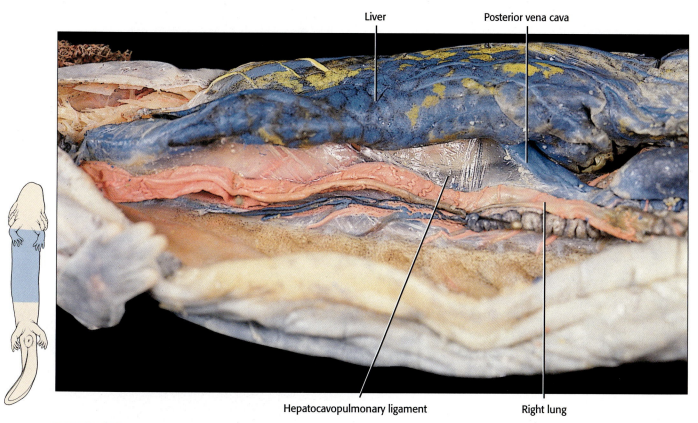

Liver   Posterior vena cava

Hepatocavopulmonary ligament   Right lung

FIGURE 27.8 Hepatocavopulmonary ligament (right side).

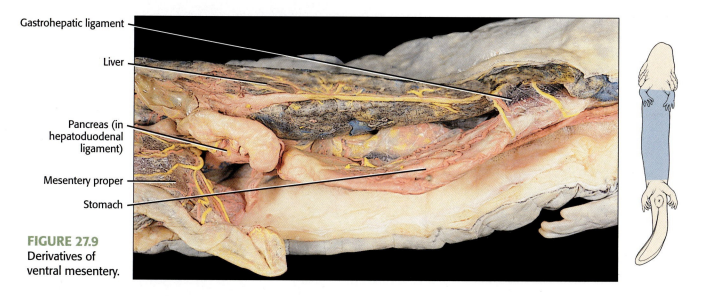

Gastrohepatic ligament

Liver

Pancreas (in hepatoduodenal ligament)

Mesentery proper

Stomach

FIGURE 27.9
Derivatives of ventral mesentery.

This is a new membrane associated with the urinary bladder of tetrapod vertebrates.

Similar to the shark, reproductive membranes associated with the reproductive organs are not derivatives of dorsal or ventral mesenteries.

In the male, the **mesorchium** (Figure 27.10) suspends the testes. In the female, the **mesovarium** suspends the ovaries from the body wall (Figure 27.11). Also in the female, the **mesotubarium** suspends the oviduct from the body wall.

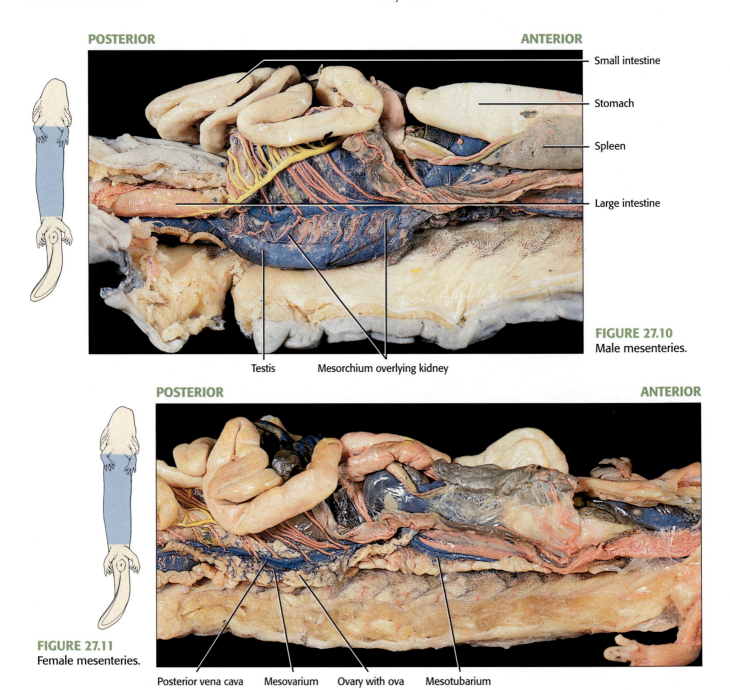

POSTERIOR   ANTERIOR

— Small intestine

— Stomach

— Spleen

— Large intestine

FIGURE 27.10
Male mesenteries.

Testis   Mesorchium overlying kidney

POSTERIOR   ANTERIOR

FIGURE 27.11
Female mesenteries.

Posterior vena cava   Mesovarium   Ovary with ova   Mesotubarium

CHAPTER TWENTY-EIGHT

Mudpuppy — Digestive and Respiratory Systems

28

Some minor changes in the digestive system occurred in *Necturus* when compared to the shark. The respiratory system, on the other hand, is a combination of the piscine gills and the terrestrial vertebrate lungs. For an overview of the evolution of these two systems, refer to Chapter 8.

To study the anterior portion of the digestive and respiratory systems, using a pair of sharp scissors, begin at the corner of the mouth and cut through the lateral side of the head and ventral to the gills to the pectoral girdle. In making this cut, you will have cut into the anterior portion of the esophagus. Angle the cut to intersect the incision made to open the pleuroperitoneal cavity (Figure 28.1). Avoid blood vessels, heart, ligaments, and other structures. (Note: Your instructor's instructions may vary from those above.)

DIGESTIVE SYSTEM

Alimentary Canal

Well-developed **lips** form the outer border of the **oral cavity**. Just interior to the lips are two rows of **teeth** in the upper jaw and a single row of teeth in the lower jaw (Figure 28.1). The teeth are conical and homodont. The **internal nares** occur as small slits just lateral and posterior to the last upper teeth in the upper jaw.

The best way to find the internal nares is to use a dissecting microscope to survey the area, as their orifice may be occluded by mucus.

A **tongue**, not much better developed than that

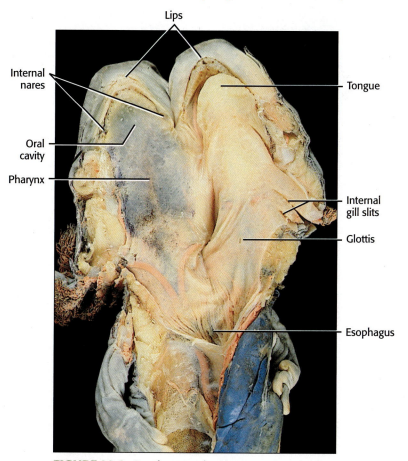

FIGURE 28.1 Oropharyngeal region.

Labels: Lips, Internal nares, Oral cavity, Pharynx, Tongue, Internal gill slits, Glottis, Esophagus

of the shark, sits just posterior to the teeth of the lower jaw (Figure 28.1). If you survey the surface of the tongue with a microscope, you may observe some distinct "bumps," perhaps containing taste buds.

Posterior to the oral cavity is the **pharynx**. Note that an internal **gill slit** occurs between the first and second gills and the second and third gills. In the midline of the floor of the pharynx on an approximate line with the second internal gill slit is the **glottis**, common to all tetrapods (Figure 28.1).

The short **esophagus** leads from the pharynx into the stomach (Figure 28.1). Notice the slightly developed folds of the esophageal wall. The **stomach** in *Necturus* is relatively straight, in contrast to the shark and most other adult vertebrates. Like the shark, rugae are present in the stomach.

 To observe the **rugae**, or folds of the stomach, carefully make a short, longitudinal slit in the mid-region of the ventral portion of the stomach. Avoid cutting blood vessels. Reflect the edges and observe the folds.

At the distal end of the stomach, the muscular **pyloric sphincter** guards the entrance into the small intestine (Figure 28.2 and Figure 28.3).

The short proximal end of the **small intestine**, the **duodenum**, protrudes anteriorly and then leads into the distal portion, which is elongate and coiled to accommodate the longer intestine (increasing digestive and absorptive surface area) in a relatively small body cavity. Make a short longitudinal incision in the ventral wall of the small intestine

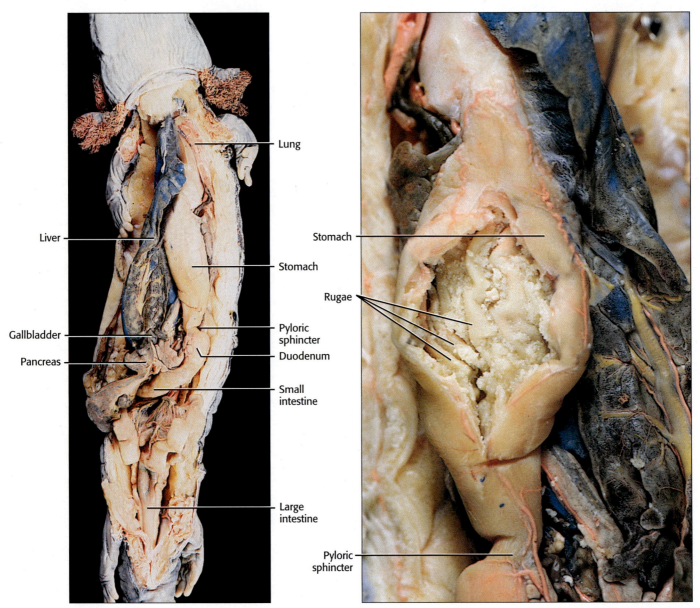

FIGURE 28.2 Digestive system.

FIGURE 28.3 Stomach.

and observe a second means of increasing surface area, the **plicae**, which are small folds in the mucosal surface. The digestive tract ends in a short **large intestine** that opens through the **anus** into the **cloaca** (Figure 28.2, Figure 28.4, and Figure 28.5).

Accessory Digestive Organs

As in all vertebrates, the liver is the largest and most prominent of the visceral organs. In contrast to the lobed liver of the shark, the **liver** in *Necturus* is weakly scalloped posteriorly. Recall that the liver is anchored by the falciform ligament to the anterior portion of the midventral body wall. A small **gallbladder** is located on the dorsal surface of the posterior edge of the middle portion of the liver near the duodenum. The gallbladder functions as an organ in which bile, synthesized in the liver, is stored. A number of **hepatic ducts** drain bile from the liver and join with a **cystic duct** leading from the gallbladder to form a **common bile duct** that empties into the **duodenum** (Figure 28.6).

 As suggested previously, a dissection microscope greatly facilitates locating the ducts.

The pinkish or white irregular mass lying in the hepato-duodenal ligament associated with the cranial loop of the duodenum is the **pancreas** (Figure 28.2). This organ forms as a result of the fusion of ventral and dorsal embryonic primordia. In the shark condition, by contrast, a ventral and dorsal pancreas persist unfused in the adult. The pancreatic ducts remain separate and open into the duodenum. These are difficult to find. The pancreas functions as an exocrine gland and also as an endocrine gland regulating metabolism.

Although it is not part of the digestive viscera, the **spleen** is obvious during the study of the contents of the pleuroperitoneal cavity (Figure 28.7). It lies to the left of the stomach in the gastrosplenic ligament. Its functions are similar to those of the shark.

RESPIRATORY SYSTEM

In most terrestrial tetrapods, the respiratory system consists of lungs where gaseous exchange occurs. In *Necturus*, an aquatic salamander, lungs are present, but they are minimally functional. Instead, respiration occurs primarily across the external gills and through the skin. More than 90% of

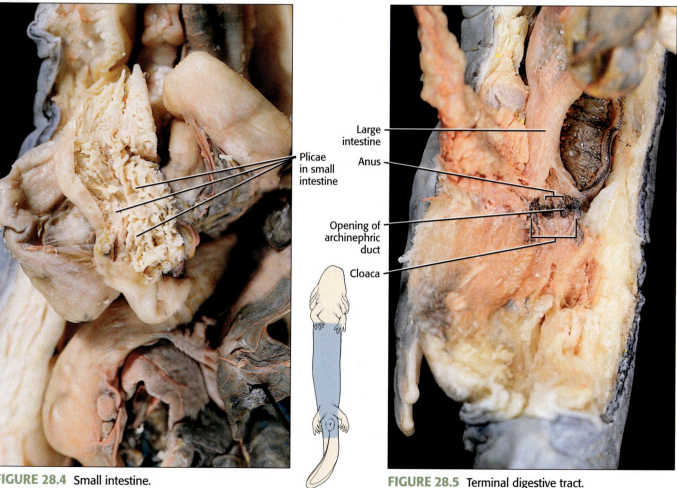

Plicae in small intestine

Large intestine

Anus

Opening of archinephric duct

Cloaca

FIGURE 28.4 Small intestine.

FIGURE 28.5 Terminal digestive tract.

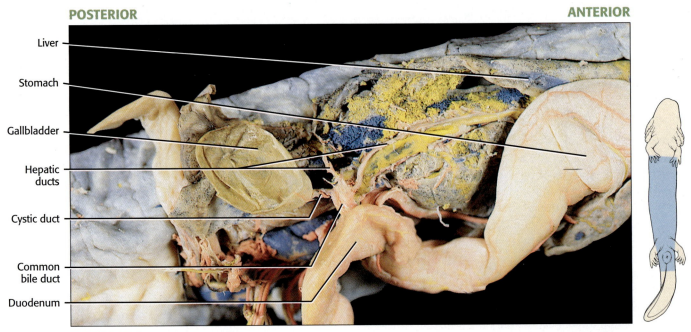

POSTERIOR   **ANTERIOR**

Liver
Stomach
Gallbladder
Hepatic ducts
Cystic duct
Common bile duct
Duodenum

FIGURE 28.6 Gallbladder and ducts.

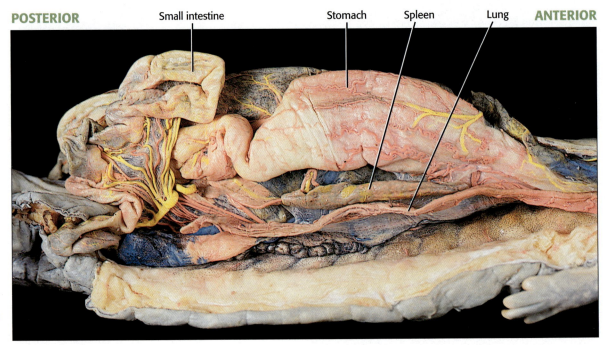

POSTERIOR   Small intestine   Stomach   Spleen   Lung   **ANTERIOR**

FIGURE 28.7
Spleen.

oxygen and carbon dioxide exchange occurs via branchial and cutaneous respiration. Slightly more than 8% is pulmonary (Guimond and Hutchinson, 1976).

Respiratory membranes must be maintained in a moist condition to permit gas exchange. Likewise, amphibian skin is specialized for gas exchange and requires a moist surface. A successful group of salamanders, members of the family Plethodontidae, have no lungs. Instead, they rely principally on cutaneous respiration. Naturalists who are interested in salamanders of the eastern United States note

that almost every species of salamander encountered belongs to this family.

The most obvious organs of respiration in *Necturus* are three pairs of bushy **external gills,** located at the posterior boundary of the head. Recall that the gills in the shark are internal, associated with the gill pouches and gill slits. Notice the tremendous vascularity of the gill filaments in injected specimens. A **gill slit** occurs between the first and second gills and the second and third gills, permitting water to be expelled from the pharyngeal cavity (Figure 28.8). Gaseous exchange

External gill slits

External gills

FIGURE 28.8 External gills and gill slits.

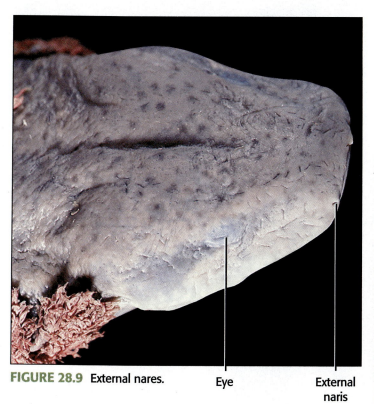

FIGURE 28.9 External nares.

Eye

External naris

Internal nares   Glottis

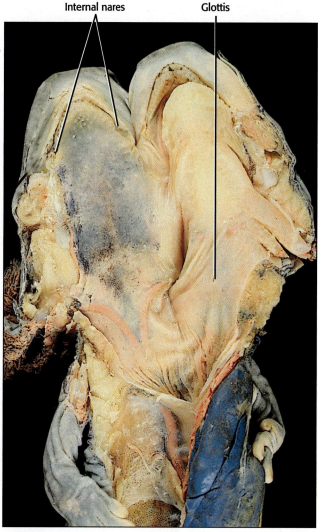

FIGURE 28.10 Respiratory anatomy I.

occurs across the membranes of the external gills, and in living specimens the gills are waved back and forth in the horizontal plane, probably ensuring a fresh supply of water.

Necturus possesses **external nares** leading into tube-like nasal chambers containing olfactory epithelia (Figure 28.9). The nasal chambers open into the mouth through a pair of small lateral slits, the **internal nares** (Figure 28.10).

The **glottis**, a small longitudinal slit, opens into a small larynx (Figure 28.10). Leading from the larynx is the short **trachea**, supported by lateral cartilages, probably remnants of posterior visceral arches (Figure 28.11). The trachea branches into a pair of short bronchi that lead into the **lungs**. The lungs of *Necturus* are elongate simple sacs that do not contain alveoli (Figure 28.12 and Figure 28.13).

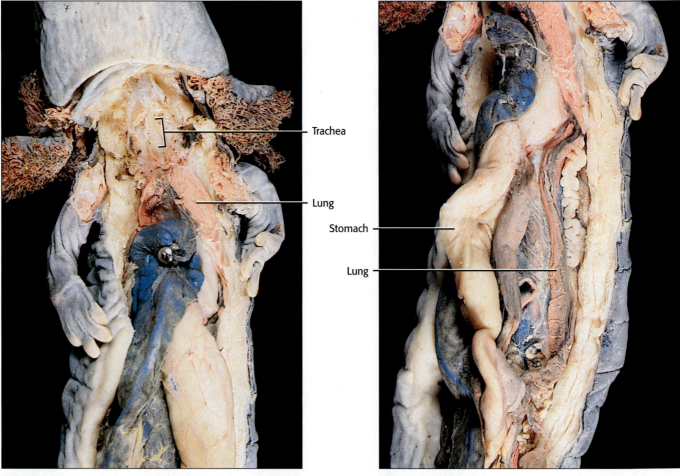

FIGURE 28.11 Respiratory anatomy II.

Trachea

Lung

Stomach

Lung

FIGURE 28.12 Respiratory anatomy III.

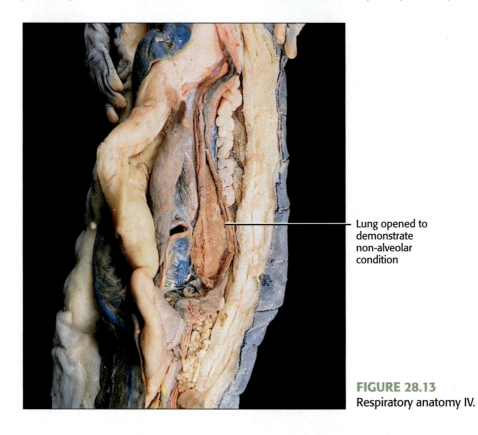

Lung opened to demonstrate non-alveolar condition

FIGURE 28.13
Respiratory anatomy IV.

CHAPTER TWENTY-NINE

Mudpuppy — Urogenital System

T he urogenital system in *Necturus* is similar to that of the shark, with minor differences. For an overview of the urinary and reproductive system, review Chapters 9 and 10.

URINARY SYSTEM

The urogenital system of *Necturus* consists of the kidney and its ducts and the reproductive system.

KIDNEY AND DUCTS

The position of the **kidney** in amphibians is an exception to the "retroperitoneal kidney" rule among most vertebrates. It projects into the pleuroperitoneal cavity and is surrounded almost completely by visceral peritoneum. The kidney is elongate, with a narrow cranial and a broad caudal end. In the male, the anterior end of the kidney has reproductive functions, whereas in the female it has degenerated. To see the extent and shape of the kidneys, lift the lateral edge and observe the dorsal aspect. Observe the **suprarenal glands** on the ventral surface of the kidneys (Figure 29.1).

POSTERIOR   **ANTERIOR**

Kidney   Collecting tubules   Suprarenal glands   Archinephric duct

FIGURE 29.1 Urinary system I.

In both sexes the kidneys are drained by the **archinephric duct**, which opens into the dorsolateral wall of the **cloaca**. In males, this duct is highly convoluted and conspicuous along the lateral surface of the kidney. In females, the duct is much smaller and not convoluted. Fine **collecting tubules** drain the kidney into the archinephric duct (Figure 29.1 and Figure 29.2).

To observe the anus, the opening of the urinary bladder, and reproductive ducts into the cloaca, carefully cut through the lateral wall of the cloaca on the side from which the cloacal gland was removed.

In terrestrial vertebrates, a new organ, the **urinary bladder**, representing a ventral evagination of the cloaca, is attached to the ventral body wall (Figure 29.2). The urinary bladder is anchored to the ventral body wall by the median ligament of the bladder. The highly vascularized urinary bladder functions as a storage bag for urine, and in amphibians is a source of water during osmoregulation. The urine is maintained in a hyposmotic condition relative to the blood plasma; *i.e.*, the urine contains more water molecules (is more dilute) than the plasma. If a frog, toad, or terrestrial salamander becomes dehydrated, as the plasma becomes more concentrated (loses water or contains a smaller number of water molecules), water follows its concentration gradient and diffuses out of the urinary bladder into the blood, thereby replenishing lost tissue water.

REPRODUCTIVE SYSTEM

Female

The paired **ovaries** appear as granular sacs suspended by the **mesovarium** from the dorsal body wall at the anterior end of the kidney. The oocytes in immature female ovaries are not well developed and appear as small granules, making it difficult to identify the ovary (Figure 29.3). In the mature female, the ovary contains large yolked eggs, which almost fill the pleuroperitoneal cavity (Figure 29.4).

The highly convoluted **oviduct**, suspended by the **mesotubarium**, originates just posterior to the pectoral limb. In contrast to the female shark, in which the two oviducts open in common through a single ventral ostium tubae, in *Necturus* the proximal funnel-shaped ends open into the pleuroperitoneal cavity as a separate left and right **ostium tubae** (Figure 29.5). The distal end of the oviduct opens as a **papilla** into the dorsal wall of the cloaca (Figure 29.6). In the dorsal wall of the cloaca is a spermatheca, homologous with a small dorsal gland in the male. It functions in the female as a sperm-storage organ where sperm can be stored for several months. Notice that the cloaca of the female has no papillae inside but has some inconspicuous folds.

Male

The somewhat elongated, solid, paired **testes** of the male are attached near the anterior end of the kidney to the dorsal body wall by means of the **mesorchium** (Figure 29.7). A number of fine **efferent ducts** permit the sperm from the testes

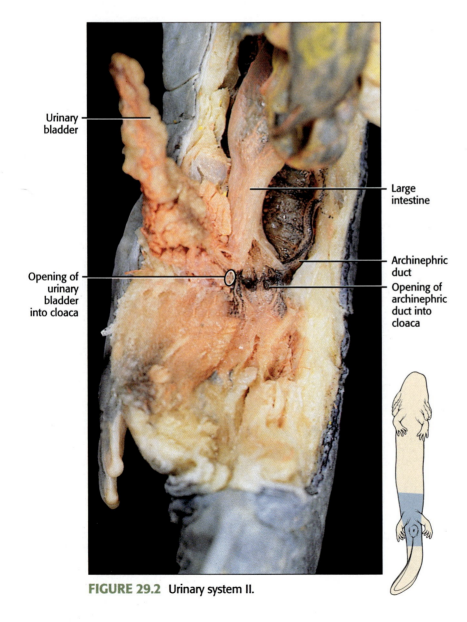

Urinary bladder

Opening of urinary bladder into cloaca

Large intestine

Archinephric duct

Opening of archinephric duct into cloaca

FIGURE 29.2 Urinary system II.

to be transported through the mesorchium to the kidney tubules of the narrow anterior end of the kidney. These kidney tubules connect to the proximal end of the **archinephric duct**, which continues posteriorly, opening into the cloaca, thereby permitting the sperm to be carried to the cloaca (Figure 29.2 and Figure 29.8). Notice the black threadlike

vestigial oviduct extending anteriorly along and beyond the proximal end of the archinephric duct (Figure 29.8).

 During dissection of the muscles in the vicinity of the male cloaca, half of the cloacal gland was removed. Observe the remaining half.

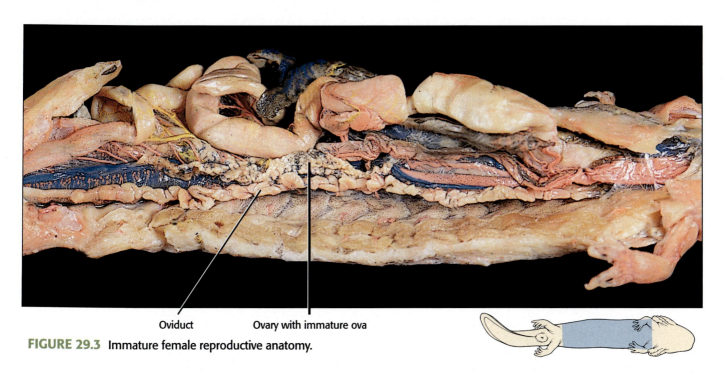

Oviduct   Ovary with immature ova

FIGURE 29.3 Immature female reproductive anatomy.

Ovary with large yolked eggs

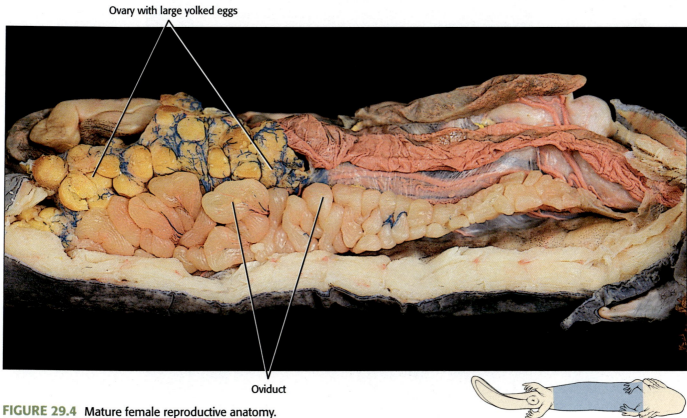

Oviduct

FIGURE 29.4 Mature female reproductive anatomy.

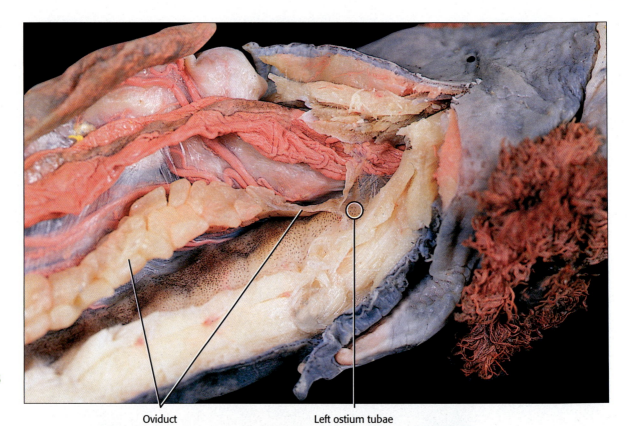

FIGURE 29.5
Female ducts.

Oviduct

Left ostium tubae

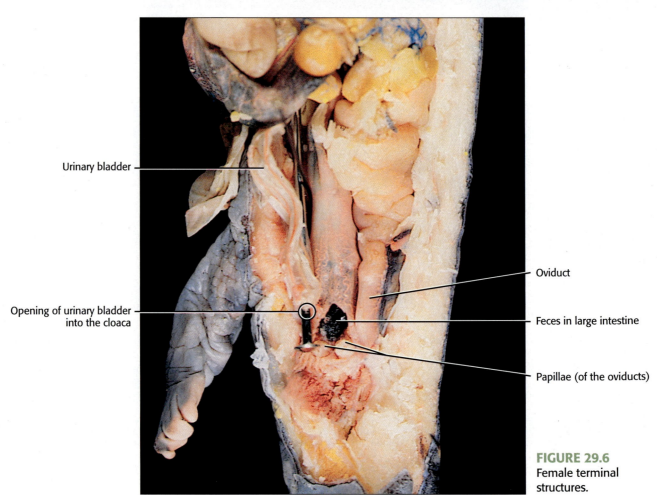

Urinary bladder

Oviduct

Opening of urinary bladder
into the cloaca

Feces in large intestine

Papillae (of the oviducts)

FIGURE 29.6
Female terminal
structures.

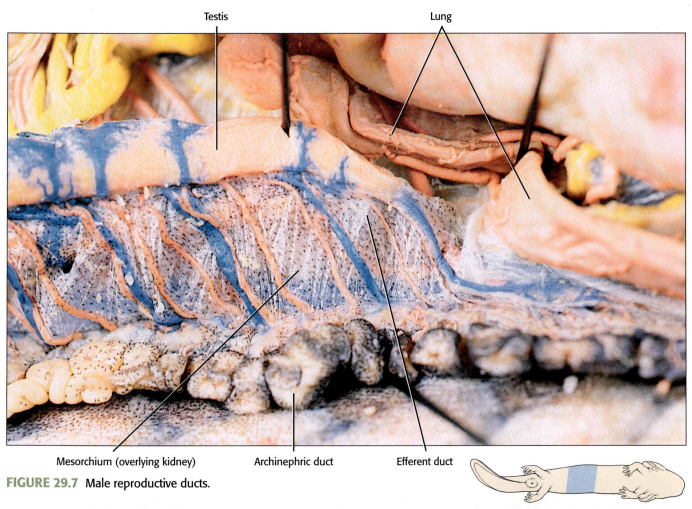

Testis

Lung

Mesorchium (overlying kidney)

Archinephric duct

Efferent duct

FIGURE 29.7 Male reproductive ducts.

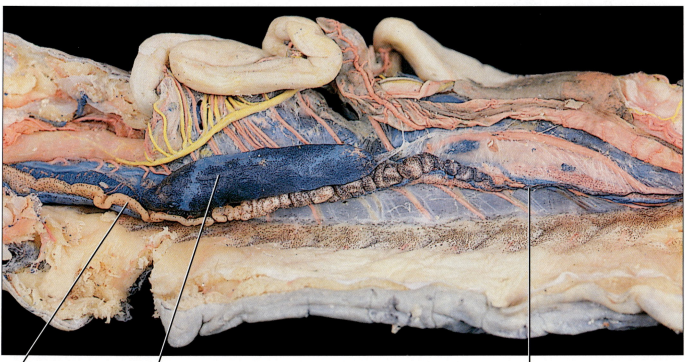

Archinephric duct

Testis

Vestigial oviduct

FIGURE 29.8 Mature male reproductive anatomy.

Secretions of the cloacal gland, combined with those of a small inconspicuous gland in the dorsal wall of the cloaca, produce spermatophores, gelatinous sperm packets. Note the presence of small **papillae** lining the cloaca, with a pair of enlarged papillae obvious at the posterior end (Figure 29.9).

Although spermatophores are deposited outside the female's body, fertilization is internal, taking place prior to egg deposition. During the breeding season in the fall, the male deposits spermatophores on the substratum of the aquatic environment. The female picks them up with her cloaca and stores them in the spermatheca until the eggs are fertilized before being deposited in the spring. Females care for their nest of oviparous eggs.

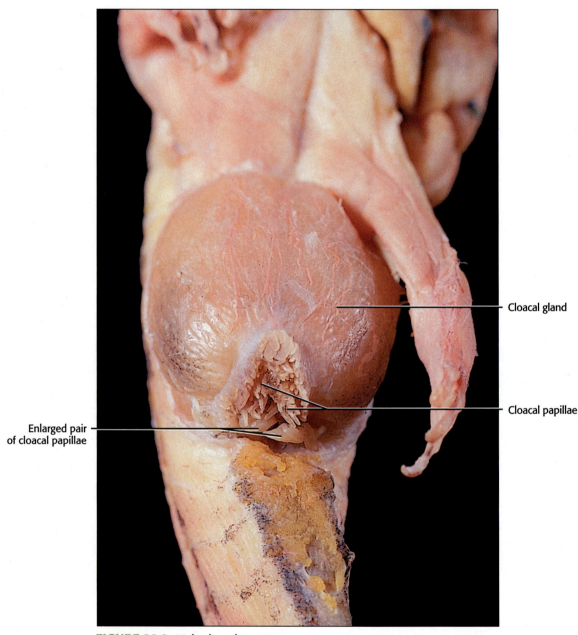

Cloacal gland

Cloacal papillae

Enlarged pair of cloacal papillae

FIGURE 29.9 Male cloacal structures.

CHAPTER THIRTY

Mudpuppy — Circulatory and Lymphatic Systems

CIRCULATORY SYSTEM

The circulatory system of *Necturus* exhibits a combination of a piscine-like and a terrestrial vertebrate-like circuit. The circulatory system, consisting of the heart and blood vessels, exhibits a great deal of variability in the branching pattern. For an overview of circulatory system evolution, see Chapter 11.

 During dissection of the circulatory system, utmost caution in handling all components of this system is essential.

HEART

 In the study of the body cavities, the heart was exposed. Now, further expose the heart and afferent branchial arteries by carefully picking away some of the surrounding tissue.

The heart projects into the **pericardial cavity**, lined with the tough **parietal pericardium**, which reflects onto the surface of the heart as the thin outer layer, the **visceral pericardium**, also known as the epicardium. Ventrally, the **ventricle**, the **left atrium** and **right atrium**, the **conus arteriosus**, and the **ventral aorta**, with its swollen base, the **bulbus arteriosus**, can be identified (Figure 30.1, Figure 30.2, and Figure 30.3).

 To view the internal anatomy, using a pair of sharp scissors, make a frontal cut through the ventricle and left atrium. With fine forceps, carefully pick out any blood or latex in these chambers. Be especially careful in working with the atrium, where the muscular strands are very delicate.

The walls of the ventricle are heavily muscled, and the myocardium exhibits well defined **muscular trabeculae**. An incomplete **interventricular septum**, consisting of a group of trabeculae lying to the left of the midsagittal plane, establishes a smaller left and a larger right area. These cavities clearly are only partially separate. Further, an incomplete **intraventricular septum** separates the right space into two (Putnam and Dunn, 1978) (Figure 30.4A and 30.4B).

Anterior to the ventricle are the left and right atria, separated by a perforated **interatrial septum**. Notice the strands of muscle extending between the walls. Reflect the ventricle to expose the **hepatic sinuses**, the thin walled, dorsal **sinus venosus**, and the **confluent pulmonary veins** opening into the left atrium (Figure 30.3).

Opening into the sinus venosus is the **posterior vena cava**, which subdivides into a pair of **hepatic sinuses** as it passes into the pericardial cavity. The hepatic sinuses merge with the sinus venosus. **Common cardinal veins (ducts of Cuvier)** draining the head, shoulder, and body wall enter the sinus venosus anteriorly (Figure 30.3).

A number of valves operate during heart function to ensure that the blood has a unidirectional flow. A flaplike **sinoatrial valve** guards the entrance of the sinus venosus into the right atrium. In contrast, the entrance of the pulmonary vein is not guarded by a valve. The opening between the atria and the ventricle is guarded by a paired atrioventricular valve (Figure 30.4B).

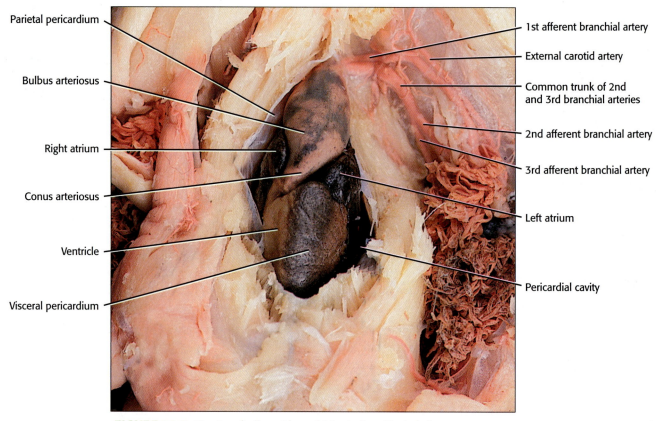

Parietal pericardium
Bulbus arteriosus
Right atrium
Conus arteriosus
Ventricle
Visceral pericardium

1st afferent branchial artery
External carotid artery
Common trunk of 2nd and 3rd branchial arteries
2nd afferent branchial artery
3rd afferent branchial artery
Left atrium
Pericardial cavity

FIGURE 30.1   Heart and afferent branchial arteries: Ventral view.

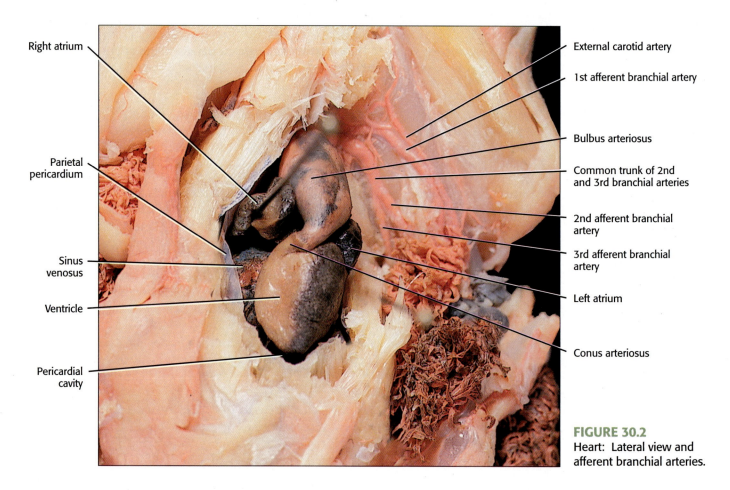

Right atrium
Parietal pericardium
Sinus venosus
Ventricle
Pericardial cavity

External carotid artery
1st afferent branchial artery
Bulbus arteriosus
Common trunk of 2nd and 3rd branchial arteries
2nd afferent branchial artery
3rd afferent branchial artery
Left atrium
Conus arteriosus

FIGURE 30.2
Heart: Lateral view and afferent branchial arteries.

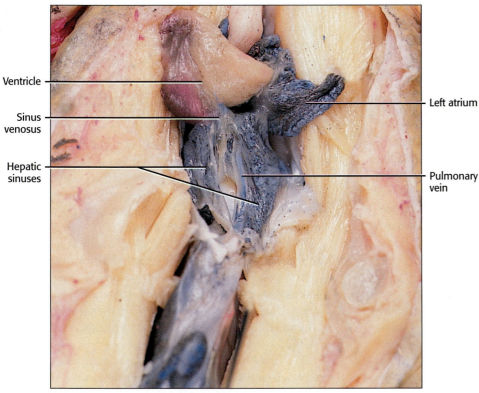

Ventricle

Sinus venosus

Hepatic sinuses

Left atrium

Pulmonary vein

FIGURE 30.3 Heart: Reflected.

To view the following interior anatomy of the conus and bulbus arteriosus, with a sharp pair of scissors carefully cut the ventral surface of both longitudinally, and reflect the cut edges.

At the base of the conus arteriosus, as it arises from the ventricle, are two rows of cup-shaped **semilunar valves**. The swollen base of the ventral aorta, the bulbus arteriosus, is subdivided by a **longitudinal partition** into two compartments (Figure 30.4A and 30.4B).

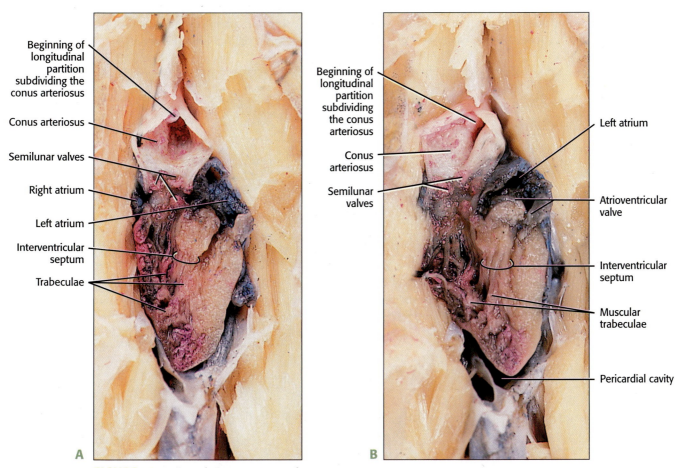

Beginning of longitudinal partition subdividing the conus arteriosus

Conus arteriosus

Semilunar valves

Right atrium

Left atrium

Interventricular septum

Trabeculae

Beginning of longitudinal partition subdividing the conus arteriosus

Conus arteriosus

Semilunar valves

Left atrium

Atrioventricular valve

Interventricular septum

Muscular trabeculae

Pericardial cavity

A

B

FIGURE 30.4 (A and B) Heart: Internal Anatomy.

ARTERIES

Ventral Aorta and Afferent Branchial Arteries

Note the similarities to the circulatory pattern in the gills of the shark. As the **ventral aorta** passes out of the pericardial cavity, it divides into a right and left vessel. Subsequently, each of these divides into two afferent arteries. The anterior artery remains single and represents the **first afferent branchial** artery associated with the third visceral arch. The posterior artery again divides to form the **second** and **third afferent branchial arteries** associated with the fourth and fifth visceral arches. The first, second, and third branchial arteries enter the first, second, and third external gills, respectively, and branch into capillaries, where gas exchange occurs (Figure 30.1 and Figure 30.2).

Efferent Branchial Arteries, Dorsal Aorta and Its Branches

From the capillaries of the external gills lead three efferent branchial arteries. The **first efferent branchial artery** arises in the first gill and remains independent. The **second** and **third efferent branchial arteries** emerge from the second and third external gills and immediately become confluent with a common stem (Figure 30.5).

> Study the branching of the efferent branchial arteries on the side opposite the cut through the gills made to expose the oral cavity and pharynx.

The first efferent branchial turns medially and gives rise to the **external carotid artery**, which branches into a number of smaller arteries supplying the surrounding musculature, thyroid gland, and tongue (Figure 30.1 and Figure 30.2). As the first efferent branchial artery continues toward the midline, it gives rise to the **internal carotid artery**, which courses anteriorly, soon disappearing into the skull. It is connected to the common stem of the second and third by a short vessel to form the **radix of the aorta**. From the right and left common stem of the second and third efferent branchial artery arises the **pulmonary artery**, which carries blood to the lung. Just medial to the origin of the internal carotid artery is the origin of the **vertebral artery**, which soon disappears into the skull (Figure 30.5).

The union of the left and right radix forms the **dorsal aorta**. Notice the paired **intersegmental arteries**, supplying body musculature, arising along the length of the dorsal aorta. Probably, the paired arteries (the subclavians, iliacs, renals, etc.) seen within the pleuroperitoneal cavity were derived from these intersegmental arteries (Figure 30.5).

The first branch of the dorsal aorta is the **subclavian artery**, which extends laterally. It branches into a prominent **cutaneous artery**, which courses posteriorly carrying blood near the skin, where gas exchange occurs. From the cutaneous artery a number of small arteries supply the shoulder. The cutaneous artery continues as the **brachial artery**, which extends into the forelimb (Figure 30.6).

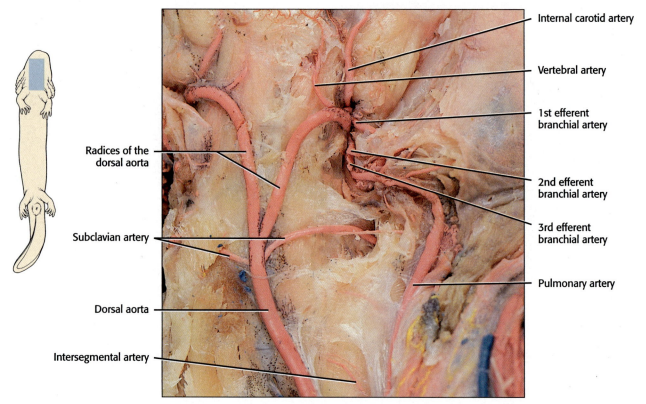

FIGURE 30.5 Efferent branchial arteries, dorsal aorta and its branches.

Just posterior to the subclavian is the unpaired **gastric artery**, which branches to form **dorsal** and **ventral gastric arteries** carrying blood to the stomach. A **splenic artery**, transporting blood to the anterior portion of the spleen, arises from one of these (Figure 30.7).

Posterior to the gastric artery, a large **coeliacomesenteric artery** arises from the dorsal aorta. It subdivides into a number of branches: a large **splenic artery** to the posterior portion of the spleen; a **pancreaticoduodenal artery** to the pancreas, duodenum, and pyloric region of the stomach; a **hepatoduodenal artery** to the liver and duodenum; and a **mesenteric artery** to the small intestine (Figure 30.8 and Figure 30.9).

A number of **mesenteric arteries,** further supplying blood to both the small and large intestines, arise from the dorsal aorta posterior to the coeliacomesenteric (Figure 30.8). At the level of the gonads are a series of paired **ovarian** or **testicular arteries** and numerous paired **renal arteries**

to the kidneys (Figure 30.10). The blood supply to the gonads is similar in both sexes.

Note: *The following dissection might be better delayed until after dissection of the ventral abdominal vein and the veins of the hindlimb.*

 Dissection must be careful, to expose the pelvic arteries and veins. Again, a dissecting microscope is most helpful during the dissection. The ventral abdominal vein generally has shifted to the right side of the specimen. To ascertain the extent of the ventral abdominal vein, carefully strip the muscles from the ventral side of the pelvic girdle. This allows you to see the yellow latex in the vein through the cartilage of the girdle. Carefully pick away the girdle, checking constantly where the

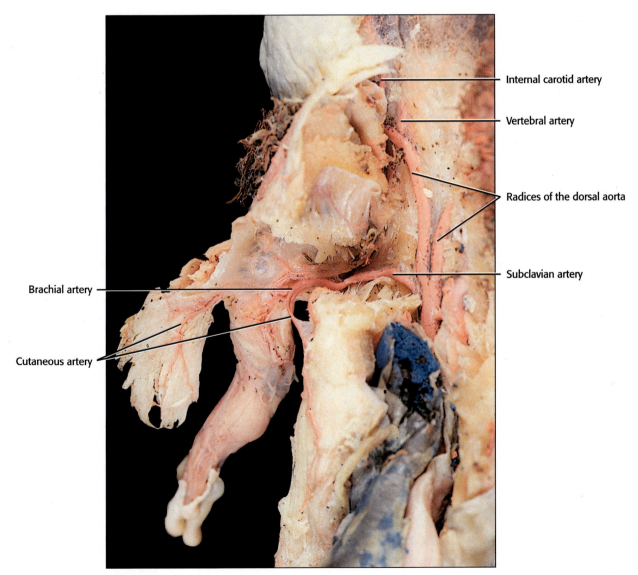

Internal carotid artery

Vertebral artery

Radices of the dorsal aorta

Subclavian artery

Brachial artery

Cutaneous artery

FIGURE 30.6 Arteries of the right shoulder and forelimb (ventral view).

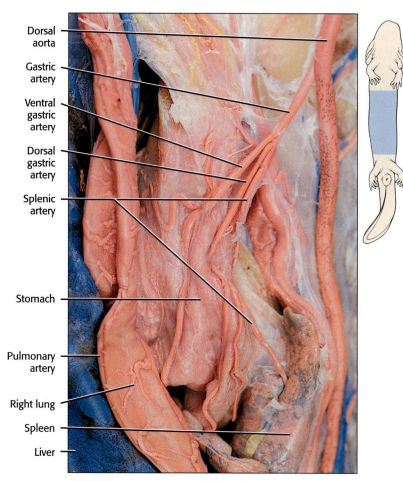

Dorsal aorta
Gastric artery
Ventral gastric artery
Dorsal gastric artery
Splenic artery
Stomach
Pulmonary artery
Right lung
Spleen
Liver

FIGURE 30.7 Gastric arteries and branches.

dissection is headed during this process. The pelvic veins lie almost directly ventral to the pelvic musculature, and the ventral abdominal vein adheres tightly to the parietal peritoneum underlying the urinary bladder and the ventral abdominal wall.

To free the ventral abdominal vein from the peritoneum, use scissors to cut between the vein and peritoneum. Like good practice in woodworking, measure twice and cut once. If you are careful, your specimen will exhibit a good venous dissection in this area, and the arterial system of this region also will be superior.

Upon reaching the posterior limit of the pleuroperitoneal cavity, the dorsal aorta gives off paired lateral **iliac arteries**. The iliac artery branches, giving rise to an **epigastric artery** supplying the anterior body wall and a **hypogastric artery** supplying the bladder and cloaca. It continues into the hindlimb as the **femoral artery**, which itself subdivides to provide blood to the tissues of the hindlimb (Figure 30.11). As the dorsal aorta passes the cloaca, it gives off a pair of **cloacal arteries**. Posterior to the cloacal arteries, the dorsal aorta continues into the tail as the **caudal artery.**

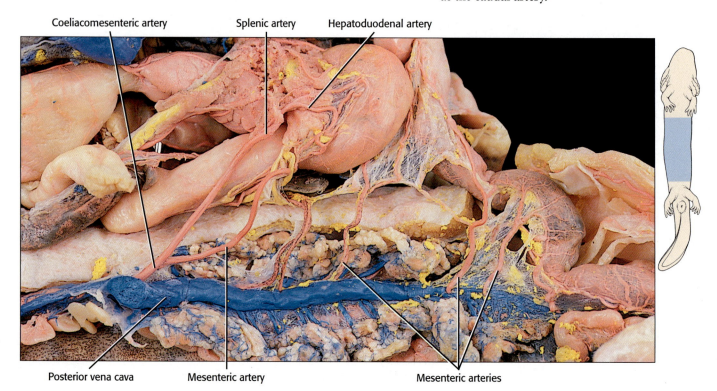

Coeliacomesenteric artery   Splenic artery   Hepatoduodenal artery

Posterior vena cava   Mesenteric artery   Mesenteric arteries

FIGURE 30.8 Branches of the coeliacomesenteric artery, mesenteric arteries I.

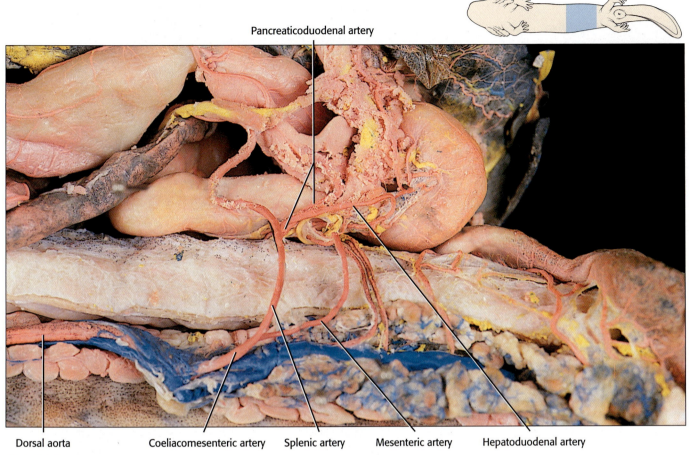

Pancreaticoduodenal artery

Dorsal aorta   Coeliacomesenteric artery   Splenic artery   Mesenteric artery   Hepatoduodenal artery

FIGURE 30.9 Branches of the coeliacomesenteric artery, mesenteric arteries II: Anterior left.

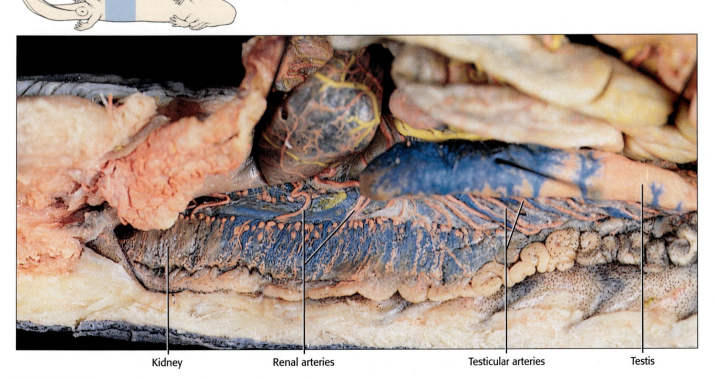

Kidney   Renal arteries   Testicular arteries   Testis

FIGURE 30.10 Testicular and renal arteries: Anterior right.

VEINS

Renal Portal System

The tail region is drained by the caudal vein, which bifurcates at the posterior end of the cloaca, forming a pair of **renal portal veins** that course along the lateral margin of the kidney (Figure 30.12). These veins connect at the anterior end of the kidney with the **posterior cardinal veins,** which extend anteriorly and empty into the **common cardinal veins (ducts of Cuvier),** which then empty into the sinus venosus (Figure 30.13).

Posterior Vena Cava

A new drainage route from the posterior region of the body evolved in tetrapods and their ancestral fish line. Portions of the embryonic paired posterior cardinal/subcardinal veins and the right hepatic vein have evolved into a single **posterior vena cava.** The posterior vena cava extends from the medial aspect of the kidneys, receiving blood from the **efferent renal veins** of the kidneys and the **ovarian veins** (or **testicular veins**) from the gonads. It continues through the liver, passing through the coronary ligament after having been joined by a number of small **hepatic veins** (Figure 30.13), as well as the large **left hepatic vein.** After passing through the transverse septum, the vena cava splits into a pair of **hepatic sinuses,** which join the **common**

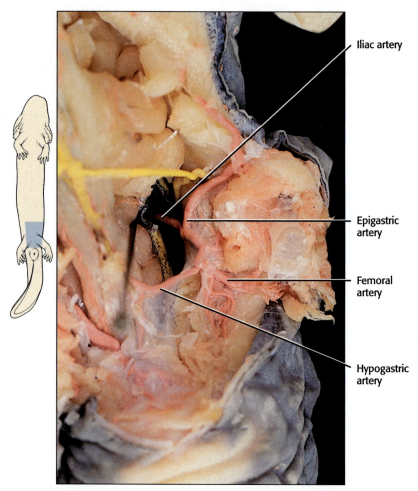

FIGURE 30.11 Vessels of the pelvic region and hindlimb.

POSTERIOR   ANTERIOR

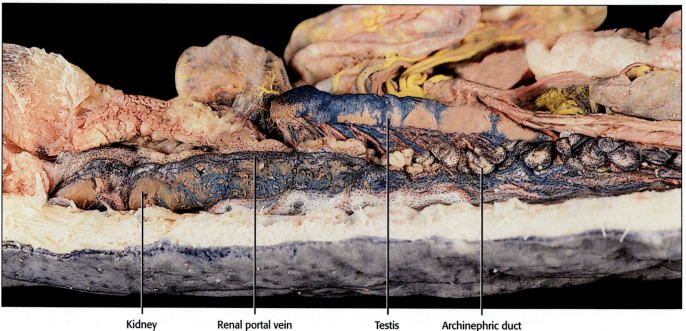

Kidney   Renal portal vein   Testis   Archinephric duct

FIGURE 30.12 Renal portal veins.

cardinal veins that empty into the sinus venosus (Figure 30.3).

A second posterior drainage route persists in the form of small subcardinal veins leaving the anterior end of the kidneys, which join the remnants of the anterior portion of the posterior cardinal veins. As the posterior cardinal veins course forward, they are joined by intersegmental veins draining adjacent body wall regions. The postcardinal veins empty into the common cardinal veins (Figure 30.13).

Pulmonary Veins

A pulmonary vein extends from the ventral side of each lung, passes dorsal to the left hepatic sinus, and unites into a single vessel that enters the left atrium of the heart (Figure 30.14). In adult tetrapods and some related fish, pulmonary veins transport oxygen-rich blood to the heart, whereas all other veins transport oxygen-poor blood.

Note: *The hepatic portal system, the ventral abdominal vein, and the veins of the hindlimb and pelvic region probably will be injected with yellow latex.*

Hepatic Portal System

As in the shark, the digestive and associated organs are drained by a venous system known as the hepatic portal. The mesenteric vein, receiving numerous intestinal veins, drains the large and small intestine. A prominent gastrosplenic vein,

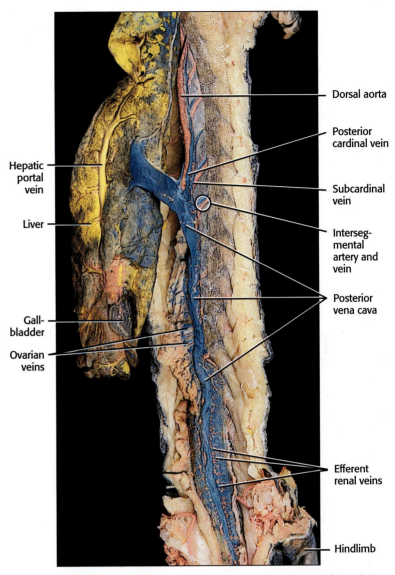

FIGURE 30.13 Posterior vena cava and hepatic portal circulation.

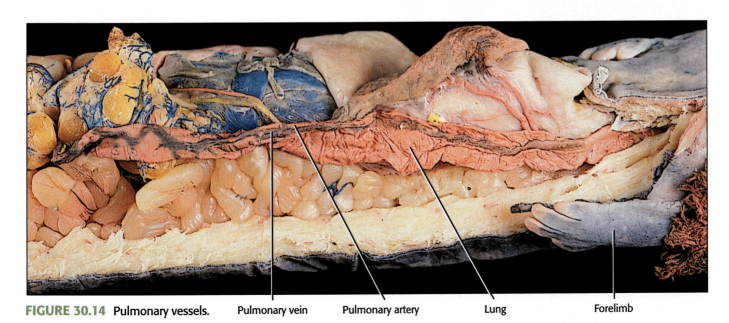

FIGURE 30.14 Pulmonary vessels.

with a **splenic vein** draining the spleen and a number of **gastric veins** draining the stomach, joins with the **mesenteric vein** in the pancreatic tissue, picking up small **pancreatic veins** to form the **hepatic portal vein**. The hepatic portal vein enters the liver and breaks up into hepatic sinuses. Hepatic veins leading from the sinuses empty into the sinus venosus (Figure 30.13 and Figure 30.15).

Ventral Abdominal Vein and Associated Veins of the Hindlimb

The ventral abdominal vein of *Necturus* seems to be homologous with the lateral abdominal veins of the shark (refer to Figure 21.23). This vein has migrated medially and now lies midventrally. Drainage of the hindlimb may occur by way of two related venous circuits. One of the routes of hindlimb drainage consists of the **femoral veins** that drain the hindlimb joining the **iliac veins** that empty into the renal portal vein. The second route consists of the femoral veins joining the **pelvic veins** that coalesce to form the **ventral abdominal vein**, which subsequently passes along the posterior edge of the falciform ligament and joins the hepatic portal vein on the surface of the liver. A **cloacal vein** drains the cloacal region and joins the ventral abdominal vein at the point at which the pelvic veins enter the ventral abdominal vein. Small **intersegmental veins** from the body wall and **vesical veins** from the urinary bladder empty into the ventral abdominal vein (Figure 30.16).

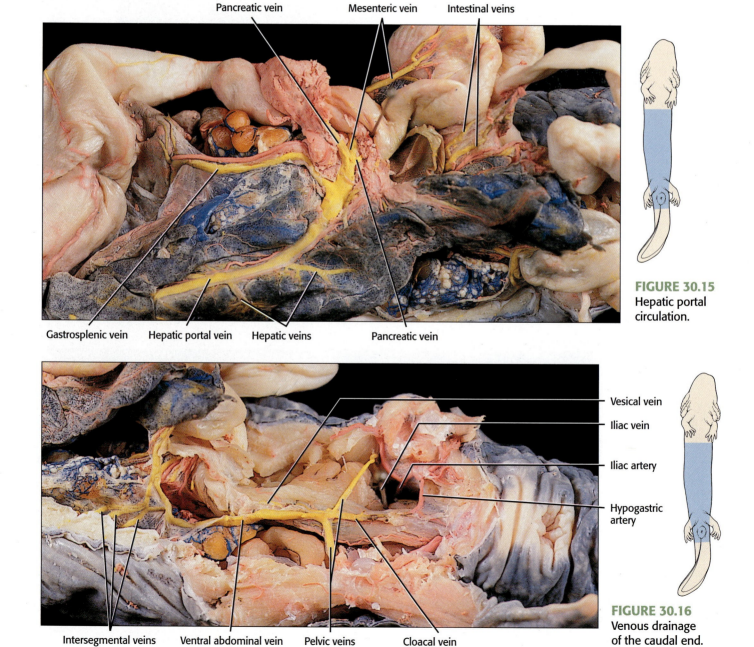

Pancreatic vein   Mesenteric vein   Intestinal veins

Gastrosplenic vein   Hepatic portal vein   Hepatic veins   Pancreatic vein

FIGURE 30.15
Hepatic portal circulation.

Vesical vein
Iliac vein
Iliac artery
Hypogastric artery

Intersegmental veins   Ventral abdominal vein   Pelvic veins   Cloacal vein

FIGURE 30.16
Venous drainage of the caudal end.

Veins of the Head and Forelimb

Tissues of the anterior region of *Necturus* are drained by veins that are partially homologous with the veins of the shark. Drainage of the head and jaws is accomplished by the internal jugular veins, homologous with the anterior cardinal vein of fishes and the **external jugular veins.** The internal jugular veins often are uninjected and therefore difficult to identify. These veins join and empty into the **common cardinal veins.** The tongue and floor of the oral cavity are drained by the **lingual vein,** homologous with the inferior jugular vein of fishes, which also drains into the common cardinal veins (Figure 30.17A, Figure 30.17B, and Figure 30.18).

The forelimbs are drained by the **brachial veins** that join the **cutaneous veins,** carrying oxygen-rich blood from the skin to form the **subclavian veins,** which empty into the common

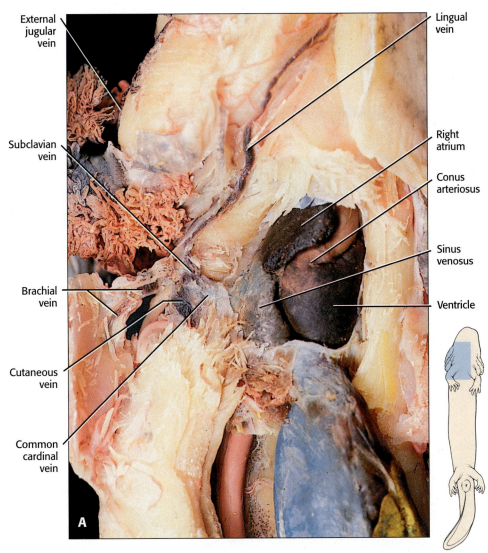

A

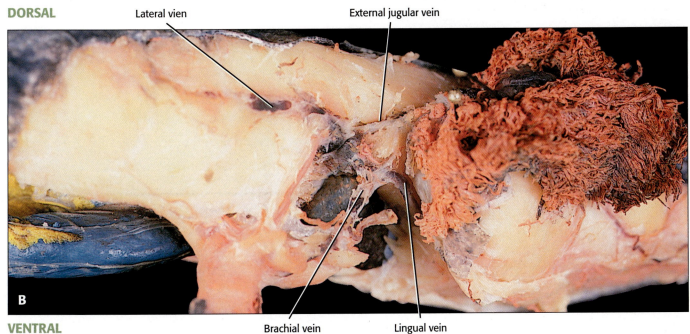

B

DORSAL

VENTRAL

FIGURE 30.17 (A) Veins of the head and forelimb I; (B) Veins of the head and forelimb II.

cardinal veins. Drainage of the forelimb of *Necturus* is almost a carbon copy of the fish fin drainage and, therefore, is homologous. A **lateral vein** drains the lateral body wall and empties into the common cardinal veins (Figure 30.17A, 30.17B, and Figure 30.18).

LYMPHATIC SYSTEM

Amphibians possess a well developed lymphatic system in the subcutaneous tissue to return fluids to the blood. The lymphatic system, however, is difficult to dissect grossly.

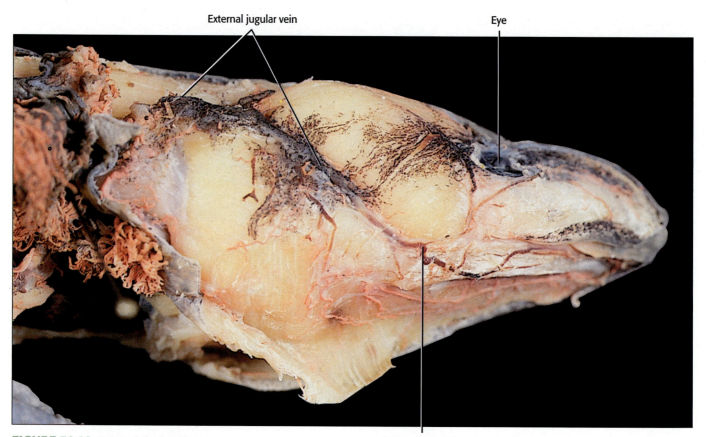

FIGURE 30.18 Veins of the head.

CHAPTER THIRTY-ONE

Mudpuppy — Nervous System and Sense Organs

NERVOUS SYSTEM

The brain and the location and distribution of cranial nerves in *Necturus* is similar to that of the fish. With the evolution of terrestrial appendages, spinal nerves to innervate them became more specialized, forming a complex called the brachial plexus associated with the pectoral limbs and a lumbosacral plexus associated with the hindlimb.

 If available, use either a dissecting microscope or a large magnifying glass for the following dissections.

Brain and Cranial Nerves

The following two dissections should be done on different specimens, as it is difficult to expose both brain surfaces without destroying the brain.

 To expose the dorsal surface of the brain, remove the dorsal musculature of the head. Carefully ease the tip of a pair of forceps under the overlying skull bones and lift, breaking off small pieces and removing them until the brain is exposed, leaving the brain in the specimen. Take special care to conserve the cranial nerves and semicircular ducts.

The complex vascularity on both brain surfaces will be obvious. To observe the underlying architecture of the dorsal surface of the brain, carefully peel off the delicate vascularized layer. In the vicinity of the paraphysis and diencephalon, the roof of the brain is extremely delicate and can be torn off. Do not be overzealous.

To expose the ventral surface of the brain, employ the same techniques as those used to dissect the efferent branchial arteries. Remove the mucosa of the roof of the oral cavity, exposing the ventral surface of the skull. If you are using a dissecting microscope, you will immediately see the outline of the ventral surface of the brain. With a pair of forceps, carefully pry the bones up, taking care to remove the pieces as you progress.

Be particularly cautious when removing skull material around the cranial nerves. Do not remove the dense vascular network covering and adhering tightly to the ventral surface of the brain.

Meninges and Ventricles of the Brain

The brain is encased in protective **meninges,** or membranes, consisting of an inner **secondary meninx** and an outer **dura mater** (Figure 31.1).

As in the shark, the cerebral hemispheres contain lateral ventricles (I and II). The ventricle of the diencephalon (III) connects with the lateral ventricles through the foramen of Monro. The mesencephalon in *Necturus* also contains a ventricle (not designated by number) that is continuous with the fourth ventricle (IV), residing between the medulla oblongata and the auricles of the cerebellum. The fourth ventricle is continuous with the spinal canal of the spinal cord.

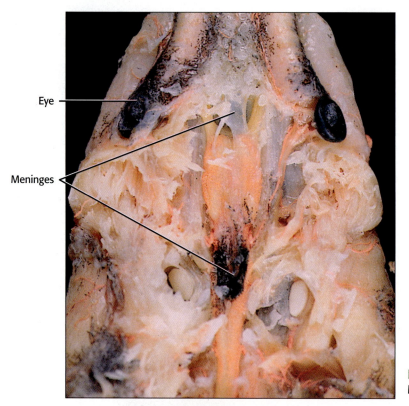

Eye

Meninges

Note: *For the following discussion, refer to Figure 31.2 and Figure 31.3.*

Telencephalon

The **telencephalon** consists of the olfactory bulbs, appearing as slight bulges at the anterior end of the elongate **cerebral hemispheres**. An **olfactory nerve** (cranial nerve I) leads from the **olfactory sac** to the **olfactory bulb**.

Diencephalon

Posterior to the telencephalon lies the **diencephalon**. A thin roof, the **epithalamus**, is covered by a **tela choroidea** similar to the shark. From the diencephalon projects the **paraphysis**, extending between the posterior ends of the

FIGURE 31.1
Meninges.

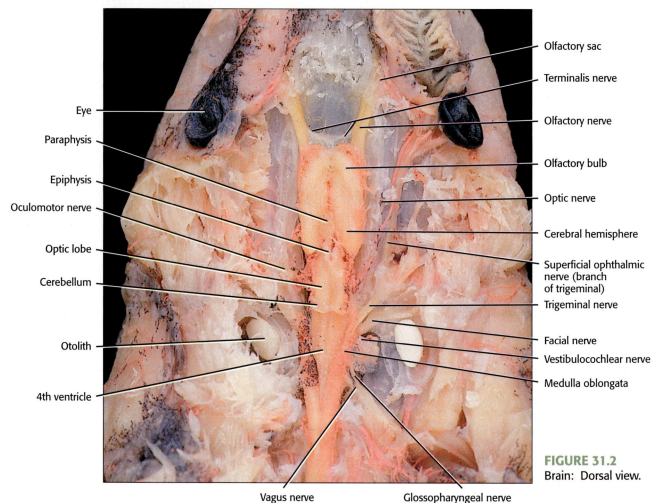

Eye

Paraphysis

Epiphysis

Oculomotor nerve

Optic lobe

Cerebellum

Otolith

4th ventricle

Olfactory sac

Terminalis nerve

Olfactory nerve

Olfactory bulb

Optic nerve

Cerebral hemisphere

Superficial ophthalmic nerve (branch of trigeminal)

Trigeminal nerve

Facial nerve

Vestibulocochlear nerve

Medulla oblongata

Vagus nerve   Glossopharyngeal nerve

FIGURE 31.2
Brain: Dorsal view.

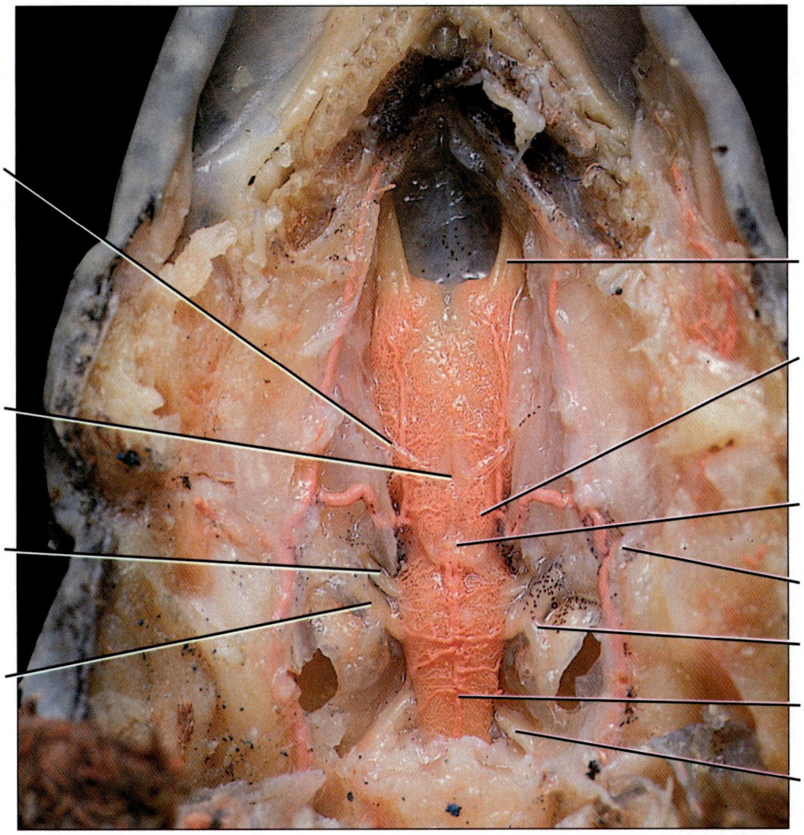

FIGURE 31.3
Brain: Ventral view.

cerebral hemispheres. Just posterior to the paraphysis is the obscure **epiphysis** (pineal gland) (Figure 31.2).

On the ventral surface of the diencephalon, observe the **optic chiasma**. Similar to the shark, the fibers of the **optic nerve** (cranial nerve II) cross to opposite sides of the brain in the optic chiasma and enter the diencephalon. Posterior to the chiasma, the large **infundibulum** connects the **hypothalamus** to the **hypophysis** (pituitary gland).

Mesencephalon

Similar to the shark, paired **optic lobes** of the **mesencephalon** occur posterior to the diencephalon. Again, similar to the shark, the **oculomotor nerve** (cranial nerve III) emerges from the mesencephalon to innervate four of the extrinsic eye muscles, and the **trochlear nerve** (cranial nerve IV) similarly emerges dorsally to innervate a fifth extrinsic eye muscle. It appears to emerge in the groove between the optic lobes and the cerebellum.

Metencephalon

In contrast to the shark, the **cerebellum** of *Necturus* is poorly developed and consists of a narrow, thin, transverse band of tissue just posterior to the optic lobes. Slightly developed auricles occur at the dorsolateral aspect of the **fourth ventricle.**

TABLE OF CRANIAL NERVES

Name	Number	Neuron Type: Sensory	Neuron Type: Motor	Distribution
Terminal**	O	X		Same as shark.
Olfactory	I	X		Same as shark.
Optic	II	X		Same as shark.
Oculomotor	III	†	X	Same as shark.
Trochlear	IV	†	X	Same as shark.
Trigeminal	V	X	X	Same as shark.
Abducens	VI	†	X	Similar to shark, innervates an extrinsic eye muscle, the lateral rectus. However, the newly evolved retractor bulbi muscle, derived from the lateral rectus, also is innervated by this nerve.
Facial	VII	X	X	Similar to shark, arises in common with C.N. VIII
Vestibulocochlear	VIII	X		Similar to shark
Glossopharyngeal	IX	X	X	External gills, branchial bars, pharynx and tongue, arises in common with C.N. X
Vagus	X	X	X	Similar to shark, two major branches of this nerve: Visceral: carries sensory impulses from and motor impulses to the heart and abdominal viscera. Lateral: carries sensory impulses from the lateral line system.

** It appears to carry impulses from the olfactory sac to the cerebrum. Its function is unknown but has been associated with chemosensory functions.

† Although cranial nerves III, IV, and VI are considered motor nerves, they carry proprioceptive fibers, which are sensory and transmit information concerning muscle status.

Myelencephalon

The **medulla oblongata**, constructed similar to that of all vertebrates, is Y-shaped and is continuous with the spinal cord. Over the roof of the fourth ventricle stretches another **tela choroidea**. Similar to the shark, cranial nerves V (trigeminal), VI (abducens), VII (facial), VIII (vestibulocochlear), IX (glossopharyngeal), and X (vagus) emerge from the ventral surface of the medulla oblongata.

SPINAL NERVES AND SPINAL CORD

Spinal nerves in *Necturus* consist of a dorsal and a ventral branch, or ramus. The roots of a spinal nerve emerge from the spinal cord and fuse. A ganglion occurs distal to the point of fusion of the rami. The dorsal ramus innervates the muscles and skin of the dorsal aspect of the body, and the ventral ramus innervates corresponding ventral regions.

In tetrapods, certain spinal nerves become associated with newly evolving regions. Spinal nerve number 1 in amphibians innervates neck musculature. This nerve contributes to the formation of C.N.XI (spinal accessory), which innervates neck and back muscles in amniotes. Neurons from spinal nerves numbers 1 and 2 in amphibians turn anteriorly and contribute to the formation of the hypobranchial nerve, which innervates the hypobranchial and tongue musculature. In amniotes, C.N. XII (hypoglossal) with a similar origin innervates the tongue.

With the appearance of limbs in tetrapods arose a complex interconnected arrangement of spinal nerves, a plexus, associated with the increased sophistication of these appendages. These plexi consist of only ventral rami of the respective spinal nerves.

Brachial Plexus

To locate the brachial plexus (see Figure 31.4), *carefully* cut through and reflect the latissimus dorsi muscle. Loosen the origin of the dorsalis scapulae, cucullaris, and levator arcuum muscles, and *carefully* pull the scapula laterally. This will expose the brachial plexus. Little additional dissection should be required.

In *Necturus*, at the level of the anterior limb, **spinal nerves 3, 4, and 5** emerge between the epaxial and hypaxial muscles and make major contributions to the brachial plexus. Spinal nerve 2 makes a minor contribution to the plexus (Figure 31.4).

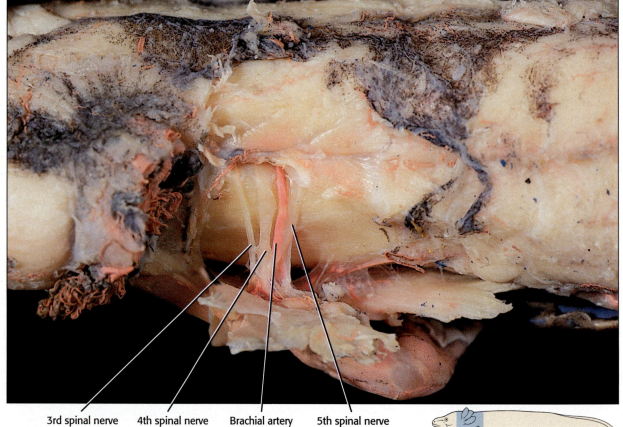

ANTERIOR

3rd spinal nerve   4th spinal nerve   Brachial artery   5th spinal nerve

FIGURE 31.4 Brachial plexus.

Lumbosacral Plexus

 To locate the lumbosacral plexus, carefully separate the epaxial and hypaxial musculature directly above the hindlimb. Carefully remove the hypaxial muscles while watching closely for the nerves of the plexus. The nerves lie directly on the lateral surface of the kidney. These nerves appear more translucent than those of the brachial plexus. They also are associated with more connective tissue that must be carefully teased away.

Spinal nerves 16, 17, and **18** make major contributions to this plexus, whereas spinal nerve 15 makes a minor contribution (Figure 31.5).

Autonomic Nervous System

The autonomic nervous system in tetrapods, in contrast to that of the elasmobranch, consists of a sympathetic division and a parasympathetic division. The sympathetic division is better developed than the parasympathetic division. This system is difficult to dissect.

SENSE ORGANS

The sense organs consist of the olfactory apparatus, lateral line system, auditory and equilibrium apparatus, and the eyes.

Olfactory Apparatus

 To expose the olfactory chamber, beginning with the external nares, carefully cut with scissors through the dorsal surface of the olfactory chamber to the posterior end. Reflect the cut edges to view the internal anatomy of the chamber. Observe the relationship of the olfactory bulb, nerve, and sac to the olfactory chamber (Figure 31.6 and Figure 31.2).

The olfactory apparatus in *Necturus* is well developed. The **olfactory chambers** are elongate tube-like organs, lined with a plicate **olfactory epithelium**. The **external nares** lead into the anterior end of the chamber (Figure 31.6). At the posterior end of the chamber, the **internal nares** open into the oral cavity just lateral to the upper teeth.

Lateral Line System

Recall that a **lateral line system** consisting of muted, elongated, dashlike indentations, occurs on the head and trunk (Figure 31.7). The lateral line system is persistent in amphibians that remain or become aquatic after becoming adults (*e.g., Cryptobranchus, Siren,* and *Amphiuma* among salamanders, some species of frogs in the family Pipidae). In most other amphibians, a lateral line system appears only in aquatic larvae. It does not appear among amniotes.

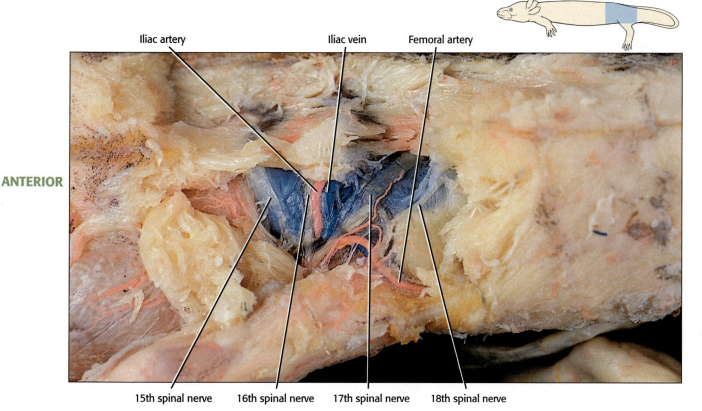

ANTERIOR

Iliac artery   Iliac vein   Femoral artery

15th spinal nerve   16th spinal nerve   17th spinal nerve   18th spinal nerve

FIGURE 31.5 Lumbosacral plexus.

Auditory and Equilibrium Apparatus

Salamanders do not possess an external or middle ear but do have an internal ear situated in the otic capsule. The equilibrium apparatus is similar to that of the shark, consisting of three semicircular ducts that communicate with the dorsal utriculus and the ventral sacculus. A conspicuous **otolith** resides in the sacculus (Figure 31.6).

The Eyes

The **eye** of *Necturus* (Figure 31.6 and Figure 31.7) is similar to other vertebrate eyes. The extrinsic eye muscles are identical to those of the shark, but for the first time a **retractor bulbi** muscle makes an appearance. It is a derivative of the lateral rectus muscle and retracts the eyeball. The lateral rectus and the retractor bulbi muscles are innervated by the abducens nerve (C.N. VI).

When you removed the skin of the head, the small, light gray spot is the cornea of the eye, which remains with the skin, as salamanders have no eyelids. Because the eye is so small, it is difficult to dissect.

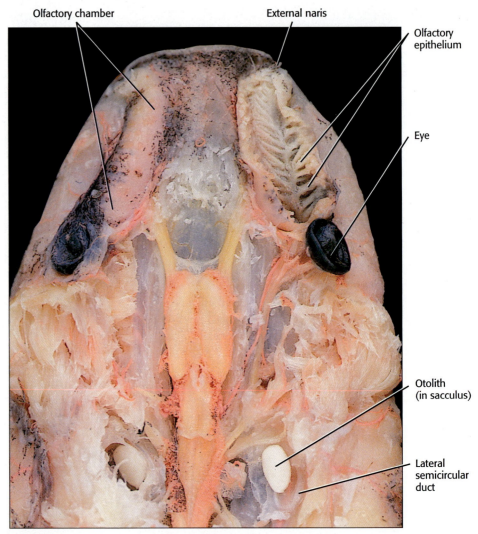

Olfactory chamber — External naris — Olfactory epithelium — Eye — Otolith (in sacculus) — Lateral semicircular duct

FIGURE 31.6 Olfactory apparatus, semicircular ducts and otolith.

Eye — Lateral line

FIGURE 31.7 Lateral line.

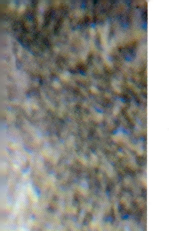

CHAPTER THIRTY-TWO

Mudpuppy — Endocrine System

32

The organs and functions of the hormones of the endocrine system in *Necturus* are similar to those of the shark.

HYPOPHYSIS OR PITUITARY

Similar to most vertebrates, the position of the **hypophysis** is on the ventral side of the brain and is suspended by the infundibulum from the hypothalamus of the diencephalon (Figure 32.1). Its formation is similar to the pituitary of the shark. Hormones secreted by the pituitary are generally similar in vertebrates, but the functions may produce effects peculiar to a specific group. For example, the hormone prolactin in mammals stimulates milk production, whereas in amphibians it stimulates the "water drive" —the urge to move toward water during the breeding season, proliferation of melanophores, etc.

Pineal Gland

The **pineal gland (epiphysis)** projects from the roof of the diencephalon (Figure 32.2). It seems to be light sensitive and plays a role similar to the pineal gland in the shark.

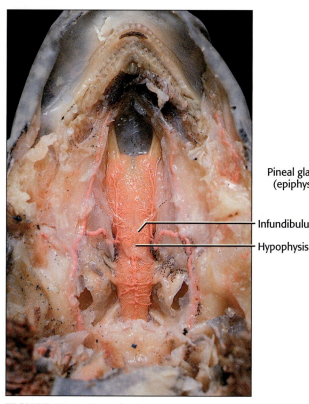

FIGURE 32.1 Hypophysis.

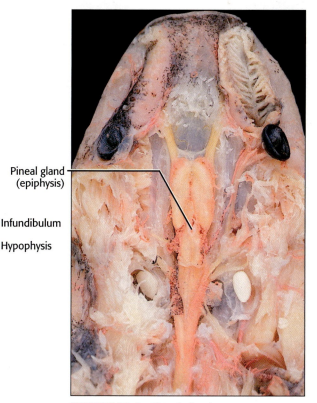

Pineal gland (epiphysis)

Infundibulum

Hypophysis

FIGURE 32.2 Pineal gland.

Thyroid Gland

The paired **thyroid glands**, derivatives of the pharyngeal pouches, lie medial to the external carotid artery (Figure 32.3). Hormones produced by this gland are similar among vertebrates, and their functions include tissue metabolism as well as integument and nervous system maintenance. Thyroid hormones play a critical role in the larval stages in amphibians that undergo metamorphosis.

Parathyroid Glands

Parathyroid glands are minute spherical bodies that develop from pharyngeal pouches and occur in tetrapods for the first time. The hormones secreted by these glands are involved in calcium metabolism. In most amphibians they are associated with the external jugular veins, but they are absent in *Necturus* (Duellman and Trueb, 1994).

Ultimobranchial Bodies

Ultimobranchial bodies are small, paired glands that originate from the last pharyngeal pouch and lie in the vicinity of the larynx. They produce a calcitonin-like hormone that affects mineral metabolism. They are extremely difficult to find.

Thymus Gland

The **thymus gland**, yet another derivative of the pharyngeal pouches, lies in the vicinity of the throat and is difficult to identify (Figure 32.4). It is a paired, trilobed gland. As in the shark, hormones stimulate development of the immune system.

Suprarenal or Adrenal Glands

In salamanders, the **suprarenal glands** appear as a series of rounded masses on the ventral side of the kidney and are easily identified in injected specimens (Figure 32.5). Secretions and functions are similar to those of the shark.

Pancreas

The **pancreas** in *Necturus*, in contrast to that of the shark, is a single gland that develops from dorsal and ventral primordia in the embryo (Figure 32.6). In tetrapods, it lies in the hepatoduodenal ligament, between the duodenum and the stomach. Its tissues and hormones are similar to those of the shark. Functions of the hormones parallel those of the shark.

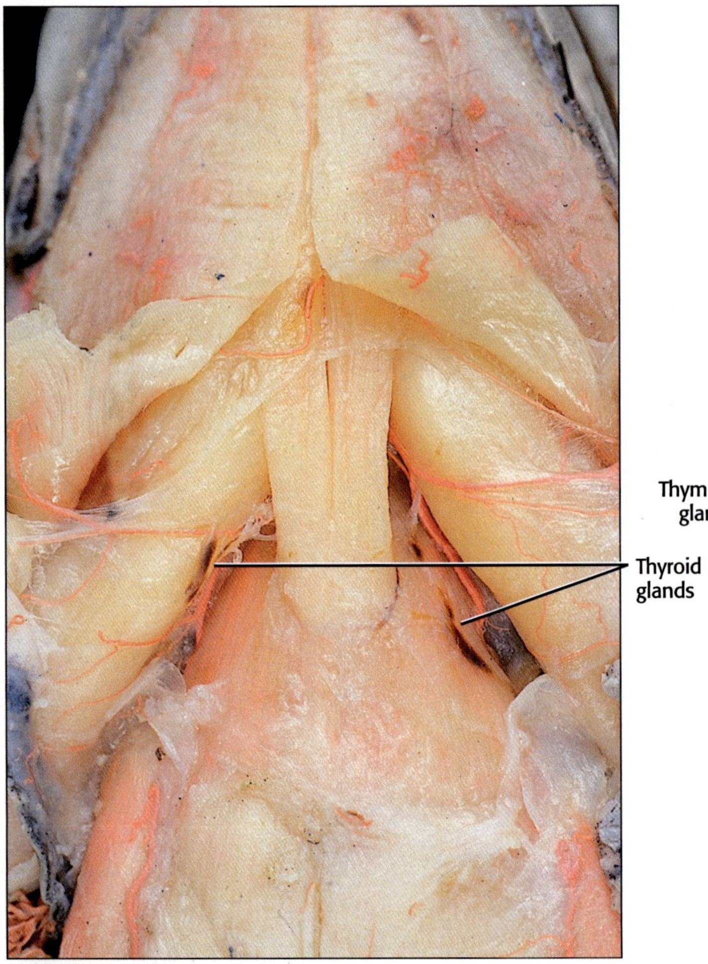

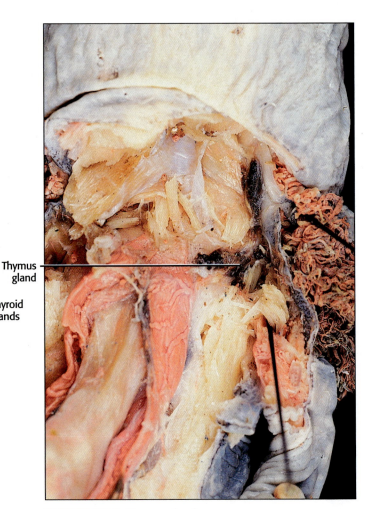

Thyroid glands

Thymus gland

FIGURE 32.3 Thyroid gland.

FIGURE 32.4 Thymus gland.

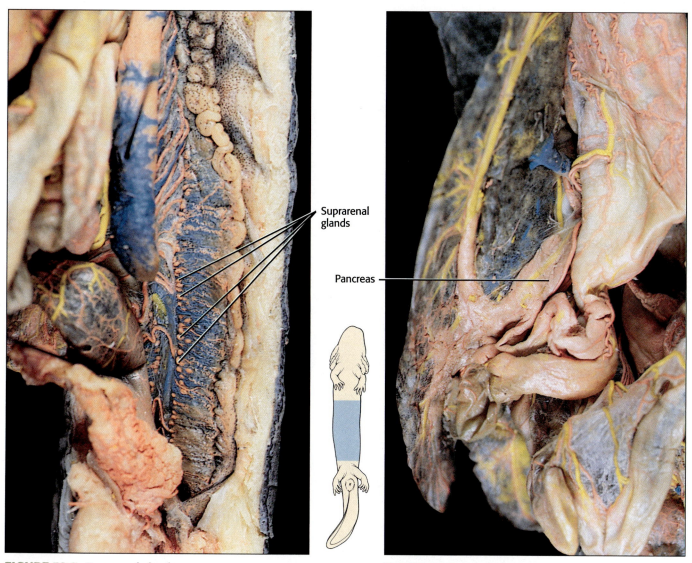

FIGURE 32.5 Suprarenal glands.

FIGURE 32.6 Pancreas.

Ovaries

The **ovaries**, female sex organs, are located ventral to the kidneys in the posterior half of the pleuroperitoneal cavity (Figure 32.7). The hormones, similar to those in the shark, stimulate and maintain sexual maturity and promote reproductive behavior.

Testes

The **testes**, male sex organs, are located ventral to the kidneys in the posterior half of the pleuroperitoneal cavity

(Figure 32.8). Similar to the shark, the hormones stimulate and maintain sexual maturity and promote reproductive behavior.

OTHER ENDOCRINE TISSUES

Other endocrine tissues and their secretions are similar to those found in the shark.

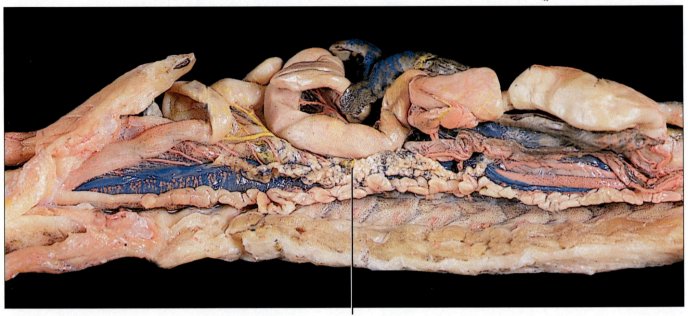

FIGURE 32.7 Ovary.   Ovary

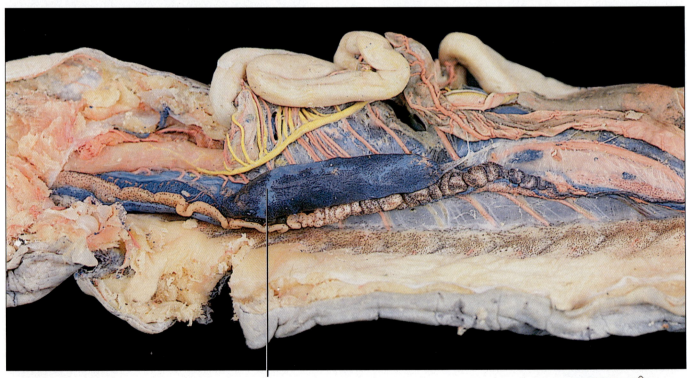

FIGURE 32.8 Testis.   Testis

PART SEVEN

Phylum Chordata / Subphylum Vertebrata

Cat

With the demise of the dinosaurs came an opportunity for some small, nocturnal, furry vertebrates to occupy the niches once dominated by those reptiles. The mammals lost little time in filling those vacancies, quickly evolving into a widely diverse fauna.

Some synapomorphies seen among these reptilian derivatives were:

1. A hairy coat covering most of the body with sensory hairs, vibrissae or whiskers occurring on the snout.

2. A metabolism based on an endothermic body temperature regulation.

3. A large brain with a thick cortex containing huge numbers of nerve cell bodies and processes with complex circuits occurring within the central nervous system as well as between the peripheral and central divisions.

4. Except for the primitive monotremes, the echidna and platypus, that are oviparous, mammals are viviparous, with the female carrying their young in their uterus for a period of time and giving birth to them.

5. A dramatic shift from the typical conical teeth of their ancestors to teeth specialized for nipping and tearing, the incisors; teeth specialized for stabbing and holding, the canines; and teeth specialized for grinding, the premolars and molars.

6. Three sound-conducting bones in the middle ear, the stapes, the incus, and the malleus.

7. Limb alignment with the legs placed toward the midline of the body and the claws associated with the digits evolving into nails and hooves or remaining as claws in some groups.

CHAPTER THIRTY-THREE

Cat — External Anatomy and Integumentary System

EXTERNAL ANATOMY

Examine the specimen assigned to your group and make the following observations. Perhaps, the most visible and obvious characteristic of most mammals is **hair** (Figure 33.1). Beneath the hair the body is divided into four major regions: **head, neck, trunk,** and **tail.** You can see a number of distinguishing features of the head, all associated with the concentration of special senses in this region. Among them are

1. the paired **external ears,** or **pinnae,**

2. the paired **eyes** with
   a. an **upper eyelid,** or **superior palpebra,** and
   b. a **lower eyelid,** known as the **inferior palpebra,**

3. paired **nostrils,** or **external nares,** and

4. tufts of coarse sensory hairs known as **whiskers,** or **vibrissae,** on either side of the face.

Note the **nictitating membrane** in the lower, medial corner of the eye.

The trunk can be divided into an anterior **thoracic** region, delineated by the rib cage, a middle **abdominal** region, and a posterior **pelvic** region. Along the ventral surface of the trunk or the belly

Pinna

Nictitating membrane
External nares

Vibrissae

Manus

Pes

Tail

FIGURE 33.1 External features of a cat.

are two rows of paired **nipples**, associated with the mammary glands—another exclusive mammalian characteristic. The nipples tend to be more prominent in females than in males, especially if the female is either pregnant or has been pregnant recently. Dorsal to the genital region in both sexes and located directly below the tail is the **anus**, the external terminal opening of the digestive system.

In mammals, tails range from being prominent to rudimentary. Examples of animals with conspicuous tails are felines (*e.g.*, housecats, lions, cheetahs). Cats use their tails for balance and communication. Other mammals, such as deer and pronghorns, use their tails for communication. When frightened, the tails of these animals become erect, displaying the prominent white underside to warn other members of their social group of danger. Some mammals (*e.g.*, herbivores) often use the tail as a switch to ward off insects. The tails of anthropoid primates (*e.g.*, chimpanzees, gorillas, humans), however, are rudimentary and, in most, are not visible externally.

Two sets of paired appendages have the same construction as *Necturus*:

1. The **forelimb**, associated with the cranial portion of the trunk, includes the **brachium** (upper arm), the **antebrachium** (forearm), and the **manus** (hand). The **carpus** (wrist) occurs in the proximal portion of the

manus. The **cubitus** (elbow joint) occurs between the brachium and the antebrachium.

2. The **hindlimb**, associated with the caudal end of the trunk, includes the **femur** (thigh), the **crus** (lower leg or shank), and the **pes** (foot). The **tarsus** (ankle) occurs in the proximal portion of the pes. The **genu** (knee joint) is between the femur and crus.

Palpate the genital area to ascertain the sex of your cat. If it is a male, you will feel the **testes** enclosed within the **scrotum**; if it is a female, note the **urogenital aperture**.

INTEGUMENTARY SYSTEM

Skin

Typical of vertebrates, the skin of mammals consists of two distinct layers, an outer **epidermis** and an inner **dermis** (Figure 33.2).

Epidermis

Mammalian skin is thick, and the epidermis is heavily keratinized. The epidermis consists of a mitotically active basal layer, the **stratum germinativum**, the middle **stratum granulosum**, where keratinization begins, and the outer **stratum**

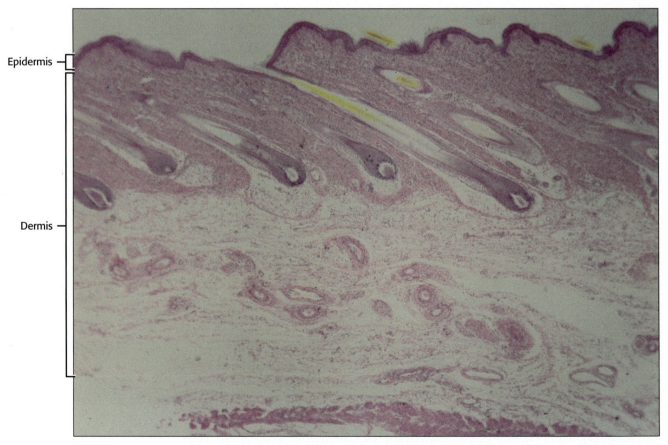

Epidermis

Dermis

FIGURE 33.2 Section through mammalian skin.

corneum, consisting of dead keratinized cells (Figure 33.3). Typically, layers of the stratum corneum are sloughed periodically. Replacement cells originating in the stratum germinativum are pushed into the stratum granulosum and ultimately into the stratum corneum by new cells continuously produced in the stratum germinativum. As they move through the epidermis, the cells become flattened because of external mechanical tension. By the time they occupy the outer layers, the cells are filled with the fibrous protein keratin.

Dermis

The dermis is thicker than the epidermis but is not as well stratified. It consists of connective tissue infiltrated with fibers and a number of inclusions: adipose tissue, blood vessels, nerves, glands, hair, and so on. The inner portion contains fewer fibers and integrates with subcutaneous connective tissue overlying the muscles.

Epidermal Derivatives

One of the most obvious and distinct characteristics is **hair**. Most mammals are covered by a complete **pelage**. Perhaps, hairs evolved as tactile organs known as **vibrissae**, or **whiskers**, associated with the nocturnal habits of mammals.

Modern mammals possess these stiff organs in the vicinity of the face and often on the legs.

Hair is an epidermal derivative. It consists of shafts that are dead and heavily keratinized with only the root containing mitotically active cells. The development of hair begins in the epidermis but sinks into the dermis, where it becomes anchored. The extensive body hair in mammals is an important insulating material, closely correlated with regulation of endothermic temperature. Attached to the sheath of a hair is an arrector pili muscle. The function of these muscles is temperature-dependent.

Cold temperatures stimulate contraction of the arrector muscle, causing erection of the hair, and thereby increasing the thickness of the insulating air next to the skin and resulting in heat conservation. Warm temperatures, conversely, cause the muscle to relax, allowing the hair to lie flat. This reduces the thickness of the air layer and permits heat loss. The emotional attitude of mammals also can be signaled through hair attitude. For example, frightened or angry cats and dogs often exhibit erect hair and an arched back. Hair color varies with pigments and air content of the hair shaft. Hair thickness, color, distribution, and the like depend upon body region, age, and seasons.

Other highly keratinized structures are characteristic of many mammals.

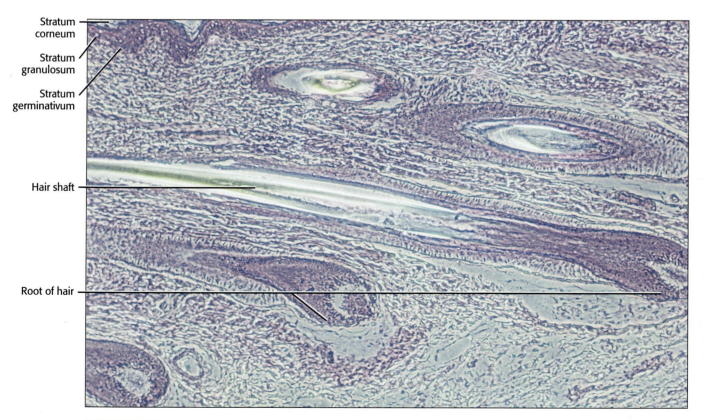

FIGURE 33.3 Section through mammalian skin: Magnified.

Claws and Nails

Claws, inherited from reptilian ancestors, extend from the terminal phalanges of the digits of many mammals. Some primates have modified claws, or **nails,** covering the upper surface of the terminal phalanges. Among ungulates another claw modification, **hooves,** are characteristic.

Horns and Antlers

Horns, consisting of a bony core covered by a keratinized sheath, are found in antelopes, cattle, and their relatives. Horns are permanent structures and are not shed. The familiar rhinoceros is a different story. Its "horn" does not have a bony corn but consists, instead, of a mass of compressed hair.

Deer and their relatives possess **antlers.** These are bony outgrowths from the skull but lack the horny sheath and are shed seasonally.

Glands

Sebaceous glands, which produce oily secretions, often are associated with hair follicles but may be present in hairless areas. The secretion lubricates the surface of the skin and the hair, preventing them from drying.

Eccrine sweat glands are not associated with hairs. In many mammal species, sweat glands occur only on bare skin surfaces. These glands are responsible for heat loss by producing watery sweat that evaporates from the hot body surface, thereby cooling it.

Apocrine glands have a different distribution and generally are confined to the axilla and anogenital region. The secretion is thick and is produced during periods of emotional arousal. Bacterial metabolism of these secretions often produces odoriferous byproducts that may be attractive to other members of the species.

Aggregations of apocrine glands or **scent glands** found in many different mammals produce chemicals that the animal uses as advertising and marking devices. These chemicals convey information such as sex, reproductive condition, and territorial boundaries.

Another unique characteristic of mammals consists of the **mammary glands.** Male and female mammals alike possess them, but they function only in lactating females, producing milk. Probably these distinctive organs are derivatives of aggregations of apocrine glands (Pough *et al.*, 1999; Feduccia and McCardy, 1991; Parsons and Romer, 1986). Although the number of mammary glands varies with the species, they generally occur somewhere along a line extending from the axilla to the groin.

Cat — Skeletal System

S imilar to *Necturus*, the mammal's skeleton also consists of bone and cartilage. Like the other two gnathostomes, the shark and the mudpuppy, the skeleton can be separated into the **axial division** and the **appendicular division** (Figure 34.1).

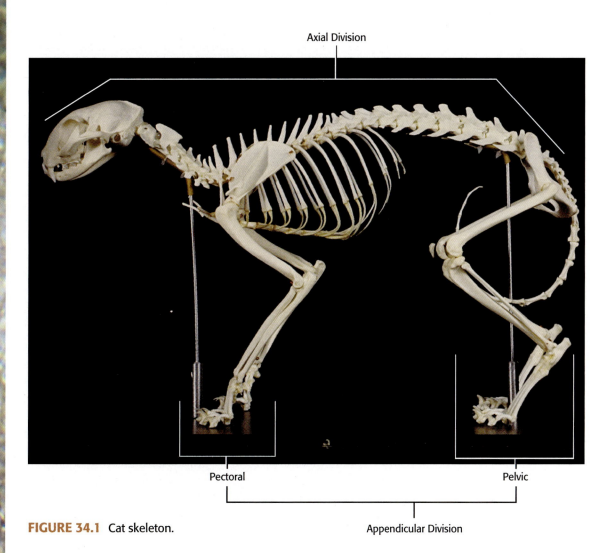

Axial Division

Pectoral

Pelvic

Appendicular Division

FIGURE 34.1 Cat skeleton.

AXIAL DIVISION

Skull

The skull consists of the cranium, whose bones surround the brain and through which the cranial nerves enter or exit through a number of openings known as **foramina, fissures,** and **canals.** Closely associated with the cranium are a number of facial bones that give the face its shape. This essential complex (cranial and facial bones) contains a number of cavities that surround and protect the major sense organs such as the eyes, portions of the auditory and balance apparatus, the olfactory organs, and with the articulated mandible, the gustatory organs.

Similar to *Necturus*, the embryonic mammalian skull consists of a chondrocranuium with paired trabeculae, parachordal bars, and a series of occipital cartilages. The occipital cartilages of *Necturus* probably are homologous with the cartilages that give rise to the vertebrae of the vertebral column and form the occipital region of the skull in mammals. Around the embryonic olfactory, optic, and auditory regions develop paired nasal, orbital, and otic cartilaginous capsules, respectively (see Figure 5.1).

Anteriorly, the trabeculae fuse, forming an ethmoid plate. The nasal capsules also fuse with this complex. Posteriorly, the trabeculae are connected by a transverse bar. The parachordal bars form a basal plate that unites with the occipital cartilages, the otic capsules, and trabeculae near the transverse bar. The resulting complex forms a trough for the brain with its associated sensory organs. The orbital cartilages remain separate from the chondrocranium.

Contrary to *Necturus*, the chondrocranium does not remain cartilaginous but, instead, becomes primarily bony in mammals. In addition, the mammalian skull consists not only of obviously homologous bones occurring in nonmammalian vertebrates but also bony complexes resulting from the fusion of nonmammalian homologs. The temporal bone of mammals consists of a combination of the squamosal; the prootic; the opisthotic; the angular with its reflected lamina, the ectotympanic; and an endochondral bone originating in mammals, the endotympanic (see Figure 5.4). Many of the dorsal and lateral bones of the skull are dermal in origin, in contrast to most of the rest of the skeleton, which is endochondral.

Evolutionary events in mammals have distinguished them from other vertebrates in their anatomy and physiology. The rate of metabolism is higher than all other vertebrates except birds. Adaptations to allow them to maintain these high metabolic rates are visible, in areas such as the integument, digestive system, and so forth, and also in the musculoskeletal system.

Some of the adaptations in the mammalian skull are related to their being endothermic or "hotblooded." To sustain an endothermic existence, mammals must find, process, digest, and assimilate large amounts of food. A number of specializations evolved to augment these activities. One of the most distinguishing features in mammalian evolution consists of changes in tooth morphology. The morphological changes correlate with new food procurement and processing functions.

The **heterodont** teeth of mammals are embedded in bony sockets of both the upper jaw and the lower jaw. Blade-like **incisors** are adapted for nipping pieces of food. The conical **canines**, probably least modified of the mammalian teeth, function as piercing and holding instruments. The molariform teeth—**premolars** and **molars**— having undergone the most modification of any, are adapted for chewing. Among carnivorous mammals (especially cats and dogs and their relatives), molariform teeth are bladelike. Further specialization among carnivores consists of a pair of lateral molariform teeth, the **carnassials**. The carnassials are used for cutting and shearing meat and tendons, cracking bones, and so forth.

The number of teeth found in a mammal is diagnostic and often distinguishes one mammalian species from another. It is true, however, that closely related species often possess the same number of teeth. The dental formulae of mammals state the tooth complement of a mammal and often indicate specialization for dietary habits. A primitive tooth formula for a placental mammal consisted of three incisors, one canine, four premolars, and three molars in each half of both the upper and lower jaw, multiplied by 2, equaling 44 teeth.

upper jaw: 3 incisors: 1 canine: 4 premolars: 3 molars

lower jaw: 3 incisors: 1 canine: 4 premolars: 3 molars

$$\times\, 2 = 44$$

The dental formula for the cat is:

$$\frac{3:1:3:1}{3:1:2:1} \times 2 = 30$$

Instead of the continuous or **polyphyodont** tooth replacement seen among nonmammalian vertebrates, mammals exhibit limited or **diphyodont** tooth replacement. The incisors, canines, and premolars erupt as the deciduous teeth (milk teeth or baby teeth) in juvenile mammals. These teeth fall out and are replaced by a permanent set of teeth consisting of the same number of incisors, canines, and premolars. In addition, the molars now erupt and become part of the permanent set.

Jaw musculature in nonmammalian vertebrates tends to be relatively unspecialized and functions primarily to close the jaws and hold prey. In mammals, jaw–muscle specialization has increased the muscle mass. To accommodate this specialization, the cheek and temporal region of the skull have been modified.

In most nonmammalian vertebrates, the roof of the oral cavity can be referred to as the primary palate. In mammals, shelves of bone grow ventrally from the premaxillae,

maxillae, and palatines to form the secondary palate, ventral to the primary palate. The secondary palate consists of a bony hard palate and a distal soft palate. The bones of the primary palate lie dorsal to the secondary palate. The air passageway in mammals is pushed posteriorly to the end of the secondary palate. Anatomically, the secondary palate separates the air (nasal) and food (oral) cavities, allowing mammals to breathe while they feed.

Surface area of the nasal cavities of mammals increased with the development of scroll-like turbinates, collectively known as the nasal conchae. The mucous membranes covering the turbinates are well-vascularized. As the air passes into the nasal cavity, it is moisturized and warmed, thereby reducing the temperature and moisture gradient between ambient air and pulmonary air. As the air is expired, it is cooled and moisture is removed, thereby conserving heat and moisture. This, then, is an energy-conserving mechanism, which is important for endothermic vertebrates. Secondarily, olfaction is enhanced by the increased surface area of the nasal cavities.

Anatomy of the Skull

As you study the cat skull, use the following figures: Figure 34.2A, Figure 34.2B, Figure 34.2C, and Figure 34.2D. It is advantageous to study disarticulated skull bones while studying articulated skulls.

The Premaxilla

Paired toothbearing **premaxillae** occur at the most anterior margin of the upper jaw. Three incisors are rooted in each premaxilla. Paired **palatine processes** contribute to the hard palate, the anterior portion of the secondary palate. The **anterior palatine foramina** (or **incisive ducts**) lie posterior to the incisors (Figure 34.3A and 34.3B). These openings permit chemical access to the vomeronasal organs.

The Maxilla

Lateral to the premaxillae are the paired **maxillae,** which complete the upper jaw and contribute to the hard palate ventrally as shelves of bone known as the **palatine processes**

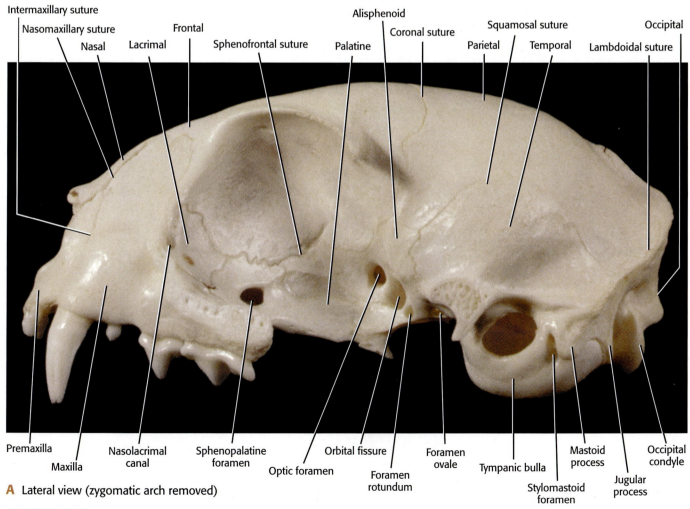

A Lateral view (zygomatic arch removed)

FIGURE 34.2A Skull.

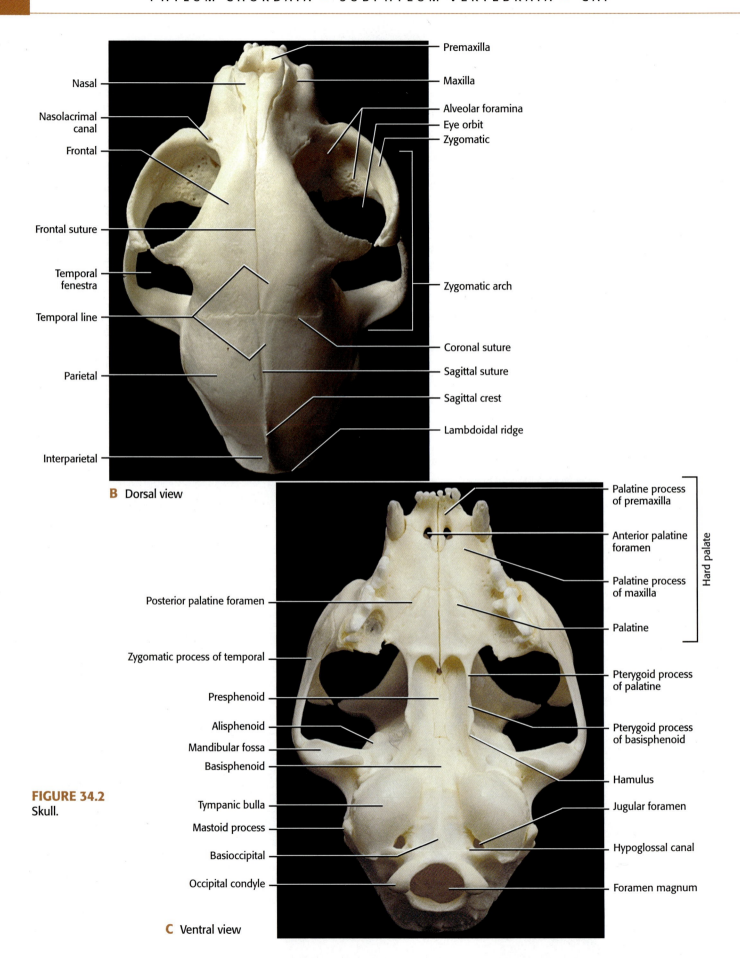

Premaxilla

Nasal

Maxilla

Nasolacrimal canal

Alveolar foramina

Eye orbit

Frontal

Zygomatic

Frontal suture

Temporal fenestra

Zygomatic arch

Temporal line

Parietal

Coronal suture

Sagittal suture

Sagittal crest

Lambdoidal ridge

Interparietal

B Dorsal view

Palatine process of premaxilla

Anterior palatine foramen

Hard palate

Palatine process of maxilla

Posterior palatine foramen

Palatine

Zygomatic process of temporal

Pterygoid process of palatine

Presphenoid

Alisphenoid

Pterygoid process of basisphenoid

Mandibular fossa

Basisphenoid

Hamulus

Jugular foramen

Tympanic bulla

Mastoid process

Hypoglossal canal

Basioccipital

Occipital condyle

Foramen magnum

FIGURE 34.2
Skull.

C Ventral view

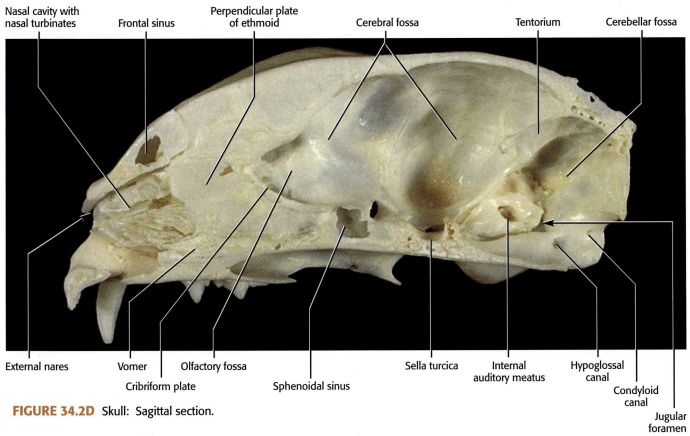

Nasal cavity with nasal turbinates

Frontal sinus

Perpendicular plate of ethmoid

Cerebral fossa

Tentorium

Cerebellar fossa

External nares

Vomer

Cribriform plate

Olfactory fossa

Sphenoidal sinus

Sella turcica

Internal auditory meatus

Hypoglossal canal

Condyloid canal

Jugular foramen

FIGURE 34.2D Skull: Sagittal section.

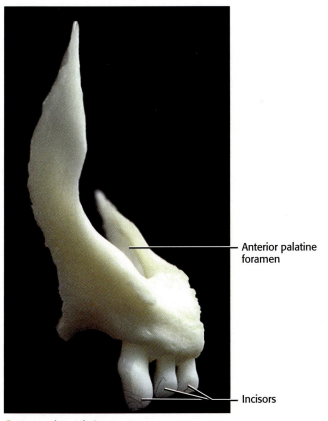

Anterior palatine foramen

Incisors

A Anterolateral view

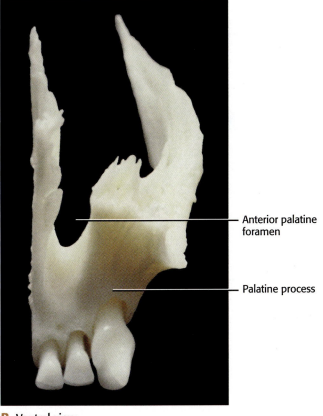

Anterior palatine foramen

Palatine process

B Ventral view

FIGURE 34.3 Right premaxilla.

of the maxillae (Figure 32.3C, Figure 34.4A, and 34.4B). Two prominent processes, the frontal process and the zygomatic process, are characteristic of this bone. The **alveolar process** accommodates the roots of the canine, premolar, and molar teeth. The **infraorbital foramen** permits the infraorbital nerve (C.N. V) and blood vessels to pass through. On the medial aspect of the maxilla exists a small lateral contribution to the nasal conchae, the **maxilloturbinate** (Figure 34.4A and 34.4B).

The remainder of the tooth complement in the upper jaw is rooted in the maxilla. At the anterior border of the maxilla is the **canine** tooth. A diastema (a space) separates the canines from the premolars. Posterior to the canine is the smallest of three **premolar** teeth. Posterior to the third premolar is a vestigial **molar** (Figure 34.4B).

The **palatine processes of the maxillae** extend back to the posterior border of the hard palate laterally (Figure 34.4A and 34.4B).

The Palatine

The palatine processes of the maxillae are joined medially by the **horizontal plates** of the palatine bones to complete the posterior border of the hard palate. The **pterygoid process of the palatine** articulates with the pterygoid process of the basisphenoid. Note the **maxillary spine**, the posterior point of articulation of the horizontal plate with

the palatine process of the maxilla. The large **sphenopalatine foramen** permits the sphenopalatine nerve (a branch of C.N. V] and corresponding blood vessel. A smaller opening, the posterior palatine canal, provides a passageway for the greater palatine nerve (a branch of C.N. V) and corresponding artery (Figure 34.5A and 34.5B).

The Nasal

Medial to the premaxilla and maxilla and cranial to the frontals are the paired **nasal** bones, which form the dorsal wall of the nasal cavity. Another small dorsal contribution to the nasal conchae, the **nasoturbinate**, is present on the ventromedial aspect of the nasals (Figure 34.6).

The Lacrimal

The small waferlike bone, the **lacrimal**, is located in the anteromedial portion of the eye orbit. An anterior notch in the lacrimal along with a groove in the frontal process of the maxilla completes the **nasolacrimal canal**, a prominent landmark of the anterior portion of the eye orbit (Figure 34.7).

The Ethmoid

The **ethmoid**, an unpaired cranial bone, is entirely associated with the nasal cavity. The major bulk of the ethmoid, the **ethmoturbinates**, composed of elaborately scrolled, thin

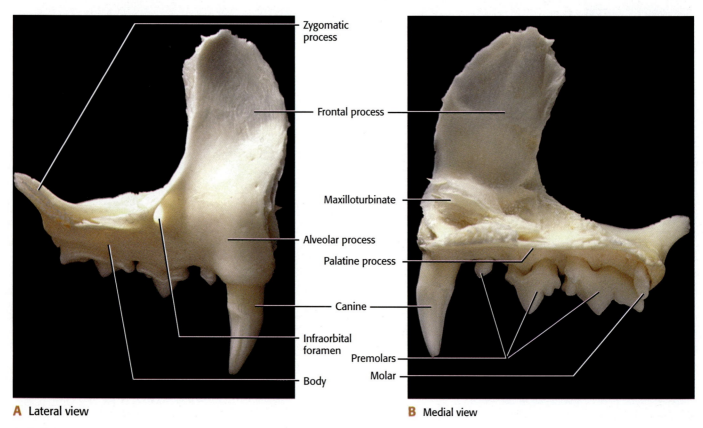

A Lateral view

B Medial view

FIGURE 34.4 Right maxilla.

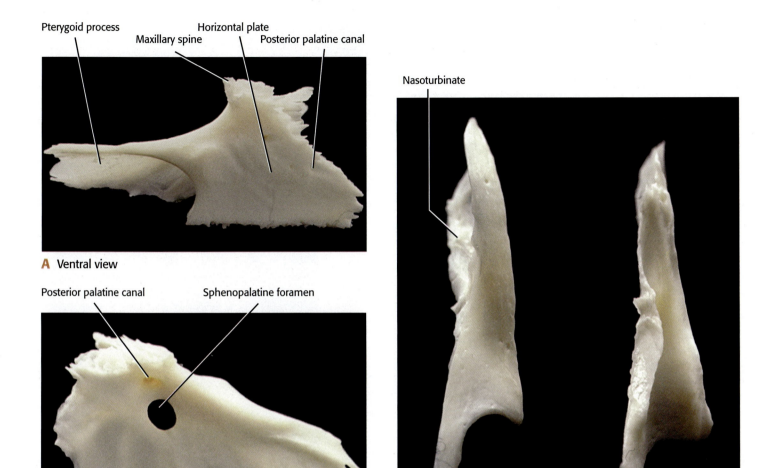

Pterygoid process
Maxillary spine
Horizontal plate
Posterior palatine canal

A Ventral view

Posterior palatine canal
Sphenopalatine foramen

B Lateral view

FIGURE 34.5 Right palatine.

Nasoturbinate

FIGURE 34.6 Nasal: right cranial, left caudal views.

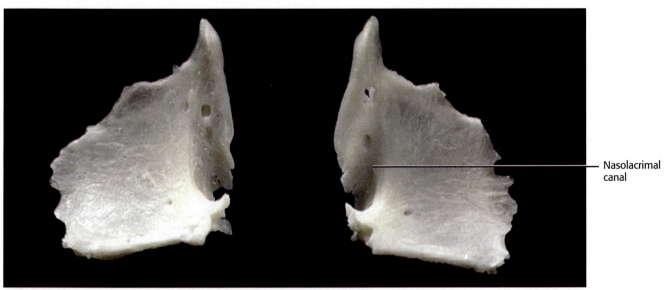

Nasolacrimal canal

FIGURE 34.7 Right and left lacrimal: lateral view.

laminate plates of bone, comprise most of the nasal conchae and fill most of the nasal cavity. The thin **perpendicular plate of the ethmoid** separates the two lateral ethmo-turbinates and, with the vomer, forms the bony portion of the nasal septum. Caudal to the perpendicular plate is the thin, concave, perforated **cribriform plate**. Through these foramina (perforations) pass fibers of the olfactory nerve (C.N. I), which synapse with neurons that lie in the olfactory bulbs situated directly above the cribriform plate (Figure 34.8A and 34.8B).

The Vomer

Ventral to the ethmoid and dorsal to the hard palate is the single **vomer**, which completes the bony nasal septum (Figure 34.9).

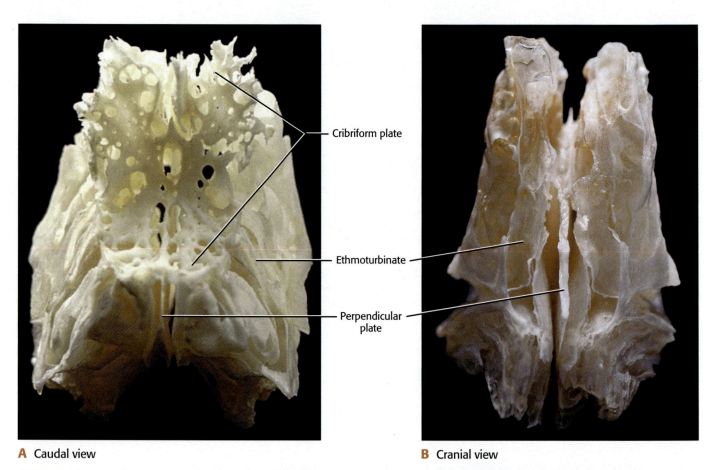

Cribriform plate

Ethmoturbinate

Perpendicular plate

A Caudal view

B Cranial view

FIGURE 34.8 Ethmoid.

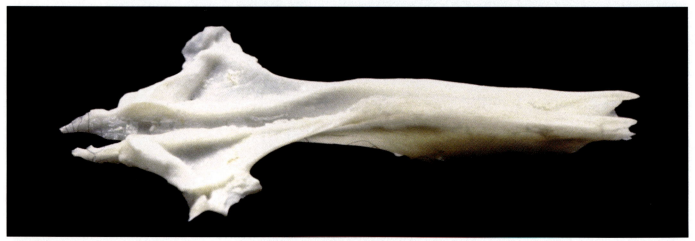

FIGURE 34.9 Vomer: dorsal view.

The Frontal

The cranial portion of the skull roof consists of the paired **frontals**. Two prominent projections, the **frontal spine** and the **orbital process** are characteristic of the frontal bone. The frontal spine articulates with the nasal and the maxilla. The orbital process delineates the upper portion of the eye orbit. Articulation of the frontal with the lacrimal, ethmoid, maxilla, presphenoid, and palatine bones completes the medial and ventral part of the eye orbit. A small **olfactory fossa** houses the olfactory lobes, and the **cerebral fossa** houses the anterior portion of the cerebrum. Identify the extensive air-filled **frontal sinus** (Figure 34.10A and Figure 34.10B).

The Zygomatic

The **zygomatic** (or **malar**) bone forms the lateral boundary of the eye orbit. The temporal fenestra (opening) inherited by mammals (*e.g.*, the cat) from their "reptilian" ancestors is occupied by muscles of mastication. The ventral **zygomatic process** of the zygomatic bone articulates with the zygomatic process of the temporal, forming the cheek prominence or **zygomatic arch**, unique in mammals. The posterior **orbital process** of the zygomatic curves upward toward the postorbital process of the frontal, forming an incomplete posterior margin of the eye orbit (Figure 34.11). Some mammals, *e.g.*, humans, retain a complete postorbital margin. A distinct ridge along the lateral surface of the zygomatic arch marks the position of the origin of the masseter muscle.

The Parietal and the Interparietal

Caudal to the frontal are paired **parietals**. In the medial view (Figure 34.12B) the **tentorium**, a curved shelf, separates the **cerebral fossa** from the anterior portion of the **cerebellar fossa** (Figure 34.12A and 34.12B). Posteriorly, the parietals diverge to articulate with the **interparietal** (Figure 34.13). A prominent medial projection is the **sagittal crest**. This small bone is represented by a pair of small postparietal bones in non-mammalian vertebrates. The interparietal may become fused with the occipital complex in mammals.

The Occipital

The **occipital** bone, representing the posterior ossified chondrocranial elements and consisting of the **basioccipital**, a pair of exoccipitals, and the supraoccipital, completes the posterior portion and base of the skull. Notice the

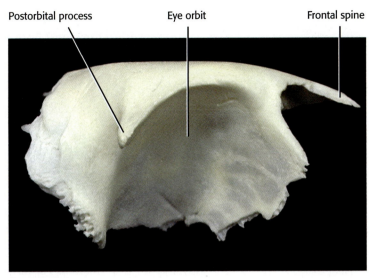

Postorbital process   Eye orbit   Frontal spine

A Lateral view

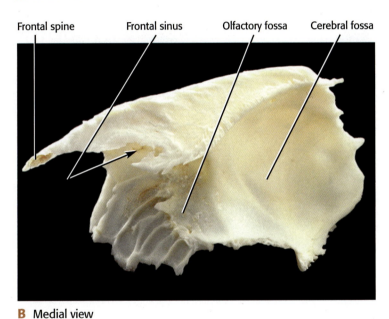

Frontal spine   Frontal sinus   Olfactory fossa   Cerebral fossa

B Medial view

FIGURE 34.10   Right frontal.

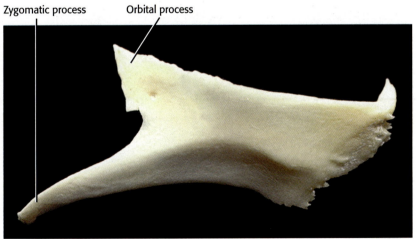

Zygomatic process   Orbital process

FIGURE 34.11   Right zygomatic: lateral view.

prominent paired **occipital condyles**, derived from a single condyle in the ancestral reptiles with which the atlas, or first cervical vertebra, articulates. A large opening, the **foramen magnum**, is situated in the ventral part of the occipital (Figure 34.14A and Figure 34.14B).

The foramen magnum marks the site of transition of the brain into the spinal cord as that portion of the central

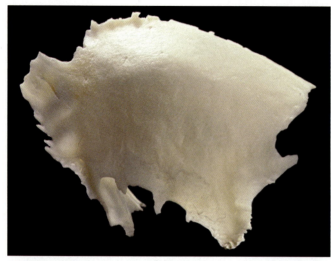

A Lateral view

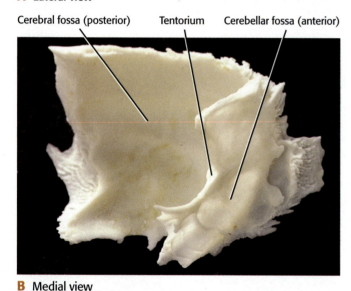

Cerebral fossa (posterior)   Tentorium   Cerebellar fossa (anterior)

B Medial view

FIGURE 34.12 Right parietal.

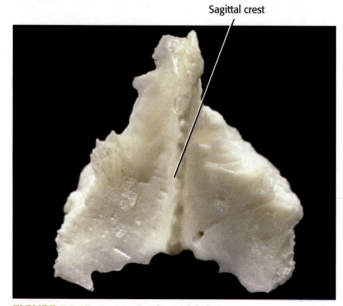

Sagittal crest

FIGURE 34.13 Interparietal: caudal view.

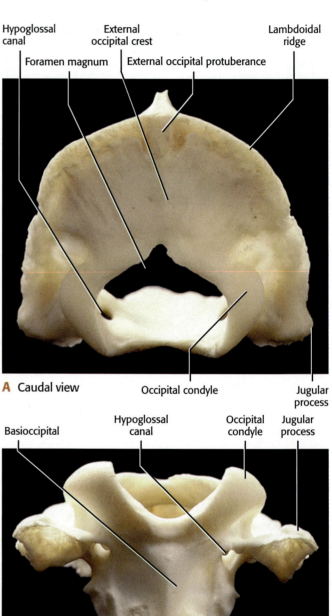

Hypoglossal canal   External occipital crest   Lambdoidal ridge

Foramen magnum   External occipital protuberance

A Caudal view   Occipital condyle   Jugular process

Basioccipital   Hypoglossal canal   Occipital condyle   Jugular process

B Ventral view

FIGURE 34.14 Occipital.

nervous system enters the vertebral canal. Lateral to each condyle and separated by a deep depression is a short, blunt projection, the **jugular process**. The **lambdoidal ridge** (nuchal crest) extends from one jugular process to the other. Through the prominent **hypoglossal canal** passes the hypoglossal nerve (C.N. XII) (Figure 34.14A and 34.14B).

The Basisphenoid

Anterior to the basioccipital sits the unpaired **basisphenoid** (Figure 34.15A). Nearly all of the mammalian basisphenoid is homologous with the basisphenoid of the nonmammalian vertebrate chondrocranium. The epipterygoid, derived from the palatoquadrate of ancestral vertebrates, is represented by the **alisphenoids**, or wings of the basisphenoid in mammals. Two anteriorly oriented projections termed the **pterygoid processes** (Figure 34.15A) have fused with the alisphenoids. In mammalian ancestors these processes were separate pterygoid bones (Figure 34.15A and 34.15B, and 34.15C).

The **hamulus**, a thin rod, extends posteriorly from the body of the pterygoid process of the basisphenoid. The dorsal surface of the basisphenoid is distinctly saddle-shaped. The anterior elevation of the saddle is the **tuberculum sellae**, whereas the more prominent and rounded posterior elevation is the **dorsum sellae**. Between the two elevations is a conspicuous depression known as the **sella turcica**, in which the hypophysis (pituitary gland) rests. Through two openings, the **foramen rotundum** and the **foramen ovale**, pass the maxillary branch and the mandibular branch of the Trigeminal Nerve (C. N. V), respectively (Figure 34.15A, 34.15B, and 34.15C).

The **orbital fissure** (Figure 34.15B) is formed by the articulation of the basisphenoid, and the presphenoid lies within the basisphenoid and the presphenoid. The Oculomotor Nerve (C.N. III), the Trochlear Nerve (C.N. IV), the ophthalmic division of the Trigeminal Nerve (C.N. V), and the Abducens Nerve (C.N. VI) pass through the orbital fissure. The anterior end of the basisphenoid articulates with the body of the presphenoid.

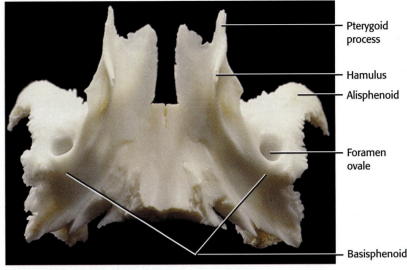

A Ventral view

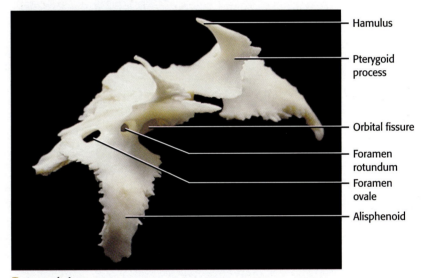

B Lateral view

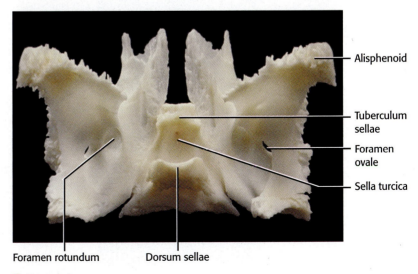

C Dorsal view

FIGURE 34.15 Basisphenoid.

The Presphenoid

The sphenethmoid region of the nonmammalian vertebrate chondrocranium is represented by a single **presphenoid,** consisting of a body and two wings resembling triangles in mammals. At the anterior end of this bone are two conspicuous **sphenoidal sinuses.** Lateral to the body and piercing the base of the triangle is the **optic foramen,** through which passes the Optic Nerve (C.N. II) and ophthalmic artery. The dorsal surface of the presphenoid is distinguished posteriorly by the **chiasmatic groove,** which extends between the optic foramina and is the site of the optic chiasma (Figure 34.16A and 34.16B).

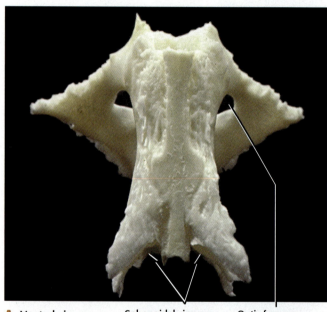

A Ventral view   Sphenoidal sinuses   Optic foramen

Chiasmatic groove   Optic foramen

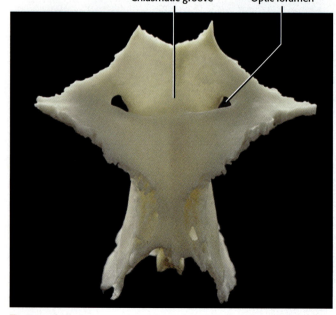

B Dorsal view

FIGURE 34.16 Presphenoid.

The Temporal

The **temporal** bone is a complex bone consisting of the **squamous,** the **petrous,** and the **tympanic** portions (Figure 34.17A and 34.17B). The squamous portion is homologous with the dermal squamosal bone of nonmammalian vertebrates. The petrous portion is homologous with the prootic and opisthotic regions of the chondrocranium of nonmammalian vertebrates. The tympanic portion consists of an auditory or tympanic bulla, which surrounds the sound transmission complex of mammals (Figure 34.18). The bulla is homologous with the dermal angular bone of nonmammalian vertebrates and the mammalian endotympanic bone.

The auditory bulla is pierced by an irregular oval opening on the lateral surface, the **external auditory meatus,** which leads into the tympanic or middle-ear cavity. A prominent, angular, nipple-shaped projection, the **mastoid process,** overlaps the bulla posterior to the external auditory meatus. The **stylomastoid foramen,** occurring anterior to the mastoid process, permits the passage of the Facial Nerve (C.N. VII). Identify the **fenestra vestibuli,** the oval window of the middle ear, and the **fenestra cochleae,** the round window of the middle ear (Figure 34.18). Within the petrous portion, the membranous acoustic and equilibrium apparatus resides in the bony labyrinth, which is slightly larger but of a similar shape (Figure 34.17A and 34.17B). A conspicuous elongated slit, the **eustachian tube passageway,** between the petrous portion and the tympanic bulla, admits the auditory tube into the middle ear cavity. The **zygomatic process** articulates with the zygomatic process of the zygomatic or malar bone completing the zygomatic arch. The condyloid process of the mandible articulates with the smooth **mandibular fossa,** a ventral groove in the zygomatic process. The small **postmandibular process** functions as a brace for the condyloid process of the mandible. Dorsal to the meatus is a small, deep fossa, the **appendicular fossa,** into which fits the small appendicular lobe of the cerebellum (Figure 34.17A and Figure 34.17B, and Figure 34.18).

A unique set of sound transmission bones evolved in mammals. Their history can be traced in nonmammalian vertebrates. All three bones, or ossicles are examples of transformations of skeletal elements that, in nonmammalian vertebrates, were associated with the jaws, either as jaw "props" or in jaw articulation, and now are located in the middle ear cavity of the temporal of mammals. The **stapes** evolved from the hyomandibula of the second visceral arch. The **incus** evolved from the quadrate of the palatoquadrate portion of the first visceral arch. The third ossicle, the **malleus,** is derived from the articular of Meckel's cartilage of the first visceral arch and the small dermal prearticular bone.

In nonmammalian vertebrates, jaw articulation occurs between the quadrate of the upper jaw, derived from the

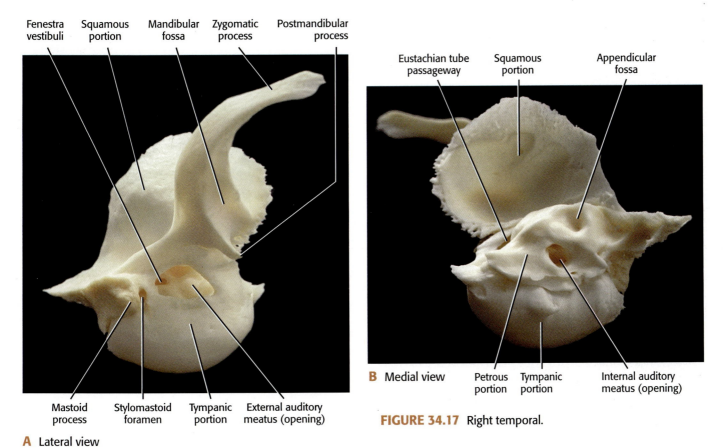

Fenestra vestibuli | **Squamous portion** | **Mandibular fossa** | **Zygomatic process** | **Postmandibular process**

Mastoid process | **Stylomastoid foramen** | **Tympanic portion** | **External auditory meatus (opening)**

A Lateral view

Eustachian tube passageway | **Squamous portion** | **Appendicular fossa**

B Medial view | **Petrous portion** | **Tympanic portion** | **Internal auditory meatus (opening)**

FIGURE 34.17 Right temporal.

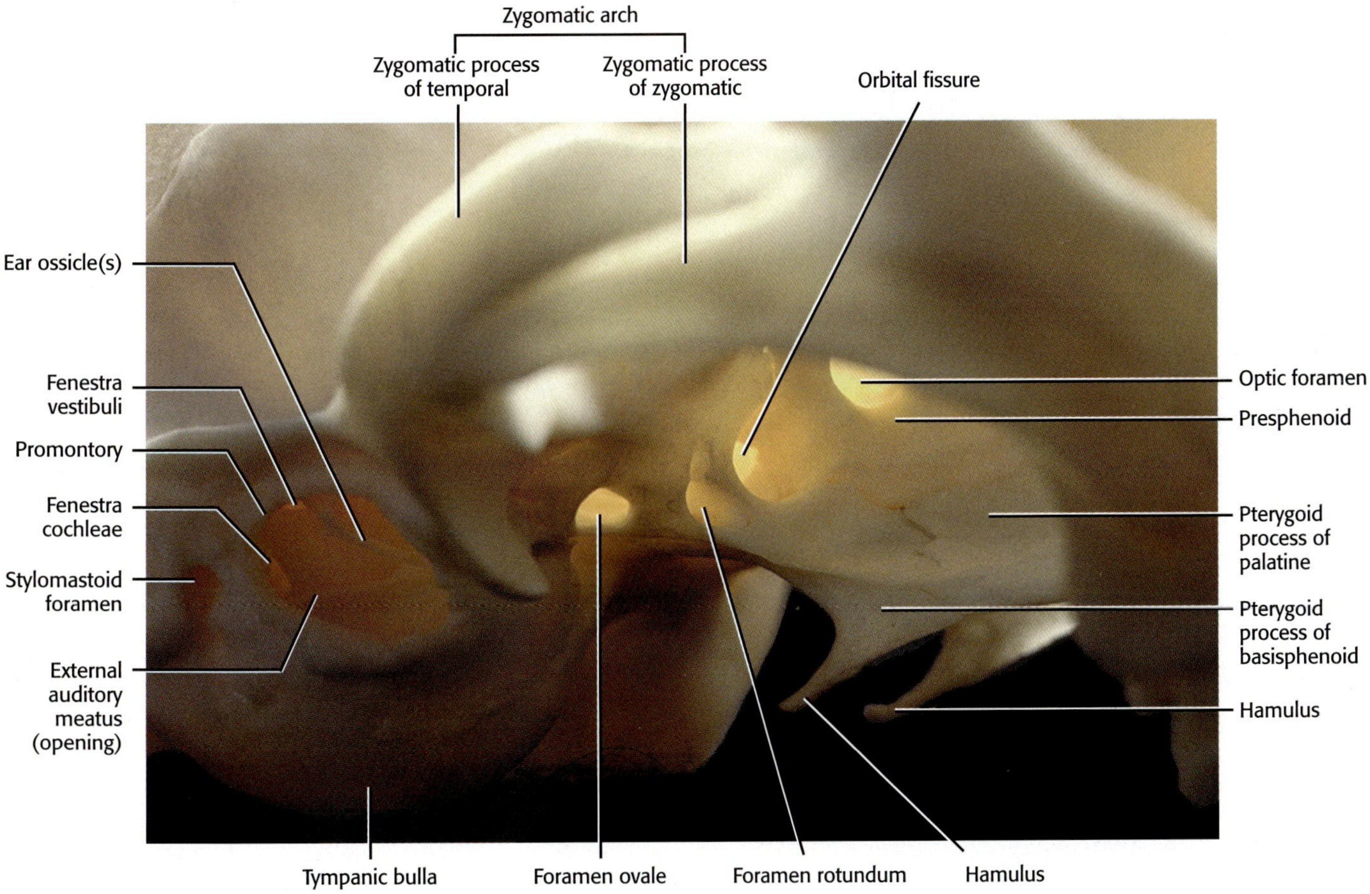

Zygomatic arch

Zygomatic process of temporal | Zygomatic process of zygomatic | Orbital fissure

Ear ossicle(s)

Fenestra vestibuli

Promontory

Fenestra cochleae

Stylomastoid foramen

External auditory meatus (opening)

Optic foramen

Presphenoid

Pterygoid process of palatine

Pterygoid process of basisphenoid

Hamulus

Tympanic bulla | Foramen ovale | Foramen rotundum | Hamulus

FIGURE 34.18 Sphenoid, presphenoid, tymbanic bulla region (enlarged).

posterior portion of the palatoquadrate and the articular, derived from the posterior portion of Meckel's cartilage. (See Figure 5.3 and Figure 5.4.) With the transformation of the quadrate into the incus and the articular into the malleus in mammals, joining the stapes appearing in amphibians, a new jaw articulation evolved in mammals. Mammals are the only living vertebrates in which jaw articulation occurs between the squamosal (squamous portion of the temporal) of the skull and the dentary of the mandible.

Surface Features of the Skull

Although the overall skull topography is fairly smooth, it has some conspicuous features such as sutures, lines and ridges, and openings. Sutures, **synarthrotic joints**, are immovable articulations found between adjacent skull bones and generally are identified using the names of the bones between which they occur.

intermaxillary suture: between the maxillary bones

nasomaxillary suture: between the nasal and maxillary bones

sphenofrontal suture: between the sphenoid and frontal bones

frontal suture: between the frontal bones

coronal suture: between the parietal and frontal bones

sagittal suture: between the parietal bones

squamous suture: between the parietal and squamous portion of the temporal bone

lambdoidal suture: caudal of the parietals, separating them from the occipital and interparietal

A *faint* dorsal feature is the **temporal line** that demarcates the origin of the temporalis muscle. It extends from the caudal margin of the postorbital process as a gentle curve over the frontal bone onto the parietal bone, meeting its opposite partner in a V-shaped pattern terminating at the cranial end of the **sagittal crest**.

Cavities and Sinuses

The **nasal cavity** houses the highly convoluted **turbinates** (**nasal conchae**) of the ethmoid, maxilla, and nasal bones. A nasal septum, consisting of the dorsal, bony perpendicular plate of the ethmoid with a cranial cartilaginous portion along with the ventrally located vomer, divides the nasal cavity into right and left halves. The **external nares** (nostrils) open anteriorly into these cavities. The caudal cribriform plate of the ethmoid, through which pass the olfactory fibers, separates the nasal cavity from the cranial cavity.

The **cranial cavity** is organized into four distinct areas.

1. The **olfactory fossa** occurs just caudal to the cribriform plate and houses the olfactory bulbs, where the olfactory fibers synapse.

2. Posterior and continuous with the olfactory fossa is the **cerebral fossa**, which houses the cerebrum, the diencephalon, and the mesencephalon (midbrain).

3. In the floor of the cranial cavity is the **sella turcica**, in which the hypophysis (pituitary) sits.

4. The **tentorium** marks the caudal end of the cerebral fossa.

Posterior to the tentorium lies the **cerebellar fossa**, which surrounds the pons, the cerebellum, and the medulla oblongata. The **internal auditory meatus** lies in the petrous portion of the temporal bone. The **foramen magnum** opens at the caudal end of the cerebellar fossa. Notice the considerable topography of the surface of the cranial cavities, which mirrors the contours of the external brain surface and also, often, the blood vessels associated with the brain.

Two air-filled spaces enclosed in certain bones of the skull are the **frontal sinus** and the **sphenoidal sinus**. Perhaps the function of these spaces is weight reduction. In the cat, the sinuses are located in the frontal and presphenoid bones.

Cranial Foramina

Mammalian skulls possess a number of openings, foramina, canals, or fissures that permit passage of cranial nerves and/or blood vessels (see Figure 34.2A, Figure 34.2B, Figure 34.2C, and Figure 34.2D). We will begin the study with the openings associated with the cranial nerves.

The olfactory nerve, cranial nerve one (C.N. I), subdivides into a number of branches that pass from the olfactory epithelium through the **cribriform foramina** of the cribriform plate of the ethmoid. A series of four foramina occur along the lateral aspect of the skull, just posterior to the eye orbit. The optic nerve (C.N. II) passes through the **optic foramen**, the first of these foramina. Through the **orbital fissure** passes the oculomotor nerve (C.N. III), the trochlear nerve (C.N. IV), the ophthalmic division of the trigeminal nerve (C.N. V), and the abducens nerve (C.N. VI). The maxillary division of the trigeminal nerve (C.N. V) passes through the third of these foramina, the **foramen rotundum**. The mandibular division of the trigeminal (C.N. V) exits through the **foramen ovale**, the fourth in the series.

A number of smaller branches of the maxillary division of the trigeminal nerve pass through the **sphenopalatine foramen**, the **anterior palatine foramen**, the **posterior palatine foramen**, and the **infraorbital foramen**. **Alveolar foramina** permit passage of a number of branches of maxillary division of the trigeminal nerve to the maxillary teeth. A branch of the mandibular division of the trigeminal nerve passes through the **mental foramina** and the **mandibular foramen** of the mandible.

Through the internal **auditory meatus** pass the facial nerve (C.N. VII) and the vestibulocochlear nerve (C.N. VIII). The facial nerve exits the skull through the **stylomastoid foramen**. The vestibulocochlear nerve is the only cranial nerve that does not exit from the skull. Through the **jugular foramen** passes the glossopharyngeal nerve (C.N. IX), the vagus nerve (C.N. X), and the spinal accessory nerve (C.N. XI). The hypoglossal nerve (C.N. XII) exits through the **hypoglossal canal**.

A **nasolacrimal canal**, a prominent landmark of the eye orbit, allows tear flow from the surface of the eye to the nasopharynx through the nasolacrimal duct. The large **foramen magnum** permits the transition of the brain to the spinal cord. As one peers through the foramen magnum, a variable **condyloid canal**, through which passes small blood vessels, can be located.

An irregular oval opening, the **external auditory meatus** leads into the tympanic cavity. In this cavity can be seen the **fenestra vestibuli** (oval window), the **fenestra cochleae** (round window), and the **auditory tube opening**, all associated with the middle ear (see Figure 34.18).

Mandible

The lower jaw, or **mandible**, completes the bones of the head. In mammals, the mandible consists of a pair of **dentary** bones that articulate cranially at the **intermandibular symphysis**. With the exception of the angular, the remaining bones found in the mandible of nonmammalian vertebrates became involved with sound transmission in mammals. At the caudal end of the mandible are a pair of **condyloid processes**, each of which articulates with an elongate groove, the mandibular fossa, in the zygomatic process of the squamous portion of the temporal (see Figure 34.18).

In lateral view, the cranial end of the mandible consists of a relatively smooth, rounded bar, the **body**, which contains two or three **mental foramina** through which pass branches of the mandibular division of the trigeminal nerve (C.N. V). The caudal end of the mandible expands into the **ramus**. The **coronoid process** is the site of insertion of the temporalis muscle, the largest and most powerful of the jaw muscles of carnivores.

The well-defined triangular-shaped depression, the **coronoid** or **masseteric fossa**, accommodates the insertion of part of the masseter muscle, another of the powerful jaw adductors. Notice the small rounded projection, the **angular process**, at the extreme caudal end of the mandible. Another portion of the masseter and pterygoid muscles insert there. The prominent **mandibular foramen** in the medial surface admits the mandibular division of the trigeminal nerve (C.N. V).

Teeth are embedded in sockets along the dorsal border of the body of the mandible. In each half are found three small **incisors**, anteriorly, followed by a single sharp **canine**. A wide space, the **diastema**, separates the canine from the two caudal **premolars** and a single **molar** (Figure 34.19A, Figure 34.19B, and Figure 34.19C).

The Hyoid

The hyoid actually is a complex of derivatives of several former ancestral gill supports. In mammals, this apparatus, located ventral to the larynx and at the root of the tongue, serves as a site of origin of tongue and larynx muscles. It is H-shaped and consists of a **body** forming the bar of the H and two **cranial** (or **lesser**) **horns**, or **cornua**, and two **caudal** (or **greater**) **horns**, or **cornua**, forming the upper and lower uprights of the H (Figure 34.20). The body and lesser horns are derivatives of the hyoid or visceral arch II, and the greater horns are derivatives of visceral arch III. A series of three small bones, also derivatives of the hyoid arch, connect the hyoid apparatus to the skull.

Vertebral Column

The vertebral column of mammals is more regionalized than *Necturus*, probably related to accompanying changes in limb placement and greater efficiency in locomotion. It consists of vertebrae separable into five regions: cervical, thoracic, lumbar, sacral, and caudal. Support of both appendicular girdles is essential to this integrated locomotory and weight-bearing unit. The pelvic girdle articulates directly with the sacral region of the vertebral column, and the pectoral girdle is associated and held in place by muscles in the thoracic region. Cushioning fibrocartilaginous **intervertebral discs** occur between each of the individual vertebrae.

Thoracic Vertebrae

In mammals, the thoracic vertebrae are the least specialized and are similar to the vertebrae of *Necturus*. Therefore, we will begin our discussion with that region (Figure 34.21). Typically, cats have 13 thoracic vertebrae.

Commonly, a vertebra consists of a solid **centrum**, or **body**, which forms the main ventral support for the spinal cord that rests in the obvious opening, the **vertebral canal**. The vertebral canal is completed by a dorsal pair of **laminae** whose dorsocaudal extensions form the **spinous process** of the vertebra. A pair of **pedicles** extends between the laterally projecting **transverse processes**, or **diapophyses**, and the centrum. Note the **articular facet** on the ventral surface of the transverse process with which the tuberculum of the rib articulates, and also the **demifacets** occurring on the centrum of two adjacent vertebrae with which the capitulum of the ribs articulates (Figure 34.21). Note that these facets are unique landmarks found only on thoracic vertebrae.

At the cranial end of each vertebra are two processes, the **prezygopophyses**, whose articulating facets face dorsally or dorsomedially and articulate with two caudal processes,

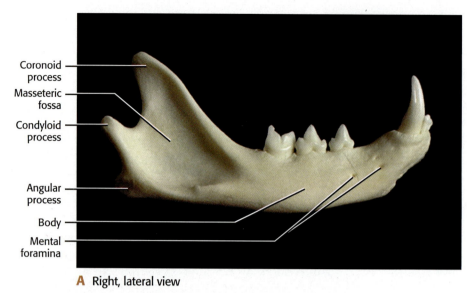

Coronoid process
Masseteric fossa
Condyloid process
Angular process
Body
Mental foramina

A Right, lateral view

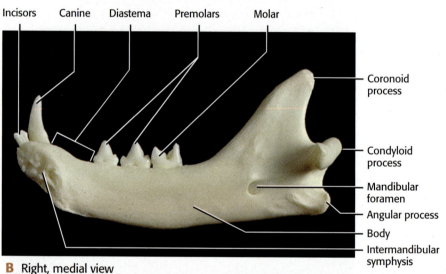

Incisors   Canine   Diastema   Premolars   Molar

Coronoid process
Condyloid process
Mandibular foramen
Angular process
Body
Intermandibular symphysis

B Right, medial view

C Skull with mandible

FIGURE 34.19 Dentary.

the **postzygapophyses**, on the posterior end of the vertebra just preceding it. The articular surface of the postzygapophysis faces ventrally or ventromedially. Notice that each vertebra possesses both pre- and postzygapophses to facilitate articulation with adjacent vertebrae (Figure 34.21).

Cervical Vertebrae

Almost all mammals, including cats, humans, whales, and giraffes, possess seven cervical vertebrae. The first and second cervical vertebrae—the atlas and axis, respectively—are distinctly structured. The modified first vertebra, or atlas, appeared in amphibians and permitted nodding movements. The second vertebra, or axis, appeared in the reptiles and permitted lateral and rotational movements. The ring-like **atlas** of mammals lacks a centrum and a distinct spine and has broad, wing-like **transverse processes.**

Typical of the first six cervical vertebrae is the **transverse foramen,** which passes through the transverse process. The vertebral arteries and veins—the blood supply of the brain—pass through these foramina. The inner articular facets of the expanded **prezygapophyses** accommodate the rounded surfaces of the occipital condyles of the skull. The **atlantal foramen,** the first intervertebral foramen, through which the first spinal nerve and vertebral vein exit and the vertebral artery enters, is located dorsal to the prezygapophysis. Notice that the **postzygapophyses** do not project as distinctly from the atlas as the prezygapophyses, but, nevertheless, bear well-defined facets for articulation with the prezygapophyses of the axis (Figure 34.22).

Study the **axis** with its pronounced spinous process overhanging the arch of the atlas. A second distinctive feature is the cranially projecting **odontoid process,** or **dens,** laterally flanked by a pair of **prezygapophyses** with smooth

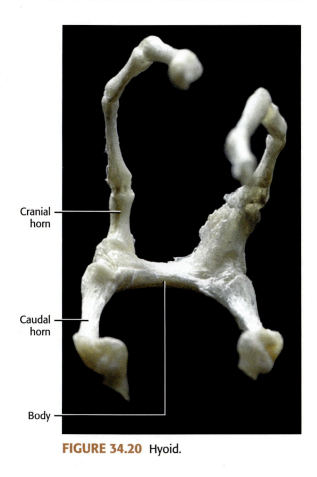

FIGURE 34.20 Hyoid.

Cranial horn

Caudal horn

Body

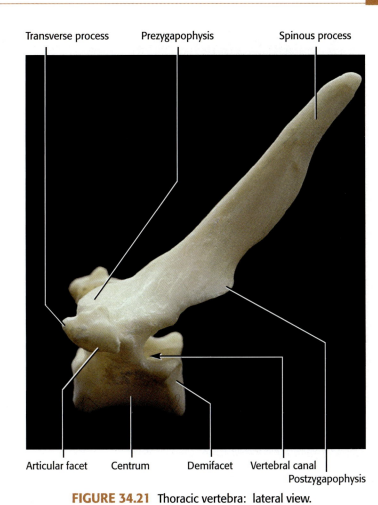

Transverse process   Prezygapophysis   Spinous process

Articular facet   Centrum   Demifacet   Vertebral canal

Postzygapophysis

FIGURE 34.21 Thoracic vertebra: lateral view.

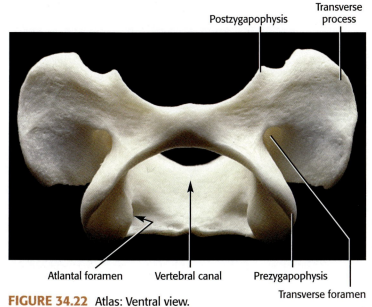

Postzygapophysis   Transverse process

Atlantal foramen   Vertebral canal   Prezygapophysis

Transverse foramen

FIGURE 34.22 Atlas: Ventral view.

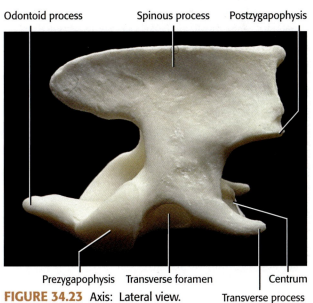

Odontoid process   Spinous process   Postzygapophysis

Prezygapophysis   Transverse foramen   Centrum

Transverse process

FIGURE 34.23 Axis: Lateral view.

articulating surfaces to articulate with the atlas. The odontoid process, the centrum of the atlas, has fused with the **centrum** and other elements of the axis to form the adult mammalian axis. Thin, caudally projecting transverse processes are present with their characteristic **transverse foramina** (Figure 34.23). The remaining cervical vertebrae, in many respects, are similar to one another and possess typical vertebral characteristics (Figure 34.24). An exception in cats is the lack of a transverse foramen in the seventh vertebra.

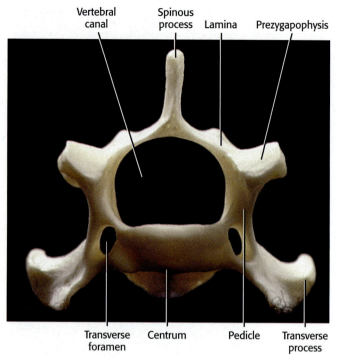

FIGURE 34.24   Vertebral canal   Spinous process   Lamina   Prezygapophysis   Transverse foramen   Centrum   Pedicle   Transverse process

FIGURE 34.24 Cervical vertebra: Anterior view.

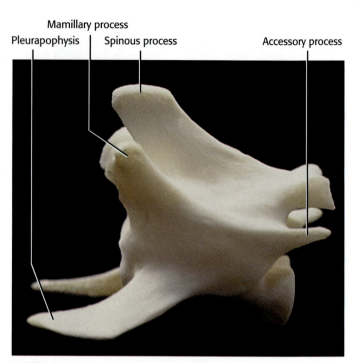

Mamillary process   Pleurapophysis   Spinous process   Accessory process

FIGURE 34.25 Lumbar vertebra: lateral view.

Lumbar Vertebrae

The seven lumbar vertebrae are the largest of the vertebrae, increasing in size toward the caudal end. Perhaps the most obvious landmark on lumbar vertebrae is the cranially projecting **pleurapophysis**, representing the transverse process with an embryonic rib fused to it. In addition, **accessory processes** are generally evident on all but the last two, and **mammillary processes** can be seen on all of the lumbar vertebrae (Figure 34.25).

Sacrum

In contrast to *Necturus*, with a single sacral vertebra, in mammals a complex consisting of three to five fused vertebrae in the adult, acting as a brace for the pelvic girdle, has evolved. Notice that the three vertebrae of the cat decrease in size, caudally. Each, however, retains most of the characteristics of the preceding lumbar vertebra. Note also the fusion of the pleurapophyses into a single lateral structure, as well as dorsal and ventral foramina to accommodate passage of spinal nerves between each of the adjacent fused vertebrae (Figure 34.26).

Caudal Vertebrae

Vertebrae of the tail are the smallest and most variable in number. The more cranial of these exhibit rather typical vertebral characteristics, whereas the more caudal vertebrae lose them rapidly and come to resemble simple cylinders representing the centra. Take note of the two small **hemal processes** on the ventral surface of the centra, with which V-shaped **hemal arches**, housing caudal blood vessels,

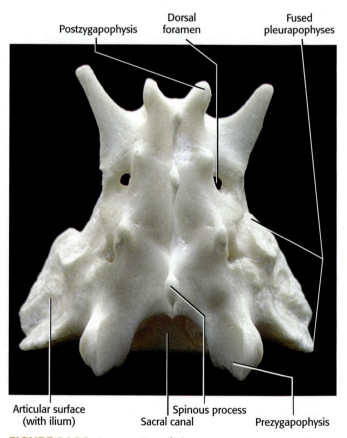

Postzygapophysis   Dorsal foramen   Fused pleurapophyses   Articular surface (with ilium)   Sacral canal   Spinous process   Prezygapophysis

FIGURE 34.26 Sacrum: Dorsal view.

articulate. Hemal arches can be identified in an intact cat but are lost during skeletal preparation (Figure 34.27).

In a number of animals (*e.g.*, some of the primates), the tail is quite abbreviated. Humans have three to five caudal

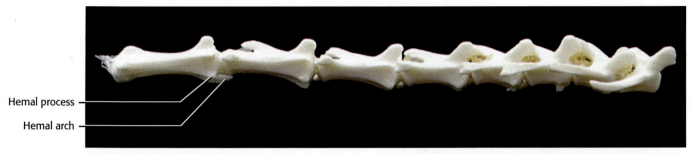

Hemal process

Hemal arch

FIGURE 34.27 Caudal vertebrae, anterior toward the right.

vertebrae, often rudimentary and sometimes fused. These vertebrae are called the *coccyx* in the human. The only time that most of us appreciate the fact that we have a tail is when we damage it and are unable to sit comfortably for some time.

Sternum

In mammals, unlike *Necturus*, the sternum is bony and elongate and is composed of a series of articulated **sternebrae**. The sternum lies ventrally in the midline of the thorax, and its posterior end lies just anterior to the diaphragm. In the cat it consists of three regions: an anterior, slightly keeled **manubrium**, which resembles the tip of a spear; the **body**, consisting of six similar, articulated sternebrae; and a posterior elongated **xiphisternum** with a distal cartilaginous end known as the **xiphoid process** (Figure 34.28).

Ribs

The cat has 13 pairs of ribs. The first nine pairs, identified as the **vertebrosternal** ribs, are considered to be true ribs because their distal ends attach individually to the sternum by means of costal cartilages. Costal cartilages associated with the vertebrosternal ribs are generally attached at intervals to the sternum. The first of these is associated with the manubrium at about its midpoint.

In some cats, a fusion line indicates the possibility of two individual sternebrae having been united to form the single manubrium. The other pairs articulate with the sternum at analogous points between adjacent sternebrae along its length. The last two sternal articulations are associated closely with one another.

The last four pairs show a great deal of variability. Generally, the next three pairs, the **vertebrochondral** ribs, are attached distally by means of cartilage to each other or to the costal cartilage of the ninth. The final pair has no sternal attachment and therefore is referred to as the **vertebral** or "floating" ribs.

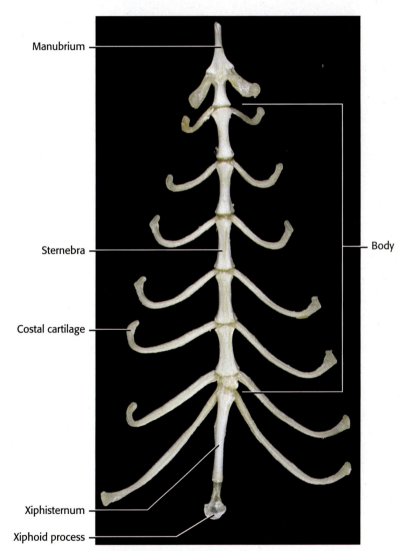

Manubrium

Sternebra

Costal cartilage

Xiphisternum

Xiphoid process

Body

FIGURE 34.28 Sternum.

Although the ribs may be of varying lengths, their morphology is similar. The basic shape is a curved, flattened rod, whose proximal end bears a **head**, or **capitulum**, which articulates with demi-facets occurring on the centrum of two adjacent thoracic vertebrae. A second projection, the **tuberculum**, bearing a smooth facet, articulates with the transverse process of a vertebra. The slightly constricted area between the capitulum and the tuberculum is known

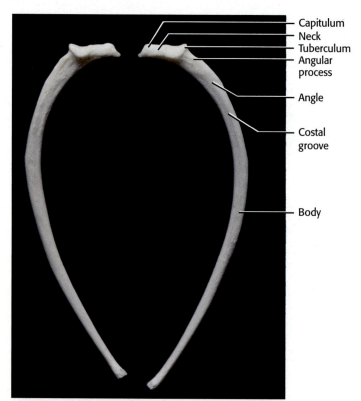

Capitulum
Neck
Tuberculum
Angular process

Angle

Costal groove

Body

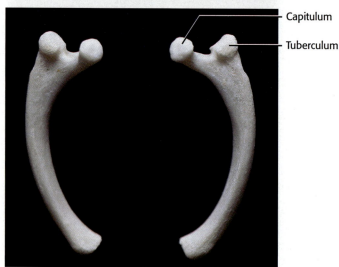

Capitulum

Tuberculum

FIGURE 34.29 Left rib, right rib: caudal view.

as the **neck**. The proximal curved portion or **angle** of the rib blends into the distal part known as the **body**. A small, pointed projection, the **angular process**, can be located on the angle of the rib (Figure 34.29).

APPENDICULAR DIVISION

In mammals, the position of the appendages has shifted from a lateral position to a position more nearly aligned with the midline of the body. With this movement came concomitant realignment and reconstruction of skeletal elements to accommodate muscle reorientation.

Pectoral Girdle and Forelimb

This pectoral girdle consists of the endochondral scapula and dermal clavicle with its associated forelimb. The mammalian scapula is a complex of the old ancestral "reptilian" scapula fused to a new anterior bony shelf to which is attached the remnants of the ancestral coracoid. The **scapula** is a flat, triangular bone that articulates with the humerus. The curved **dorsal border** continues anteriorly as the **cranial border** and posteriorly as the **caudal border**. The scapula in a cursorial mammal such as the cat is oriented in line with the axis of the limb rather than at a right angle, as in more primitive mammals. (See Figure 5.17 and Figure 5.18.)

At the vertex of the triangle is a ventral concave surface known as the **glenoid fossa**, with which the head of the humerus articulates. A small beaklike projection, the **coracoid process**, extends medially from the anterior border of the glenoid fossa. Lateral to it is the **supraglenoid tubercle** (Figure 34.30).

A prominent lateral ridge, the **scapular spine**, separates the scapula into a posterior **infraspinous fossa** representing the old ancestral portion, and the anterior **supraspinous fossa**, the "new" shelf added during mammalian evolution. The spine terminates in a pointed **acromion process**. The upper curved, thicker portion of the free edge of the spine is identified as the **tuberosity of the spine**, and the caudally projecting **metacromion** occurs just dorsal to the acromion. On the medial surface is the **subscapular fossa** (Figure 34.30).

In many mammals the **clavicle** is robustly constructed, and its articulation reflects a more primitive mammalian condition, acting as a brace between the sternum and the pectoral girdle. The clavicle in cursorial (running) mammals may be lost or reduced drastically and often embedded in the muscle of the shoulder. This anatomy permits the scapula to be aligned with the forelimbs, thereby effectively lengthening the leg and stride. In the cat, the clavicle is a slender, rod-like, curved bone with the sternal end slightly enlarged (Figure 34.31).

The proximal long bone of the brachium is the **humerus**. Typical of a long bone, the humerus consists of a central diaphysis and proximal and distal epiphyses. It articulates proximally with the scapula, and distally with the radius and ulna. The proximal **head** articulates with the glenoid fossa of the scapula. Medial to the head is the smaller **lesser tuberosity**, and lateral to the head is the larger **greater tuberosity** for muscle attachment. The **bicipital groove** between them accommodates the tendon of the biceps brachii muscle. At the distal end is a pair of prominent condyles—a larger medial **trochlea** and a smaller lateral **capitulum**, with which the ulna and radius, respectively, articulate (Figure 34.32).

Medial and proximal to the trochlea is the prominent **medial epicondyle**, and lateral and proximal to the capitulum is the less obvious **lateral epicondyle**. The conspicuous

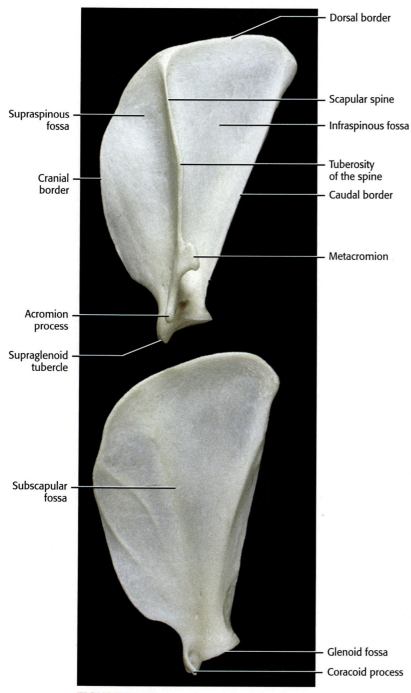

Dorsal border

Scapular spine

Infraspinous fossa

Tuberosity
of the spine

Caudal border

Metacromion

Supraspinous
fossa

Cranial
border

Acromion
process

Supraglenoid
tubercle

Subscapular
fossa

Glenoid fossa

Coracoid process

FIGURE 34.30 Scapula: Left lateral, right medial views.

During the evolution of locomotory changes in mammals, a rotation at the wrist brought the manus forward and resulted in the radius being rotated over the ulna. The natural position of the **radius** in the antebrachium of the cat extends from the lateral proximal humeral end to the medial distal carpal end, and therefore crosses over the ulna in the forearm. The proximal **head** is slightly concave to accommodate its articulation with the capitulum of the humerus. Below the head is the **neck**. Just distal to the neck on the posterolateral surface of the diaphysis is the **bicipital tuberosity** for insertion of the tendon of the biceps brachii muscle. The distal end is characterized by the medial **styloid process** (Figure 34.33).

The second bone of the forearm is the **ulna**. The proximal end, better known as the **olecranon**, articulates with the olecranon fossa of the humerus. Furthermore, this process functions as the insertion site of the tendon of the powerful forearm extensor, the triceps brachii. On the anterior surface, just distal to the olecranon, is an unusually shaped, articulating surface, the deeply excavated **semilunar notch**. The trochlea of the humerus articulates with the notch (Figure 34.34).

An anterior projection, the **coronoid process**, forms the medial margin of a lateral concave articulating surface, the **radial notch**, for articulation of the head of the radius. The distal end of the ulna is distinguished by the peg-like lateral **styloid process**. The medial surface articulates with the cuneiform and pisiform of the carpals. On the medial surface of the ulna, just proximal to the styloid process, observe the small articulating surface with which its radial counterpart articulates (Figure 34.34).

A series of seven small, irregularly shaped bones, the **carpals** or wrist bones, is organized into two rows—a proximal set of three and a distal set of four—and comprises the proximal end of the forefoot or manus. The most medial and largest carpal of the proximal row is the **scapholunate**, which in many mammals remain separate as the scaphoid and lunate. In the middle of this row is the **cuneiform**. The most lateral of the proximal set is called the **pisiform**. This row articulates proximally with the radius and ulna and distally with the second row of carpals (Figure 34.35).

In the distal or second row, from medial to lateral, are the **trapezium, trapezoid, capitate,** and **hamate**. Distally, these bones articulate with the five **metacarpals**, which articulate with the proximal **phalanges** of the five digits. The thumb has two phalanges, and each of the other four toes

ovoid slit occurring proximal to the medial epicondyle is the **supracondyloid foramen**. Through the foramen pass the median nerve and the brachial blood vessels. On the posterior side of the humerus is the deep, prominent **olecranon fossa**, with which the olecranon of the ulna articulates. A prominent rugosity, the medial **pectoral ridge** and the sharp, crest-like, lateral **deltoid ridge**, converge anteriorly at about midshaft. These surface irregularities mark the positions of muscle attachments (Figure 34.32).

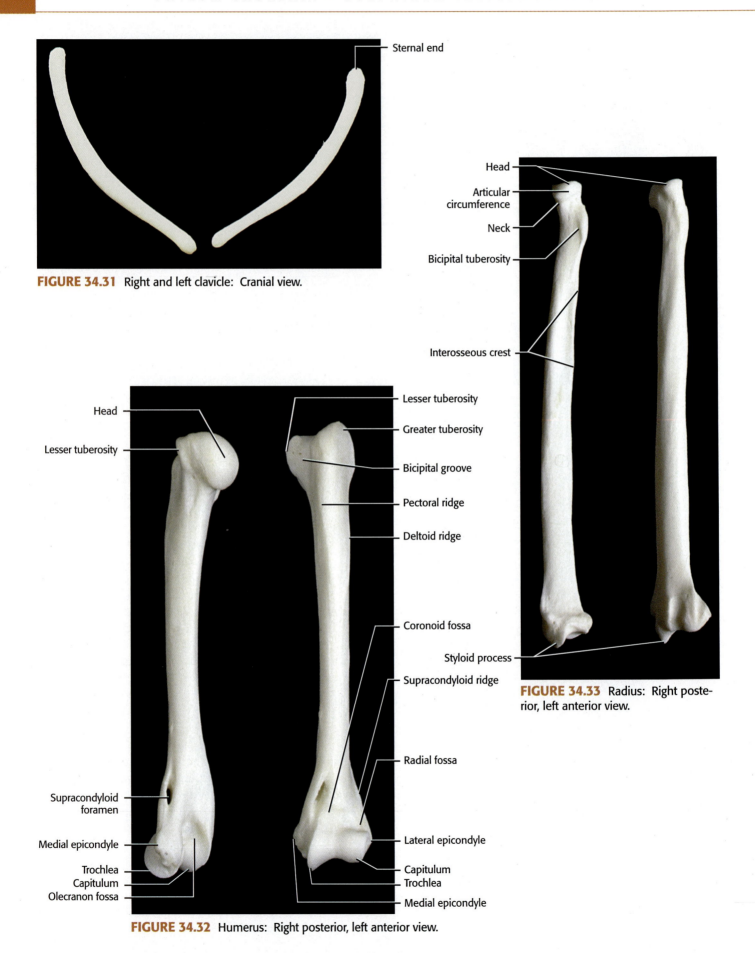

FIGURE 34.31 Right and left clavicle: Cranial view.

Sternal end

Head
Articular circumference
Neck
Bicipital tuberosity
Interosseous crest
Styloid process

FIGURE 34.33 Radius: Right posterior, left anterior view.

Head
Lesser tuberosity
Lesser tuberosity
Greater tuberosity
Bicipital groove
Pectoral ridge
Deltoid ridge
Coronoid fossa
Supracondyloid ridge
Radial fossa
Supracondyloid foramen
Medial epicondyle
Lateral epicondyle
Trochlea
Capitulum
Capitulum
Trochlea
Olecranon fossa
Medial epicondyle

FIGURE 34.32 Humerus: Right posterior, left anterior view.

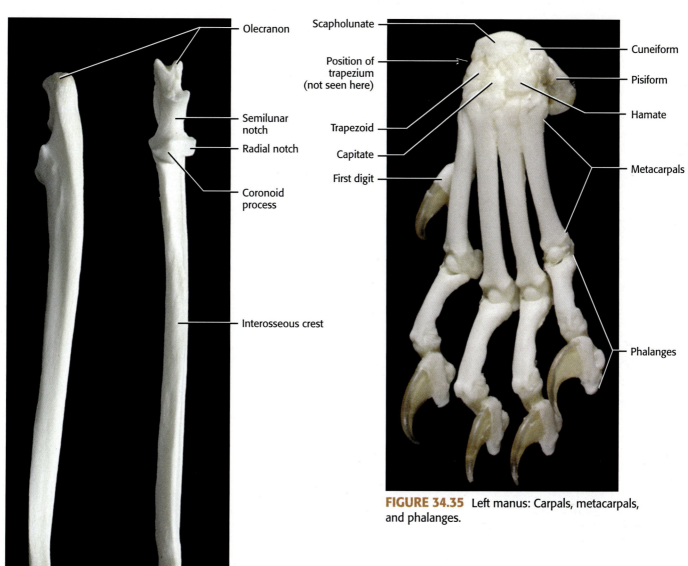

FIGURE 34.34 Ulna: Right posterior, left anterior view.

FIGURE 34.35 Left manus: Carpals, metacarpals, and phalanges.

has three. Sesamoid bones—facilitating smooth gliding of muscle tendons over the rough surfaces of adjacent bones—occur at the junctions of the metacarpals and the phalanges. Typical of most feline carnivores, the cat has well-developed retractile claws, an obvious adaptation among these predators to aid in capturing and holding frisky prey (Figure 34.35).

Pelvic Girdle and Hindlimb

The pelvic girdle of mammals consists of paired **os coxae** (Figure 34.36). The **os coxa**, an irregularly shaped, elongated, complex, has three individual bones: the ilium, the ischium, and the pubis.

Accompanying the medial shift in limb alignment in mammals has been an anterior reorientation of the ilium to accommodate the origin of important locomotory muscles. The cranially projecting **ilium** is articulated medially with the sacrum rather firmly. The articulation between the sacrum and the os coxae stabilizes the position of the pelvic girdle and its appendages and provides the posterior weight-bearing surface. The dorsal edge is called the **crest**. The posterior part of the os coxa consists of the **ischium**, with the roughened, thickened **tuberosity of the ischium**.

The third bone that contributes to the formation of the os coxa is the **pubis**. The paired halves of the pubis and the ischium meet in the ventral midline as the **pubic symphysis** and **ischiadic symphysis**, respectively. A lateral cuplike depression, the **acetabulum**, occurring at the junction of the ilium, ischium, and pubis, accommodates the articulation of the head of the femur.

In the cat, the thin shell-like medial wall of the socket is formed by the **acetabular bone**, a minor component of the os coxa. The prominent oval opening in the os coxa, the **obturator foramen**, is surrounded by the ischium and

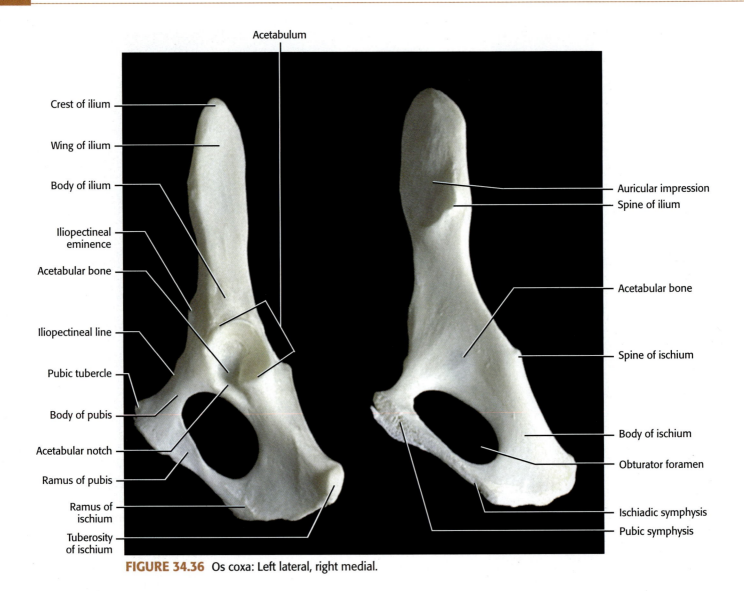

FIGURE 34.36 Os coxa: Left lateral, right medial.

pubis. Among many living reptiles, including mammalian ancestors, the pelvic girdle was pierced by a large opening, the obturator fenestra, as well as a small obturator foramen. Through this foramen passes the obturator nerve to muscles of the leg. In mammals, the obturator foramen was incorporated into the larger fenestra—hence, the large obturator foramen.

The proximal bone of the upper portion of the hindlimb is the **femur** (Figure 34.37A and Figure 34.37B). Because of the medial shift in limb position, a prominent ball-like projection, the **head**, with an irregular depression, the **fovea capitis**, articulates with the acetabulum of the os coxa. The fovea capitis is the site of attachment of a ligament that helps anchor the femur to the ventral portion of the acetabulum.

Equally prominent on the lateral aspect of the femur is the **greater trochanter**, where the hip muscles attach. A **lesser trochanter** is situated on the posterior aspect of the femur, again for the attachment of hip muscles. A

depression, the **trochanteric fossa**, is located between the two trochanters. Notice the prominent ridge extending from the greater trochanter and along the posterolateral aspect of the femur, crossing from lateral to medial at midshaft. This ridge is joined by a much less conspicuous line extending from the lesser trochanter to form the **linea aspera**. These ridges provide sites for muscle attachment.

The distal end of the femur is distinguished by two prominent projections, the **medial condyle** and the **lateral condyle**, whose smooth, rounded surfaces articulate with the proximal end of the tibia. A deep posterior notch, the **intercondyloid fossa**, separates the condyles. Take note of the two irregularly shaped prominences, the **medial epicondyle** and the **lateral epicondyle**, located above the condyles, which provide sites for muscle attachment. The anterior tongue-shaped, smooth surface, the **patellar surface**, is joined by the **patella** ("kneecap") to form a smooth surface over which major extensor muscle tendons slide (Figure 34.37A and 34.37B).

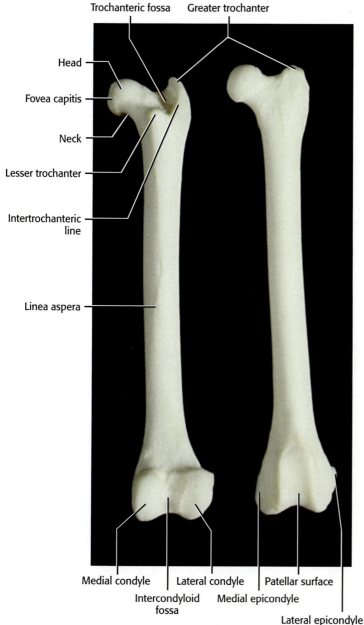

Trochanteric fossa   Greater trochanter

Head

Fovea capitis

Neck

Lesser trochanter

Intertrochanteric line

Linea aspera

Medial condyle   Lateral condyle   Patellar surface
Intercondyloid fossa   Medial epicondyle
Lateral epicondyle

FIGURE 34.37A Femur: Right posterior, left anterior.

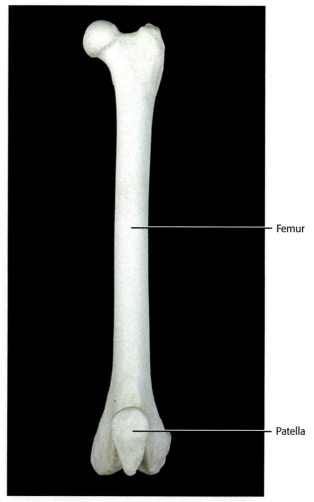

Femur

Patella

FIGURE 34.37B Patella *in situ:* Left anterior.

Of the two shank bones (Figure 34.38), the medial **tibia** is the longer and more robust and is the primary weight-bearing bone of the hindlimb. A **medial condyle** and a **lateral condyle** articulate with the articular surfaces of the femoral condyles. On either side of the condyles are a **medial tuberosity** and a **lateral tuberosity**. The rough anterior protuberance at the proximal end is the **tibial tuberosity**. Continuous with the tuberosity is the **tibial crest**. Beneath the lateral condyle is the **articular facet** for the head of the fibula. The distal end extends as the **medial malleolus**, with which the tarsals articulate (Figure 34.38).

The slender, lateral long bone of the crus is the **fibula**. The proximal **head** articulates with the lateral surface of

the tibia. The distal end is distinguished by the **lateral malleolus**. Grooves to accommodate lower leg muscle tendons are visible laterally and ventrally (Figure 34.38).

The pes consists of the tarsals, **metatarsals**, and **phalanges** (Figure 34.39). The number of tarsal bones (seven) is identical to the number of carpals. In contrast to the carpals, however, they vary in size. This is true particularly of the medially located **talus**, whose size and shape probably relate to the fact that it is the bone with which the tibia and fibula articulate and, therefore, represents the major weight-bearing bone of the ankle set.

The **calcaneus**, laterally located, is the longest bone of the ankle and forms the heel. The groove on the exposed surface of the calcaneus accommodates the tendon of Achilles. Distal to the talus lies the **navicular**. A row of four tarsal bones, identified from medial to lateral, consists of the **medial cuneiform**, **intermediate cuneiform**, **lateral cuneiform**, and **cuboid**. Five **metatarsals** articulate with the distal row of four tarsal bones. With the exception of the first, which has been greatly reduced, the other four are elongated and each articulates with a series of three **pha-langes**, terminating in a well-developed retractile claw (Figure 34.39).

During mammalian evolution, changes in stance played a significant role in locomotion and three divergent foot postures emerged. The least specialized is the **plantigrade** foot, in which the entire foot is placed on the ground during locomotion. Mammals such as primates and bears have plantigrade feet. In canine and feline carnivores evolved a **digitigrade** foot, in which the heel is held off the ground with only the digits making contact.

A third posture is the **unguligrade foot**, in which the lower foot surface is pulled even higher off the ground, with the animal walking on the tips of its toes. Two distinct groups of ungulates evolved. The **artiodactyls**—cows, pigs, sheep, antelopes, and deer—walk on an even number of toes. The **perissodactyls**—rhinoceroses, tapirs, and horses—walk on an odd number of toes. Digitigrade and unguligrade foot postures are correlated with cursorial habits. Lifting part of the foot from the ground reduces friction and lengthens the stride, thereby affecting the speed of the runner. See Figure 5.20.

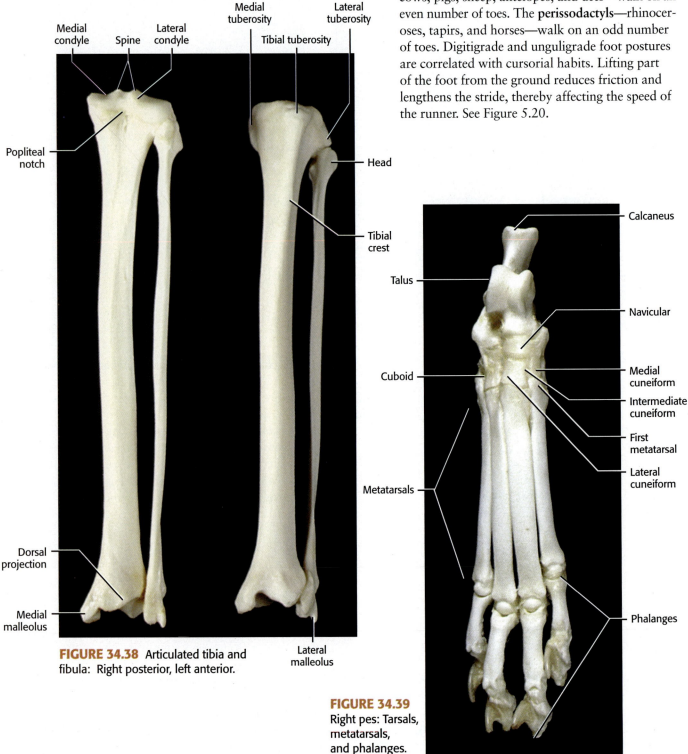

FIGURE 34.38 Articulated tibia and fibula: Right posterior, left anterior.

FIGURE 34.39
Right pes: Tarsals, metatarsals, and phalanges.

CHAPTER THIRTY-FIVE

Cat — Muscular System

35

SKINNING

In spite of what you may have heard, there is only one way to skin a cat as far as we're concerned. Cat specimens usually are packed individually in plastic bags containing both the cat and some preservative fluid to aid in maintaining a moist environment. Therefore, you will have to remove the specimen from the bag carefully, retaining the fluid to keep the cat moist when it is returned to the bag for storage.

Lay the cat on its dorsal surface on a large dissecting tray. The cat will be skinned on *one side only*, and the skin should be kept in a *single piece* so it can be wrapped around the skinned surface when the cat is not being worked on actively. Before attempting to remove the skin, observe the areas on the body where skin may have been removed to facilitate the injection of blood vessels with latex—*e.g.*, the neck region, the forearm, the hindleg, and the hepatic portal system. A major factor affecting your determination of which side of the cat to skin should be the injection site of the hepatic portal system, as the incision in the abdominal area usually is stapled or sutured shut, making it difficult to skin around this area. Therefore, skin the specimen to avoid this injection site.

 Because the success of the skinning process is closely correlated with your ability to complete clean, precise cuts, a new blade in your scalpel is essential.

Make a careful, shallow incision, just deep enough to break the skin, beginning at the base of the neck 1.25 cm left or right of the midventral line to avoid any muscles whose origin or insertion is on the midline. Determine whether it is possible to pull the skin away from the underlying tissue. Using your fingers, a pair of forceps, or a scalpel with the blade held parallel to the underlying muscle or toward the skin, sever the connective tissue from the skin.

If your specimen is a female that has been pregnant recently, as you skin the thoracic and abdominal regions, you may encounter the mammary glands, which will appear as flattened, tan masses that you might mistake for muscle. It is preferable to remove these glands with the skin. Continue caudally to a level approximately 5 cm anterior to the cranial edge of the hindlimb. Now angle your incision along the midline of the hindlimb, continuing to a point just proximal to the digits, where you will make an encircling incision around the pes.

If your specimen is a male, exercise extreme caution at this juncture, as the spermatic cord is imbedded within the fat and connective tissue of the groin area and directly beneath the skin. Also, the leg skin is very thin and a major superficial vein, the saphenous, lies directly under the skin. Therefore, carefully sever the fascia from the skin in this area.

Return to the thoracic region and begin an incision opposite the forelimb, continuing down the medial aspect and encircling the manus just proximal to the digits. Be exceptionally cautious when skinning the radial side of the lower forelimb, as a thin, narrow muscle band, the brachioradialis muscle, a nerve and blood vessels, adhere closely to the skin and may be severed mistakenly. In addition, along the lateral aspect of the forelimb, from the wrist to the shoulder, courses the cephalic vein. Again, this vein can be removed easily along with the skin—but it should not be!

Carefully skin the body, hindlimb and forelimb. As you skin the trunk, you will encounter an extensive dermal muscle, the **cutaneus maximus** muscle, especially prominent in the axillary, pectoral, and abdominal regions. This is the muscle, derived from ancestral pectoralis muscles, that allows twitching for horses to get rid of flies, dogs to shake water out of their coats, and so on. It is best to remove this muscle with the skin, exerting great care in the axillary region. In the dorsal shoulder region, take care not to cut through a heavy white connective tissue (aponeurosis) between the paired acromiotrapezius muscles.

Concentrate now on the neck and head regions. Be careful of superficial blood vessels in the ventrolateral position in the neck. During your dissection of this area, note another dermal muscle, the **platysma** muscle, derived from ancestral hyoid constrictors, which adheres closely to the neck and head muscles. Again, it is desirable to remove the platysma with the skin. Notice that the skin in the head region is much thicker than other areas of the body. Extend the ventral incision to the base of the mandible, outlining the mouth, the nose, and the eye, continuing the incision to the midline of the forehead. As you loosen the skin from the underlying tissues, make a circular incision around the ear and reflect the skin. Continue the skinning process to approximately 1.25 cm past the middorsal line along the entire length of the cat.

Preparing the Cat for Muscle Dissection

To dissect and appreciate muscle relationships, extraneous tissues that tend to adhere to the surface of the superficial muscles must be removed. Carefully remove fatty tissue lying on external muscle surfaces. Usually the groin area has a heavy deposit of fat. If your specimen is a male, exercise extreme caution while removing fat in this region because the spermatic cord, a thin, small-diameter tube, lies in close proximity to this fat deposit. Another area where fat may accumulate is in the region between the scapulae on the surface of the aponeurosis connecting the two acromiotrapezius muscles. While removing this fatty tissue here, be careful not to destroy the aponeurosis.

Heavy sheets of dense, connective tissue covered by fatty tissue may occur in other areas on the body surface (*e.g.*, the lumbosacral region, the insertion end of the tensor fascia latae muscle, biceps femoris muscle).

 Always take care to avoid damaging the important areas of fascia and any other underlying structures. Fascia associated with muscle insertions should *not* be removed.

After skinning the cat, pieces of the dermal muscles (platysma and cutaneus maximus) may remain, adhering to the muscle surfaces. These should be removed carefully. The epitrochlearis muscle on the medial surface of the brachium appears very much like a piece of cutaneus maximus. Do not remove this muscle or the thin aponeurosis by which it inserts in the vicinity of the elbow.

Direction of Muscle Fibers

During the dissection of your specimen, it is advantageous to be able to distinguish where one muscle ends and an adjacent or overlapping muscle begins. To identify individual muscles, look for the direction of muscle fiber orientation. For example, in muscles such as the abdominals, consisting of three sheetlike layers superimposed one on the other, it becomes essential to detect changes in fiber direction. Furthermore, most muscles are individually wrapped in layers of connective tissue called fascia, and the areas where these layers abut one another often can be observed as distinct lines between muscles. Your ability to distinguish and separate contiguous muscles will be enhanced greatly by training yourself to appreciate these relationships—and will be greatly appreciated by your instructor.

 Because you will be identifying superficial and deep muscles, you must cut the superficial muscles before dissecting the underlying muscles. Isolate the muscle from origin to insertion, freeing it from contiguous muscles. Lift the isolated muscle and cut through the midpoint (belly) of the muscle. Always cut the muscle perpendicular to its fibers. Muscles to be cut will be identified as:

Cut this muscle.

SUPERFICIAL THORACIC MUSCLES

 Muscles in the superficial thoracic muscle group have a tendency to adhere tightly to one another. Therefore, exercise care when separating them. Watch for the changes in muscle-fiber orientation and the subtle white lines created by the connective tissue surrounding each muscle, indicating the extent of individual muscles.

Pectoral muscles in mammals, in contrast to the simple pectoral muscles of *Necturus*, have diversified, giving rise to a number of individual muscles that adduct or pull the forelimb toward the midline and retract or pull the forelimb backward. The change in limb position and the increased mobility of mammals probably contributed to the differentiation.

Pectoantebrachialis

The **pectoantebrachialis** is the most superficial of the chest or pectoral muscles, homologous with a portion of the pectoralis muscles of *Necturus*. It is a narrow, *thin* band extending from the midline of the body to the upper portion of the forelimb (Figure 35.1).

Origin: Manubrium of the sternum

Insertion: Flat tendon into the superficial fascia of the antebrachium above the elbow

Action: Draws the forelimb toward the midline

 Cut this muscle.

Pectoralis Major

Pectoralis major is a diagonally oriented band partially covered by **clavotrapezius** and clavobrachialis muscles and, further, partially hidden by **pectoantebrachialis**. Like pecto-antebrachialis, this muscle is homologous with a portion of the pectoralis muscles of *Necturus* (Figure 35.1). To see the entire extent of pectoralis major, the clavotrapezius and clavobrachialis muscles should now be dissected (see pages 313–314).

 Exercise care while separating clavotrapezius to avoid damaging the underlying pectoralis major. Furthermore, take care in isolating pectoralis major to avoid damaging the underlying pectoralis minor.

Origin: Cranial half of the sternum and midventral raphe

Insertion: Proximal two-thirds of the shaft of the humerus

Action: Draws the forelimb toward the midline and turns the manus forward

 Cut this muscle.

Pectoralis Minor

A thick band of muscle, **pectoralis minor**, also homologous with a portion of the pectoralis muscles of *Necturus*, extends caudal to and beneath pectoralis major (Figure 35.1). Exert care to preserve **xiphihumeralis**, which passes dorsal to pectoralis minor. In addition, with great care, separate the **latissimus dorsi** from the lateral border of pectoralis minor and xiphihumeralis.

Origin: From the six sternebrae and sometimes the xiphoid process, resulting in the appearance of several slips that appear to be separate muscles

Insertion: Ventral border of the humerus

Action: Draws the forelimb toward the midline

 Cut this muscle.

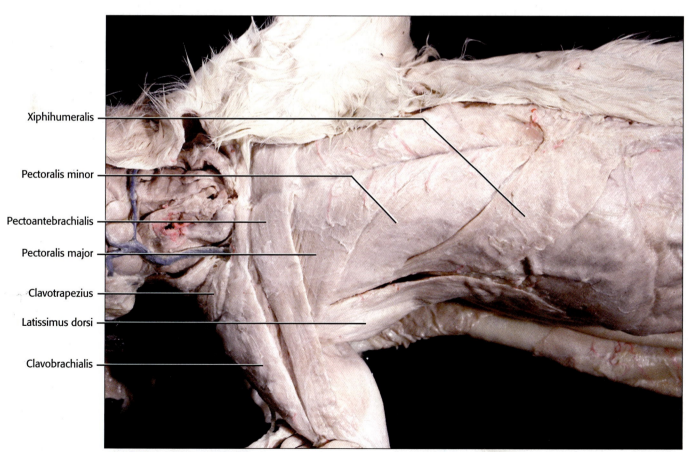

Xiphihumeralis

Pectoralis minor

Pectoantebrachialis

Pectoralis major

Clavotrapezius

Latissimus dorsi

Clavobrachialis

FIGURE 35.1 Superficial thoracic muscles (anterior left).

Xiphihumeralis

A long, very thin, narrow band of muscle, also homologous with a portion of the pectoralis muscles of *Necturus*, **xiphi-humeralis** lies along the posterior border of pectoralis minor. According to some anatomists, it is actually a part of that muscle (Figure 35.1).

Origin: Median raphe in the vicinity of the xiphoid process

Insertion: Along the ventral border of the humerus

Action: Synergistic with the pectoralis minor in drawing the forelimb toward the midline

 Cut this muscle.

ABDOMINAL MUSCLES

Three sheetlike muscles and a longitudinal, bandlike muscle, homologous with similar muscles in *Necturus*, make up the group of abdominal muscles. Note that the left and right portions of the abdominal muscles are separated by a longitudinal white line of connective tissue known as the **linea alba**. The sheetlike muscles are thin and quite extensive, supporting the entire abdominal area and a portion of the ventral thoracic region. These muscles are layered and adhere closely to one another by means of fascia. The direction of fibers within each sheet is distinctive, and this feature is used as a tool to identify the individual muscles.

 To facilitate dissection of these sheets, make a three-sided opening, 2.5 cm on each side, in the flank (Figure 35.2). Carefully separate and identify the muscles.

External Oblique

In the **external oblique, the** most superficial of the three sheetlike abdominal muscles, the direction of the fibers extends craniodorsally (Figure 35.2).

Origin: Lumbodorsal fascia and the last 9 or 10 ribs

Insertion: Median raphe of distal portion of sternum and linea alba from sternum to pubis

Action: Compresses the abdominal region

Internal Oblique

The **internal oblique** lies directly beneath the external oblique, and its fibers extend caudodorsally (Figure 35.2).

Origin: Lumbodorsal fascia in common with the external oblique, and dorsal iliac border

Insertion: Linea alba in common with the external oblique and transversus abdominis

Action: Compresses the abdominal region

Transversus Abdominis

The deepest muscle sheet, the **transversus abdominis**, lies directly beneath the internal oblique. Its fibers are oriented

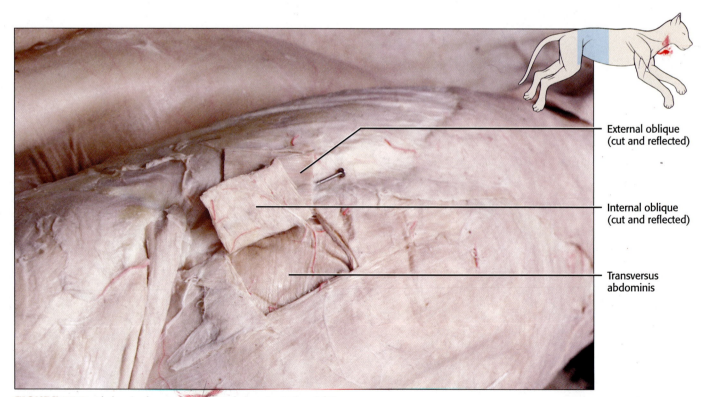

External oblique (cut and reflected)

Internal oblique (cut and reflected)

Transversus abdominis

FIGURE 35.2 Abdominal muscles: Lateral view (anterior right).

almost transversely between the origin and the insertion (Figure 35.2).

Origin: From the costal cartilages of the posterior ribs, transverse processes of lumbar vertebrae and ventral border of the ilium

Insertion: Linea alba in common with the two obliques

Action: Compresses the abdomen

Rectus Abdominis

Rectus abdominis occurs as a longitudinal band of muscle on either side of the linea alba and is encased in a sheath formed by the aponeuroses of the other three abdominal muscles (Figure 35.3).

Origin: Pubis

Insertion: First and second costal cartilage and proximal end of sternum

Action: Compresses the abdominal region, pulls sternum and ribs caudally, causing flexion of the trunk

SUPERFICIAL BACK MUSCLES

Clavotrapezius

Clavotrapezius is a wide, flat muscle that covers most of the lateral portion of the neck and is homologous with a portion of the cucullaris of *Necturus* (Figure 35.4). Observe that levator scapulae ventralis passes below clavotrapezius and must be isolated from clavotrapezius.

Origin: Lambdoidal ridge, middorsal raphe over spine of the axis

Insertion: Clavicle and raphe between clavotrapezius and clavobrachialis

Action: Protracts the humerus

 Cut this muscle.

Clavobrachialis

Clavobrachialis, a homolog of a portion of procoraco-humeralis of *Necturus*, appears to be a continuation of clavotrapezius onto the forelimb. Some anatomists consider it to be the cranial portion of the deltoid, called the clavodeltoid (Figure 35.4).

Origin: Clavicle and raphe between clavotrapezius and clavobrachialis

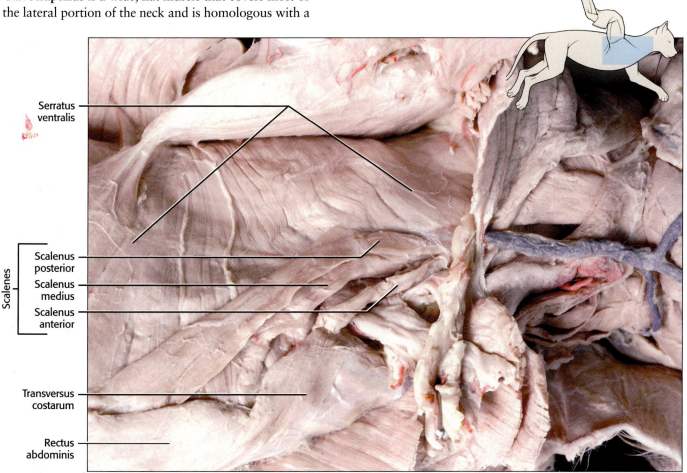

Serratus ventralis

Scalenes — Scalenus posterior

Scalenus medius

Scalenus anterior

Transversus costarum

Rectus abdominis

FIGURE 35.3 Cranial deep thoracic muscles. Note: This cat had an unusually long scalenus posterior (anterior right).

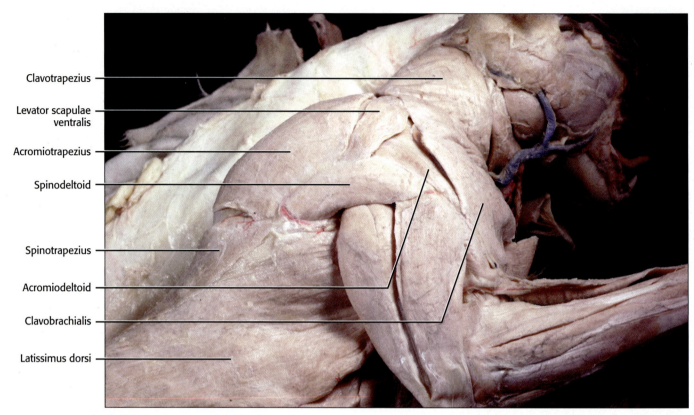

Clavotrapezius

Levator scapulae
ventralis

Acromiotrapezius

Spinodeltoid

Spinotrapezius

Acromiodeltoid

Clavobrachialis

Latissimus dorsi

FIGURE 35.4 Superficial back muscles.

Insertion: Commonly inserted with the brachialis on the medial surface of the ulna distal

Action: Flexes the forearm

 Cut this muscle.

Acromiotrapezius

The **acromiotrapezius** is a *thin* trapezoidal muscle, homologous with a portion of the cucullaris of *Necturus*, lying over the scapulae (Figure 35.4). Extreme care must be exercised while dissecting this muscle, especially to prevent damage to the whitish middorsal tendon and fascia that hold the left and right acromiotrapezius muscles together over the vertebral column.

Origin: Middorsal line from the spine of the axis to the spinous process of the fourth thoracic vertebra

Insertion: Metacromion process and spine of the scapula

Action: Adduct and stabilize the position of the scapulae

 Cut this muscle, *not* the aponeurosis!

Spinotrapezius

The triangular trapezius, the **spinotrapezius**, also homologous with a portion of the cucullaris of *Necturus*, is the

most posterior of the trapezius group (Figure 35.4). With great care, dissect this muscle from the craniodorsal surface of **latissimus dorsi.**

Origin: May originate from the spinous processes of most of the thoracic vertebrae

Insertion: Fascia of supraspinatus and infraspinatus muscles on either side of the spine

Action: Pulls the scapula dorsally and caudally

 Cut this muscle.

Latissimus Dorsi

Latissimus dorsi is a large, thick, flat, triangular muscle, homologous with latissimus dorsi of *Necturus*. It lies just posterior to the trapezius group and is covered craniodorsally by **spinotrapezius** (Figure 35.4). As mentioned previously, in the dissection of pectoralis minor, exercise care while separating the lateral edges of latissimus dorsi and pectoralis minor.

Origin: Neural spines of the fourth or fifth thoracic to the sixth lumbar vertebrae

Insertion: Medial surface of humerus at the proximal end

Action: Pulls forelimb dorsocaudally

 Cut this muscle.

DEEP THORACIC MUSCLES

Serratus ventralis

A large fan-shaped muscle, homologous with thoraciscapularis and a portion of the external oblique of *Necturus*, **serratus ventralis** consists of individual straplike slips that extend between the thorax and the scapula. Notice that these individual slips are more conspicuous at the caudal end of the muscle (Figure 35.3 and Figure 35.5).

Origin: From the surface of the first 9 or 10 ribs and the transverse processes of the last five cervical vertebrae

Insertion: Vertebral border of the scapula

Action: Draws the scapula toward the thoracic wall and helps to support the scapula

Scalenes

The **scalenes**, homologous with a portion of the external oblique of *Necturus*, are separable into three bands lying at an oblique angle along the lateral aspect of the thorax and cranially uniting into a single band or bundle (Figure 35.3).

Origin: From the second through the ninth ribs

Insertion: Transverse processes of cervical vertebrae

Action: Bends the neck and pulls ribs cranially

Transversus Costarum

A thin, bandlike muscle extending from the sternum and covering the cranial portion of the rectus abdominus muscle, the **transversus costarum** is still another homolog of a portion of the external oblique of *Necturus* (Figure 35.3).

Origin: From the side of the sternum

Insertion: First rib and costal cartilage

Action: Pulls the ribs cranially

Intercostalis Externus

Note that the fibers of this outer layer of muscles, the **intercostalis externus**, lying in the intercostal spaces between adjacent ribs, are oriented craniodorsally, similar to the external oblique layer. Not surprisingly, they are homologous

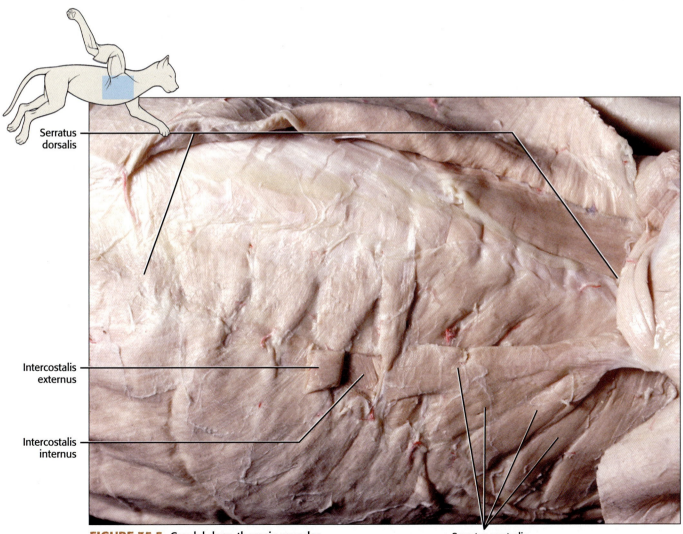

Serratus dorsalis

Intercostalis externus

Intercostalis internus

FIGURE 35.5 Caudal deep thoracic muscles.

Serratus ventralis

with a portion of the external oblique of *Necturus* (Figure 35.5).

Origin: From a cranial rib

Insertion: To an adjacent caudal rib

Action: Protracts the ribs

Intercostalis Internus

The **intercostalis internus** layer lies directly medial to the external intercostals. Its fibers are oriented caudodorsally, similar to the internal oblique layer, and are homologous with a portion of the internal oblique of *Necturus* (Figure 35.5).

Origin: From a caudal rib

Insertion: To an adjacent cranial rib

Action: Retracts the ribs

Transversus Thoracis

An incomplete third layer, the **transversus thoracis**, lies beneath part of the internal intercostals. It represents the thoracic portion of the transversus abdominis with which it is homologous in *Necturus* (Figure 35.6).

Origin: Dorsolateral border of the sternum between the third and the eighth rib

Insertion: Costal cartilages near the junction of the ribs

Action: Moves ribs

Serratus Dorsalis

A thin layer of muscle, homologous with a portion of the external oblique of *Necturus*, **serratus dorsalis** appears as slips extending along the dorsal part of the thorax and neck beneath the latissimus dorsi muscle (Figure 35.5).

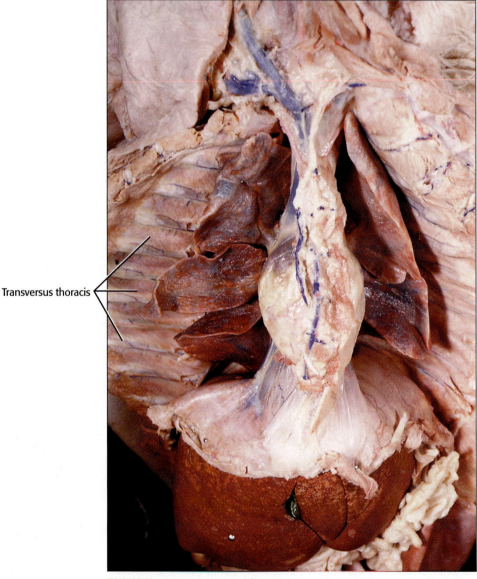

Transversus thoracis

FIGURE 35.6 Transversus thoracis.

Origin: From the middorsal fascia

Insertion: On the outer surface of the ribs

Action: Anterior portion draws the ribs cranially while the posterior portion draws the ribs caudally.

LOWER BACK MUSCLES: LUMBAR AND THORACIC

 To begin dissection of the lower back and thoracic muscles, use a pair of forceps to hold up the lumbodorsal fascia, make a small hole in the double-layered fascia, and then cut a three-sided window approximately 5 cm on a side (Figure 35.7). Depending upon the size of the cat, the dimensions of the window may vary.

The following lumbar and thoracic muscles are homologous with portions of dorsalis trunci of *Necturus*.

Multifidus Spinae

The **multifidus spinae** is an extensive muscle consisting of many bundles of fibers that can best be distinguished abutting the vertebral column in the lumbar region (Figure 35.7).

Origin: Primarily transverse processes of vertebrae

Insertion: Neural spines of more anterior vertebrae

Action: When both sides contract simultaneously, extends the vertebral column; when contracted unilaterally, bends the vertebral column toward that side

Longissimus Dorsi

Longissimus dorsi is an extensive muscle occupying the space between the neural spines and transverse processes and extending from the prominent lumbar region to the much less bulky thoracic and cervical regions. There are a number of distinguishable bundles in the lumbar region—a **medial bundle** and a **lateral bundle**, further subdivided by fascia into upper and lower lateral portions. These do not have to be dissected. Note that longissimus capitis is the cervical extension of longissimus dorsi (Figure 35.7).

Origin: From the ilium and neural spines of the vertebrae in the vertebral column

Insertion: Onto various processes of more anterior vertebrae of the vertebral vertebrae

Action: Extends the vertebral column

Spinalis Dorsi

Spinalis dorsi results from a medial subdivision of longissimus dorsi in the thoracic region (Figure 35.7).

Origin: From the neural spines of more posterior thoracic vertebrae

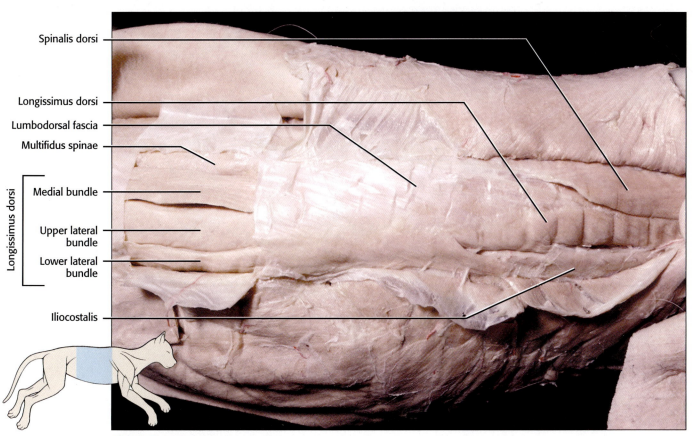

FIGURE 35.7 Caudal back window: Lumbar and thoracic back muscle complex (anterior right).

Insertion: Transverse processes of more cranial vertebrae

Action: Extends the vertebral column

Iliocostalis

A thin muscle confined to the thoracic region, **iliocostalis** consists of a number of bundles lying lateral to longissimus dorsi over the dorsal aspect of the ribs (Figure 35.7).

Origin: Lateral surface of the ribs

Insertion: Onto the lateral surface of more cranial ribs

Action: Pulls the ribs together

MUSCLES OF THE NECK

Sternomastoid

The origin of the **sternomastoid**, a paired bandlike muscle, homologous with a portion of the cucullaris of *Necturus*, forms the apex of a V just cranial to the anterior end of the sternum, while the arms of the V continue toward the base of the ear (Figure 35.8A and 35.8B).

Origin: Cranial end of the manubrium

Insertion: Lambdoidal ridge and mastoid portion of the temporal bone

Action: As a pair—flexion of the head; individually—turns the head

Cleidomastoid

Cleidomastoid is a flat bandlike muscle, also homolgous with a portion of the cucullaris of *Necturus*, which extends between the clavicle and the temporal region of the skull; lies dorsolateral to sternomastoid (Figure 35.8B).

Origin: Clavicle

Insertion: Mastoid process of the temporal bone

Action: When the clavicle is stationary, turns the head; when head is stationary, moves the clavicle anteriorly

Sternohyoid

A slender bandlike muscle, homologous with a portion of rectus cervicis of *Necturus*, the **sternohyoid** lies along either side of the midventral line of the neck (Figure 35.9A and Figure 35.9B).

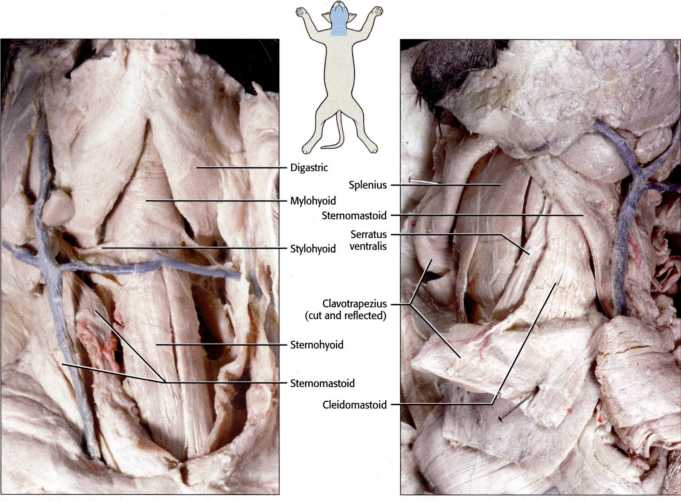

Digastric	Splenius
Mylohyoid	Sternomastoid
Stylohyoid	Serratus ventralis
Sternohyoid	Clavotrapezius (cut and reflected)
Sternomastoid	Cleidomastoid

A Superficial neck muscles (ventral view)

B Sternomastoid and cleidomastoid (lateral view)

FIGURE 35.8 Muscles of the neck.

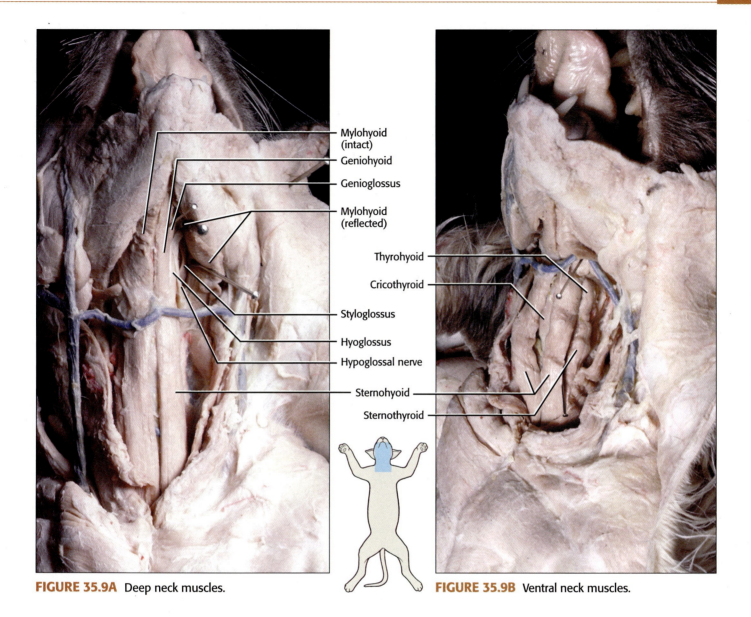

Mylohyoid (intact)
Geniohyoid
Genioglossus
Mylohyoid (reflected)
Thyrohyoid
Cricothyroid
Styloglossus
Hyoglossus
Hypoglossal nerve
Sternohyoid
Sternothyroid

FIGURE 35.9A Deep neck muscles.

FIGURE 35.9B Ventral neck muscles.

Origin: First costal cartilage

Insertion: Hyoid bone

Action: Retracts the hyoid

Sternothyroid

Another slender band-like muscle of the ventral neck, **sternothyroid**, homologous with a portion of rectus cervicis of *Necturus*, lies somewhat dorsal to the sternohyoid and lateral to the trachea (Figure 35.9B).

Origin: First costal cartilage

Insertion: Thyroid cartilage of the larynx

Action: Retracts the larynx

Thyrohyoid

The **thyrohyoid** is a short, bandlike muscle, homologous with a portion of rectus cervicis of *Necturus*, which lies along the lateral aspect of the larynx (Figure 35.9B).

Origin: Lateral portion of the thyroid cartilage of the larynx

Insertion: Hyoid bone

Action: Protracts the larynx

Stylohyoid

Stylohyoid is a slender band, homologous with a portion of the interhyoideus/sphincter colli muscles of *Necturus*, which stretches horizontally across the posterior surface of the digastric muscle (Figure 35.8A).

> Great care must be exercised in dissecting this muscle, as it is overlain with connective tissue and this, combined with its minute size, makes it especially vulnerable to destruction.

Origin: Stylohyal, a segment of the lesser horn of the hyoid

Insertion: Body of the hyoid

Action: Elevates the hyoid

Mylohyoid

The thin, roughly triangular **mylohyoid** muscle is homologous with a portion of intermandibularis of *Necturus*. It lies between the two dentary bones of the mandible and consists of a pair of muscles with distinct transverse fibers connected by a thin, tendinous median raphe (Figure 35.8A).

 With a sharp scalpel, make an incision through the median raphe. Reflect one of these thin muscles toward the mandible, taking care not to shred its fibers.

Origin: Medial surface of the mandible

Insertion: Median raphe

Action: Elevates the floor of the mouth

Geniohyoid

The **geniohyoid**, a narrow, elongated muscle, homologous with the geniohyoid of *Necturus*, lies along the median raphe dorsal to the mylohyoid (Figure 35.9A).

Origin: Ventral surface of the mandible

Insertion: Body of the hyoid

Action: Protracts the hyoid

Genioglossus

A second narrow, elongated muscle, the **genioglossus**, lies dorsolateral to the geniohyoid (Figure 35.9A). It is homologous with the genioglossus of *Necturus*.

Origin: Ventral surface of the mandible near the symphysis and dorsal to the geniohyoid

Insertion: Tongue

Action: Draws tip of the tongue backward and the root forward

Hyoglossus

Lateral to the geniohyoid lies the roughly rhomboidal **hyoglossus** muscle (Figure 35.9A). It can be identified readily because the hypoglossal nerve (C.N. XII) lies over its surface. The hyoglossus is homologous with a portion of the geniohyoid of *Necturus*.

Origin: Body of the hyoid

Insertion: Tongue

Action: Retracts and depresses the tongue

Styloglossus

Yet another bandlike muscle homologous with a portion of the geniohyoid of *Necturus*, the **styloglossus** lies lateral to the hyoglossus and parallel to the digastric (Figure 35.9A).

To observe the styloglossus, carefully pull the digastric muscle laterally.

Origin: Mastoid process

Insertion: Tongue

Action: Retracts and elevates the tongue

Intrinsic muscles of the tongue are entirely contained within the body of the tongue and are known as the **lingualis proprius**. They are homologous with a portion of the genioglossus of *Necturus*. They make up the bulk of the tongue and assist in tongue movement.

DEEP NECK AND BACK MUSCLES

Rhomboideus Capitis

The rhomboideus complex is homologous with a portion of the external oblique of *Necturus*. The cranial portion, **rhomboideus capitis**, consists of a narrow, thin, flat, lateral band (Figure 35.10).

Origin: Lambdoidal ridge

Insertion: Dorsal border of the scapula

Action: Rotates and pulls scapula cranially

Rhomboideus Cervicis and Thoracis

The **rhomboideus cervicis** and **thoracis** portion of the complex is a thick trapezoidal muscle separated into an anterior cervical and posterior thoracic part (Figure 35.10).

Origin: Vertebral spines of the first four thoracic vertebrae

Insertion: Vertebral border and outer surface of the scapula

Action: Adducts scapulae

Splenius

The **splenius**, a large, flat muscle, homologous with a portion of the dorsalis trunci of *Necturus*, lies along the dorsolateral aspect of the neck beneath rhomboideus capitis (Figure 35.11).

Origin: Middorsal line of neck and adjacent fascia

Insertion: Along the lambdoidal ridge of the occipital bone

Action: The joint action of the left and right splenius muscles elevate or extend the head; individually, each muscle flexes the head laterally

Longissimus Capitis

Longissimus capitis, a narrow, straplike muscle, is a cranial continuation of longissimus dorsi. It is homologous with a portion of dorsalis trunci of *Necturus* and lies along the ventral edge of splenius with which it has a tendency to fuse (Figure 35.11).

Origin: By several tendons from the prezygapophyses of cervical vertebrae 4–7

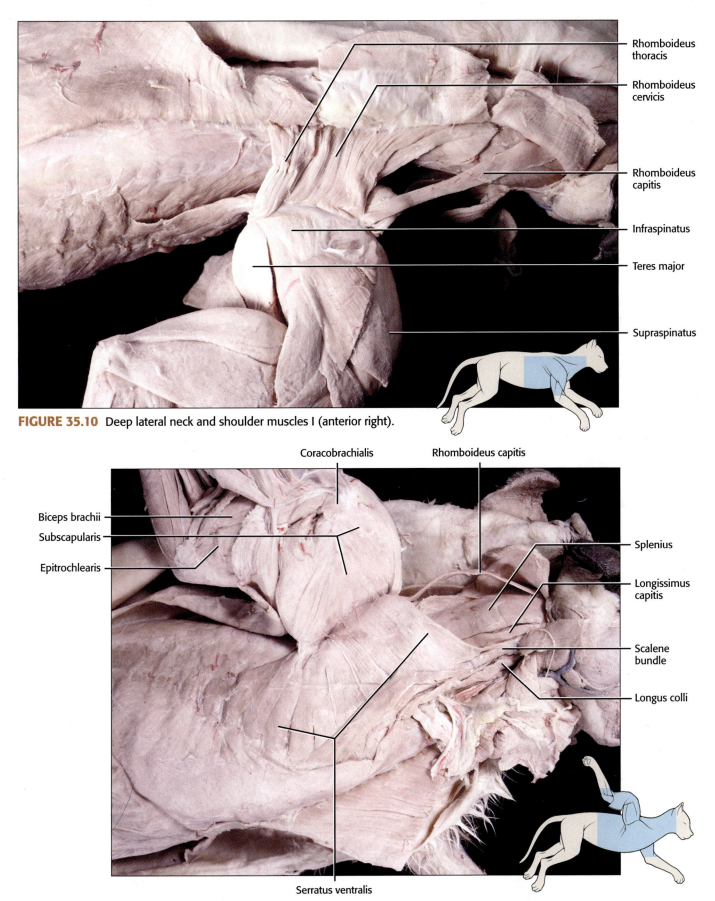

FIGURE 35.10 Deep lateral neck and shoulder muscles I (anterior right).

FIGURE 35.11 Deep lateral neck and shoulder muscles II (anterior right).

Insertion: Mastoid process of the temporal bone

Action: Lateral flexion of the head

The next two muscles, homologous with a portion of dorsalis trunci of *Necturus*, are sometimes identified individually as **semispinalis cervicis** and **semispinalis capitis**. These are flat muscles lying beneath **splenius** (Figure 35.12). To identify the semispinalis muscles, it will be necessary to bisect splenius. First longissimus capitis has to be separated from splenius.

 Bisect splenius at right angles to its fibers and reflect the halves.

Semispinalis Cervicis

Origin: Spinous processes of cervical vertebra 7 and the first three thoracic vertebrae

Insertion: Lambdoidal crest

Action: Elevates the head

Semispinalis Capitis

Origin: Prezygapophyses of cervical vertebrae 3–7 and first three thoracic vertebrae

Insertion: Lambdoidal crest

Action: Elevates the head

Longus Colli

Lying along the lateral aspect of the neck, **longus colli** is a narrow band of muscle that is homologous with a portion of the subvertebralis muscles of *Necturus* (Figure 35.11).

Origin: From the ventral surfaces of the first six thoracic vertebrae, and ventral surfaces of bodies and transverse processes of the cervical vertebrae

Insertion: Slips from the thoracic vertebrae unite and insert commonly onto the ventral portion of the sixth cervical transverse process; slips from the cervicals extend cranially to insert on the midline of the centra of more anterior cervical vertebrae

Action: Bends the neck both ventrally and laterally

MUSCLES OF THE HEAD

Ventral hyoid constrictors, interhyoideus, and sphincter colli of nonmammalian vertebrates have evolved into a dermal muscle sheet, the platysma and muscles of facial expression producing the eye, ear, nose, and lip movements that we associate with mammals. The muscles of facial expression are not included in this discussion.

The dramatic changes that took place among mammalian ancestors to produce the distinctive food-handling

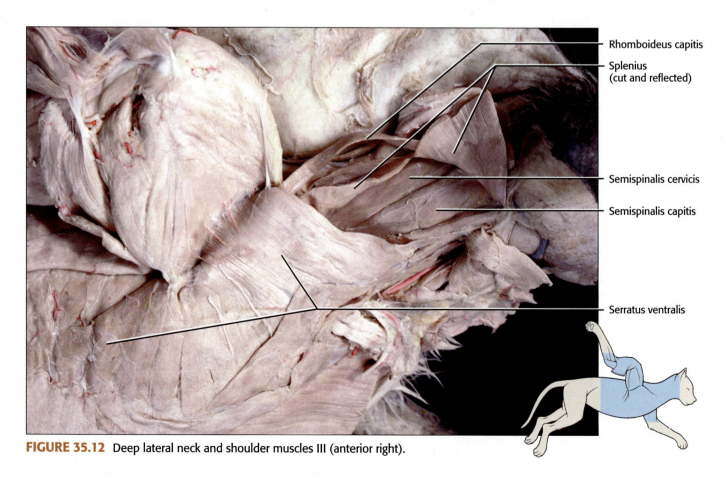

Rhomboideus capitis

Splenius (cut and reflected)

Semispinalis cervicis

Semispinalis capitis

Serratus ventralis

FIGURE 35.12 Deep lateral neck and shoulder muscles III (anterior right).

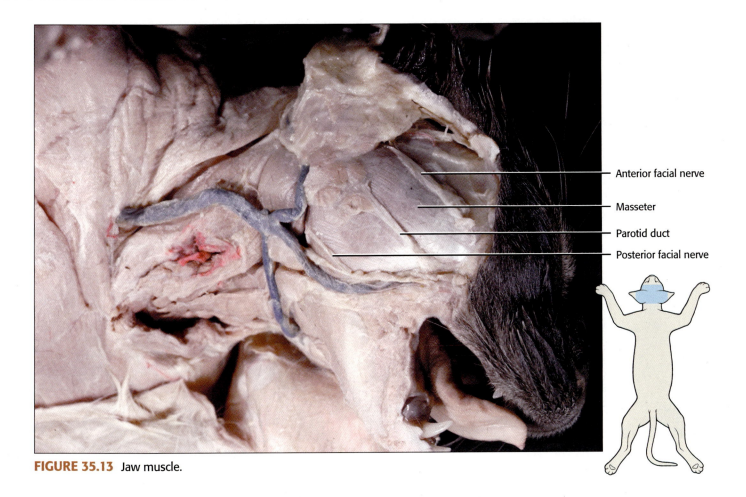

FIGURE 35.13 Jaw muscle.

Labels: Anterior facial nerve, Masseter, Parotid duct, Posterior facial nerve

characteristics of mammals were accompanied by specialization of their simple jaw adductor musculature. The partitioning of the muscles into three distinct groups—**masseters, temporals,** and **pterygoids**—permitted more sophisticated capturing and processing tools.

Temporal muscles play a primary role in holding the jaws shut, and hence are more massive in carnivores, while masseters and pterygoids are more important in the grinding action of mastication and therefore constitute a larger proportion of the jaw musculature among herbivores and ominvores. The jaw "closing" muscles—the masseter, temporal, and pterygoids—are homolgous with the adductor mandibulae of *Necturus.*

Masseter

The heavy muscle projecting prominently beneath and posterior to the eye and making up the cheek region in the cat is the **masseter** (Figure 35.13). Although this muscle's construction consists of three separate layers of fibers whose directions are distinct, we will not attempt to dissect them.

Origin: From the zygomatic arch

Insertion: Masseteric fossa and adjacent portions of the mandible

Action: Elevation of mandible

Temporalis

A massive muscle, the **temporalis** occupies the temporal fossa of the skull (Figure 35.14).

Origin: Most fibers originate from the temporal bone and a few from the zygomatic arch

Insertion: Coronoid process of the mandible

Action: Elevates mandible

Pterygoideus Externus

Pterygoideus externus, a muscle of mastication, is located ventral to the temporalis and is not illustrated.

Origin: From the external pterygoid fossa that extends from the sphenopalatine foramen of the palatine to the foramen rotundum of the basisphenoid

Insertion: Ventral border of the medial aspect of the mandible

Action: Elevates the mandible

Pterygoideus Internus

A second muscle of mastication, **pterygoideus internus**, is located posterior to the previous muscle and, likewise, is not illustrated.

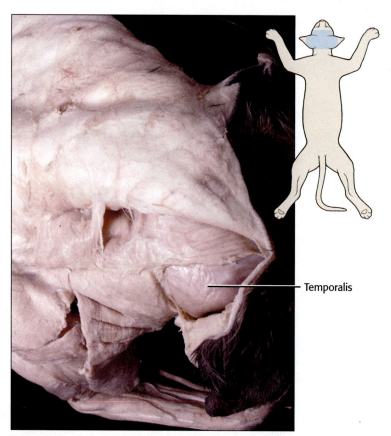

FIGURE 35.14 Temporalis: Dorsal view.

Origin: From the internal pterygoid fossa lying along the lateral surface of the pterygoid process and hamulus of the basisphenoid

Insertion: Angular process of the mandible and pterygoideus externus

Action: Synergistic with pterygoideus externus in elevating the mandible

Digastric

The **digastric** is a thick muscle lying along the medial ventral border of the mandible (Figure 35.8A). It has an unusual homology, in that the anterior portion evolved from intermandibularis and the posterior portion evolved from the interhyoideus/sphincter colli of *Necturus*.

Origin: From the mastoid and jugular processes

Insertion: Medial ventral border of the mandible

Action: Depresses the mandible

MUSCLES OF THE SHOULDER

Supraspinatus

Supraspinatus is a thick muscle, homologous with a portion of supracoracoideus of *Necturus*, that lies in the supraspinous fossa of the scapula (Figure 35.10).

Origin: From the entire surface of the supraspinous fossa

Insertion: Greater tuberosity of the humerus

Action: Protracts the humerus

Infraspinatus

Another thick, but somewhat smaller, muscle, also homologous with a portion of supracoracoideus of *Necturus*, **infraspinatus** fills the infraspinous fossa (Figure 35.10).

Origin: From the surface of the infraspinous fossa

Insertion: Lateral surface of the greater tuberosity of the humerus

Action: Rotates the humerus laterally

Teres Major

The **teres major** is a thick triangular muscle homologous with a portion of latissimus dorsi in *Necturus*. It occupies the caudal border of the scapula (Figure 35.10).

 Carefully separate teres major from infraspinatus.

Origin: Cranial border of the scapula

Insertion: Proximal end of the humerus

Action: Retracts humerus and rotates it medially

Levator Scapulae Ventralis

Levator scapulae ventralis, a bandlike muscle homologous with levator scapulae of *Necturus*, can be seen lying between clavotrapezius and acromiotrapezius muscles (Figure 35.4).

Origin: By two heads from the ventral surface of the transverse process of the atlas and from the basioccipital near the tympanic bulla

Insertion: The two heads unite, forming a flat band that inserts onto the ventral border of the metacromion of the scapula and into the infraspinous fossa

Action: Pulls scapula cranially

Acromiodeltoid

A flat muscle homologous with a portion of dorsalis scapulae of *Necturus*, **acromiodeltoid** is positioned ventral to levator scapulae ventralis and caudal to clavobrachialis (Figure 35.4).

Origin: Acromion of the scapula

Insertion: Surface of the spinodeltoid muscle

Action: Retracts the humerus and rotates it laterally

Spinodeltoid

Spinodeltoid is yet another muscle that is homologous with a portion of dorsalis scapulae of *Necturus*. It lies ventral to acromiotrapezius and levator scapulae ventralis and caudal to acromiodeltoid (Figure 35.4).

Origin: Spine of the scapula

Insertion: Deltoid ridge of the humerus

Action: Synergistic action with the acromiodeltoid in retracting the humerus and rotating it outward

Teres Minor

Teres minor is a small, somewhat triangular muscle, homologous with a portion of procoracohumeralis of *Necturus*. It is located between infraspinatus and the long head of triceps brachii and beneath the spinodeltoid (Figure 35.15).

Origin: Cranial border of the scapula near the glenoid fossa

Insertion: Greater tuberosity of the humerus

Action: Synergistic action with the infraspinatus in rotating the humerus laterally

Subscapularis

The **subscapularis** is a large, medial, triangular muscle located in the subscapular fossa. It is homologous with subcoracoscapularis of *Necturus* (Figure 35.11 and Figure 35.16).

Origin: Subscapular fossa

Insertion: Lesser tuberosity of the humerus

Action: Adducts the humerus

MUSCLES OF THE UPPER FORELIMB, OR BRACHIUM

Coracobrachialis

A very short bandlike muscle, homologous with coracobrachialis of *Necturus*, **coracobrachialis** lies on the medial aspect of the shoulder joint in close proximity to the insertion of subscapularis and the origin of biceps brachii (Figure 35.16).

Origin: Coracoid process of the scapula

Insertion: Proximal end of humerus

Action: Adducts the humerus

Epitrochlearis

Epitrochlearis is a delicate, thin, flat muscle, homologous with a portion of triceps brachii of *Necturus*. It appears on the medial surface of the brachium, where it partially overlies triceps brachii (see Figure 35.20A).

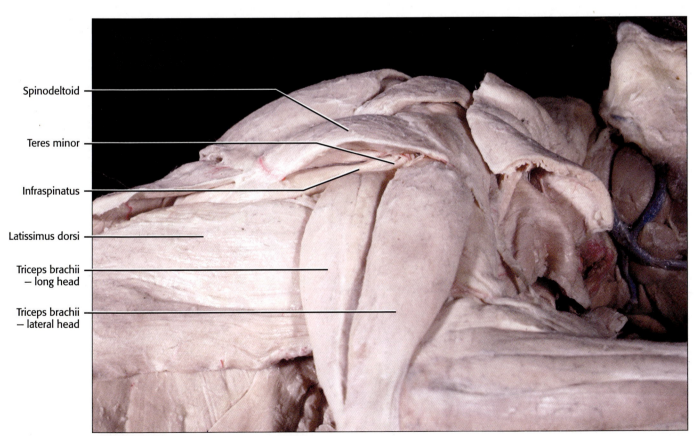

Spinodeltoid

Teres minor

Infraspinatus

Latissimus dorsi

Triceps brachii — long head

Triceps brachii — lateral head

FIGURE 35.15 Teres minor (anterior right).

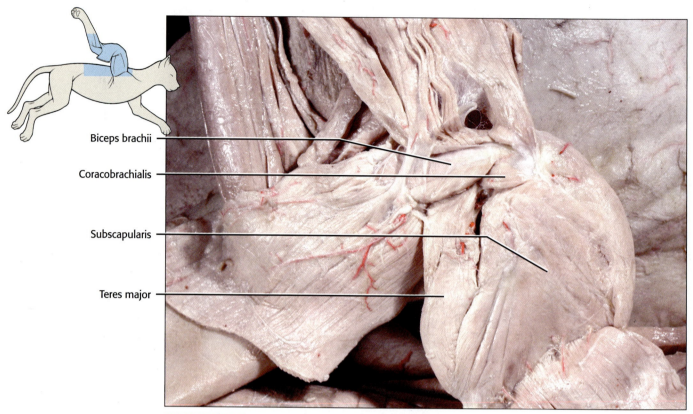

FIGURE 35.16 Muscles of the scapular region: Medial view.

Origin: From the lateral border of latissimus dorsi

Insertion: By a thin aponeurosis continuous with the antebrachial fascia onto the olecranon of the ulna

Action: Acts synergistically with the triceps brachii in extending the antebrachium

Biceps Brachii

An important arm flexor lying on the cranial surface of the humerus, **biceps brachii** is a thick muscle that is homologous with a portion of the humeroantebrachialis/coracoradialis muscles of *Necturus* (Figure 35.16). In the human, this muscle has two heads—a long and a short head—whereas the cat has only the homolog of the long head.

Origin: By a tendon above the glenoid fossa of the scapula

Insertion: By a tendon on the radial tuberosity

Action: Flexes the forearm synergistically with the brachialis; tends to supinate the manus; and stabilizes the shoulder joint

Triceps Brachii

A very large, lateral muscle, homologous with triceps of *Necturus*, **triceps brachii** consists of three heads that originate from separate sites, but all of which have a common insertion (Figure 35.17).

Origin: (1) Lateral head—deltoid ridge of proximal end of humerus; (2) Long head—near glenoid fossa of axillary

border of scapula; (3) medial head— consists of three parts, all of which originate from the humerus

Insertion: By a common strong tendon onto the surface of the olecranon of the ulna

Action: Extends the forearm

Anconeus

A small triangular muscle, homologous with a portion of triceps in *Necturus*, **anconeus** covers the lateral surface of the elbow (Figure 35.18).

Origin: Lateral epicondyle

Insertion: Ulna

Action: Acts synergistically with the triceps brachii in extending the forearm

Brachialis

Brachialis is an important lateral flexor located along the cranial surface of the humerus and lying partially obscured by the long head of the triceps brachii. It is homologous with a portion of humeroantebrachialis of *Necturus* (Figure 35.18).

Origin: Lateral surface of humerus

Insertion: Lateral surface of ulna near semilunar notch

Action: Flexes the forearm or antebrachium and is synergistic with the biceps brachii

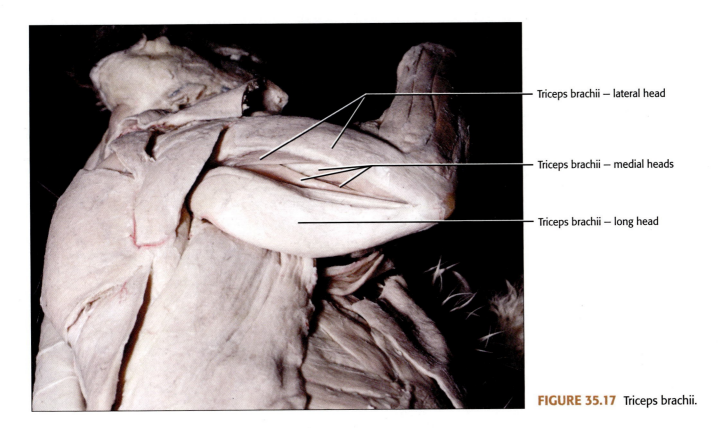

FIGURE 35.17 Triceps brachii.

MUSCLES OF THE LOWER FORELIMB, OR ANTEBRACHIUM

Notice that the antebrachium is covered with a two-layered connective tissue. This is called the **antebrachial fascia**. The outer layer is loose and is a continuation of the subcutaneous fascia. The inner layer is in close contact with the underlying muscles and extends between the dorsal or extensor muscles and adheres closely to their tendons. This sheet is continuous on the ventral or flexor surface and is closely attached to the pronator teres and radius. In the carpal area, the fascia thickens to form a dorsal transverse ligament, the **extensor retinaculum** (Figure 35.18 and Figure 35.19), and a ventral transverse ligament, the **flexor retinaculum**, to hold the tendons of these muscles in place. It further continues as the dorsal fascia of the manus; however, on the palmar surface, it unites with the pad and is continuous with the tendon sheaths of the flexors.

 Carefully remove the antebrachial fascia, being especially cautious in the region of the tendons, retinacula, and the brachioradialis muscle. This is an area in which one can be destructively creative. Take great care in separating the muscles while maintaining their integrity. Do not split muscles.

The following mammalian forelimb extensors and flexors are homologous with the forelimb extensors and flexors of *Necturus*, respectively.

Brachioradialis

The **brachioradialis** biceps brachii, a narrow, bandlike muscle, extends along the radial border of the antebrachium in company with blood vessels and a nerve (Figure 35.18).

Origin: Humerus

Insertion: Styloid process of the radius

Action: Supinates the manus

Extensor Carpi Radialis Longus

A slender muscle whose main mass lies on the radial side of the antebrachium, **extensor carpi radialis longus** lies deep to the brachioradialis (Figure 35.18).

Origin: Humerus

Insertion: Base of the second metacarpal

Action: Extends the manus

Extensor Carpi Radialis Brevis

Extensor carpi radialis brevis, a somewhat shorter muscle, lies just medial to extensor carpi radialis longus and adheres closely to it (Figure 35.18). *Carefully separate the two extensor muscles with a steel probe.*

Origin: Humerus

Insertion: Base of the third metacarpal

Action: Extends the manus

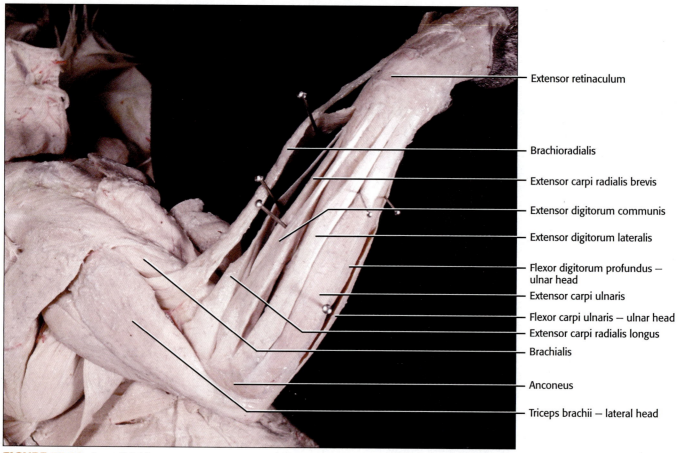

- Extensor retinaculum
- Brachioradialis
- Extensor carpi radialis brevis
- Extensor digitorum communis
- Extensor digitorum lateralis
- Flexor digitorum profundus — ulnar head
- Extensor carpi ulnaris
- Flexor carpi ulnaris — ulnar head
- Extensor carpi radialis longus
- Brachialis
- Anconeus
- Triceps brachii — lateral head

FIGURE 35.18 Superficial forearm extensors: Lateral view.

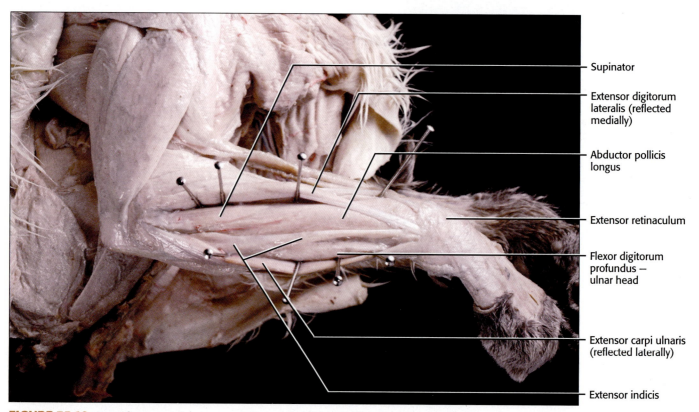

- Supinator
- Extensor digitorum lateralis (reflected medially)
- Abductor pollicis longus
- Extensor retinaculum
- Flexor digitorum profundus — ulnar head
- Extensor carpi ulnaris (reflected laterally)
- Extensor indicis

FIGURE 35.19 Deep forearm extensors, supinator, and abductor policis longus: Lateral view.

Extensor Digitorum Communis

Extensor digitorum communis, a long, slender dorsal muscle, partially overlies extensor carpi longus and brevis (Figure 35.18).

Origin: Humerus

Insertion: Tendon divides into four slips that insert on the dorsal surface along the medial aspect of the three phalanges of the second, third, fourth, and fifth digits

Action: Extension of second, third, fourth, and fifth digits

Extensor Digitorum Lateralis

Extensor digitorum lateralis is a long, slender, dorsal muscle that lies just lateral to extensor digitorum communis (Figure 35.18).

Origin: Humerus

Insertion: Division of the tendon similar to that of the extensor digitorum communis that inserts along the dorsolateral surface of the phalanges of digits 3, 4, and 5, or 2, 3, 4, and 5 respectively

Action: Extends the digits with the extensor digitorum communis, synergistically

Extensor Carpi Ulnaris

The last of the superficial extensors, the extensor carpi ulnaris is a long, slender muscle lying along the ulnar side of the antebrachium (Figure 35.18).

Origin: Lateral epicondyle of humerus

Insertion: Base of the fifth metacarpal

Action: Extension of carpals of ulnar side

Supinator

The **supinator** is a flat muscle that surrounds the proximal end of the radius and lies deep to extensor digitorum communis and lateralis (Figure 35.19).

Origin: From stabilizing elbow ligaments and the lateral epicondyle of the humerus

Insertion: Fibers pass obliquely to insert on the proximal third of the radius

Action: Supinates the forearm synergistically with the brachioradialis

Abductor Pollicis Longus

Abductor pollicis longus is a deep, flat muscle located distal to the supinator. Its oblique fibers extend between the radius and the ulna (Figure 35.19).

Origin: Ventrolateral surface of the ulna and dorsal surface of the radius

Insertion: On radial side of first metacarpal

Action: Extends and abducts the pollex (thumb, digit I)

Extensor Indicis

This slender muscle, the **extensor indicis**, lies deep to extensor carpi ulnaris. It may consist of a single muscle, or two separate muscles with a common origin (Figure 35.19).

Origin: Ulna

Insertion: If the muscle is single with a single tendon, it inserts on the second phalanx of the second digit. If the muscle is single with a divided tendon, both may insert on the second phalanx of the second digit or one may insert on the second digit and one on the pollex (digit I). If there are two muscles, an extensor digiti I inserts by means of a tendon on the pollex and an extensor digiti II inserts on digit II.

Action: Extends the first and second digits

Flexor Carpi Ulnaris

Flexor carpi ulnaris is a muscle having a humeral and an ulnar head. It lies along the ulnar side of the ventral aspect of the lower forelimb (Figure 35.20A and Figure 35.20B).

Origin: Humerus near the medial epicondyle (humeral head), lateral surface of the olecranan of the ulna (ulnar head)

Insertion: Pisiform bone of the carpals

Action: Wrist flexor

Flexor Digitorum Superficialis

Flexor digitorum superficialis is a two-part muscle, with (1) a superficial head that is the widest of the flat, bandlike muscles of the ventral surface of the lower forelimb and (2) a deep head that lies on the surface of the tendon of flexor digitorum profundus (Figure 35.20A and Figure 35.20B).

Origin: Superficial head—medial epicondyle of the humerus; deep head—tendon of two humeral heads of the flexor digitorum profundus

Insertion: The tendons of the two heads pass under the flexor retinaculum and insert on the middle phalanx of digits 2–5

Action: Flexes the digits

Flexor Carpi Radialis

Flexor carpi radialis is a thin muscle that extends from the humerus to the manus (Figure 35.20A).

Origin: Medial epicondyle of the humerus

Insertion: Bases of the second and third metacarpals

Action: Flexes the wrist

Pronator Teres

Pronator teres is a ventral muscle oriented obliquely over the upper surface of the forearm (Figure 35.20A and Figure 35.20B).

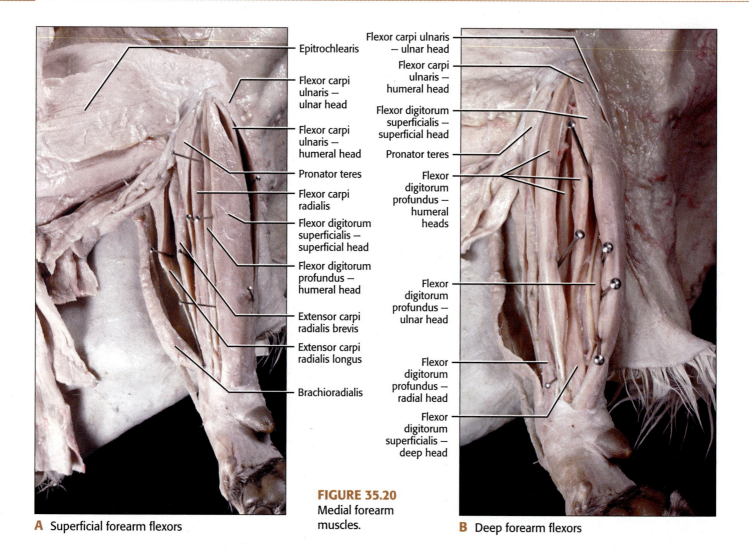

A Superficial forearm flexors

FIGURE 35.20
Medial forearm muscles.

B Deep forearm flexors

Origin: Medial epicondyle of the humerus

Insertion: Medial border of the radius

Action: Pronation of the manus by rotating the radius

Flexor Digitorum Profundus

Flexor digitorum profundus is a deep, five-headed muscle whose tendons are united at the wrist (Figure 35.20B).

Origin: Radial border of the ulna (ulnar head), radius, interosseous ligament between the radius and ulna, and ulna (radial head), medial epicondyle of the humerus (three humeral heads)

Insertion: Five tendons join at the wrist to form a strong, wide, white, glistening band that subdivides into five tendons to insert on the bases of the distal phalanx of digits 1–5

Action: Flexes all digits

Pronator Quadratus

A deep muscle, **pronator quadratus** has fibers that extend obliquely between the distal ends of the ulna and the radius (Figure 35.20C).

 To find this muscle, separate the tendon and radial head of flexor digitorum profundus and the tendon of flexor carpi radialis. Look for a flat, bluish-purple muscle covered by shiny fascia. Carefully slit the fascia to reveal the oblique fibers of this muscle.

Origin: Ulna

Insertion: Radius

Action: Rotates the radius synergistically with the pronator teres

MUSCLES OF THE MANUS

Lumbricales

These **lumbricales** are small, intrinsic muscles of the manus that will not be dissected. In addition, a number of small intrinsic muscles associated with the metacarpals and digits will not be treated here.

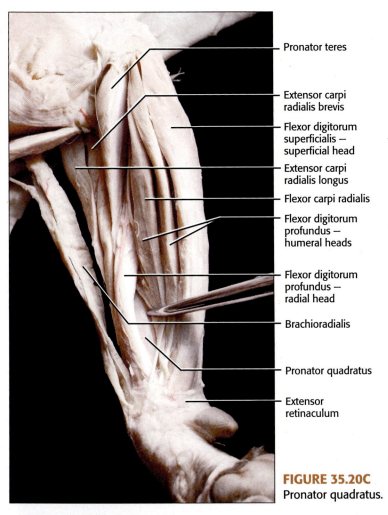

- Pronator teres
- Extensor carpi radialis brevis
- Flexor digitorum superficialis — superficial head
- Extensor carpi radialis longus
- Flexor carpi radialis
- Flexor digitorum profundus — humeral heads
- Flexor digitorum profundus — radial head
- Brachioradialis
- Pronator quadratus
- Extensor retinaculum

FIGURE 35.20C
Pronator quadratus.

MUSCLES OF THE THIGH

Sartorius

Sartorius is a muscle that appears as a narrow, thick band laterally and continues as a thin band extending almost halfway across the medial surface of the cranial aspect of the thigh (Figure 35.21). It is homologous with iliotibialis of *Necturus*.

 Use caution in dissecting this muscle, as it often has fat and extensive connective tissue associated with it.

Origin: Crest and ventral border of the ilium

Insertion: Patella, tibia, and fascia of the knee

Action: Adducts and rotates the femur; extends the shank

 Cut this muscle.

Gracilis

A thin muscle, **gracilis** occupies the caudal half of the medial surface of the thigh (Figure 35.21). It is homologous with puboischiotibialis of *Necturus*. Because the insertion is an aponeurosis, do not be overzealous in cleaning the surface of this muscle and thereby destroy it.

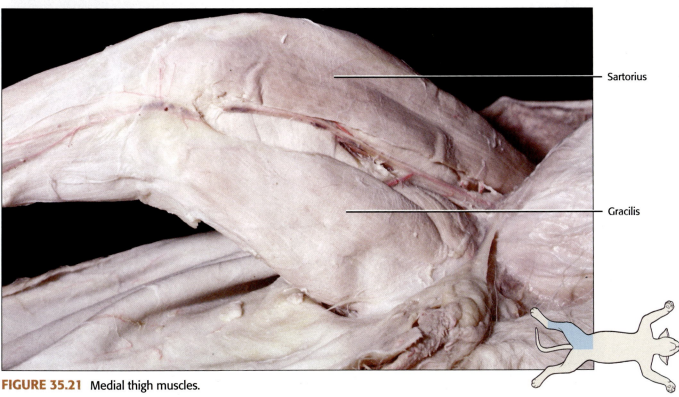

- Sartorius
- Gracilis

FIGURE 35.21 Medial thigh muscles.

Origin: Symphysis of the ischium and the pubis

Insertion: A thin aponeurosis on the medial surface of the tibia and continuous with the fascia of the shank

Action: Adducts and retracts the leg

 Cut this muscle.

Biceps Femoris

Biceps femoris is a large, thick muscle covering almost three-fourths of the lateral surface of the thigh (Figure 35.22 and Figure 35.23). Some anatomists suggest that this muscle may be homologous with ischioflexorius of *Necturus*, but others disagree.

Origin: Ischial tuberosity

Insertion: Tibia and patella

Action: Abducts thigh and flexes the shank

 Cut this muscle after reading the following.

With reference to Figure 35.22 and Figure 35.23, when bisecting the biceps, the three major concerns are:

1. The delicate tendon of the **caudofemoralis** is closely applied to the undersurface of the cranial aspect of the biceps.

2. The **tenuissimus** lies just beneath the caudal edge of the biceps.

3. The **ischiadic nerve** is positioned under and approximately in the middle of the biceps.

 Separate each from biceps femoris to avoid cutting them along with it.

Tenuissimus

Tenuissimus is an extremely slender muscle that strongly adheres to biceps femoris (Figure 35.23). This muscle may have the same debatable homology as biceps femoris.

Origin: Transverse process of second caudal vertebra

Insertion: In common with the biceps femoris

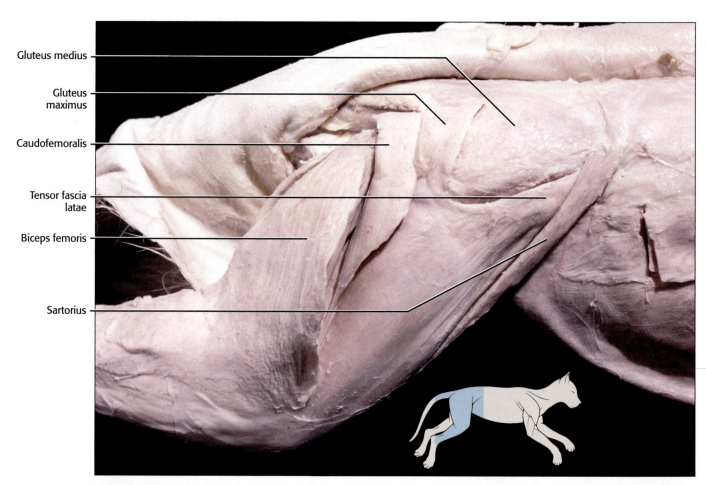

FIGURE 35.22 Lateral thigh muscles.

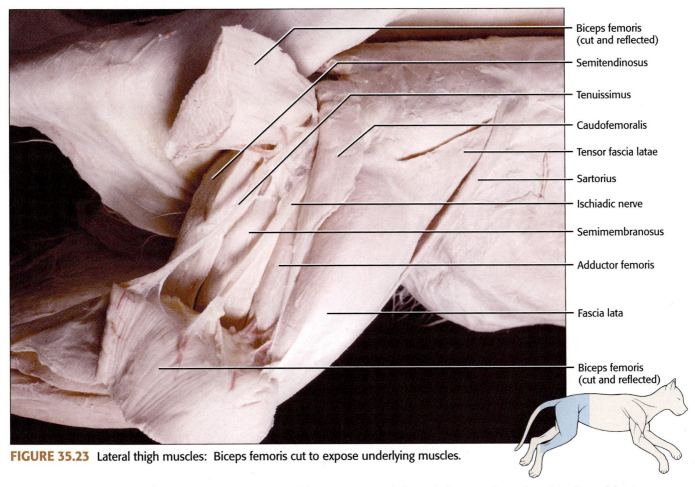

- Biceps femoris (cut and reflected)
- Semitendinosus
- Tenuissimus
- Caudofemoralis
- Tensor fascia latae
- Sartorius
- Ischiadic nerve
- Semimembranosus
- Adductor femoris
- Fascia lata
- Biceps femoris (cut and reflected)

FIGURE 35.23 Lateral thigh muscles: Biceps femoris cut to expose underlying muscles.

Action: Synergistically assists the biceps femoris in abducting the thigh and flexing the shank

Caudofemoralis

The major mass of **caudofemoralis** lies beneath and just cranial to biceps femoris (Figure 35.22 and Figure 35.23). This muscle is homologous with caudofemoralis of *Necturus*. Again, exercise care when dissecting this muscle, as it often is associated with fat and connective tissue and is inserted by way of a thin, narrow tendon that adheres closely to the medial surface of biceps femoris.

Origin: Transverse processes of second and third caudal vertebrae

Insertion: Thin tendon along the lateral border of the patella

Action: Abducts the thigh and extends the shank

Semitendinosus

Semitendinosus is a long muscle that forms the caudal border of the thigh (Figure 35.23 and Figure 35.24). This muscle is homologous with a portion of ischioflexorius in *Necturus*. It adheres strongly to the caudal edge of biceps femoris and should be separated from it by picking away

carefully with forceps along the white line of fascia separating the two.

Origin: Ischial tuberosity

Insertion: Tibia

Action: Flexes the shank

Semimembranosus

A thick muscle, homologous with a portion of ischioflexorius of *Necturus*, **semimembranosus** lies on the medial aspect of the thigh (Figure 35.23 and Figure 35.24).

Origin: Ischial tuberosity and adjacent area of the ischium

Insertion: Medial epicondyle of the femur and adjacent medial surface of the tibia

Action: Extends the thigh

Adductor Femoris

Adductor femoris is a broad muscle lying cranial to and partially covering semimembranosus (Figure 35.23 and Figure 35.24). This muscle is homologous with pubotibialis of *Necturus*.

Origin: Pubis and ischium

Insertion: Femur

Action: Adducts thigh

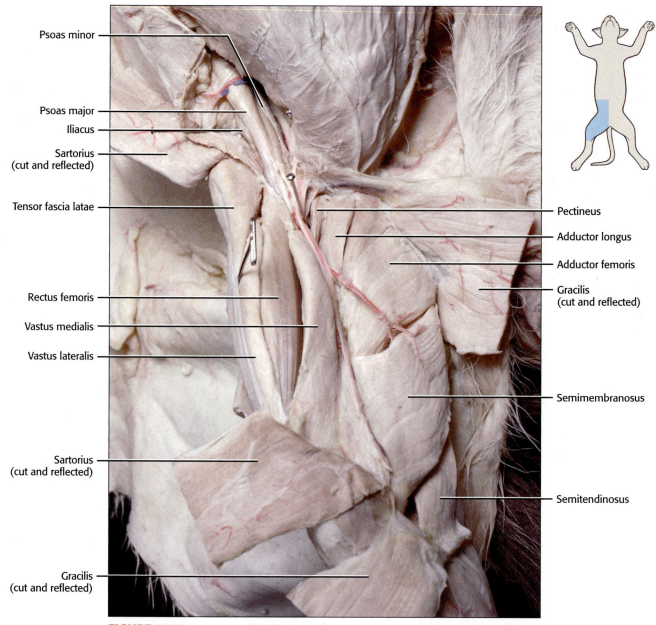

Psoas minor

Psoas major

Iliacus

Sartorius
(cut and reflected)

Tensor fascia latae

Rectus femoris

Vastus medialis

Vastus lateralis

Sartorius
(cut and reflected)

Gracilis
(cut and reflected)

Pectineus

Adductor longus

Adductor femoris

Gracilis
(cut and reflected)

Semimembranosus

Semitendinosus

FIGURE 35.24 Anteriomedial thigh muscles.

Adductor Longus

Cranial to the **adductor femoris** is the thin triangular adductor longus (Figure 35.24). It is homologous with adductor femoris of *Necturus*.

Origin: Pubis

Insertion: Linea aspera of the femur

Action: Adducts thigh

Pectineus

A small, triangular muscle, **pectineus** lies just beneath the femoral vessels and saphenous nerve (Figure 35.24). This muscle is homologous with a portion of puboischio-femoralis internus of *Necturus*. Often, fat obscures the pectineus and will have to be removed carefully.

Origin: Pubis

Insertion: Proximal end of femur

Action: Adducts the thigh

To observe the following three muscles, cut anteriorly through the abdominal muscles by making a 5 cm incision beginning at the posterior edge of the lateral body wall just anterior to the thigh.

Iliacus

Iliacus is the most lateral and slightly dorsal of these muscles (Figure 35.24). It is homologous with a portion of puboischiofemoralis internus of *Necturus*.

Origin: Ilium

Insertion: Lesser trochanter of the femur

Action: Flexes and rotates the thigh

Psoas Major

Psoas major is the largest of the three muscles and occurs in the middle (Figure 35.24). It is also homologous with a portion of puboischiofemoralis internus of *Necturus*.

Origin: Last thoracic and all of the lumbar vertebrae

Insertion: Lesser trochanter of the femur

Action: Flexes and rotates the thigh

Psoas Minor

Psoas minor is a thin muscle, homologous with a portion of the subvertebral muscles of *Necturus*, occurring medial to psoas major and distinguished by its long, narrow, glistening tendon (Figure 35.24).

Origin: Last thoracic and first few lumbar vertebrae

Insertion: Pubis, by a long, narrow, conspicuous glistening tendon

Action: Flexes vertebral column

Tensor Fascia Latae

Tensor fascia latae is a rather thick, triangular muscle that largely covers the cranial portion of vastus lateralis and abuts the caudal border of sartorius and the cranioventral border of gluteus medius (Figure 35.22). It is homologous with a portion of iliofemoralis of *Necturus*.

> Take extreme care in loosening the extensive fascia lata from the caudal edge of vastus lateralis. Do not shred or destroy the fascia lata.

Origin: Illium, fascia of surrounding hip muscles

Insertion: By fascia lata. The fascia lata on the surface of the patella.

Action: Helps to extend the shank

Quadriceps Complex

The following four muscles make up the quadriceps complex, which lies on the craniolateral and craniomedial aspects of the thigh and is a powerful extensor of the shank (Figure 35.24,

Figure 35.25, and Figure 35.26). The vasti muscles are homologous with puboischiofemoralis externus of *Necturus*, and rectus femoris is homologous with ilioextensorius of *Necturus*.

Vastus Medialis

The **vastus medialis** is the most medial of the four muscles.

Origin: Femur

Insertion: Crosses the patella and inserts by means of the patellar ligament on the tibial tuberosity

Action: Extends the shank

Rectus Femoris

Rectus femoris is a spindle-shaped muscle that rests between vastus medialis and vastus lateralis and is distinguished by a shiny covering of fascia.

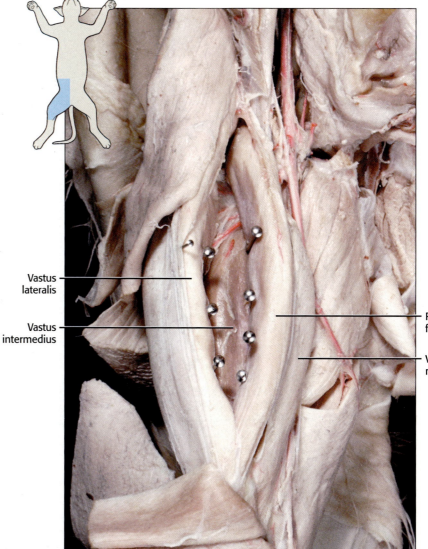

FIGURE 35.25 Quadriceps femoris complex.

Origin: Ilium near the acetabulum

Insertion: In common with the vastus medialis and lateralis

Action: Extends the shank

Vastus Lateralis

A large, flat muscle, **vastus lateralis** covers the cranial and lateral surface of the thigh.

Origin: Greater trochanter and adjacent area of the femur

Insertion: In common with the vastus medialis and rectus femoris

Action: Extends the shank

Vastus Intermedius

The deepest of the quadriceps muscles, **vastus intermedius** can be seen by carefully separating rectus femoris and vastus lateralis.

Origin: Femur

Insertion: In common with the other three members of this complex

Action: Extends the shank

MUSCLES OF THE SHANK

Beware of the tough fascia surrounding the shank muscles. Carefully remove this tissue and, while doing so, avoid destruction of the tendons of insertion of the thigh muscles.

The shank muscles are roughly divided into extensors and flexors. A distinct band of connective tissue, the **extensor retinaculum,** holds the cranial extensor muscle tendons in place (Figure 35.26).

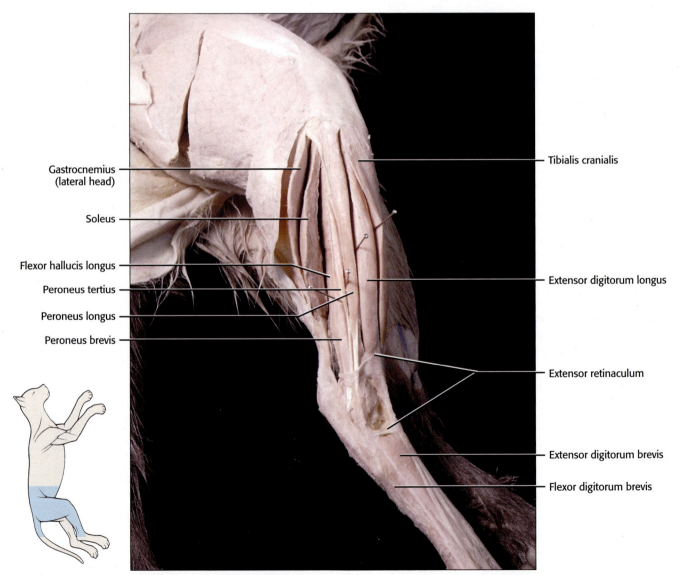

FIGURE 35.26 Lateral shank muscles.

Labels: Gastrocnemius (lateral head); Soleus; Flexor hallucis longus; Peroneus tertius; Peroneus longus; Peroneus brevis; Tibialis cranialis; Extensor digitorum longus; Extensor retinaculum; Extensor digitorum brevis; Flexor digitorum brevis

The following mammalian hindlimb extensors and flexors are homologous with the hindlimb extensors and flexors of *Necturus*, respectively.

With regard to the actions of the shank muscles, extension of the pes is synonymous with plantar flexion, and flexion of the pes is synonymous with dorsiflexion. Flexion of a shank muscle decreases the angle of the joint at the ankle by pulling the pes closer to the shank, and extension of a shank muscle increases the angle of the joint at the ankle by pulling the pes away from the shank.

Tibialis Cranialis

Tibialis cranialis is a large, flat, meaty muscle that lies on the craniolateral aspect of the tibia (Figure 35.26).

Origin: Proximal end of the tibia and the fibula

Insertion: Medial surface of the first metatarsal after passing beneath the extensor retinaculum

Action: Flexes the pes

Extensor Digitorum Longus

A large muscle lying beneath tibialis cranialis, **extensor digitorum longus** appears as a narrow strip caudal to it along the lateral surface of the shank (Figure 35.26).

 Carefully loosen this muscle from tibialis cranialis.

Origin: Lateral epicondyle of the femur

Insertion: After passing under the extensor retinaculum, the tendon, inserts on the dorsal surface of the second and third phalanges of digits 2–5

Action: Extends the digits and flexes the pes

Peroneus Longus

Peroneus longus is the most superficial of a group of three peroneus muscles on the lateral surface of the shank (Figure 35.26).

Origin: Head and lateral surface of the fibula

Insertion: Proximal ends of all five metatarsals

Action: Extends the pes

Peroneus Tertius

A slender muscle, the **peroneus tertius** lies beneath peroneus longus and must be separated from it (Figure 35.26).

 Trace its long tendon of insertion, separating it from underlying muscle with forceps.

Once the tendon has been isolated, the muscle usually become evident.

Origin: Lateral surface of the fibula

Insertion: Its tendon lies in the groove of the lateral malleolus and passes along the lateral margin of the foot and inserts on the first phalanx of the fifth digit

Action: Flexes pes and abducts and extends the fifth digit

Peroneus Brevis

The shortest of the three peroneus muscles, **peroneus brevis** lies posterior to the other two (Figure 35.26).

Origin: Fibula

Insertion: Note that the tendon of peroneus brevis passes in common with the tendon of peroneus tertius within the groove of the lateral malleolus and inserts on the base of the fifth metatarsal

Action: Extends the pes

Popliteus

Popliteus is a triangular muscle that wraps obliquely around the posterior aspect of the knee from the femur to the tibia (Figure 35.27). A clue to its position is that the insertion fascia of semitendinosus lies over popliteus and must be trimmed carefully to reveal it.

Origin: Lateral epicondyle of the femur

Insertion: Medial aspect of the proximal end of the tibia

Action: Flexes and medially rotates the leg

Flexor Digitorum Longus

Flexor digitorum longus is a long, slender muscle on the medial aspect of the shank just posterior to the tibia (Figure 35.27).

 Separate its conspicuous tendon by working forceps beneath the tendon and gently sliding them toward the muscle until it is free from tibialis caudalis.

Origin: Fibula and tibia

Insertion: A common, broad tendon dividing into four discrete tendons that insert on the base of the terminal phalanx of each toe

Action: Flexes toes and pes

Flexor Hallucis Longus

Flexor hallucis longus is a somewhat larger muscle than flexor digitorum longus, lying lateral to flexor digitorum longus on the posterior aspect of the shank (Figure 35.26 and Figure 35.27).

Origin: Fibula and tibia

Insertion: In common with the flexor digitorum longus

Action: Flexes toes and pes

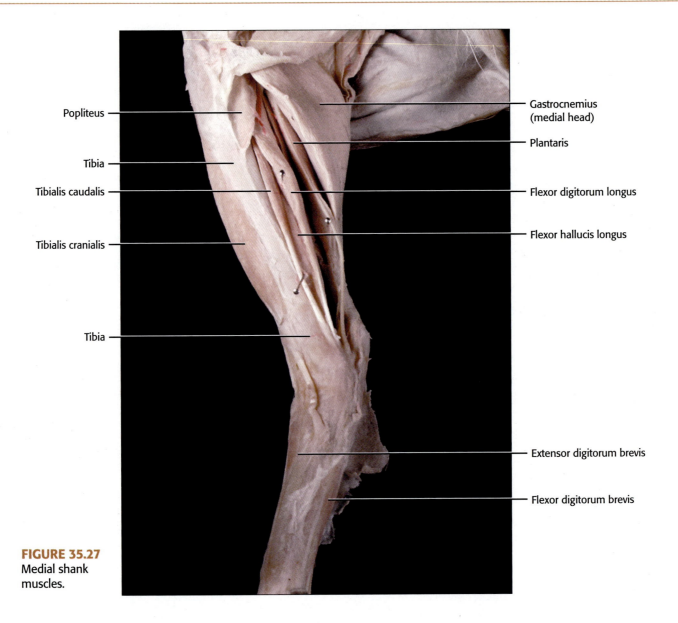

Popliteus

Tibia

Tibialis caudalis

Tibialis cranialis

Tibia

Gastrocnemius
(medial head)

Plantaris

Flexor digitorum longus

Flexor hallucis longus

Extensor digitorum brevis

Flexor digitorum brevis

FIGURE 35.27
Medial shank
muscles.

Tibialis Caudalis

A flat, slender muscle, **tibialis caudalis** lies beneath flexor digitorum longus and between flexor digitorum longus and flexor hallucis longus (Figure 35.27). Notice the prominent tendon of this muscle.

 To isolate tibialis caudalis, use the same technique as described for flexor digitorum longus.

Origin: Fibula, tibia and adjacent aponeurosis

Insertion: Plantar surface of the navicular and medial cuneiform

Action: Extends the pes

Gastrocnemius

Gastrocnemius comprises most of the bulky posterior muscle mass of the shank known as the calf (Figure 35.26

and Figure 35.27). It has two heads, a lateral and a medial head.

Origin: (1) Patella, the superficial fascia of the shank, the sesamoid bone located above the lateral epicondyle of the femur, and an aponeurosis from the plantaris and adjacent tibia. (2) The medial head originates from the sesamoid bone above the medial epicondyle and adjacent area of the femur

Insertion: By means of a common powerful tendon, the Achilles tendon, formed by the individual tendons of the gastrocnemius, the soleus, and the plantaris muscles, that inserts on the proximal end of the calcaneus

Action: Extends the pes

Plantaris

Plantaris is medial, lies beneath gastrocnemius, and can be seen protruding between the proximal ends of the heads of this muscle (Figure 35.27). Take care in dissecting this muscle away from the two heads of the gastrocnemius.

Origin: From the sesamoid above the lateral epicondyle of the femur and the patella

Insertion: By way of the Achilles tendon in common with the gastrocnemius

Action: Acts synergistically with the gastrocnemius and the soleus to extend the pes

Soleus

Soleus is a lateral flat muscle located beneath plantaris (Figure 35.26).

Origin: Fibula

Insertion: In common with the tendon of the gastrocnemius and contributes to the formation of the Achilles tendon

Action: Synergistic extension of the pes with the gastrocnemius and plantaris

Triceps Surae

Gastrocnemius and soleus sometimes have been considered a single calf muscle with three heads, the triceps surae.

MUSCLES OF THE PES

Extensor Digitorum Brevis

A thin muscle, the **extensor digitorum brevis** covers the dorso-lateral surface of the tarsus and metatarsus (Figure 35.26 and Figure 35.27).

Origin: From the proximal ends of metatarsals 3–5

Insertion: By three tendons that terminate on the dorsal and lateral surface of the first phalanx

Action: Extends the toes

Flexor Digitorum Brevis

Flexor digitorum brevis is a muscle lying on the plantar surface of the foot (Figure 35.26 and Figure 35.27).

Origin: Plantaris tendon

Insertion: By four tendons to the second phalanx of digits 2–5

Action: Flexes the toes

In addition, a number of small muscles associated with the tarsus, metatarsus, and phalanges will not be described individually. They produce various intricate movements.

MUSCLES OF THE HIP

 Carefully remove the tough fascia covering the hip region.

Re-identify tensor fascia **latae**, **caudofemoralis**, and **biceps femoris** (Figure 35.22).

Gluteus Maximus

Gluteus maximus is a thin trapezoidal muscle lying just anterior to caudofemoralis (Figure 35.22). This muscle is homologous with iliofibularis of *Necturus*.

Origin: Last sacral and first caudal vertebrae, as well as adjacent fascia

Insertion: Greater trochanter of the femur

Action: Abducts the thigh

After gluteus maximus has been identified and isolated, slide a probe under the muscle perpendicular to the fibers in the region of the belly, and with a sharp scalpel cut along the top of the probe, using the probe as a hard surface to prevent cutting into important muscles beneath it.

Gluteus Medius

In the cat, **gluteus medius** is a very thick muscle sandwiched between the cranial tensor fascia latae and the caudal gluteus maximus (Figure 35.22). This muscle is homologous with a portion of iliofemoralis of *Necturus*.

Origin: Crest and lateral surface of the ilium, last sacral and first caudal vertebrae and adjacent fascia

Insertion: Greater trochanter of the femur

Action: Abducts the thigh

 Extreme care must be taken in isolating gluteus medius from important underlying musculature with which you will be dealing shortly. After cutting and reflecting the gluteus maximus muscle, you should be able to identify the posterior portion of pyriformis extending from beneath the caudal edge of the gluteus medius muscle. Note the ischiadic nerve passing obliquely under the pyriformis muscle. The cranial edge of gluteus medius is very thick and must be loosened from the muscular part of tensor fascia latae.

Lift the cranial edge of gluteus medius and observe the shiny, spindle-shaped gluteus minimus muscle. Insert a probe beneath the caudal edge of gluteus medius and perpendicular to the direction of the muscle fibers, continuing until it emerges beneath the cranial edge. Be sure that the probe remains dorsal to pyriformis and gluteus minimus. With a sharp scalpel, cut through the belly of gluteus medius along the dorsal edge of the probe.

Pyriformis

A fan-shaped muscle, the **pyriformis** lies under both of the superficial gluteus muscles (Figure 35.28A and Figure 35.28B). It is homologous with a portion of iliofemoralis of *Necturus*.

Origin: Last two sacral and first caudal vertebrae

Insertion: Greater trochanter of the femur

Action: Abducts the thigh

 Insert a probe perpendicular to the fibers of pyriformis approximately 3/4 of the distance from the point of origin, beneath the caudal edge, until it emerges cranially. Take care not to pick up the sciatic nerve with the muscle. With a sharp scalpel, cut through pyriformis, following the top edge of the probe.

Gluteus Minimus

Gluteus minimus consists of an anterior shiny, spindle-shaped portion and a small, flat, fan-shaped posterior portion (Figure 35.28A, Figure 35.28B, and Figure 35.28C). It is homologous with a portion of iliofemoralis of *Necturus*.

Origin: Lateral surface of the ilium

Insertion: Greater trochanter of the femur

Action: Abducts and rotates the thigh outward

Articularis Coxae

Articularis coxae is ia small, flat muscle lying beneath gluteus minimus and between the heads of vastus lateralis and rectus femoris (Figure 35.28C).

Origin: Ilium cranial to the acetabulum

Insertion: Dorsal surface at the proximal end of the femur

Action: Assists in flexion and rotation of the thigh

Gemellus Cranialis

Closely abutting the caudal border of gluteus minimus is **gemellus cranialis**, a fan-shaped muscle lying below the pyriformis (Figure 35.28.B). Gemellus cranialis is homologous with a portion of ischiofemoralis of *Necturus*.

Origin: Dorsal ilium and ischium

Insertion: Greater trochanter of the femur

Action: Abducts and rotates the femur

Coccygeus

Coccygeus is a deep hip muscle lying dorsal to gemellus cranialis and often mistaken as part of it (Figure 35.28B).

Origin: In common with gemellus cranialis, the dorsal edge of the ilium and ischium

Insertion: Proximal caudal vertebrae

Action: Contributes to the pelvic wall and bends the tail laterally

Obturator Internus

Obturator internus is located directly caudal to gemellus cranialis (Figure 35.28B). It is homologous with a portion of ischiofemoralis of *Necturus*.

Origin: Medial surface of schium from the symphysis to the tuberosity

Insertion: Trochanteric fossa of the femur

Action: Abducts the thigh

Gemellus Caudalis

A flat muscle lying caudal to gemellus cranialis, **gemellus caudalis** is overlapped by obturator internus (Figure 35-28B). Gemellus caudalis is homologous with a portion of ischio-femoralis of *Necturus*.

Origin: Dorsolateral surface of the ischium

Insertion: In common with the tendon of the obturator internus into the trochanteric fossa

Action: Abducts the thigh

Quadratus Femoris

The last of the visible deep hip muscles is the short, thick **quadratus femoris** (Figure 35.28B). It is homologous with a portion of puboischiofemoralis externus of *Necturus*.

Origin: Ischial tuberosity

Insertion: Greater and lesser trochanters

Action: Retraction of the femur

Obturator Externus

 To observe this deep muscle, carefully separate gemellus caudalis and quadratus femoris.

Obturator externus is homologous with a portion of pubo-ischiofemoralis externus of *Necturus*.

Origin: Lateral surface of the pubis and ischium near the border of the obturator foramen

Insertion: Trochanteric fossa of the femur

Action: Rotates and retracts the thigh

TAIL MUSCLES

A number of intrinsic muscles perform the typical tail movements of a feline mammal— straight carriage, lateral bending, and the nervous twitching and lashing motions prior to pouncing upon some unsuspecting prey. Several of these muscles are continuations of the lumbar muscles discussed previously (*e.g.*, the extensor caudae medialis, a continuation of the multifidus spinae, raises the tail).

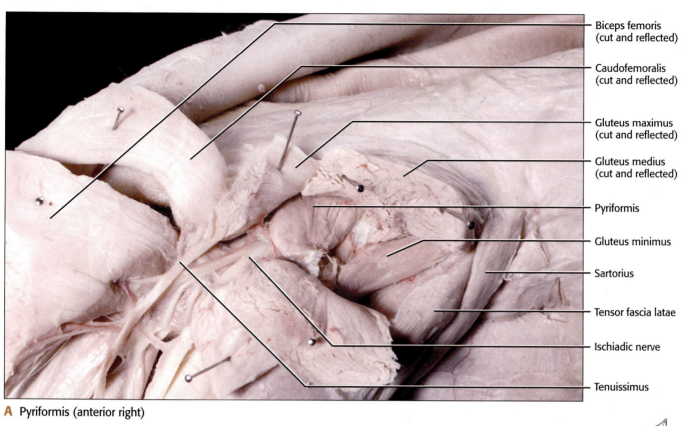

Biceps femoris
(cut and reflected)

Caudofemoralis
(cut and reflected)

Gluteus maximus
(cut and reflected)

Gluteus medius
(cut and reflected)

Pyriformis

Gluteus minimus

Sartorius

Tensor fascia latae

Ischiadic nerve

Tenuissimus

A Pyriformis (anterior right)

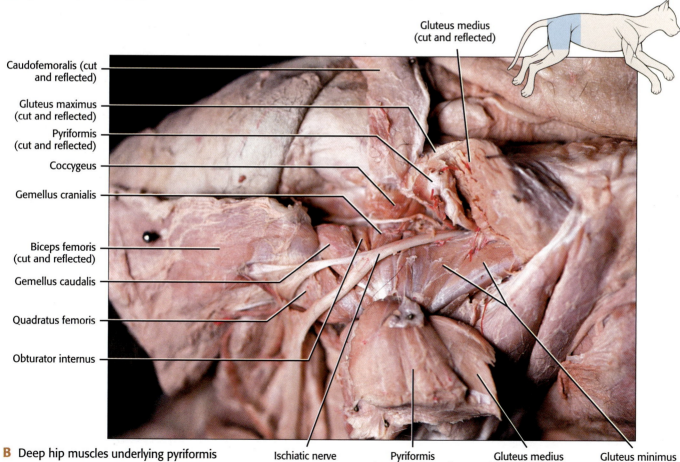

Gluteus medius
(cut and reflected)

Caudofemoralis (cut
and reflected)

Gluteus maximus
(cut and reflected)

Pyriformis
(cut and reflected)

Coccygeus

Gemellus cranialis

Biceps femoris
(cut and reflected)

Gemellus caudalis

Quadratus femoris

Obturator internus

B Deep hip muscles underlying pyriformis

Ischiatic nerve

Pyriformis
(cut and reflected)

Gluteus medius
(cut and reflected)

Gluteus minimus

35.28 Deep hip muscles.

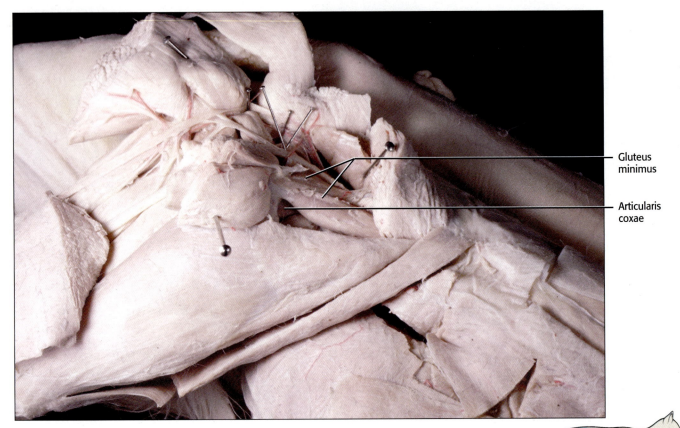

Gluteus minimus

Articularis coxae

C Articularis coxae

FIGURE 35.28 Deep hip muscles.

CHAPTER THIRTY-SIX

Cat — Body Cavities and Mesenteries

OPENING THE CAT

Place the cat ventral side up on the tray. Palpate the sternum to locate the xiphoid process. Remember that the sternum in the cat is elongate, so do not panic if it appears that you are too far posterior. Also, note that the xiphoid process is cartilaginous and will feel softer than the sternum. A good muscle dissection should have left you with the xiphihumeralis muscle in the chest area. The xiphoid process occurs approximately 1.25 cm cranial to the posterior edge of this muscle. During the following procedure, refer to Figure 36.1.

All of the following instructions apply to the skinned side of the cat.

■ With your scalpel make a 2.5 cm horizontal incision through all muscle layers and connective tissue about .75 cm posterior to the tip of the xiphoid process (Incision 1). Take care to cut through only the muscles and not damage underlying organs. Insert your finger into the incision to determine whether you have penetrated the body cavity and to locate the diaphragm. There is a high probability that you have made your incision just posterior to the diaphragm. The diaphragm is a muscular

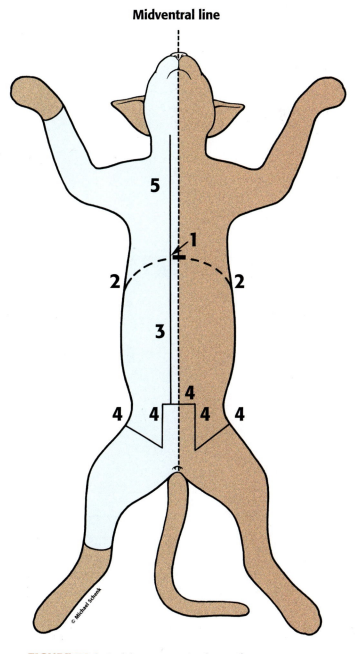

Midventral line

FIGURE 36.1 Incisions to expose internal organ systems.

partition separating the thoracic and abdominal cavities that feels similar to a taut balloon.

With scissors, extend the incision laterally on both sides toward the back or dorsal aspect of the cat along the curved contour of the diaphragm about 7.5–10 cm on each side (Incision 2).

With a scalpel, carefully loosen the diaphragm from the body wall, allowing it to come to rest on the liver. From the original horizontal incision, make a caudally directed longitudinal incision 1.25 cm to one side of the midline. While you are making this incision, pull up the abdominal wall and watch for a midventral mesentery attaching the urinary bladder to the wall (Incision 3).

When within about 1.25 cm of this mesentery, *stop*. Laterally, cut about 2.5 cm to either side of the midline and then continue these cuts to the caudal end of the abdominal cavity to form a small doorlike structure (Incision 4).

If your specimen is a male, be careful not to destroy the spermatic cord, the inguinal canal,

or the vas deferens. The cord can be recognized because it contains not only the vas deferens but also injected blood vessels lying on the medial aspect of the leg and passing over the brim of the pelvis, continuing through the inguinal canal and through the abdominal muscles. Extend the incisions laterally from the caudal cuts (above), again avoiding the inguinal canal and spermatic cords.

To expose the organs of the thoracic cavity, with scissors cut from the xiphoid process cranially (toward the head). To ensure cutting through the costal cartilages and avoiding the ribs, your incision should be 1.25 cm from the midventral line of the skinned side of the specimen (Incision 5). As you approach the neck region, you will be cutting through muscles in which major blood vessels and nerves occur (arteries are red, veins are blue, nerves are white). To avoid destroying these structures, cut through one muscle at a time.

Firmly grasp the thoracic wall of the skinned side and reflect it back until you hear a cracking

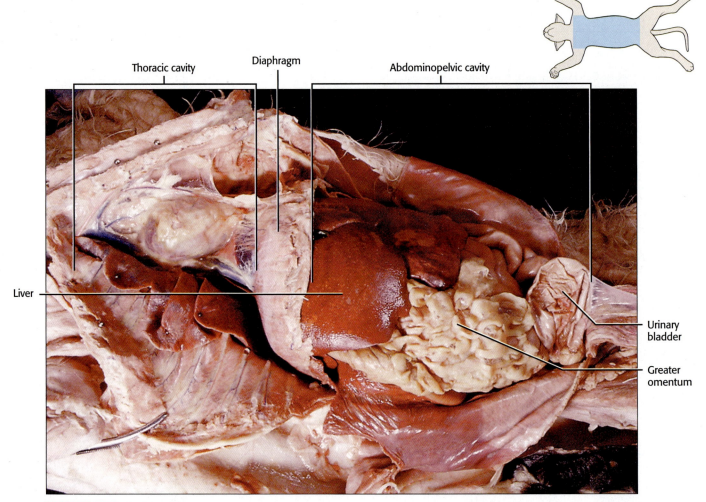

FIGURE 36.2 Body cavities.

sound as the ribs break. Gently lift the opposite side with the skin intact and observe the mesentery adhering to the midventral body wall. Carefully, with your scalpel release the mesentery from the body wall, allowing it to fall on the surface of the underlying organs. Just lateral to the attachment of this mesentery, on either side, observe an artery and vein adhering to the inner surface of the ventral thoracic wall. Again, with your scalpel, release these vessels from this surface. If necessary, remove some of the muscle. Now, with both hands grasp the rib cage on the unskinned side and quickly reflect back the thoracic wall until you hear a cracking sound.

Until you have read the next sections, *do not remove or disturb any tissues that might appear fatty or membranous, particularly the greater omentum (Figure 36.2). To preserve these delicate membranes, it is highly recommended that the mesenteries be studied before the viscera.*

BODY OR COELOMIC CAVITIES

Body plans of major invertebrate phyla and vertebrates are quite similar. In both invertebrates and vertebrates, bodies have evolved with body cavities lined with mesenteries allowing organs that are suspended within them to move independently of the general body surface. In general, this has allowed animals to develop a number of important characteristics and to become larger and more mobile.

Vertebrate body cavities typically are lined with a shiny membrane called a **serosa.** The serosa lining the body wall is known as the **parietal** layer, and the outer covering reflected over the surface of most organs suspended in these cavities is a continuation of this membrane called the **visceral** serosa. In reality, what we refer to as a cavity in the living animal is a potential space containing a small amount of lubricating fluid, which allows free sliding movement of the organs. Mesenteries are membranes that suspend organs within the cavities and extend between the parietal and visceral layers.

To illustrate the relationship of the parietal and visceral membranes associated with cavities, imagine a ball with a fist thrust into it. The fist represents any organ projecting into the cavity. The layer of the ball surrounding the fist is analogous with the visceral membrane and the outer surface of the ball is analogous with the parietal membrane. The space between the two ball surfaces is analogous with a body cavity or coelomic cavity (Figure 36.3A).

In mammals, the **coelomic cavity** is divided into a **thoracic cavity** and an **abdominopelvic** (peritoneal) **cavity** by the unique muscular **diaphragm** (Figure 36.3A, Figure 36.3B, and Figure 36.3C). The thoracic cavity is subdivided

further into a more or less central **pericardial cavity** surrounding the heart and paired lateral **pleural cavities** containing the lungs. Between the two pleural cavities is a potential space, the **mediastinum**, in which is located the heart and its cavity, the esophagus, the trachea, major blood vessels, nerves, and some endocrine organs, all of which are held in place by loose connective tissue (Figure 36.3C).

The abdominopelvic cavity is divided arbitrarily by an imaginary line drawn from the iliac crest to the pubic rim, defining a cranial abdominal space and a caudal pelvic space. In the abdominal cavity are suspended primarily digestive organs, and in the pelvic cavity are urinogenital organs and the distal organs of the digestive tract (Figure 36.3B).

MESENTERIES OF THE THORACIC CAVITIES

The pleural cavities are delimited by the **parietal pleura** as a glistening membrane that continues as the reflected **visceral pleura**, the outer covering of the lungs. The joined left and right medial parietal pleurae form the **mediastinal septum**. It is located both ventral and dorsal to the pericardium and continues anterior to the heart and adheres to the organs residing in the mediastinum. Note that the ventral portion of the mediastinal septum is the membrane that you loosened from the mid-ventral thoracic body wall when you "opened the cat." The right posterioventral portion of the mediastinal septum forms a pocket, the **caval fold**, into which projects the small accessory lobe of the right lung. The caval fold takes its name from the large blue blood vessel, the posterior vena cava, adhering to the medial portion of this mesentery. With your fingers, carefully pull one of the lungs medially and observe the stretched mesentery, the **pulmonary ligament** (Figure 36.4A and 36.4B).

The serosal membranes associated with the heart are similar in their relationships to those of the pleural membranes and their association with the lungs. With a pair of scissors, make a slit in the ventral portion of the pericardial sac to appreciate the **pericardial cavity**. The **parietal pericardium** defines the potential space, called the pericardial cavity, and then reflects over the surface of the heart as the **visceral pericardium** (Figure 36.4C). In contrast to the pleura, the parietal pericardium is intimately associated externally with other tissues forming the pericardial sac.

MESENTERIES OF THE ABDOMINOPELVIC CAVITY

Three ligaments can be identified between the diaphragm and the liver. The most prominent of these is the **falciform ligament,** found between the left and right halves of the liver and extending to the ventral abdominal wall. The thickened free margin of this membrane is the **round**

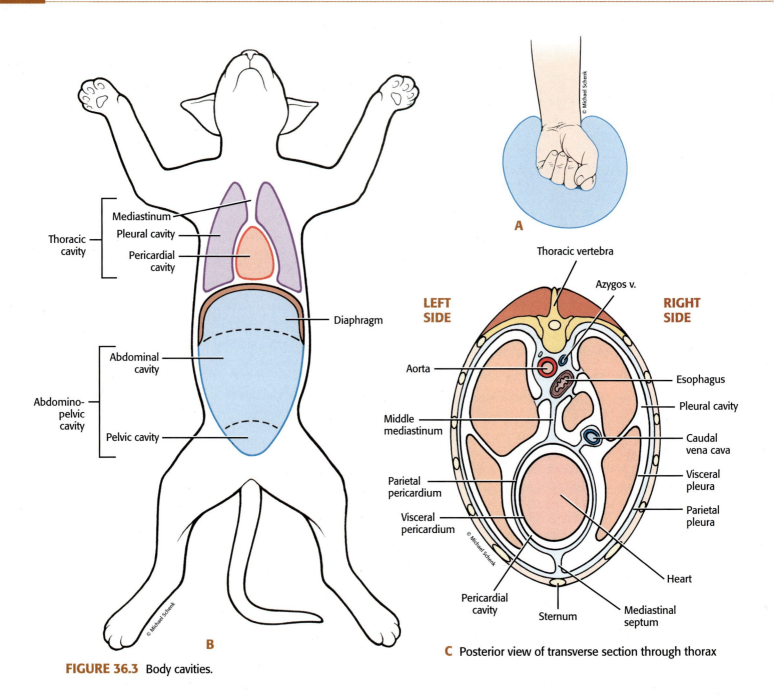

A

FIGURE 36.3 Body cavities.

C Posterior view of transverse section through thorax

ligament, a remnant of the umbilical vein, a fetal blood vessel. The continuation of the falciform ligament on either side of the central tendon of the diaphragm is the **coronary ligament** (Figure 36.5).

The very conspicuous and fragile **greater omentum** must be handled with care (Figure 36.6). It lies over most of the visceral organs, holding them in place. Because it is the site of fat storage, it is liberally laced with fat. Quite often, the greater omentum is tucked among the coils of the small intestine and may even be attached by strands of serosa and connective tissue to the dorsal parietal peritoneum.

 Carefully, with your fingers loosen the omentum from the intestinal loops, clipping, if necessary, the strands anchoring it dorsally.

The greater omentum of the mammal is a double-layered, apron-like mesentery extending from the greater curvature of the stomach and attaching to the dorsal wall of the peritoneal cavity. During the embryology of the mammalian digestive tract and its suspensory mesenteries, the stomach rotates around its longitudinal axis and also around its vertical axis. This results in a double-layered greater

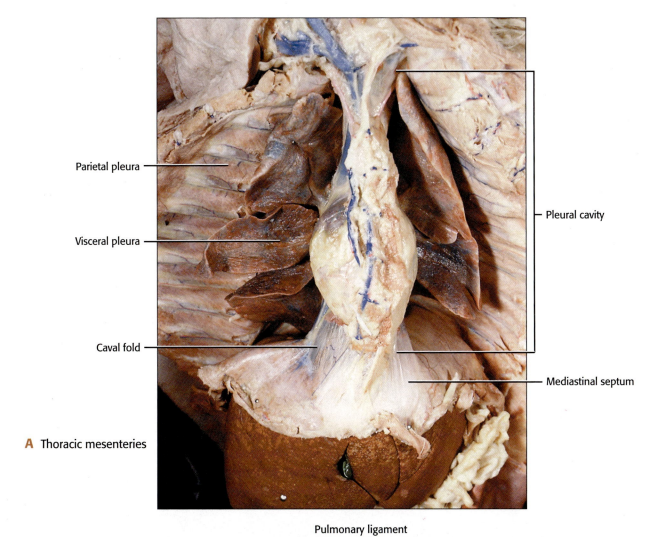

A Thoracic mesenteries

Parietal pleura

Visceral pleura

Caval fold

Pleural cavity

Mediastinal septum

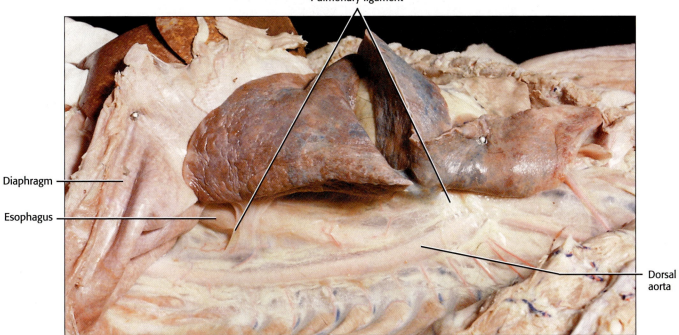

Pulmonary ligament

Diaphragm

Esophagus

Dorsal aorta

B Thoracic mesenteries (anterior right)

FIGURE 36.4 Thoracic mesenteries and pericardia.

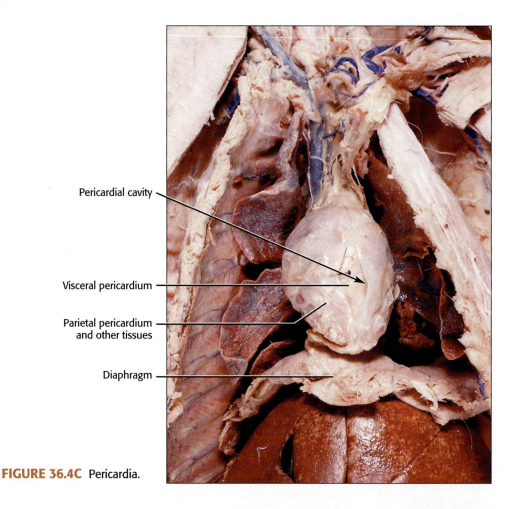

Pericardial cavity

Visceral pericardium

Parietal pericardium
and other tissues

Diaphragm

FIGURE 36.4C Pericardia.

Round ligament

Falciform ligament

Coronary ligament

Central tendon
of the diaphragm

Diaphragm

Liver

FIGURE 36.5 Ligaments associated with the liver and diaphragm (anterior right).

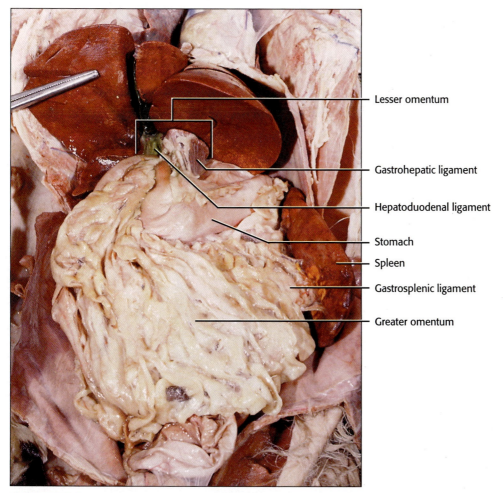

Lesser omentum

Gastrohepatic ligament

Hepatoduodenal ligament

Stomach

Spleen

Gastrosplenic ligament

Greater omentum

FIGURE 36.6 Greater and lesser omentum.

omentum enclosing a space or purse-like bursa that communicates through a restricted opening into the peritoneal cavity.

The **ventral layer of the greater omentum**, attached to the greater curvature of the stomach and extending to the pelvic region, turns back on itself toward the stomach as the **dorsal layer** (Figure 36.7). It then passes dorsal to the stomach, incorporating the tail of the pancreas and ultimately attaches to the dorsal peritoneal wall. The portion of the ventral layer of the greater omentum, extending from the stomach to the spleen, is called the **gastrosplenic ligament** (Figure 36.6). The potential space between the dorsal and ventral layers is the **omental bursa** (Figure 36.7).

 To demonstrate this cavity, carefully separate the two layers, and with your scissors cut along the bottom of the apron. Slowly pull the two layers apart and observe the saclike cavity.

The opening of this sac, the **epiploic foramen**, opens dorsally into the peritoneal cavity, thereby establishing continuity between the omental bursa and the peritoneal cavity (Figure 36.7). The foramen occurs just to the right of the lesser omentum, discussed below.

A much less extensive mesentery, extending from the liver to the lesser curvature of the stomach and the duodenum, is the **lesser omentum**. The portion of this mesentery between the liver and the stomach is known as the **gastrohepatic ligament**, and the continuation of this mesentery between the liver and the duodenum is the **hepatoduodenal ligament** (Figure 36.6). Part of the caudate lobe of the liver projects ventrally into this mesentery. Blood vessels, the common bile duct, nerves, and lymphatic vessels pass on the right through the free surface of this mesentery.

A short, somewhat inconspicuous mesentery, the **hepatorenal ligament**, stretches from the caudate lobe of the liver to the parietal peritoneum covering the *right* kidney (Figure 36.8). Do not destroy this ligament as the left side has no corresponding ligament.

Also on the right side is an often more conspicuous and fragile triangular mesentery, the **duodenocolic fold**, extending from the duodenum to the mesocolon (Figure 36.9).

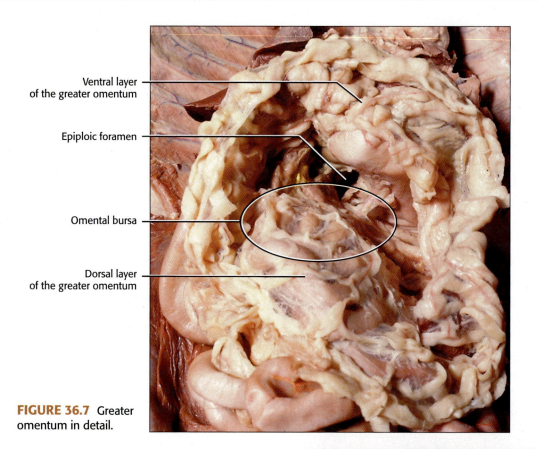

Ventral layer
of the greater omentum

Epiploic foramen

Omental bursa

Dorsal layer
of the greater omentum

FIGURE 36.7 Greater
omentum in detail.

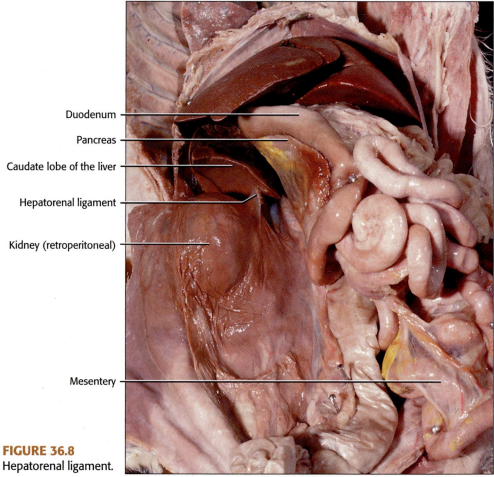

Duodenum

Pancreas

Caudate lobe of the liver

Hepatorenal ligament

Kidney (retroperitoneal)

Mesentery

FIGURE 36.8
Hepatorenal ligament.

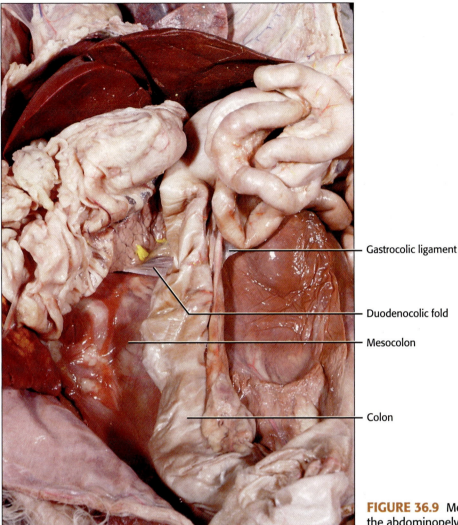

Gastrocolic ligament

Duodenocolic fold

Mesocolon

Colon

FIGURE 36.9 Mesenteries of the abdominopelvic cavity I.

On the left side of the peritoneal cavity and extending from the dorsal layer of the greater omentum to the opposite side of the mesocolon is another triangular **gastrocolic ligament**. From the dorsal body wall, the large intestine (colon) is suspended by the a part of the dorsal mesentery, the **mesocolon** (Figure 36.9).

 To appreciate its extent, gently pull the colon ventrally.

At the same time, observe that the duodenocolic fold and gastrocolic ligament merge with the mesocolon at opposing acute angles.

Gently lift the coils of the small intestine and observe the extensive mesentery suspending and supporting this organ. Notice the wealth of blood vessels coursing through this mesentery. The portion of this mesentery that supports the duodenum and the head of the pancreas is called the **mesoduodenum**, and the remainder, which supports the jejunum and ileum, is the **mesentery** proper (Figure 36.10).

Three prominent mesenteries anchor the urinary bladder in the abdominopelvic cavity. The **median vesical ligament** extends from the ventral surface of the urinary bladder to the ventral body wall. Both sides of the urinary bladder are anchored to the lateral body wall by the **lateral vesical ligaments**. Along the free edge of the lateral ligaments runs the **round ligament**, which is the remnant of the fetal umbilical arteries that carried oxygen-poor blood to the placenta (Figure 36.11). A generous rounded pad of fat is attached to each of the lateral vesical ligaments and often masks them.

If your specimen is a female, an elongated **broad ligament** supports the internal reproductive structures (Figure 36.12). The most extensive part of the broad ligament is the **mesometrium,** supporting the uterus. The portion of the broad ligament, much less extensive and supporting the ovary, is the **mesovarium** and the least extensive, supporting the uterine tubes (oviducts), is the **mesosalpinx**. The cranial end of the ovary is attached to the body wall by a thickened **suspensory ovarian ligament**. The caudal end of the ovary is held in place by a second thickened band called

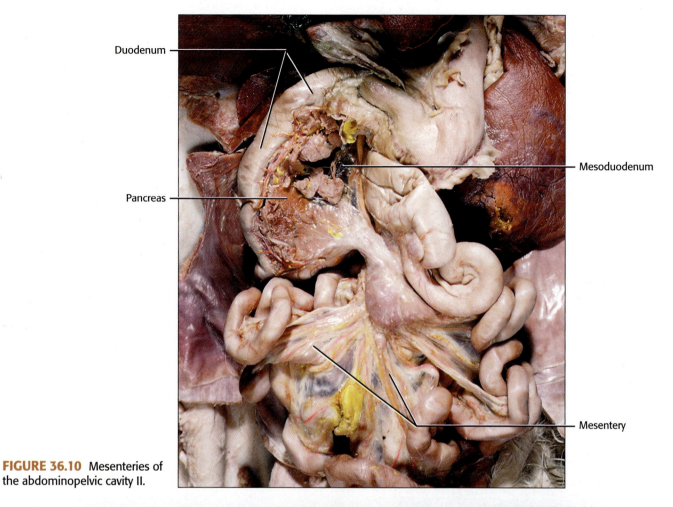

Duodenum

Mesoduodenum

Pancreas

Mesentery

FIGURE 36.10 Mesenteries of the abdominopelvic cavity II.

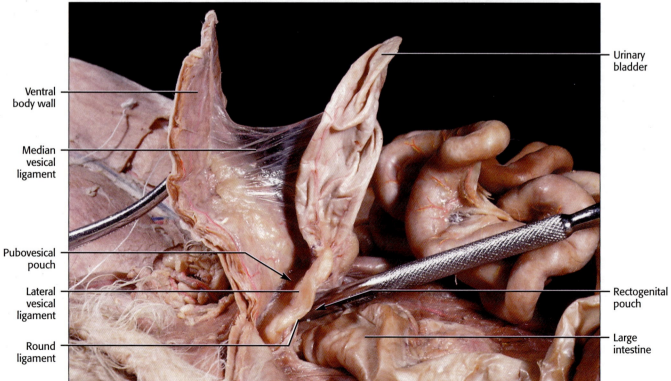

Urinary bladder

Ventral body wall

Median vesical ligament

Pubovesical pouch

Lateral vesical ligament

Round ligament

Rectogenital pouch

Large intestine

FIGURE 36.11 Male abdominopelvic region (anterior right).

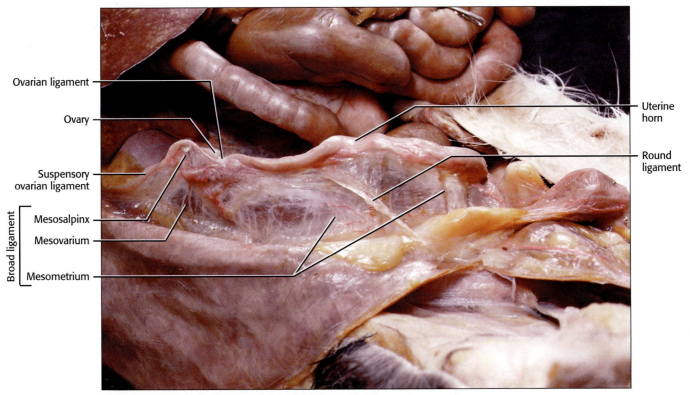

FIGURE 36.12 Mesenteries of the female reproductive system.

the **ovarian ligament**, extending between the ovary and the cranial end of the uterine horns.

Almost perpendicular and lateral to the mesometrium is the **round ligament** (Figure 36.12). It is the female counterpart of the gubernaculum in the male, which will be discussed in the reproductive system. If your specimen is a male, mesenteries associated with the reproductive system are largely outside the main body cavities and will be discussed with the reproductive system in Chapter 38.

The abdominopelvic cavity extends into the pelvic region in both sexes. In the male, the part of the cavity that

extends between the large intestine and the urinary bladder is the **rectogenital pouch** and the cavity that extends between the urinary bladder and the ventral body wall is the **pubovesical pouch** (Figure 36.11). In contrast, females, because the uterus passes between the large intestine and the urinary bladder, have three spaces versus the two in males. The **rectogenital pouch** is between the large intestine and the uterus, the **vesicouterine** pouch lies between the uterus and the urinary bladder, and the **pubovesical** pouch is in the same position as in the male, between the urinary bladder and the ventral body wall (Figure 36.13).

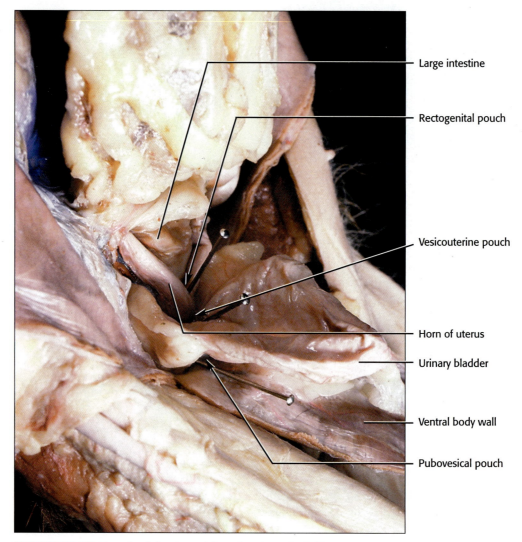

FIGURE 36.13 Female abdominopelvic region (anterior left).

Large intestine

Rectogenital pouch

Vesicouterine pouch

Horn of uterus

Urinary bladder

Ventral body wall

Pubovesical pouch

Cat — Digestive and Respiratory Systems

DIGESTIVE SYSTEM

During mammalian evolution the digestive system underwent a number of changes. A distinctive characteristic of mammals is the great specialization of teeth correlated with the need to handle large amounts of nutritious food to sustain their mobile endothermic bodies. The tongue became more muscular and mobile, also associated with food handling, grooming, and so forth. In addition, the tongue became the focus of taste receptors. As in several other groups of vertebrates, mucous glands were aggregated into salivary glands.

As pointed out with other vertebrates, efficient digestion requires maximizing exposure to enzymes in the intestine. Similar demands occurred in mammals. Herbivorous mammals tend to have longer intestines than carnivores because vegetation is more difficult to digest than meat. Coiling permitted mammals, in general, to accommodate their longer intestines. Further, increases in surface area were accomplished by increasing the internal topography of the intestine.

With the exception of the primitive monotremes—the spiny anteater and the duckbilled platypus—mammals do not have a cloaca. Among marsupial and placental mammals, feces are eliminated through an anus, and the urinary and reproductive systems open independently, apart from the digestive system.

ALIMENTARY CANAL

Salivary Glands and Ducts

 To examine the salivary glands, carefully remove surface connective tissue along the side of the head and neck and the surface of the masseter muscle, being especially cautious not to destroy blood vessels, nerves, and nervelike structures that may be salivary ducts (Figure 37.1A and 37.1B).

The parotid gland and its duct are particularly vulnerable to destruction while attempting to clean the area below the ear and over the masseter muscle.

Paired salivary glands are located along the lateral surface of the head beneath connective tissue and skin. The largest of these paired glands, located ventral to the ear is the **parotid gland**. It is a large, diffuse, lobulated structure that is intimately associated with overlying connective tissue and can be easily misidentified and removed while trying to expose the gland. The parotid duct emerges from approximately the midpoint of the anterior surface of the gland, crosses the **masseter** muscle, and enters the vestibule (space between the teeth and inside of the lip) of the oral cavity, where it opens opposite the third upper premolar tooth. Two similar-appearing bands, one above the parotid duct (the anterior branch of the facial nerve) and one below the parotid duct (the posterior branch of the facial nerve) serve as reference points in identification of the duct.

The **mandibular gland**, also referred to as the submaxillary, is located just ventral to the parotid and posterior to the angular process of the mandible. It, too, is lobular, but the lobes are less diffuse, giving the impression that this gland is smoother and better defined. A duct can be identified emerging from beneath the anterior edge of the gland and continuing laterally and beneath the digastric and

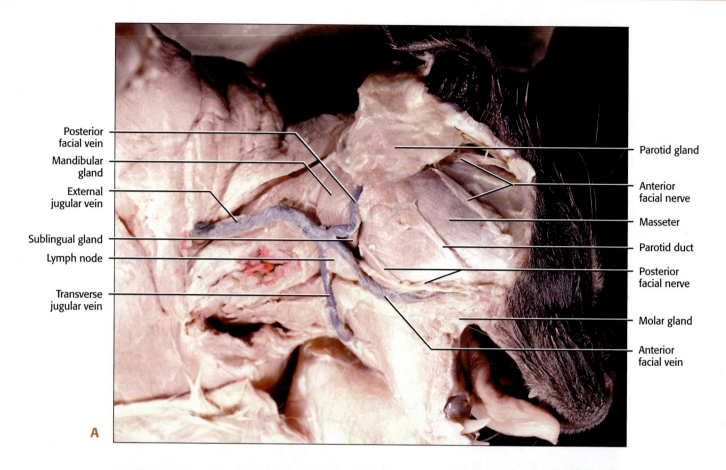

Posterior facial vein
Mandibular gland
External jugular vein
Sublingual gland
Lymph node
Transverse jugular vein

Parotid gland
Anterior facial nerve
Masseter
Parotid duct
Posterior facial nerve
Molar gland
Anterior facial vein

A

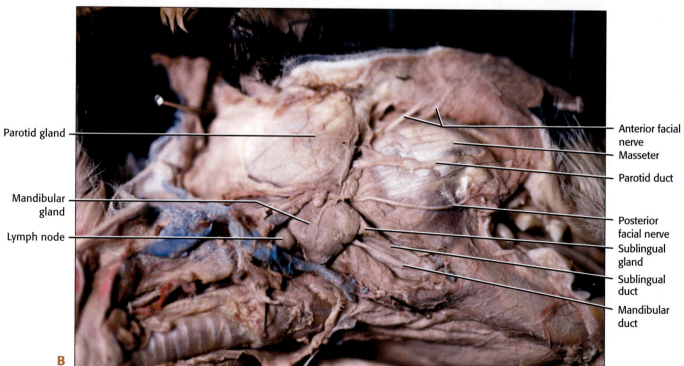

Parotid gland
Mandibular gland
Lymph node

Anterior facial nerve
Masseter
Parotid duct
Posterior facial nerve
Sublingual gland
Sublingual duct
Mandibular duct

B

FIGURE 37.1 Salivary glands.

under the mylohyoid muscles. The duct enters the floor of the oral cavity and opens at the base of a small papilla just anterior to the lingual frenulum. In the vicinity of this gland are generally one or two lymph nodes that can be easily mistaken for the mandibular. These may vary in size and, when large, may be particularly confusing.

The **sublingual gland** is the smallest of the three salivary glands. It is conical and is the smoothest, and often adheres to the anterior surface of the mandibular gland. The sublingual gland wraps around the proximal end of the mandibular duct. The sublingual duct is inconspicuous, running parallel to the mandibular duct, and also opens in the floor of the oral cavity in the vicinity of the mandibular duct.

In the cat, two other glands—a molar and a zygomatic —are considered part of the salivary system. The **molar gland** occurs at the angle of the jaw. It is located immediately beneath the skin and is embedded in the surrounding connective tissue. It has a brownish-gray, granular appearance, and varies in prominence. Several inconspicuous, small ducts open on the inner surface of the cheek. The **zygomatic** or infraorbital gland lies in the floor of the orbit of the eye. It opens by means of a small duct into the posterolateral portion of the roof of the mouth. This gland is difficult to find without removing an eye. We do not recommend that this be done.

Mouth or Oral Cavity

To facilitate observation of the oral cavity, the pharynx, and the larynx, we describe two methods of exposing these areas. One results in a specimen in which the oral cavity, pharynx, and larynx are presented as dorsal and ventral halves. The other is a bisected left and right half. Each has its own advantages. The first method that we describe is performed with an electrical craniotomy saw and may best be done as a demonstration for each class, perhaps with a cat whose dissection was begun but abandoned for one reason or another.

To prepare the cat for this operation, make an incision with a scalpel through the skin from the nose to the back of the neck. Peel the skin back from the incision about 5 cm (Figure 37.2). The next steps probably are best performed by the instructor and a laboratory assistant. While the assistant holds the head steady on the dissecting tray, generally locking the thumbs around the ears and holding either side of the head tightly, the instructor uses the craniotomy saw to bisect the entire head, including the mandible. The person doing the cutting and the holding assistant both must keep their eyes on the saw at all times and wear safety glasses. This really is not as dangerous as it might sound, but both individuals should be cautious.

The completed bisection may require some minor cutting of the tongue and the laryngeal area with a scalpel. This technique permits viewing of various head organs (*e.g.,* brain, pituitary gland, nasal conchae, and sinuses, as well as the relationship of cavities such as the oral and the pharyngeal and other difficult-to-demonstrate organs and openings such as the palatine tonsils and the opening of the auditory tube into the nasopharynx (Figure 37.3).

The second dissection is more difficult to perform. With a sharp scalpel, cut through the masseter muscle to the ramus of the jaw on both sides, avoiding the parotid duct and posterior facial nerve, if at all possible. Use a pair of bone shears to cut through the ramus of the jaw. An audible crunch will be heard. When completed, it should result in your being able to depress the lower jaw. It will be necessary to cut through the juncture of the palatoglossal arches and soft palate to gain full access to the pharyngeal area (Figure 37.4).

This dissection has the advantage of allowing observation of the entire hard and soft palates, the tongue, lingual and labial frenula, opening of the nasopalatine duct, and the

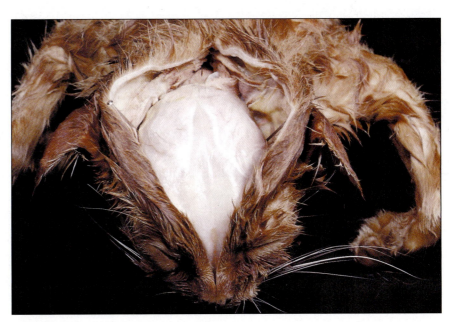

FIGURE 37.2 Head prepared for bisection.

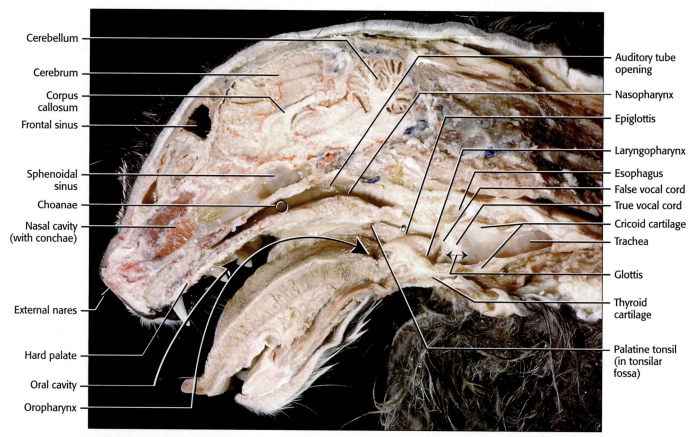

Cerebellum
Cerebrum
Corpus callosum
Frontal sinus
Sphenoidal sinus
Choanae
Nasal cavity (with conchae)
External nares
Hard palate
Oral cavity
Oropharynx

Auditory tube opening
Nasopharynx
Epiglottis
Laryngopharynx
Esophagus
False vocal cord
True vocal cord
Cricoid cartilage
Trachea
Glottis
Thyroid cartilage
Palatine tonsil (in tonsilar fossa)

FIGURE 37.3 Bisected head.

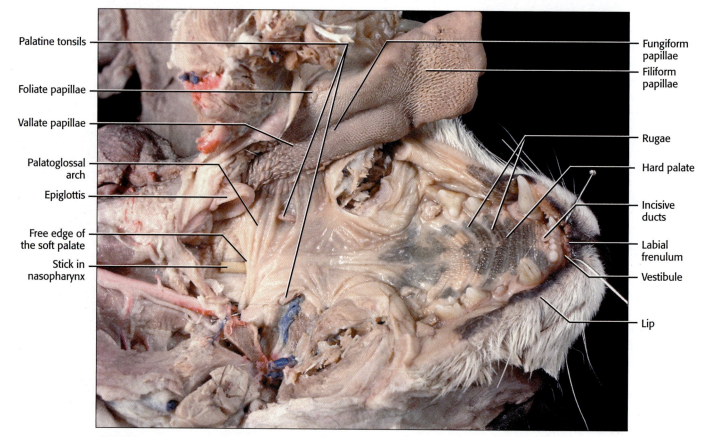

Palatine tonsils
Foliate papillae
Vallate papillae
Palatoglossal arch
Epiglottis
Free edge of the soft palate
Stick in nasopharynx

Fungiform papillae
Filiform papillae
Rugae
Hard palate
Incisive ducts
Labial frenulum
Vestibule
Lip

FIGURE 37.4 Dorso-ventrally dissected head.

like. It further avoids some damage to the circulatory and nervous systems that may result from the bisection.

The mouth or oral cavity is defined externally by the lips along its border. The movable **lips** are a pair of folds whose inner surface is covered by a mucous membrane and whose outer surface is hairy. The space between the lips and the teeth is called the **vestibule**. In the vestibule, the **labial frenulum**, a fold of tissue, connects the upper and lower lips at the midline to their respective gumlines (Figure 37.4). The **oral cavity** proper is the area of the mouth extending from the lingual side of the teeth to the entrance of the oropharynx (Figure 37.3).

The **tongue** is a mobile, muscular organ that plays an important and versatile role in the life of the cat. It is used as an organ of food manipulation, swallowing, drinking, and grooming. The **lingual frenulum** anchors the tongue in the anterior floor of the oral cavity and is one of the most obvious structures there. The frenulum becomes obvious when lifting the tongue (Figure 37.5).

On the surface of the tongue are four types of projections known as **papillae** (Figure 37.6).

1. The most numerous are the **filiform papillae**, which in cats are located on most of the surface of the tongue. Anteriorly, they appear spiky and are used in grooming or as rasping devices to remove tissue from bones. Posteriorly, they are less pointed.

2. **Fungiform papillae** are mushroom-shaped, fewer in number, and scattered among the filiform papillae.

3. **Vallate papillae** are larger, round papillae isolated by shallow grooves and arranged in a V configuration near the root of the tongue. The apex of the V is oriented toward the pharynx. Frequently, these papillae are difficult to distinguish.

4. The **foliate papillae** are leaf-shaped and are located on the posterolateral aspect of the tongue.

Taste buds, microscopic structures important in detecting chemicals identifiable in tasting food, are located in fungiform, vallate, and foliate papillae.

The roof of the oral cavity is formed anteriorly by the **hard palate** and posteriorly by the **soft palate**. The hard palate consists of a bony shelf constructed of the palatine processes of the premaxilla, the palatine processes of the maxilla, and the palatines. The hard palate is covered by tissue formed into a series of folds known as **rugae**. The soft palate extends from the caudal end of the hard palate and terminates in a free border. At the anterior end of the hard palate and directly posterior to the incisors is a pair of ducts, the **incisive ducts**, whose openings are distinguished by a small nipplelike structure (Figure 37.4). These ducts lead to vomeronasal organs, whose function intensifies olfaction in mammals. Vomeronasal organs first appeared among the amphibians and continue to occur among other vertebrates.

The pharynx, a space shared by the digestive and respiratory systems, extends from the oral cavity to the larynx. It is arbitrarily subdivided into three regions—the **nasopharynx**, the **oropharynx**, and the **laryngopharynx** (Figure 37.3). The dorsal nasopharynx extends from the **internal nares**, termed **choanae**, to the free border of the soft palate. Air moves through the nasopharynx to the **trachea**. In the

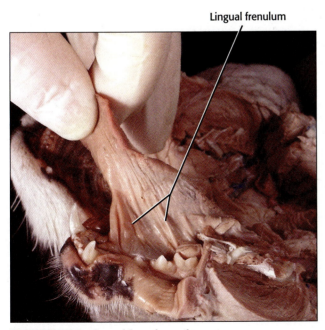

Lingual frenulum

FIGURE 37.5 Lingual frenulum of tongue.

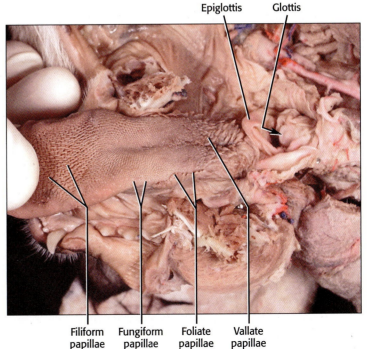

Epiglottis   Glottis

Filiform papillae   Fungiform papillae   Foliate papillae   Vallate papillae

FIGURE 37.6 Papillae of tongue.

lateral walls of this portion of the pharynx are the paired **auditory tube openings**, better seen in the bisected specimen. These tubes connect the air-filled middle ear cavity with the nasopharynx and are important in equalizing air pressure.

The oropharynx is the space bounded laterally by the **palatoglossal arches** and extending from the root of the tongue to the free border of the soft palate (Figure 37.3 and Figure 37.4). The **fauces**, the space between the arches, marks the transition between the oral cavity and the pharynx. Air, food, and liquids pass through the oropharynx on their way to the trachea and the esophagus, respectively. Two small lymphoid masses, the **palatine tonsils**, lying in shallow depressions, the **tonsilar fossae**, are located in the dorsolateral walls of the oropharynx (Figure 37.3 and Figure 37.4).

The **laryngopharynx** is that part of the pharynx continuing from the tip of the **epiglottis** to the **glottis**, an opening into the larynx (Figure 37.3).

Movement of food and liquids from the pharynx into the esophagus involves swallowing. During swallowing, the larynx is elevated and pulled up against the epiglottis, covering the glottis or opening into the larynx, thereby preventing food or liquids from entering it and facilitating passage of these materials into the esophagus. The epiglottis is unique to mammals.

The Digestive Tube

The **esophagus** is a collapsible muscular tube, capable of considerable distension, that passes through the mediastinum dorsal to the trachea. Wavelike muscular movements (peristalsis) in the walls of the esophagus propel the food through this tube to the stomach. In mammals, the thoracic and abdominopelvic cavities are separated by a muscular partition known as the diaphragm, which plays an essential role in breathing and through which the esophagus passes. At its caudal end, the esophagus terminates in a ring of smooth muscle, the **cardiac sphincter**, permitting movement of material from the esophagus into the stomach and functioning primarily in preventing reflux of the bolus (food mixed with saliva and oral enzymes) back into the esophagus (Figure 37.7).

The stomach is a J-shaped organ lying mainly on the left side of the body. Its left margin is convex and is known as the **greater curvature**. Its right margin is concave and is known as the **lesser curvature**. That portion of the stomach below the cardiac sphincter is the **cardiac** end, and the narrow portion connected to the intestine is the **pyloric** region. A muscular valve, the **pyloric sphincter (pylorus)**, located at the distal end of the pyloric region, regulates the movement of the stomach contents into the intestine. The large inflated portion between these two ends consists of the upper **fundus** and lower **body** (Figure 37.7). Some interesting variation

occurs in the stomachs of mammals, such as compartmentalization among ruminant herbivores.

 Carefully make an incision along the greater curvature from the fundus to the pyloric area of the stomach, avoiding the greater omentum.

If the stomach is full, carefully remove some of the material to observe the **gastric rugae**, or folds in this organ, which allow it to expand when food is eaten.

From the stomach, the contents of the alimentary canal move into the small intestine. This is the major site of digestion and absorption of nutrients and water in the digestive tract. This lengthy portion of the tract actually occupies a minimal abdominal volume because it is tightly coiled. The small intestine is regionally subdivided into three areas: the **duodenum**, the **jejunum** and the **ileum** (Figure 37.8).

The short proximal part of the small intestine known as the duodenum, about 12 cm to 18 cm in length in an adult cat, extends from the pylorus to the position of the duodenocolic fold. The distal-most portion of the small intestine is the ileum, and the middle portion is the jejunum. At the junction of the ileum and the large intestine, on the righthand side of the body, is a doughnut-shaped, muscular **ileocecal valve**, which, like the pylorus, regulates movement of the contents of the small intestine into the large intestine. It also prevents reflux of the contents into the small intestine (Figure 37.9).

 With a sharp scalpel, carefully make a small incision in the wall of the cecum opposite the site of the entry of the ileum. To observe the valve, you may have to clear any contents in this area.

The luminal surface of the small intestine in mammals exhibits fingerlike villi and microvilli, greatly increasing the surface area.

Further absorption of water occurs in the large intestine, along with fermentation, rotting of undigested material, and vitamin synthesis by resident colonies of bacteria. The end result is the production of gases and semisolid feces, eliminated from the body through the anus.

Anatomically, the proximal part of the large intestine is a blind diverticulum, the **cecum** (Figure 37.8 and Figure 37.9). In mammals, the cecum varies considerably in size, depending upon their food preference. Generally, the cecum of herbivores is large to accommodate digestion of large amounts of indigestible plant food. Some mammals have more than one cecum. A few mammals possess an elongate, fingerlike tube, the appendix, extending from the cecum. Cats do not have an appendix. The function of the appendix is unknown.

From the cecum, the colon continues cranially on the right side of the body as the **ascending colon**, makes

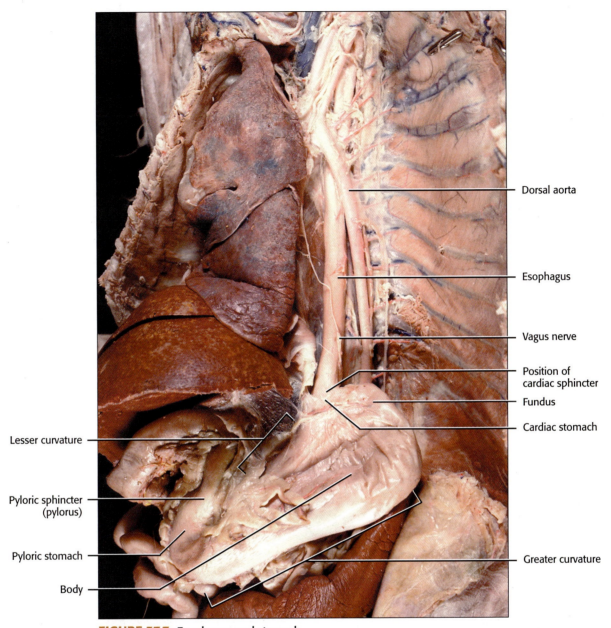

Dorsal aorta

Esophagus

Vagus nerve

Position of
cardiac sphincter

Fundus

Cardiac stomach

Lesser curvature

Pyloric sphincter
(pylorus)

Pyloric stomach

Body

Greater curvature

FIGURE 37.7 Esophagus and stomach.

a lefthand turn, and crosses the abdominal cavity to the left side as the **transverse colon**, where it curves caudally and continues as the **descending colon**, terminating as the **rectum**. A mesentery, the mesocolon, suspends the large intestine from the parietal peritoneum of the dorsal wall. The **anus** is the caudal opening of the digestive tract and is surrounded by sphincter muscles (Figure 37.10).

Accessory Digestive Organs

The largest internal organ in most mammals is the dome-shaped **liver**, which rests directly below the diaphragm (Figure 37.11). Among its many functions are chemical syntheses, detoxification of potentially harmful chemicals, storage of metabolic products, and bile production. If you

cannot pinpoint an organ where a particular function occurs, a good guess is the liver.

In the cat, the prominent, reddish-brown liver is divided into six lobes. A deep cleft, from which the falciform ligament extends from the liver to the ventral body wall, separates the left and right halves. Identify a **left medial lobe** and a **left lateral lobe**. Adjacent and to the right of the falciform ligament is the small **quadrate lobe**, which is partially united with the **right medial lobe**. The gallbladder is located in a semicircular depression, the cystic fossa, between the quadrate and right medial lobes.

Just posterior to the right medial lobe is the **right lateral lobe**. The **caudate lobe**, just posterior to the right lateral lobe and sometimes appearing like a subdivision of that

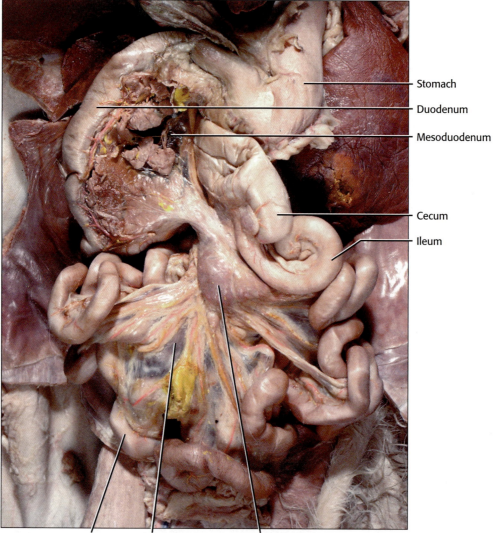

Stomach

Duodenum

Mesoduodenum

Cecum

Ileum

FIGURE 37.8 Regions of the small intestine.

Jejunum   Mesentery   Mesenteric lymph node

Ileocecal valve   Cecum

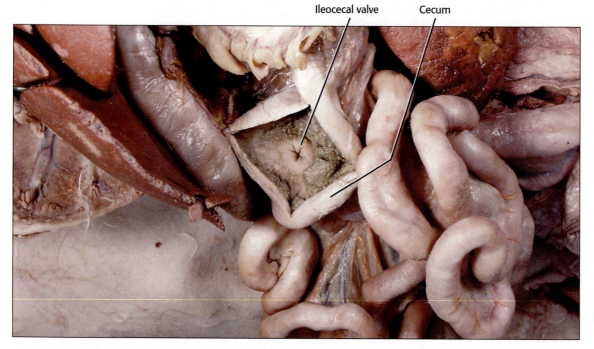

FIGURE 37.9
Ileocecal valve.

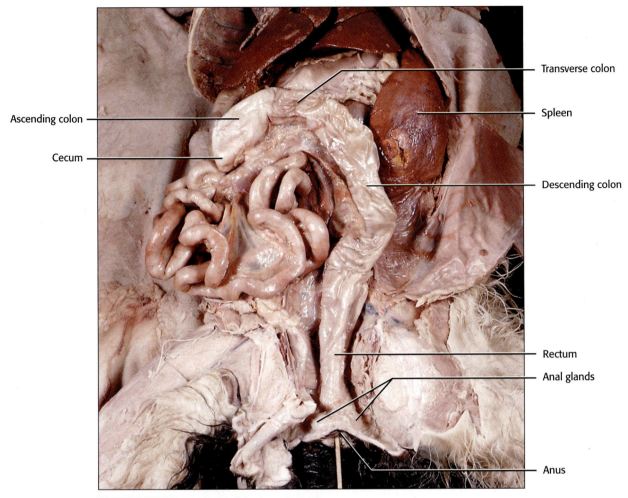

Ascending colon

Cecum

Transverse colon

Spleen

Descending colon

Rectum

Anal glands

Anus

FIGURE 37.10 Regions of the large intestine.

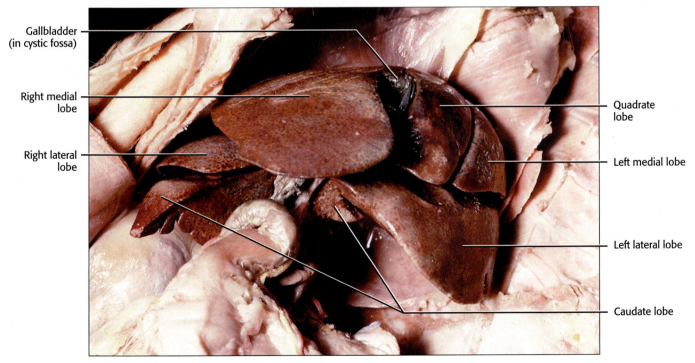

Gallbladder
(in cystic fossa)

Right medial
lobe

Right lateral
lobe

Quadrate
lobe

Left medial lobe

Left lateral lobe

Caudate lobe

FIGURE 37.11 Lobes of the liver.

lobe, extends into the lesser omentum (Figure 37.11). The point at which the liver and the diaphragm actually contact one another is the site of the central tendon of the diaphragm.

An important digestive function of the liver is to secrete a solution containing bile salts, which are important in emulsifying fats (reducing the size of fat globules, and subsequently increasing the surface area of the fat to facilitate enzymatic digestion. In addition, bile is involved during fat absorption.

Bile is secreted by liver cells and stored in the "bag" called the **gallbladder**. A number of ducts are associated with the liver and gallbladder. Many small hepatic ducts converge to form two or more quite prominent **hepatic ducts** leading from the left lobes and right lateral lobe of the liver that join the **cystic duct**, draining the gallbladder, forming the **common bile duct** that empties into the duodenum (Figure 37.12A and 37.12B).

To observe the duct system, with forceps, carefully isolate the common bile duct (a tannish, flat structure) within the border of the hepatoduodenal ligament, beginning at its distal end, where it joins the main pancreatic duct.

Both open into the duodenum at a site distinguished by a small pimplelike bump known as the **hepatopancreatic ampulla** (ampulla of Vater) (Figure 37.12A and 37.12B). The common bile duct is fragile and can be torn easily. From the ampulla it extends toward the liver, where it is joined by the hepatic ducts from various lobes of the liver, as well as the often greenish cystic duct leading from the gallbladder.

The **pancreas** is a gland with aggregates of endocrine cells producing essential hormones that control various metabolic activities throughout the body, and an exocrine portion producing essential digestive enzymes and a buffering solution of sodium bicarbonate. The pancreas of mammals is derived from a dorsal and a ventral primordium that fuse to form a single organ.

The appearance of this organ is lobulated and glandular. It is the tannish, elongated organ whose **head** (duodenal portion) lies within the mesoduodenum and whose tail (gastrosplenic portion) lies within the dorsal part of the greater omentum near the greater curvature of the stomach (Figure 37.13). The **main pancreatic duct** (Figure 37.12A and 37.12B) is associated with the duodenal part. This duct joins the common bile duct in the hepatopancreatic ampulla on the duodenum. An accessory duct often drains the gastrosplenic portion independently.

To observe the main duct, carefully and gently pick away pancreatic tissue along the duodenal border of the duodenal portion of the pancreas, starting from the ampulla and working toward the head end, taking special care to preserve all red-injected and yellow-injected blood vessels in the process. This duct is white and delicate, with smaller branches feeding into it. The accessory duct often is difficult to locate and is absent in some cats.

The **spleen** is a large lymphoid organ often discussed with the digestive system because it occurs in the peritoneal cavity along with the viscera. This large, tonguelike organ lies on the left side of the body. It is anchored in the ventral layer of the greater omentum, the **gastrosplenic ligament** (Figure 37.13).

RESPIRATORY SYSTEM

Gaseous exchange of oxygen and carbon dioxide is a life-sustaining function of the respiratory system. In mammals, the organs facilitating this exchange are the lungs. They are generally found within the thoracic cavity of the body. The diffusion of gases occurs across the moist lungs into capillaries. The spongy lungs are made up of many gas-filled small spheres, or alveoli, where this diffusion takes place. The vast number of alveoli makes it possible to maximize the surface area in a reasonably sized organ such as the lung.

A system of tubes—the trachea, various bronchi, and bronchioles—conveys the oxygen-rich air from the atmosphere to the lungs and the oxygen-poor air back to the atmosphere. The diameter of these tubes decreases while the number increases as they extend from the nasal cavity to the alveoli. The overall effect is to increase the respiratory surface area dramatically.

Among some reptiles and all mammals, a secondary palate described in the discussion of the digestive system separates the oral and the nasal cavities. In the digestive system of mammals, the palate serves as the roof of the oral cavity and as the floor of the nasal cavity of the respiratory system. This "shelf" allows a mammal to chew and hold food items in the mouth and breathe at the same time.

The pathway by which air moves from the atmosphere to the alveoli begins through the nostrils, or **external nares,** and into the **nasal cavities**. In these cavities the air is warmed, moistened, filtered, and smelled. From the cavities it passes through the internal nares, or **choanae,** and into the **nasopharynx**—one part of a more extensive region called, simply, the pharynx, a space shared with the digestive system. The nasopharynx extends from the internal nares to the free border of the soft palate. The air then moves into the **laryngopharynx**, extending from the hyoid to the esophagus and larynx. From this space the air moves into the larynx (see Figure 37.3).

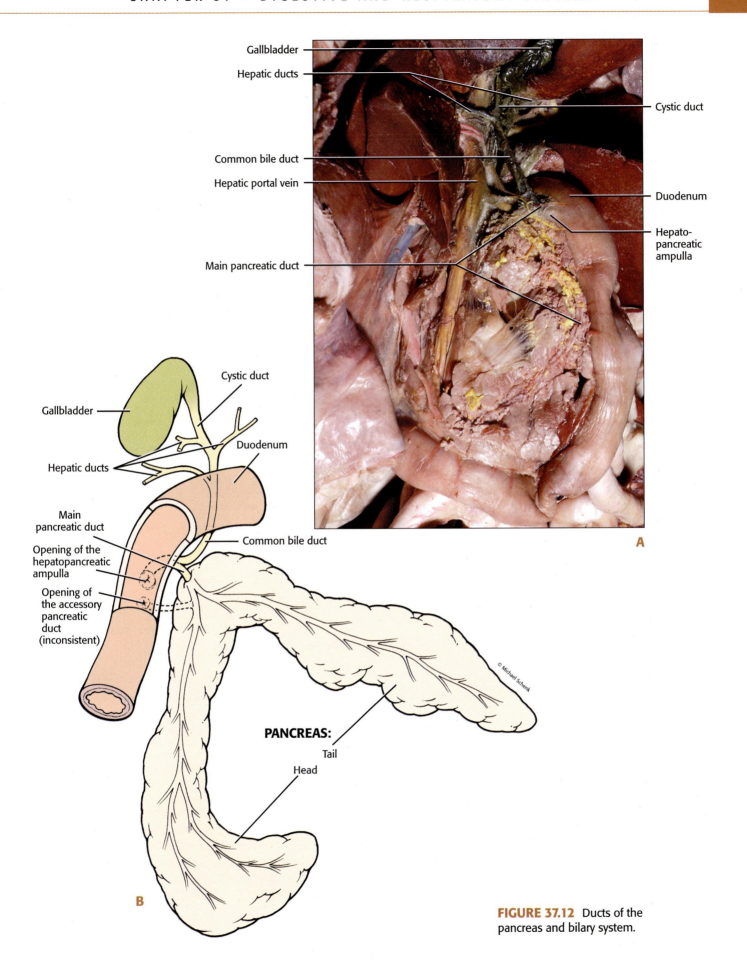

Gallbladder

Hepatic ducts

Cystic duct

Common bile duct

Hepatic portal vein

Duodenum

Hepato-pancreatic ampulla

Main pancreatic duct

A

Cystic duct

Gallbladder

Duodenum

Hepatic ducts

Main pancreatic duct

Opening of the hepatopancreatic ampulla

Opening of the accessory pancreatic duct (inconsistent)

Common bile duct

© Michael Schenk

PANCREAS:

Tail

Head

B

FIGURE 37.12 Ducts of the pancreas and bilary system.

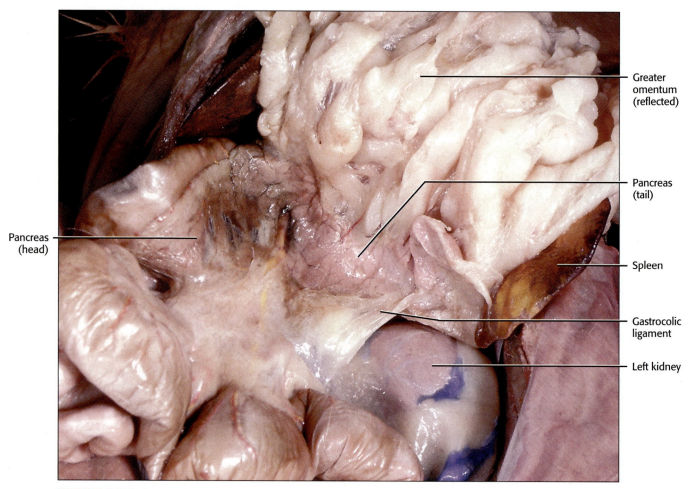

Greater omentum (reflected)

Pancreas (tail)

Spleen

Gastrocolic ligament

Left kidney

Pancreas (head)

FIGURE 37.13 Pancreas.

 Before the larynx is dissected, locate the paired thyroid glands situated on either side of the trachea, just posterior to the larynx (Figure 37.14).

The larynx is cartilaginous and consists of several single and paired elements. One of the single elements is the **epiglottic cartilage**, which strengthens the **epiglottis**, a unique structure in mammals. The function of the epiglottis is to prevent the contents of the oral cavity from entering the trachea during swallowing. As the material is swallowed, muscles pull the larynx anteriorly, causing the epiglottis to block off the **glottis**, a slit between the vocal cords. Air passes into the larynx through this slit. To view the epiglottis in a bisected specimen, see Figure 37.3.

 With a specimen dissected into a dorsal and a ventral half, pull the tongue forward while peering into the pharyngeal region, and observe the epiglottis (Figure 37.4).

The most conspicuous single cartilage of the larynx, the **thyroid cartilage**, is located ventrally (Figure 37.3 and

Figure 37.14). Observe that this cartilage is continuous dorsally.

 To see this structure better, you should remove the musculature obscuring it.

In humans, this cartilage, commonly called the Adam's apple, exhibits a distinct sexual dimorphism; that is, in males it is much more prominent.

Posterior to the thyroid cartilage is the ring-shaped **cricoid cartilage** (Figure 37.3 and Figure 37.14). The dorsal wall of the larynx consists primarily of the broad part of this cartilage. You also may want to clear musculature from this cartilage.

Small, paired cartilages, the **arytenoid cartilages** (Figure 37.14), sit on the dorsal rim of the cricoid cartilage and abut the dorsal projections of the thyroid cartilage. Some other inconspicuous cartilages, part of the larynx, are difficult to see.

With a sharp scalpel, make an incision in the midventral wall of the larynx, perhaps extending it

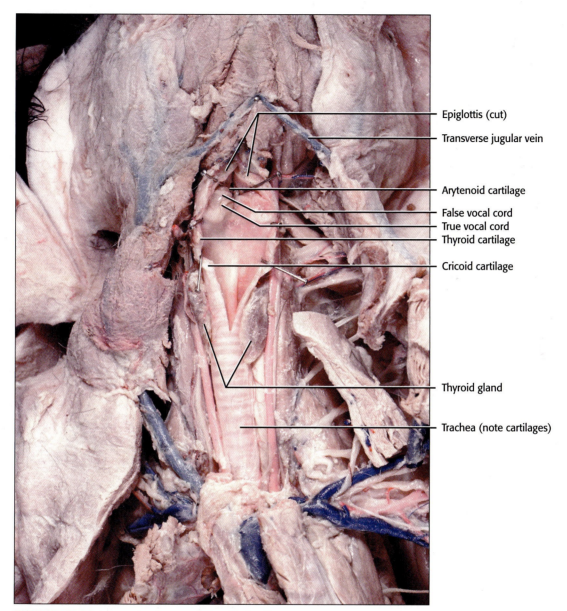

— Epiglottis (cut)

— Transverse jugular vein

— Arytenoid cartilage

— False vocal cord
— True vocal cord
— Thyroid cartilage

— Cricoid cartilage

— Thyroid gland

— Trachea (note cartilages)

FIGURE 37.14 Larynx, trachea, and thyroid glands.

into the trachea posteriorly. With scissors, continue the cut anteriorly, cutting the epiglottis. These directions apply only to dorsoventral head dissection. Carefully reflect the walls of the larynx.

Observe two pairs of tissue folds at the cranial end of the larynx. The anterior folds, the **false vocal cords**, often appearing tannish, extend from the arytenoid cartilages to the epiglottis. The posterior folds, the **true vocal cords**, appearing as whitish bands, extend from the arytenoid cartilages to the thyroid cartilage (Figure 37.3 and Figure 37.14). In cats, vocalization is the result of air movement across the true vocal cords.

The **trachea** extends posteriorly from the larynx to the root of the lungs (Figure 37.3 and Figure 37.14). This large-diameter respiratory tube is the main passageway for air from the larynx to the lungs. Observe the C-shaped cartilages that reinforce the wall of the trachea, with the dorsal open space of the "C" reinforced with muscle. This anatomical configuration allows compression during the act of swallowing but prevents collapse during respiration. The lobulated, glandular tissue, the **thymus**, lying along the ventral aspect of the trachea and heart, should be conserved (Figure 41.5).

The trachea is divided within the tissue of the lungs, with a resulting interconnected network of air-conducting tubes often referred to as a bronchial or respiratory tree. The first division of the trachea forms the **primary bronchi** (Figure 37.15). These are subdivided sequentially into secondary and tertiary bronchi, bronchioles, and alveolar

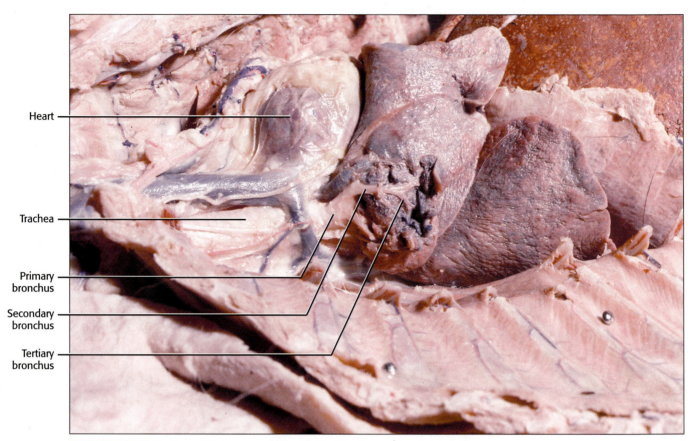

Heart

Trachea

Primary
bronchus

Secondary
bronchus

Tertiary
bronchus

FIGURE 37.15 Trachea: Primary, secondary, and tertiary bronchi.

ducts, terminating in alveoli, illustrating the decreasing diameter but increasing surface area of the respiratory tree. The supporting cartilage of these tubes decreases proportionately to the increased surface area of the tree.

 To observe the primary bronchi, reflect the lungs medially and carefully pick away lung tissue at the level of the root of the lungs. Be careful to avoid the pulmonary vessels in this area that enter and leave the lung at the hilus.

The bronchi appear as a whitish, shiny tube containing cartilage. Because the rest of the tree is enclosed within the lungs, we do not recommend further dissection. Your instructor may wish to provide a demonstration dissection, however.

The lungs are spongy primarily because of the terminal structures of the respiratory tree, called alveoli. Air moves through the tubular network into the alveoli—small, thin

terminal air sacs—where gas exchange occurs. The left lung is subdivided into three lobes, the **anterior**, the **medial**, and the **posterior**. The right lung is divided similarly, with the exception of the posterior, which is further subdivided, forming an **accessory lobe**. Remember that the accessory lobe of the lung projects into a mesentery pocket called the caval fold, and its lefthand mesentery is the **mediastinal septum** (Figure 37.16). In addition, remember that the lungs are suspended within the pleural spaces by the pulmonary ligaments (Figure 36.4B).

The often mentioned muscular partition between the thoracic and abdominopelvic cavities is the **diaphragm** (Figure 37.16). Mammals are the only vertebrates in which these two cavities are completely separated. This partition is partially formed from the transverse septum, homologous with the same structure of the shark and *Necturus*. In addition, membranous tissues and muscles derived from the cervical region contribute to formation of the diaphragm. Through the diaphragm pass the esophagus, the posterior vena cava, and the aorta.

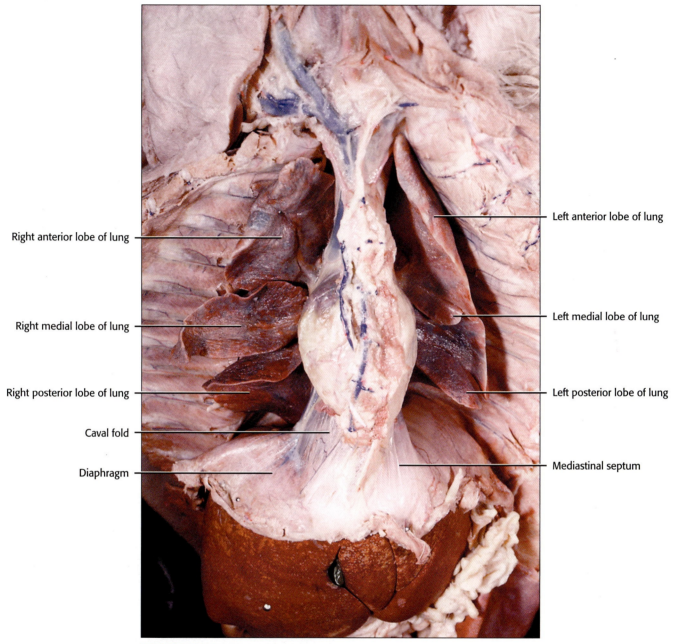

Right anterior lobe of lung

Right medial lobe of lung

Right posterior lobe of lung

Caval fold

Diaphragm

Left anterior lobe of lung

Left medial lobe of lung

Left posterior lobe of lung

Mediastinal septum

FIGURE 37.16 Lobes of the lungs.

CHAPTER THIRTY-EIGHT

Cat — Urogenital System

I n mammals, as in many other vertebrates, the urinary and reproductive systems are closely interrelated not only anatomically, but also embryologically. For a review of the urinary reproductive systems, see Chapter 9 and Chapter 10. In particular, most of the terminal ductwork in males is common to both systems. Under the influence of testosterone, normal development of the reproductive system of male mammals progresses, whereas female reproductive structures develop in the absence of testosterone.

With the exception of the monotreme mammals (the spiny anteater and the platypus), mammals no longer possess a cloaca. Therefore, the ureters (urinary drainage tubes) enter the urinary bladder, which is drained by a single new tube, the urethra, to the outside. The structure and function of the urinary system in the two sexes is virtually identical from the kidneys to the urethra. In males, the pre-prostatic urethra is purely urinary and the rest of the urethra functions as part of both the urinary and reproductive systems. In females the entire urethra is urinary.

The functions of the testes and the accessory glands of the male reproductive system in mammals include sex hormone and semen production (containing sperm and associated glandular fluids). Hormones are important to maintain "maleness" and continued stimulation of semen production. While urine passes from the urinary bladder to the outside by way of the entire length of the urethra, semen travels only along that part of the urethra distal to the prostate gland to the outside. One of the functions of the penis is to deposit seminal fluid into the female reproductive tract. Although reptiles and mammals alike have a penis, the anatomy is not identical.

The functions of the ovaries in the female reproductive system are production of sex hormones and oocytes and secretions by accessory glands. The system of "tubes" in the female reproductive apparatus of mammals is adapted to receive semen from the male during copulation, movement of the sperm to effect fertilization, and subsequent possible implantation and continued nourishment and maintenance of the developing embryo in the uterus.

URINARY SYSTEM

Kidney and Ducts

The glomerular kidney is the fundamental organ of this system. Similar to the kidney of the shark and *Necturus*, the primary functional filtering units of the kidney are the nephrons. The nephron consists of a cuplike Bowman's capsule with an elongate tubule extending from it. A specialized arterial capillary, the glomerulus, is intimately associated with Bowman's capsule. Also associated with the tubules of the nephron are other capillaries, numerous blood vessels, and nerves.

The kidneys are paired, retroperitoneal ("behind" the parietal peritoneum), organs surrounded by fat deposits in the dorsal portion of the lumbar region. Notice that the position of the right **kidney** is slightly more posterior than the left because of the posterior extension of the caudate lobe of the liver on the right side. Take note of the **hepatorenal ligament** extending between the caudate lobe and the right kidney. It has no counterpart on the left side; therefore, be careful to conserve this membranous structure.

To observe the gross internal anatomy of the kidney, make a slit through the parietal peritoneum covering the left kidney. Carefully separate the kidney from the surrounding fat, taking care to expose it sufficiently to allow you to make a mid-frontal cut through the kidney.

In life, the kidney resembles a large kidney bean in color and in shape. The medial indentation is the **hilus** (Figure 38.1). Through this region passes the expanded proximal end of the **ureter** (renal pelvis) renal arteries and veins and nerves. Notice the tough, whitish, fibrous connective tissue encapsulating the kidney, the **renal capsule**.

 To better view this capsule, carefully peel it back, without removing it.

The outer narrow band of lighter tissue in the kidney section is the granular **cortex**, and the central darker region is the striated **medulla** (Figure 38.2). The glomerulus and portions of the nephron tubule are found in the cortex, whereas other tubular regions of the nephron, as well as collecting ducts, are found in the medulla. The medulla of the kidney in the cat consists of a single pyramid with its base abutting the cortex, and its vertex, the **papilla**, opening into the **renal pelvis,** the expanded proximal end of the ureter. The renal pelvis is located within the **renal sinus**, the potential space surrounding the renal pelvis (Figure 38.2). Fat and blood vessels may be seen in the renal sinus.

New urine drainage tubes, the ureters, appear among the amniotes (the reptiles, birds, and mammals). Ureters are tubes with muscular walls leading from the kidney to the urinary bladder. The **ureter** begins as the renal pelvis located within the sinus of the kidney, courses posteriorly in a retroperitoneal position, and passes through the lateral ligament of the **urinary bladder** to enter the urinary bladder (Figure 38.1 and Figure 38.3). The ureteral openings along with the urethral opening delineate a triangular area in the base of the bladder.

 Carefully pick away the connective tissue and fat covering the left ureter, exposing it from the hilus to the urinary bladder. Do not destroy the entire left lateral ligament of the bladder. In males, the vas deferens coils around the ureter at the base of the urinary bladder. Do not damage the vas deferens (Figure 38.3).

The saclike **urinary bladder** is a reservoir for urine. The free domed end of the urinary bladder is the **vertex**, and the attached caudal end is the **fundus** (Figure 38.3). Remember that the bladder is held in position ventrally by the median ligament and on each side by the lateral ligaments.

Urine leaves the urinary bladder by way of the tubular **urethra.** Like the rest of the urinary tract, the walls of this tube are similarly constructed. Because this tube lies primarily in the pelvic canal, it will be seen during dissection of the reproductive system (Figure 38.4 and Figure 38.5).

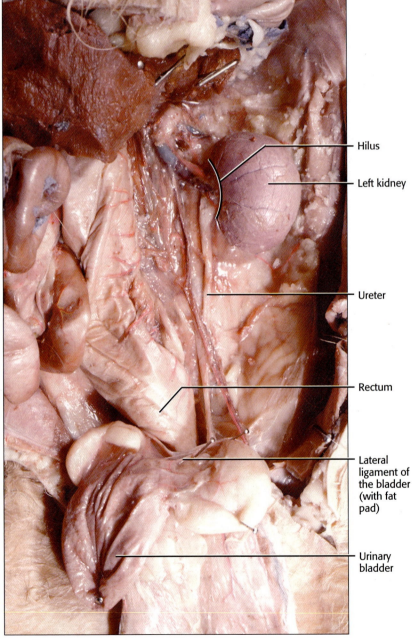

Labels: Hilus, Left kidney, Ureter, Rectum, Lateral ligament of the bladder (with fat pad), Urinary bladder

FIGURE 38.1 Overview of excretory system.

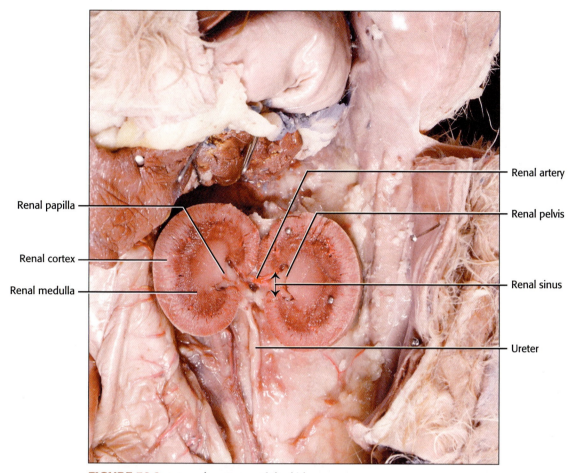

FIGURE 38.2 Internal anatomy of the kidney.

Renal papilla

Renal cortex

Renal medulla

Renal artery

Renal pelvis

Renal sinus

Ureter

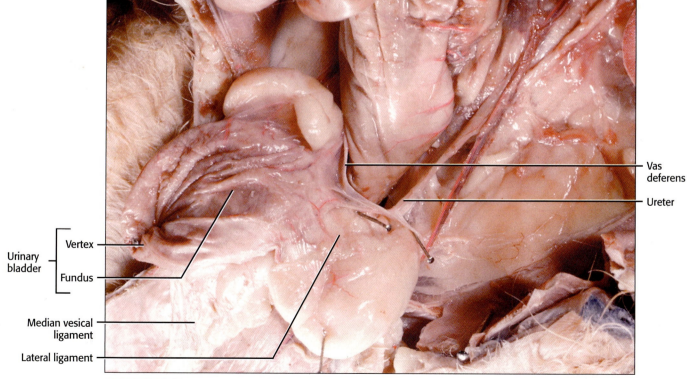

FIGURE 38.3 Ureter and vas deferens in the male.

Vas deferens

Ureter

Urinary bladder

Vertex

Fundus

Median vesical ligament

Lateral ligament

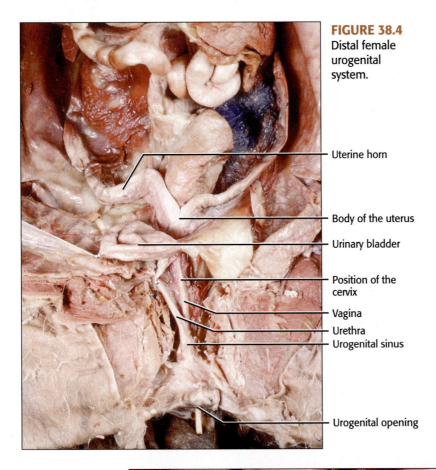

FIGURE 38.4
Distal female urogenital system.

Uterine horn

Body of the uterus

Urinary bladder

Position of the cervix

Vagina

Urethra

Urogenital sinus

Urogenital opening

REPRODUCTIVE SYSTEM

Female

The primary reproductive organs are the **ovaries,** which are small, oval organs suspended anteriorly from the dorsal body wall by the **suspensory ovarian ligament** and posteriorly by the **ovarian ligament** connecting it to the cranial end of the uterine horns. Remember that the ovaries are further supported by a portion of the broad ligament, the **mesovarium** (Figure 38.6A and Figure 38.6B).

The ovaries are not physically connected to the tubes associated with the female reproductive system but, rather, rest in the peritoneal cavity. When the oocytes rupture through the serosa of the ovary, they are swept into the **oviducts,** or **uterine tubes.** The expanded proximal end of the oviduct, the **infundibulum,** is hoodlike, wraps around the ovary laterally, and opens medially by way of the **ostium tubae.** Along its edges are small, fingerlike projections, the **fimbriae,** whose movements are responsible for sweeping the oocyte into the ostium tubae. The small, coiled

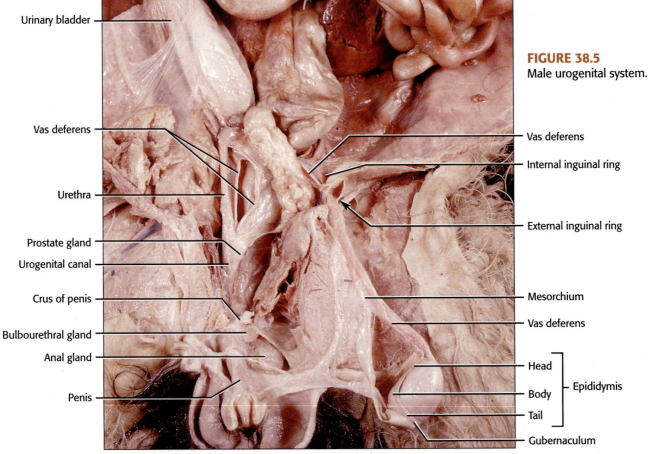

Urinary bladder

Vas deferens

Urethra

Prostate gland

Urogenital canal

Crus of penis

Bulbourethral gland

Anal gland

Penis

FIGURE 38.5
Male urogenital system.

Vas deferens

Internal inguinal ring

External inguinal ring

Mesorchium

Vas deferens

Head

Body ⎤ Epididymis

Tail

Gubernaculum

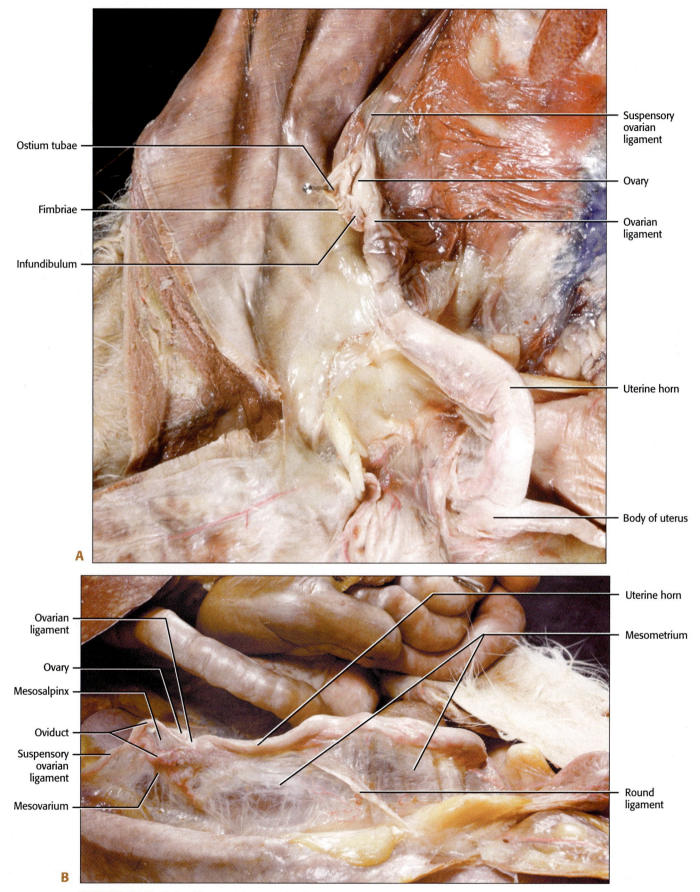

FIGURE 38.6 Right side: Female reproductive organs.

oviducts lie lateral to the ovary and are suspended by another portion of the broad ligament, the **mesosalpinx** (Figure 38.6A and Figure 38.6B).

 To observe these structures, carefully reflect the ovary and uterine tubes laterally (Figure 38.6A and Figure 38.6B). To appreciate the ostium, carefully separate the edges of the infundibulum and insert a probe between them.

In mammals, fertilization occurs in the upper third of the oviducts. Cilia, lining the oviducts, move oocytes or zygotes (fertilized eggs) toward the uterus.

The distal ends of the oviducts are enlarged and fused to form a uterus, which projects into the vagina, a muscular tube that opens into the urogenital sinus independent of the urinary system. The uterine tubes merge into the larger-diameter **horns of the uterus,** or **uterine horns.** The two uterine horns fuse to form the **body of the uterus** and give the impression of a Y-shaped organ (Figure 38.4 and Figure 38.6A).

Successful fertilization results in implantation of early embryos in the uterine horns, where development continues for 9 to 10 weeks, leading to the birth of kittens. The mammalian uterus is lined with a highly vascularized lining that sustains the developing fetus. Recall that the mesometrium of the broad ligament supports the horns of the uterus. Note the round ligament present in the mesometrium, which is homologous with the gubernaculum in the male.

To view the distal portion of the reproductive system, the pelvic cavity has to exposed.

 Avoiding the median ligament of the bladder and blood vessels on the undersurface of the abdominal window, carefully make about a 2.5 cm centrally located horizontal cut along the pelvic rim. Insert your index finger and palpate with your thumb externally to ascertain the position of the pubic symphysis, which will feel like a shallow V-shaped groove.

This should correspond with the juncture of the medial muscles of the hindlegs that appears as a line overlying the symphysis.

 With a sharp scalpel, make an incision between the leg muscles, following the line formed by the muscle juncture.

You now should be able to see the symphyses.

 Begin at the pubic groove and press down with the scalpel blade until it separates. Likely, pressure will have to be applied anteriorly between the pubes and posteriorly between the ischia.

Often, the pubes and the ischia separate suddenly, so be ready to remove the pressure of the scalpel blade immediately. If a clean separation of these bones does not occur, cut through the symphysis of the os coxae with a pair of bone cutters. Now grasp the legs and break through the symphysis, exposing the organs of the pelvic cavity. Carefully clean the exposed area of fat and connective tissue, being especially careful not to remove associated blood vessels and nerves (Figure 38.4).

To observe the anatomy of the distal portion of the urogenital system, beginning at the urogenital aperture, make a cut through the lateral wall of the canal. While making this short cut, continue until you observe the opening of the urethra and vagina into the urogenital sinus (Figure 38.7). Do not cut further.

The body of the uterus tapers distally to form the **cervix,** the necklike region of the uterus, which protrudes into the vagina. The cervix can be palpated externally as a sphincterlike region.

 With forceps, gently grasp the external surface of the vagina above the point of entry of the urethra into the urogenital sinus and slide the forceps toward the body of the uterus.

Almost always the position of the cervix is indicated when the forceps abut the tip of the cervix as it projects into the vagina. The **vagina** extends from the cervix to the **urogenital sinus.** The **urethra** extends from the urinary bladder to the urogenital sinus, a space into which opens the **urethral orifice** and the **vaginal orifice,** and thereby serves as a common canal for the reproductive and urinary systems (Figure 38.7 and Figure 38.8).

 To observe these openings and other features of the urogenital system, reflect the cut ventral half of the canal.

The urogenital sinus is quite long in the cat and opens to the outside through the **urogenital aperture.** In the cat, the **labia** are slight skinfolds situated laterally around the urogenital aperture and are not easily identifiable. Notice the small papillate **clitoris** resting in a shallow, midventral depression (Figure 38.7 and Figure 38.8). This organ is partially homologous with the penis.

While describing the anatomy of the digestive system, the terminal portion of the rectum was not observable until the pelvic canal was exposed. Notice its position dorsal to the uterus as it continues to the outside through the anus. The **anal glands** are located on either side of the anus and open into the rectum (Figure 38.8). These glands are derivatives of the integumentary system. Cats use the secretions of

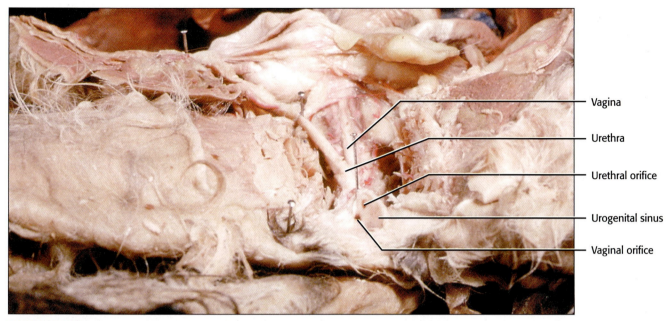

FIGURE 38.7 Female urogenital sinus with associated openings.

Vagina

Urethra

Urethral orifice

Urogenital sinus

Vaginal orifice

these glands to mark their territories and also to advertise their sex and sexual condition.

If a female cat is pregnant, has just given birth, or is suckling young, the **mammary glands**, unique in mammals, extending along either side of the midventral line, will be prominent (Figure 38.9 and Figure 38.10).

Male

The primary reproductive organs in the male are the **testes**, egg-shaped organs that are suspended externally within the **scrotum**, an obvious hairy sac protruding posteriorly just ventral to the anus (Figure 38.11). In most male mammals the testes are carried in the **scrotum**, an external sac that maintains temperatures below the higher body temperatures. (In most species, high temperatures are damaging to sperm production.) Take note also of the position of the sheath of the penis with its opening ventral to the scrotum, typical of most placental mammals (Figure 38.12). In some mammals (*e.g.*, marsupials), however, the penis is dorsal to the scrotum.

One of the major challenges encountered when studying the intact male reproductive system is to avoid severing the **spermatic cords** (Figure 38.12). Because they pass from the abdominal cavity through the abdominal wall and lie in the fat of the inguinal (groin) area, they can be cut or destroyed in the wink of an eye. This is the reason you were warned to avoid removing fatty tissue in this region during the skinning and initial connective tissue-cleaning process.

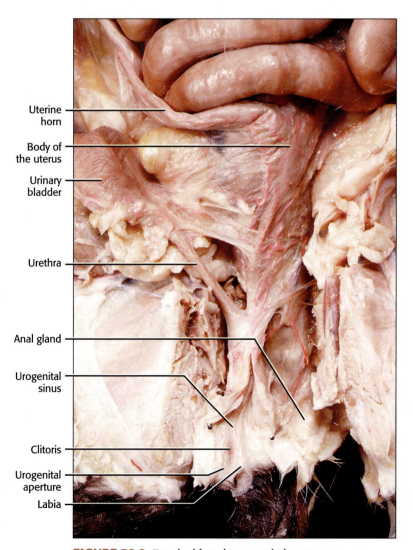

Uterine horn

Body of the uterus

Urinary bladder

Urethra

Anal gland

Urogenital sinus

Clitoris

Urogenital aperture

Labia

FIGURE 38.8 Terminal female urogenital system.

Nipples

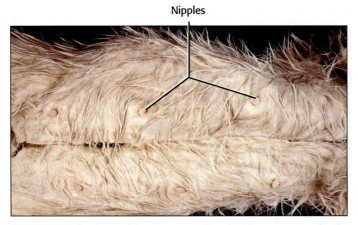

FIGURE 38.9 External features associated with mammary glands.

Mammary gland tissue

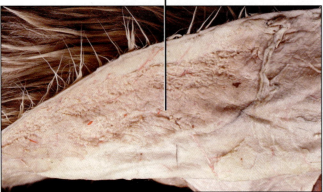

FIGURE 38.10 Internal mammary glands.

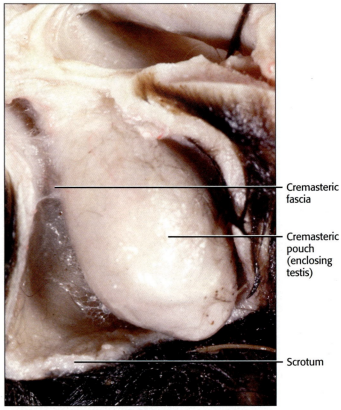

Cremasteric fascia

Cremasteric pouch (enclosing testis)

Scrotum

FIGURE 38.11 Testis in scrotum.

Dissection of the male reproductive system is somewhat more difficult and time-consuming than dissection of the female.

Begin by cautiously making a small cut (5 mm) through the skin of the posterior wall of scrotum. Carefully work the scissors under the skin and continue the incision to the anterior limit of the testis. Repeat the process on the other side. Now, carefully peel the scrotum laterally to expose the testis on both sides.

Notice the abundant connective tissue, the **cremasteric fascia,** stretching between the inner wall of the scrotum and the white sac, the **cremasteric pouch,** surrounding the testis (Figure 38.11). Observe that the pouch narrows abruptly into a tubelike structure, the **spermatic cord,** at its anterior end. Within the spermatic cord lies the vas deferens, the spermatic artery and vein, lymphatic vessels, and nerves.

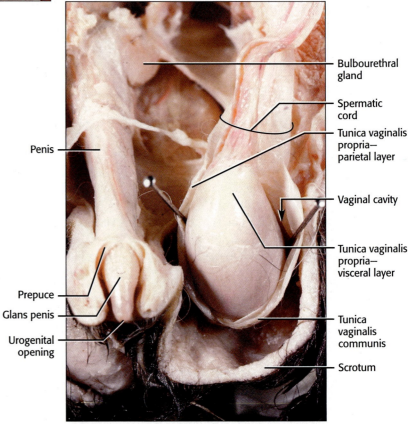

Penis

Prepuce

Glans penis

Urogenital opening

Bulbourethral gland

Spermatic cord

Tunica vaginalis propria— parietal layer

Vaginal cavity

Tunica vaginalis propria— visceral layer

Tunica vaginalis communis

Scrotum

FIGURE 38.12 Testis with associated tunics.

Leave one of the testes held in place by cremasteric fascia within the scrotum, but gently remove the other by loosening the cremasteric fascia. Do not cut the spermatic cord and detach the testis from the body of the cat. Leave it attached.

The organ you now observe is the **testis**, enclosed within the **cremasteric pouch**. The pouch and its contents have the same relationship as any of the major ventral body cavities to the body proper (*i.e.*, the abdominopelvic cavity). The tough, outer layer of the cremasteric pouch, the **tunica vaginalis communis**, is analogous to the body proper (Figure 38.11 and Figure 38.12).

 Carefully nip through this outer covering and then continue the cut to expose the structures suspended within the sac.

Notice the space between these enclosed structures and the sac, the **vaginal cavity**. It is analogous to the potential space of the peritoneal cavity. The inner lining of the pouch, the **tunica vaginalis propria–parietal layer**, is analogous to the parietal layer of the peritoneum (Figure 38.12). This layer comes together to form the mesenterylike **mesorchium**, and

then spreads over the surface of the structures suspended within the vaginal cavity as the **tunica vaginalis propria–visceral layer**, analogous to the visceral peritoneum of the abdominopelvic cavity (Figure 38.12 and Figure 38.13).

The mesorchium can be demonstrated by gently grasping the structures of the spermatic cord and pulling them medially. Notice the thin, mesenterylike membrane anchoring these structures in place within the cord. The two layers of the tunica vaginalis propria can be seen as the inner shiny layer of the cremasteric pouch (parietal layer) and the outer shiny layer covering the testis (visceral layer) (Figure 38.12).

A band of highly convoluted tubules, the **epididymis**, adheres to the dorsal portion of the testis. Observe the free anterior end, the **head**, the middle portion, the **body**, and the posterior **tail** of the epididymis. The **vas deferens** begins its journey as a somewhat convoluted tube from the tail of the epididymis and enters the spermatic cord, held in place by the **mesorchium** (Figure 38.13).

The testes in most male mammals have an interesting developmental history. The testes begin their life in the abdominal cavity, but attached to the inner portion of the scrotum by a long band of connective tissue called the **gubernaculum**, which passes through the lower abdominal wall on its way to the scrotum. As the fetus matures, the

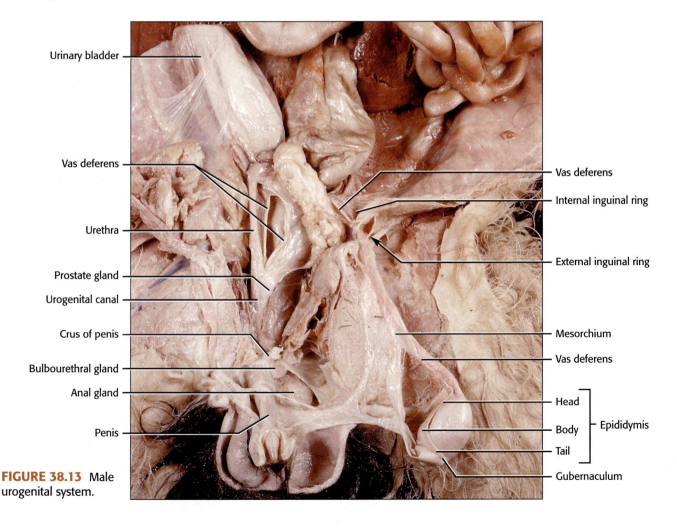

FIGURE 38.13 Male urogenital system.

gubernaculum contracts, pulling the testes into the scrotum. As the testes move into the scrotum, they are accompanied on their journey by a small bit of the peritoneal cavity (vaginal cavity) along with its parietal and visceral layers. So, really, it is not surprising that the organs suspended within the vaginal cavity have relationships similar to the space and membranes of the peritoneal cavity.

> To see the gubernaculum, while holding the testis, gently pull the cremasteric pouch wall at the posterior end of the testis and observe the short, tough, fibrous band holding the tail of the epididymis tightly against the inner wall of the pouch (Figure 38.13).
>
> Now, carefully separate the spermatic cord from the surrounding tissues and trace it anteriorly to an opening piercing the abdominal wall, the external inguinal ring.

The cord passes through a short **inguinal canal**, and the vas deferens enters the abdominal cavity through the **internal inguinal ring**. Notice also that the spermatic artery and vein enter and exit the spermatic cord at this point. Follow the vas deferens as it courses dorsally around each of the ureters and the urinary bladder (Figure 38.3 and Figure 38.13).

The next dissection should be completed prior to further work.

> Grasp the opening of the sheath of the penis with a pair of forceps, and make a cut through the anterior wall of the sheath with scissors (Figure 38.13). Continue the cut through the connective tissue along the shaft of the penis to expose it. Take care to avoid cutting into the penis itself.

To observe the urethra, glands, and other associated parts of the system, cut through the symphysis of the os coxae.

> As in the female, carefully make about a 2.5 cm cut through the abdominal wall along the rim of the pelvis, avoiding the median ligament of the bladder and blood vessels associated with the inner portion of the abdominal window that was cut when the cat was opened. Carefully palpate the slight depression between the pubes and, with a sharp scalpel, cautiously make an incision, beginning with the depression and following the raphe between the gracilis muscles. Remember that the cut through the raphe will be quite deep.
>
> Now grasp the hindlegs and reflect them laterally to complete the separation of the two os coxae. You may have to complete the separation by carefully cutting through the ischiatic symphysis

with the scalpel. At all times, be careful not to cut into organs lying within the pelvic canal.

With some luck and good dissection, you should be looking into the canal.

> Begin cleaning connective tissue from the tube leading from the urinary bladder, being careful to conserve associated blood vessels and nerves in the process.

As you proceed distally, you will encounter an enlarged, whitish mass, the **prostate gland**, which creates prostatic fluid, a major component of semen. Dorsally, this is the point at which the **vasa deferentia** connect to the urethra. The part of the tube extending from the urinary bladder to the prostate, the **urethra**, carries only urine. The part of the tube posterior to the prostate, the **urogenital canal**, continues to the tip of the penis, where it opens to the outside through the urogenital aperture. The urogenital canal carries both urine and semen (Figure 38.13).

The distal portion of the urogenital canal is surrounded by a column of erectile tissue, the corpus spongiosum, which lies along the posterior surface of the penis when it is not erect. The distal portion of this erectile tissue caps the tip of the penis as the conical **glans penis**. The **urogenital aperture** opens at the tip of the glans. Notice the pocket of skin, the **prepuce**, that encloses the glans (Figure 38.12).

A pair of erectile columns of tissue, the corpora cavernosa, lie along the anterior surface of the penis when it is not erect. All three of these columns possess blood sinuses that bring about erection under the influence of sexual excitement. The proximal ends of the corpora cavernosa are attached to the ischia by bands of tough connective tissue called the **crura** of the penis (singular: **crus**). Each crus is covered ventrally by the ischiocavernosus muscle. Dorsal to each of the crura is a **bulbourethral gland** (Cowper's gland), which secretes lubricating fluid into the urogenital canal during sexual excitement (Figure 38.13).

> To expose the bulbourethral glands, it is necessary to use scissors to cut through the crus on one side. Pull the cut end ventrally and observe the gland. It often is necessary to clean connective tissue from its surface.

As in the female, the male possesses **anal glands** situated near the terminal end of the large intestine, near the anus (Figure 38.13). These serve a similar function in males. Perhaps you have seen large cats, such as lions and tigers, in those wildlife documentaries on television as they back up to a bush or shrub and spray fluid on them. The fluid is ejected forcefully from their anal glands. Unneutered male domestic cats do something similar on furniture and other objects in their home territories, leading to odoriferous surroundings.

CHAPTER THIRTY-NINE

Cat — Circulatory and Lymphatic Systems

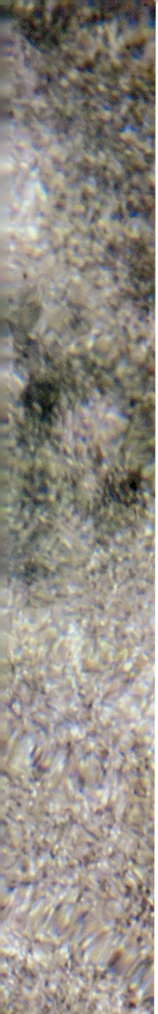

Although mammalian hearts are all similar in morphology, the anatomy described is that of the adult sheep in the discussion that follows. Our recommendation is that the heart be cut with a sharp kitchen knife into dorsoventral halves, preferably by your instructor. Review Chapter 11.

THE HEART

External Anatomy

The mammalian heart consists of two muscular pumps, the **right atrium** and the **right ventricle**, which receive and pump oxygen-poor blood from the body to the lungs and the **left atrium** and **left ventricle**, which receive oxygen-rich blood from the lungs and pump it throughout the body (Figure 39.1 and Figure 39.2).

The heart is shaped somewhat like a blunt wedge, with a flattened **base** and more acute **apex** (Figure 39.1). In the living mammal the heart projects into a membranous sac, the pericardium, in a potential space called the mediastinum. The outer layer of the pericardium is fibrous and may be covered by the parietal pleura, as in the cat. Lining the potential pericardial space into which the heart projects is the serous membrane known as the parietal pericardium. The parietal pericardium reflects over the heart surface as the visceral pericardium.

The walls of the heart consist of three layers. The outer covering is a single layer of squamous epithelium that is really nothing more than the visceral layer of pericardium, also known as the epicardium or outermost layer of heart tissue. The muscular middle layer, or myocardium, makes up almost the entire tissue of the heart. The inner lining, continuous with the inner lining of circulatory vessels, also consists of a single layer of squamous epithelium known as the endothelium.

Before beginning to study the heart, determine which side is dorsal and which is ventral. Remember that the heart, as it rests in the normal living sheep's body, has a dorsal and ventral aspect. In mammals, no matter how they walk, the **pulmonary trunk,** a major blood vessel that exits from the **right ventricle**, stretches from right to left across the ventral side of the heart. Another vessel, the **aorta,** generally of greater diameter and slightly dorsal to the pulmonary trunk, also curves left. From the aorta branches the **brachiocephalic artery** (Figure 39.1).

 Carefully pick away fat between the pulmonary trunk and aorta until you discover a short ligamentous band, the ligamentum arteriosum, connecting the two vessels.

This represents the remnant of a fetal circulatory bypass, the ductus arteriosus, which allowed most of the blood pumped into the pulmonary trunk during intrauterine life to be shunted into the systemic circulation and away from the nonfunctional lungs (Figure 39.1). If you examine the dorsal side, you will probably notice that it has more vessels—the four **pulmonary veins** and the **anterior vena cava** and **posterior vena cavas** (Figure 39.2). In many specimens, the dorsal vessels sometimes are cut close to the surface of the heart, leaving only openings or holes.

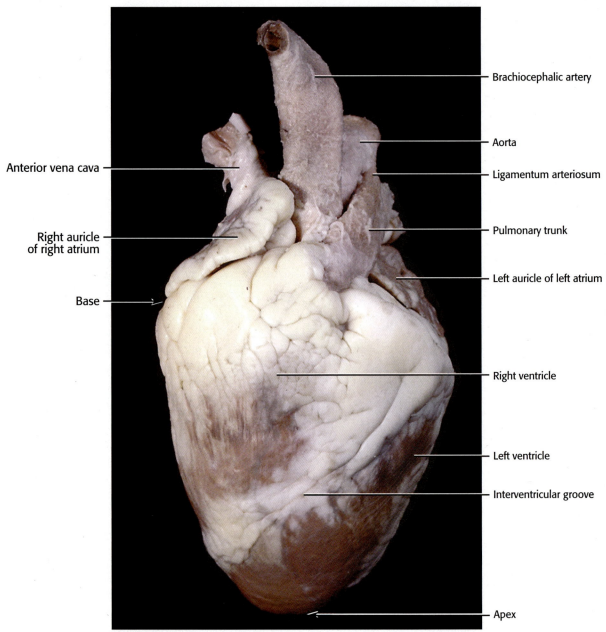

FIGURE 39.1 Heart: Ventral view.

A shallow **atrioventricular groove**, or **coronary sulcus**, marks the separation of the heart into the two anterior atria and the two posterior ventricles (Figure 39.2). Encircling the heart in this groove is the main venous drainage of the heart, the **coronary sinus**. Venous blood returns from heart muscle to the right atrium through this channel. **Interventricular grooves** on the ventral and dorsal surfaces demarcate the division of the ventricle into right and left chambers (Figure 39.1 and Figure 39.2). Major coronary blood vessels lie in these grooves, generally buried in a considerable amount of fat.

Note that these grooves are oriented almost parallel to the left margin of the heart, giving the appearance that the left ventricle is larger. This is not an optical illusion. The thicker muscle in the left ventricle is correlated with the higher pressures observed in the systemic circulation. The resistance in the systemic circulation is far greater than in the pulmonary circuit, and high pressures are needed to pump oxygen-rich blood throughout the systemic circuit. This disparity in ventricular thickness is obvious upon examination of the cut surface of the muscular walls of both chambers. Also obvious are two earlike flaps, the **auricles**, attached to the atria (Figure 39.1 and Figure 39.2). They increase the volume of each atrium somewhat.

Internal Anatomy

In our treatment of the internal morphology, we will describe the anatomy with respect to the incoming blood

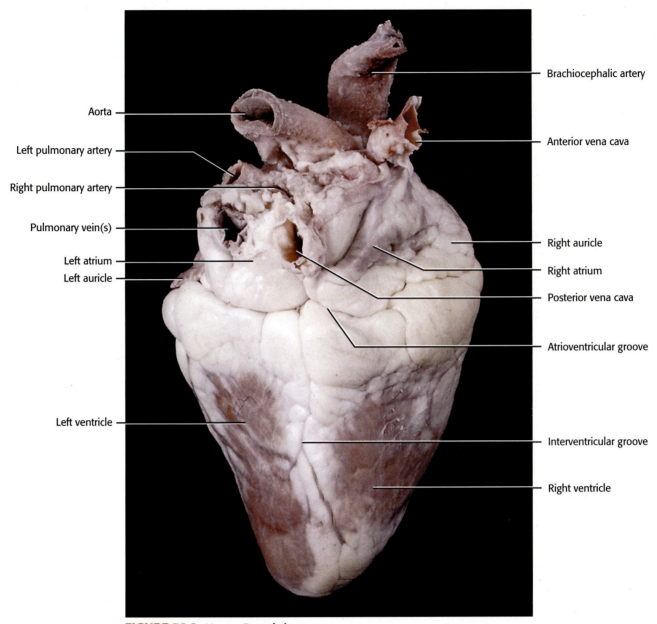

FIGURE 39.2 Heart: Dorsal view.

flow through the right pump, followed by the outgoing blood flow through the left pump. Be aware that both pumps function at the same time.

Blood enters the **right atrium** from the cranial region of the body by way of the **anterior vena cava** and from the caudal region of the body via the **posterior vena cava** (Figure 39.3). Locate the openings of these two major vessels in the inner wall of the right atrium. If a substantial section of these vessels remain attached to your specimen, notice that the posterior vena cava curves along the dorsal aspect of the lower left side of the heart, turning cranially to enter the right atrium.

In addition, locate the **opening of the coronary sinus**, situated caudal to the **opening of the posterior vena cava** (Figure 39.3). Venous blood from the heart itself drains

into the right atrium through this opening. A depression, somewhat lighter in color than the surrounding tissue, sits just ventral to the entrance of the posterior vena cava in the **interatrial septum**. This area, the **fossa ovalis**, represents the position of an opening between the right and left atria, the **foramen ovale**, which functions as a shunt during embryonic development (Figure 39.3).

Notice that the atrial wall is thin compared to the ventricle wall. The inner wall of the auricle appears honeycombed because of the presence of muscular ridges, the **musculi pectinati** (Figure 39.4). These muscular ridges also extend into the atrium.

Blood flows from the right atrium into the right ventricle through the **right atrioventricular** or **tricuspid valve**, taking its name from three flaplike structures or cusps: the

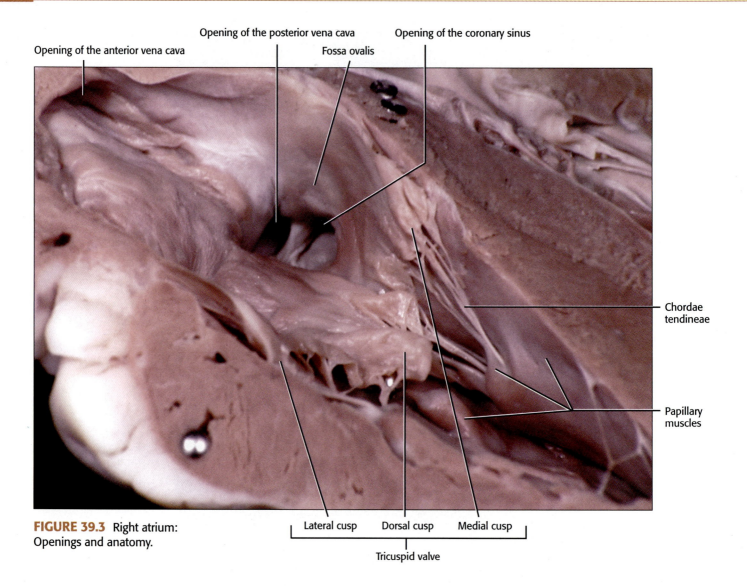

FIGURE 39.3 Right atrium: Openings and anatomy.

Labels: Opening of the anterior vena cava · Opening of the posterior vena cava · Fossa ovalis · Opening of the coronary sinus · Chordae tendineae · Papillary muscles · Lateral cusp · Dorsal cusp · Medial cusp · Tricuspid valve

dorsal cusp attached to the dorsal wall, the **lateral cusp** attached to the lateral wall and the **medial cusp** attached to the medial wall (Figure 39.3 and Figure 39.4). Attached to the apex of each cusp are cordlike structures, the **chordae tendineae**, extending to cone-shaped **papillary muscles** in the wall of the right ventricle (Figure 39.3 and Figure 39.4). During ventricular filling, the ventricle is relaxed and the tricuspid valve is open. Therefore, most of the ventricular filling is passive, but the final ventricular volume results from atrial contraction.

As the right atrium relaxes and the right ventricle contracts, high pressure in the ventricle has a tendency to promote backflow of blood into the atrium, causing the valve to slam shut, but contraction of the papillary muscles pulling on the chordae tendineae prevents eversion of the cusps into the atrium. Notice that the inner walls of the ventricle are crisscrossed with muscular bands, the **trabeculae carneae**. A slender band of tissue, the **moderator band**, extends between the lateral and medial wall of the right ventricle (Figure 39.5). Not all mammals have a moderator

band. The moderator band probably prevents over-expansion of the ventricle.

Note that the **myocardium** of the right ventricle is thinner than that of the left ventricle, since pressures are lower in the right ventricle because blood pumped by the right side travels only to the lungs located close to the heart (Figure 39.6).

As the right ventricle contracts, blood is pumped through the **pulmonary semilunar valve** into the pulmonary trunk. Three membranous pockets, each in the shape of a half-moon, form this valve. Following contraction, pressure in the ventricle decreases as it relaxes, causing the blood under high pressure in the pulmonary trunk to backwash into the ventricle. This backwash is prevented by blood filling the pockets of the semilunar valve, which now slams shut. Blood is transported to the lungs, where gaseous exchange occurs.

From the lungs, blood returns through the **pulmonary veins** to the left atrium. The two veins from each lung become confluent as they enter the atrium, sometimes giving

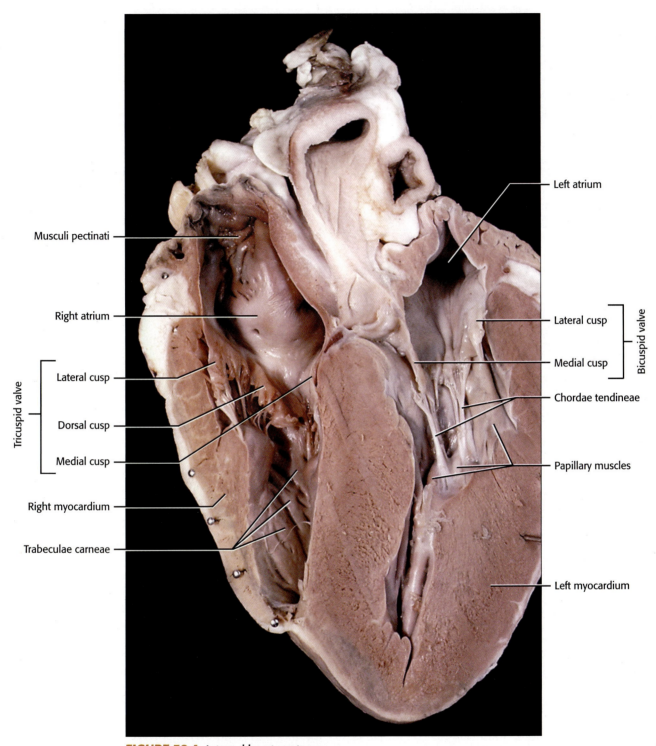

FIGURE 39.4 Internal heart anatomy.

the appearance of only two openings. Locate the openings of the pulmonary veins on the posterior aspect of the atrium (Figure 39.2). Often, the pulmonary veins are removed close to the heart and appear only as holes in the left atrium.

The inner appearance of the left atrium is similar to the right atrium, with the exception that the musculi pectinati seem to be less extensive. They are prominent in the left auricle, however. As in the right pump, blood flows passively from the left atrium into the left ventricle—in this case, through the two cusps of the **left atrioventricular** or **bicuspid valve**. A **medial cusp** and a **lateral cusp** are attached to those respective walls (Figure 39.6). In a manner similar to construction of the tricuspid valve, chordae tendineae extend between the margins of the cusps to papillary muscles

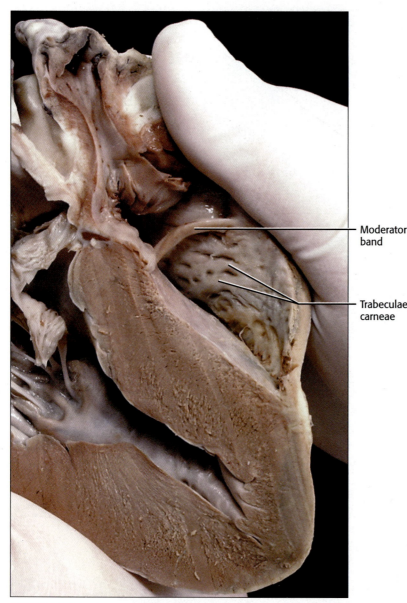

FIGURE 39.5 Right ventricle: Moderator band.

protruding from the muscular walls of the ventricle. Contraction of the left atrium accomplishes the final filling of the ventricle.

As the atrium relaxes and the ventricle contracts, the pressure of the engorged ventricle supersedes that of the atrium and blood backwash again is prevented when the cusps of the bicuspid valve slam shut. Contraction of the papillary muscles resulting in a pull on the edges of the cusps prevents eversion of the cusps into the left atrium.

The obviously thicker myocardium of the left ventricle is one of its most distinguishing features. Although it cannot be observed readily, an **interventricular septum** separates the myocardium of the two ventricles, indicated externally by the interventricular grooves. **Trabeculae carneae** again are a feature of the inner surface of the chamber (Figure 39.5). Contraction of the left ventricle forces blood past

the **aortic semilunar valve** into the **aorta** (Figure 39.6). The morphology of this semilunar valve is identical to that of the pulmonary semilunar valve. As the ventricle relaxes, the blood pressure in the aorta exceeds that of the ventricle and blood backwashes toward the ventricle, filling the membranous pockets and causing them to slam shut.

Continuous pumping of the heart throughout the life of a mammal demands a constant supply of highly oxygenated blood to this hard-working muscular organ. This requirement is met by delivery of the most highly oxygenated blood by way of the **left** and **right coronary arteries**, which originate immediately above the **aortic semilunar valve**. Examine the medial and lateral walls of the aorta, where you will observe the openings of these two vessels (Figure 39.6). You may see a number of small openings in the cut ventricular walls, which represent **coronary vessels**.

Blood Vessels

The architecture of the circulatory system exhibits a consistent pattern of divergence or branching in the arterial portion to supply the tissues of the body, and convergence or confluence in the venous portion of the system to drain those tissues. Another common theme is that arteries and veins are paired when the organs they supply or drain are paired (*e.g.,* each kidney is supplied by an artery and a vein). You also will notice that a vessel that is continuous from its origin to its destination may have different names as it passes from one area to another (*e.g.,* the subclavian becomes the axillary as it enters the armpit but then is called the brachial as it enters the arm, and so on). This is common in the terminology of the circulatory system. To identify blood vessels correctly, it is necessary to trace them to or from the tissues they supply.

In a triply injected specimen, arteries are red, veins are blue, and the hepatic portal system, consisting of the veins of most of the abdominal organs, have been injected with yellow latex. Interspecific as well as intraspecific variability in branching patterns of arteries and veins occurs in mammals.

 Therefore, you must examine all of the specimens in your class.

Pulmonary Circulation

Pulmonary circulation involves arterial vessels carrying oxygen-poor blood to the lungs and venous vessels carrying

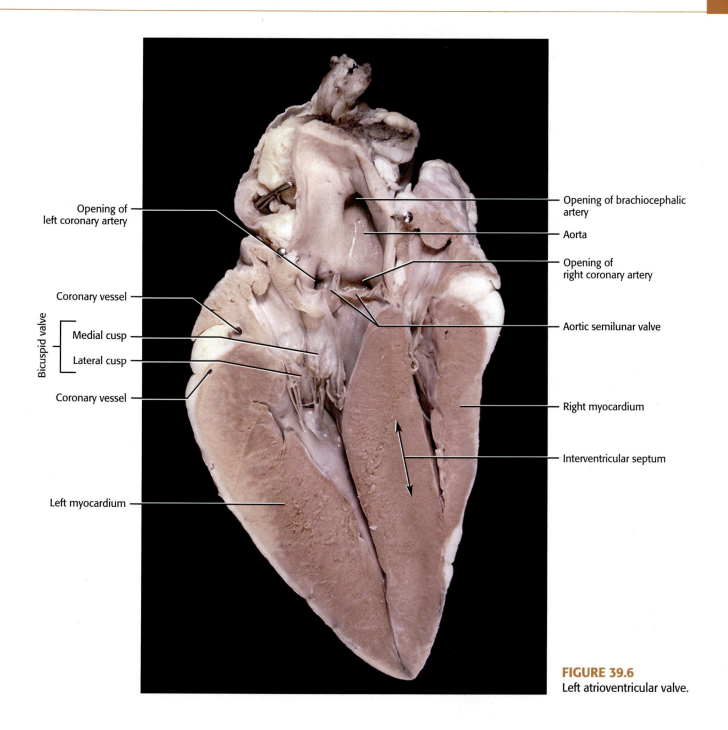

Opening of
left coronary artery

Coronary vessel

Bicuspid valve
- Medial cusp
- Lateral cusp

Coronary vessel

Left myocardium

Opening of brachiocephalic
artery

Aorta

Opening of
right coronary artery

Aortic semilunar valve

Right myocardium

Interventricular septum

FIGURE 39.6
Left atrioventricular valve.

oxygen-rich blood back to the heart. Note that these vessels follow the anatomical rule in which blood is carried back to the heart by veins and away from the heart by arteries. In contrast to the blood vessels of the rest of the body of an adult, however, the oxygen level of the blood carried in the pulmonary arteries is low and it is high in the pulmonary veins. These arteries are injected with red latex, even though the blood transported by them is low in oxygen.

To observe these vessels, cut the pericardial sac with a pair of scissors.

Identify the **pulmonary trunk** of the cat exiting from the right ventricle. Dorsal to the heart, the pulmonary trunk bifurcates into **right** and **left pulmonary arteries**. Trace these vessels into the root of their respective lungs, where they branch further, supplying the lobes of the lungs. In the root of the lung, identify the **pulmonary veins** as they leave the lungs and enter the left atrium (Figure 39.7C).

Systemic Circulation

Systemic circulation carries oxygen-rich blood from the heart to the body in arteries and returns oxygen-poor blood to the heart in veins (Figure 39.7A and 39.7B). To appreciate

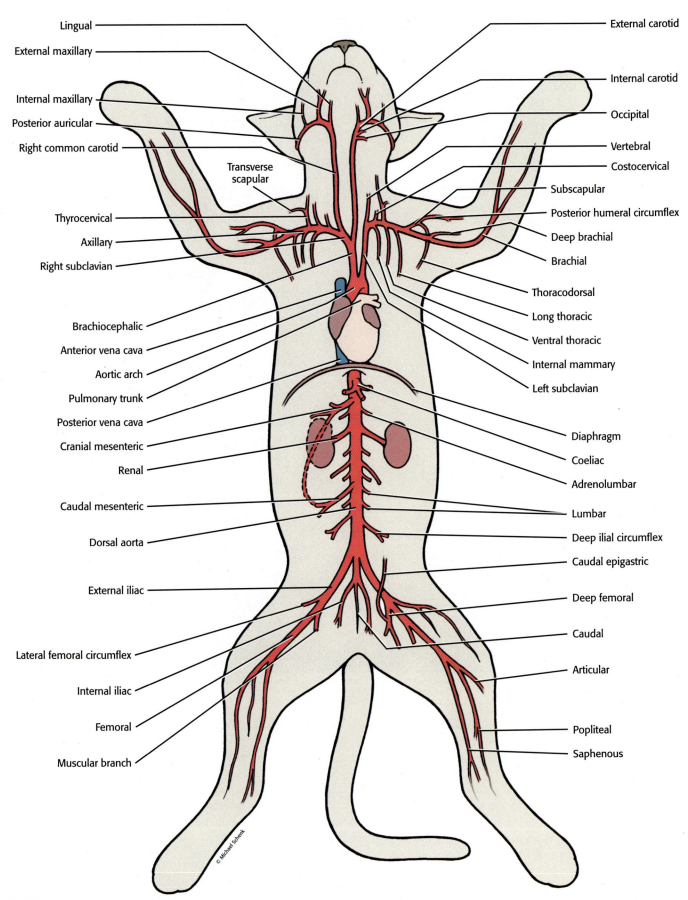

FIGURE 39.7A Arterial schematic.

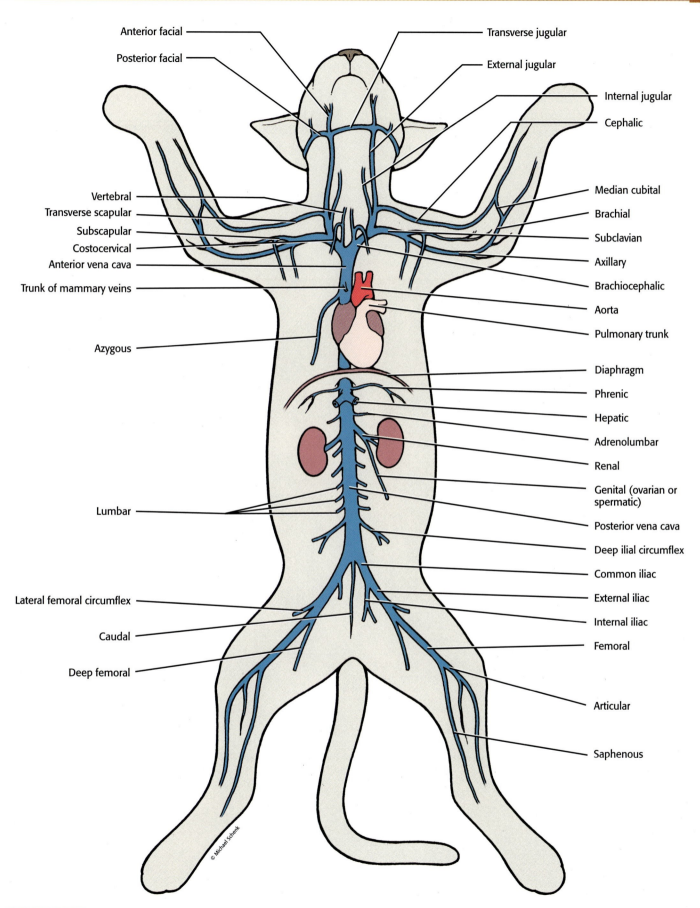

Anterior facial
Posterior facial
Transverse jugular
External jugular
Internal jugular
Cephalic
Vertebral
Transverse scapular
Subscapular
Costocervical
Anterior vena cava
Trunk of mammary veins
Azygous
Median cubital
Brachial
Subclavian
Axillary
Brachiocephalic
Aorta
Pulmonary trunk
Diaphragm
Phrenic
Hepatic
Adrenolumbar
Renal
Genital (ovarian or spermatic)
Posterior vena cava
Deep ilial circumflex
Common iliac
External iliac
Internal iliac
Femoral
Articular
Saphenous
Lumbar
Lateral femoral circumflex
Caudal
Deep femoral

© Michael Schenk

FIGURE 39.7B Venous schematic.

the enormity and complexity of the circulatory system, we will begin our exploration of the arterial component of this circuit by identifying the **aorta** as it leaves the left ventricle ventrally and curves dorsally to the left as the **aortic arch** (Figure 39.7C). The aorta then continues posteriorly, passing through the diaphragm into the peritoneal cavity, generally following the middorsal line of the body.

The portion of the aorta in the thoracic region may be identified as the **thoracic aorta** and the segment in the peritoneal area may be identified as the **abdominal aorta.** In mammals the left fourth aortic arch persists and is functional, whereas the right aortic arch disappears and is represented only as the base of the subclavian artery. By contrast, in birds the right aortic arch is the functional arch (see Figure 11.7).

As the aorta begins to curve dorsally, identify the **brachiocephalic artery,** followed by the **left subclavian artery** branching from the arch, supplying the neck, head, and

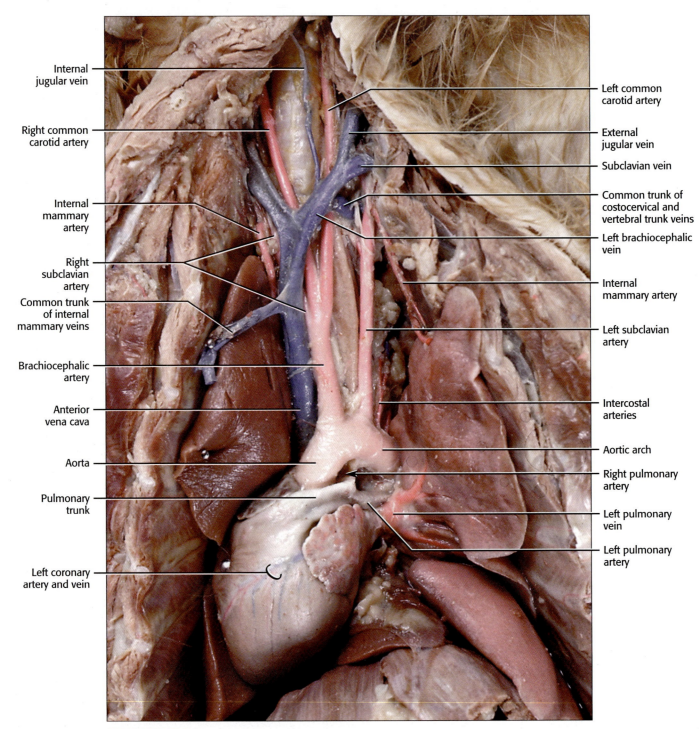

FIGURE 39.7C Vessels of the thoracic area.

forelimbs. The brachiocephalic artery, at about the level of the second rib, subdivides into the **right subclavian artery**, the **right common carotid artery** and the **left common carotid artery** (Figure 39.7C). In other mammals the aortic arch may have different branching patterns.

Four small arteries branch from the left subclavian artery before it emerges from the thoracic cavity. In general, dissection of this portion of the circulatory system is minimal and consists mainly of picking connective tissue and fat from the surface of the vessels and nerves.

> As usual, this activity is to be done carefully. Note that nerves often accompany the blood vessels and appear as shiny cream-colored strands of tissue. These can be destroyed easily, so be careful!

The first branch, the **internal mammary artery,** leaves ventrally and enters the ventral thoracic body wall (Figure 39.8).

As it courses posteriorly, it gives off branches to nearby muscles, pericardium, mediastinum, and diaphragm, and then anastomoses with the caudal epigastric arteries (described later).

The second branch, the **vertebral artery,** arises as a dorsal artery, continues cranially and dorsally, enters the transverse foramen of the sixth vertebra, giving off branches to nearby neck muscles and spinal cord segments as it passes through the transverse foramina of the cervical vertebrae.

As it nears the foramen magnum, it joins with the left vertebral artery to form the basilar artery, which courses along the midventral aspect of the medulla oblongata.

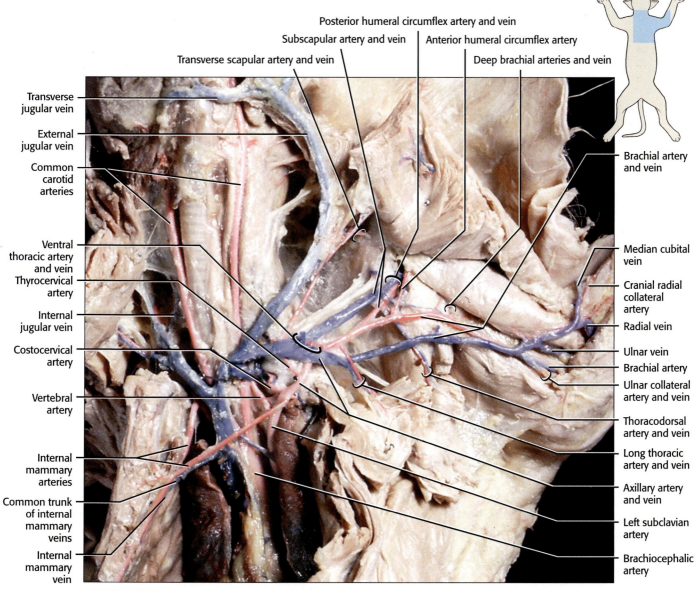

FIGURE 39.8 Vessels of the thoracic and brachial area (left side).

The third branch, the **costocervical artery**, also emerges from the dorsal surface of the subclavian artery, curves dorsally, and subdivides to send branches to neighboring neck, back, and intercostal muscles.

The last branch of the subclavian artery, the **thyrocervical artery**, leaves the anterior aspect of that vessel, travels anteriorly and laterally, and as it approaches the shoulder, becomes known as the **transverse scapular artery.** These vessels supply blood to adjacent muscles of the region, particularly the shoulder.

As the subclavian artery curves around the first rib and enters the axilla, it becomes known as the **axillary artery.** From this artery, several others branch. The **ventral thoracic artery** emerges from the ventral surface of the axillary artery and courses posteriorly to supply pectoral muscles. Notice that the anterior ventral thoracic nerve of the brachial plexus accompanies this artery. A short distance laterally, the **long thoracic artery** branches and, with the posterior ventral thoracic nerve of the brachial plexus, supplies the pectoral muscles, the thoracic mammary glands, and the latissimus dorsi muscle.

The relatively large **subscapular artery** leaves the axillary as it nears the shoulder. The **posterior humeral circumflex artery** usually branches from the subscapular and, accompanied by the axillary nerve of the brachial plexus, travels beneath the biceps brachii muscle carrying blood to several muscles of the medial aspect of the scapula, the shoulder, and triceps brachii muscle.

A second branch of the subscapular artery, the **thoracodorsal artery**, arises dorsal to the brachial plexus, giving off branches to the teres major and latissimus dorsi muscles (Figure 39.8). The subscapular artery continues deep to supply dorsal shoulder muscles.

Beyond this point, the axillary artery becomes known as the **brachial artery** as it passes into the arm. The **anterior humeral circumflex artery** arises almost immediately as a branch of the brachial artery to bring blood to the biceps brachii muscle. At about the midpoint of the biceps brachii, the **deep brachial artery** originates and, with the radial nerve of the brachial plexus, passes under the belly of that muscle to supply blood to the triceps brachii, latissimus dorsi, and other muscles in the area. Occasionally, the deep brachial artery originates from the subscapular artery.

A number of muscular branches emerge from the brachial artery as it runs parallel with the median nerve of the brachial plexus until it nears the elbow, where a pair of arteries, the cranial **radial collateral** and **ulnar collateral**, are given off. The radial collateral artery turns laterally and supplies the arm extensor muscles while the ulnar collateral artery turns medially and supplies blood to the muscles of the elbow (Figure 39.8). With the median nerve, the brachial artery passes through the supracondyloid foramen. Below this point it subdivides into an ulnar, a radial, and an interosseous artery.

The left common carotid artery can be seen closely associated with the internal jugular vein (which may not be well-injected), the vagus nerve, and the sympathetic trunk, and is bound to them by a fibrous sheath. Follow the common carotid cranially to the level of the thyroid gland and find a medial branch, the **cranial thyroid artery,** supplying the thyroid gland, parathyroid glands and neck muscles, and directly opposite it, find the **muscular branch** passing laterally carrying blood to deep neck muscles (Figure 39.9).

The next medial branch is the **laryngeal artery**, supplying muscles in the laryngeal area. A variable number of branches may originate from the lateral side, carrying blood to muscles and lymph nodes in the area.

 To view the cranial branches of the common carotid artery, remove the sternomastoid and cleidomastoid muscles on one side.

Portions of the digastric muscle, mandibular gland, and parotid gland and lymph nodes usually must be dissected away from the vessels. The hypoglossal nerve runs across the surface of the common carotid just anterior to the next pair of vessels that branch from it, and serves as a landmark for the **occipital** and **internal carotid arteries** (Figure 39.9) These vessels usually are very small, hairlike branches and may leave the dorsal part of the common carotid individually or may leave as a common stem that subsequently branches into two arteries.

Care must be exercised when probing for them because there is a great deal of connective tissue in the area and these vessels often are not much more than 2–3 mm apart. The occipital artery originates first (more posteriorly) and passes caudal to the tympanic bulla carrying blood to the neck and occipital region. The internal carotid artery passes cranial to the tympanic bulla and transports blood to the base of the diencephalon. Along with several other arteries, this artery contributes to the major arterial blood supply of the brain in the form of the Circle of Willis.

The common carotid now continues as the **external carotid artery,** following the branching of the internal carotid artery. A large branch, the **lingual artery**, leaves the external carotid. This artery carries blood to muscles of the hyoid and pharynx while continuing into the tongue as its major supply. The more dorsal **external maxillary artery** branches from the external carotid and carries blood to the outer facial region (Figure 39.9).

As the external carotid nears the posterior boundary of the masseter muscle, the **posterior auricular artery** emerges and supplies the outer ear area. The external carotid artery continues along the margin of the masseter, gives off the **superficial temporal artery** that delivers blood to the masseter and outer ear, and continues deep to the masseter as the **internal maxillary artery**, which supplies internal

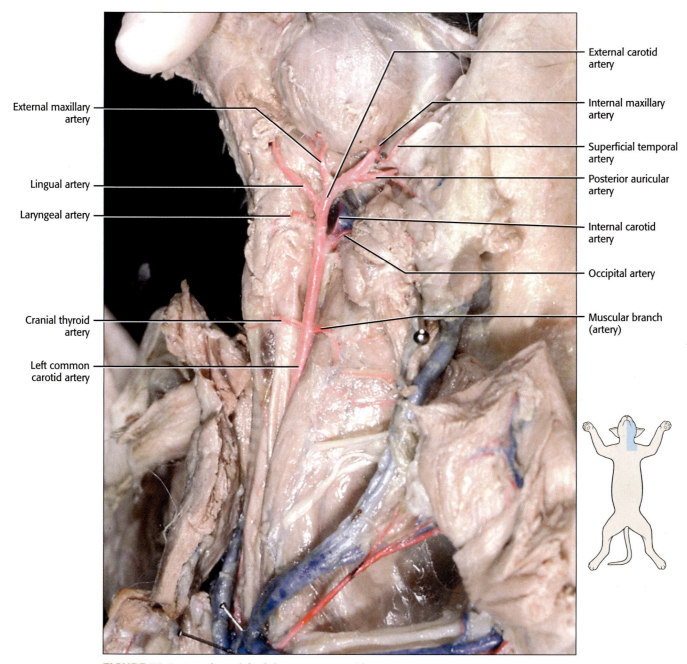

External maxillary artery

Lingual artery

Laryngeal artery

Cranial thyroid artery

Left common carotid artery

External carotid artery

Internal maxillary artery

Superficial temporal artery

Posterior auricular artery

Internal carotid artery

Occipital artery

Muscular branch (artery)

FIGURE 39.9 Branches of the left common carotid artery.

structures and tissues associated with the upper and lower jaw, nose, eyes, and surrounding regions.

We have just described the circulation of the left subclavian artery and the left common carotid artery. The branches of the right common carotid artery and the right subclavian artery are similar.

For the most part, arteries and veins lie in close proximity. Therefore, as an artery is identified, the corresponding vein, often with the same name, should be identified. In general, the venous system parallels the arterial system in this portion of the body with the following exceptions (Figure 39.7C, Figure 39.8, and Figure 39.10).

▪ Tributaries draining the brain, auricular, labial, lingual, palatal, and dental regions of the head form **anterior** and **posterior facial veins** that coalesce to form the **external jugular vein.**

▪ In the hyoid region, a **transverse jugular vein** draining that area connects the anterior facial veins.

▪ As the external jugular vein approaches the heart, it is joined by the **transverse scapular vein**, draining the shoulder region.

▪ The brain and surrounding tissues are drained by tributaries of the **internal jugular vein** joining the external jugular. This vein lies close to the trachea and it may not be well-injected.

Drainage of the forelimb involves deep and superficial vessels. The superficial veins include the **cephalic vein,** which drains the lateral aspect of the forelimb and is connected to the **brachial vein** by the **median cubital vein.** The **cephalic vein** continues superficially to join with the **posterior humeral circumflex vein** and the **transverse scapular vein.**

The deep veins of the forelimb parallel the arterial circulation (Figure 39.8 and Figure 39.11).

The brachial-axillary-subclavian vein joins the external jugular vein to form the **brachiocephalic vein** (Figure 39.7C and Figure 39.8). In contrast to the arterial supply, there are two brachiocephalic veins. The **costocervical** and **vertebral veins** merge to form a common trunk that empties into the brachiocephalic vein. Note the absence of a thyrocervical vein. The left and right brachiocephalic veins join to form

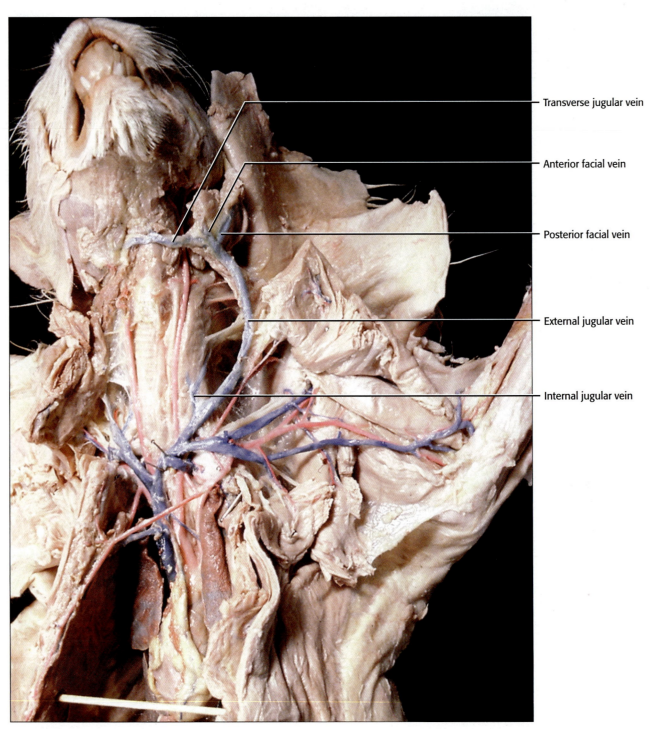

Transverse jugular vein

Anterior facial vein

Posterior facial vein

External jugular vein

Internal jugular vein

FIGURE 39.10 Veins of the head, neck, and thorax.

the **anterior vena cava**. The **left** and **right internal mammary veins** merge to form a common trunk and join the ventral surface of the anterior vena cava. Although the pattern of the venous and arterial circulation differs somewhat, the regions served are similar.

In *Necturus*, drainage of the anterior portion of the body is accomplished by jugular veins that drain into anterior cardinal veins, draining into common cardinal veins. In the mammal, drainage occurs via jugulars that empty into a short anterior vena cava representing the common cardinal and anterior cardinal veins of *Necturus*. Some mammals have a left and a right anterior vena cava.

Examine the inner thoracic wall and observe the segmental **intercostal arteries** and **veins** associated with the

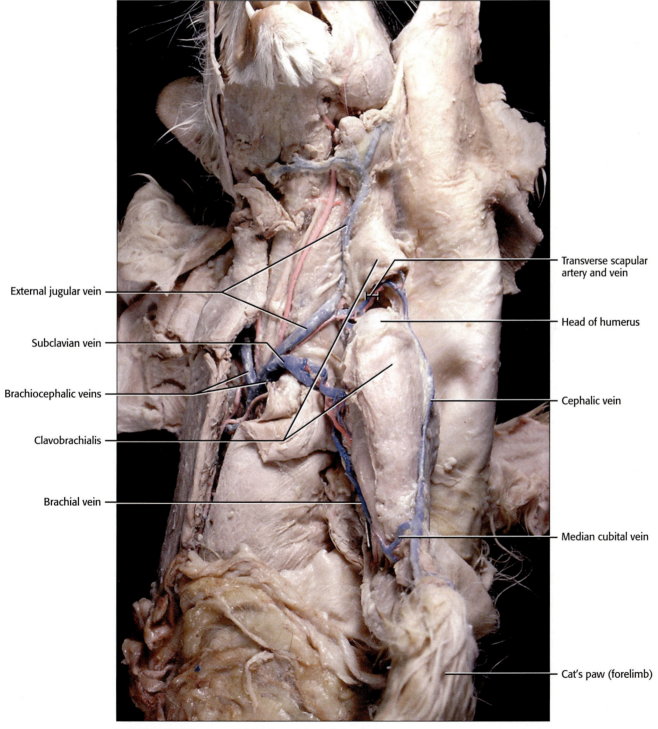

FIGURE 39.11 Superficial veins of the left forelimb.

External jugular vein

Subclavian vein

Brachiocephalic veins

Clavobrachialis

Brachial vein

Transverse scapular artery and vein

Head of humerus

Cephalic vein

Median cubital vein

Cat's paw (forelimb)

blood supply to the tissues between the ribs. The fairly large **azygous vein** runs along the right side of the aorta in the thoracic region and receives blood from most of the intercostal veins on both sides (Figure 39.12). It empties into the dorsal region of the anterior vena cava. The origin of the azygous vein is actually in the dorsal abdominal area, but it is best seen in the thoracic region and therefore is discussed here.

Some mammals not only have an azygous vein on the right side but also a hemiazygous vein on the left side, with a number of interconnecting vessels. An accessory hemizygous vein also is present. The azygous and hemiazygous veins are remnants of the anterior portions of the posterior cardinal veins.

As the aorta travels through the abdominopelvic area, it gives rise to three unpaired major arterial branches—the **coeliac artery**, the **cranial mesenteric artery**, and the **caudal mesenteric artery**—to visceral organs of the region in most vertebrates and in all mammals (Figure 39.13A, Figure 39.13B, and Figure 39.13C). In *Necturus*, the branching pattern of the visceral arteries is different, although it supplies the same organs as in mammals. A major difference is the combination of the coeliac and a mesenteric trunk, the coeliacomesenteric artery, whose branches largely parallel the visceral blood supply of the coeliac and cranial mesenteric arteries seen in mammals.

 To observe the complex subbranching of the coeliac artery in the cat, carefully cut through the diaphragm to the esophagus, avoiding vessels associated with the diaphragm. Free the esophagus from connective tissue anchoring it in place, and cut across the esophagus at a point about 2 cm anterior to the stomach. Pull the stomach ventromedially, carefully freeing it, with scissors, from the diaphragm and surrounding tissues.

To begin dissecting the coeliac and its branches, first locate the coeliac artery, then carefully pick away the connective tissue to expose its three major branches. The most difficult to locate is the **hepatic artery**. The first branch of the coeliac artery is covered by tough connective tissue. It ascends toward the liver, where it subdivides into several hepatic branches that supply the lobes of the liver and the **cystic artery** lying on the surface of the cystic duct. Use great care in removing connective tissue to expose these delicate branches.

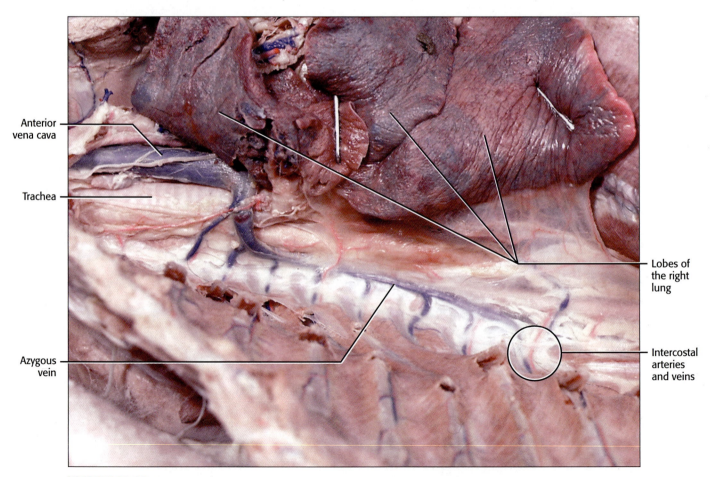

FIGURE 39.12 Azygous vein.

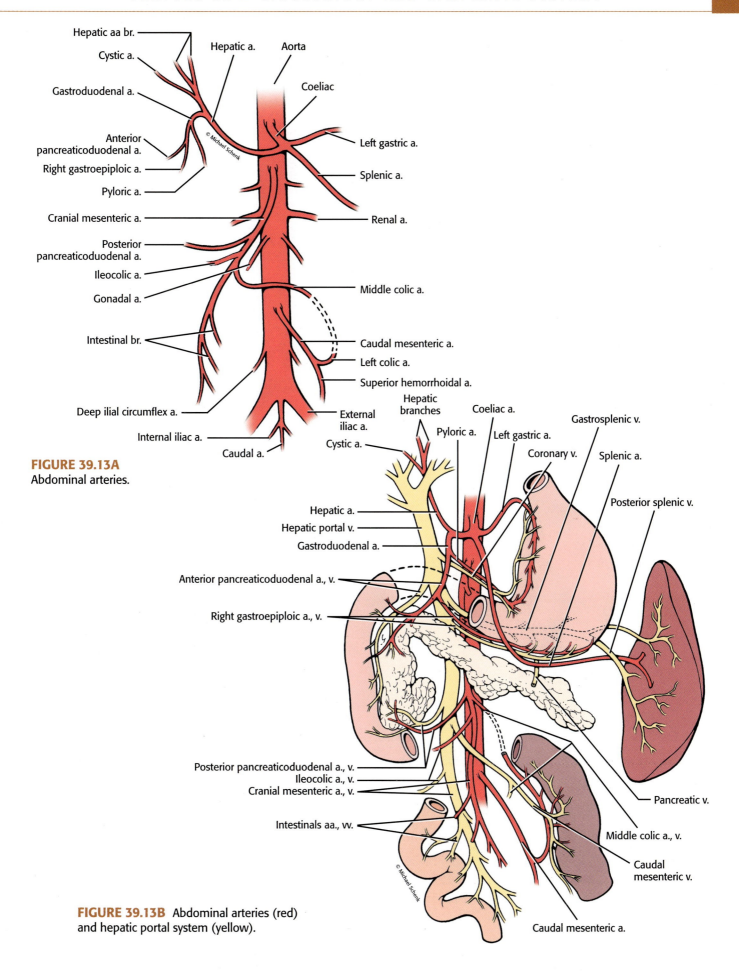

FIGURE 39.13A
Abdominal arteries.

FIGURE 39.13B Abdominal arteries (red)
and hepatic portal system (yellow).

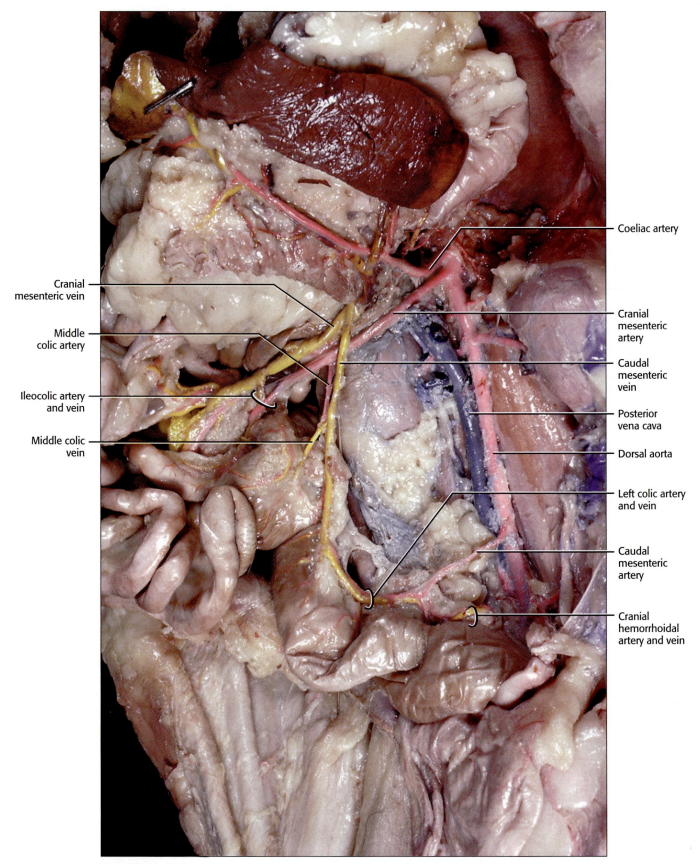

Coeliac artery

Cranial mesenteric vein

Middle colic artery

Ileocolic artery and vein

Middle colic vein

Cranial mesenteric artery

Caudal mesenteric vein

Posterior vena cava

Dorsal aorta

Left colic artery and vein

Caudal mesenteric artery

Cranial hemorrhoidal artery and vein

FIGURE 39.13C Distribution of the three abdominal arteries (organs reflected to cat's right).

As the hepatic artery travels toward the liver, observe a short **gastroduodenal artery** curving posteriorly and almost immediately giving off 1–3 delicate **pyloric artery(ies)** that supplies blood to the pyloric region of the stomach, continuing along the lesser curvature of the stomach and anastomosing with the left gastric artery. A few millimeters from the pyloric artery, the gastroduodenal artery bifurcates into the **right gastroepiploic artery** supplying the greater curvature of the stomach and greater omentum and the **anterior pancreaticoduodenal artery** supplying the head of the pancreas and the duodenum (Figure 39.14B). These arteries are generally more robust than the pyloric artery, but you still must exercise great care in exposing them.

The **left gastric artery** leaves the coeliac artery carrying blood to the lesser curvature of the stomach. A **gastric artery** transporting blood to the greater curvature of the stomach arises close to the left gastric artery. The largest and most obvious branch of the coeliac is the **splenic artery** leading toward the spleen. Notice the numerous small branches supplying the tail of the pancreas and the body of the stomach.

The main portion of the artery travels along the length of the spleen, giving off an **anterior splenic artery** and continuing as the **posterior splenic artery** (Figure 39.14A and Figure 39.14B). A continuation of the posterior splenic artery, the **left gastroepiploic artery**, follows along the greater curvature of the stomach and anastomoses with the right gastroepiploic artery (Figure 39.15).

Although the description of the coeliac artery and its branches is applicable to most mammals, including the human, the pattern may vary, with additional small branches occurring.

The second unpaired, somewhat larger major abdominal artery is the **cranial mesenteric artery**, extending farther than the coeliac artery before supplying blood to the intestines and pancreas. The first branch is the **posterior pancreaticoduodenal artery** supplying the pancreas and duodenum and anastomosing with the anterior pancreaticoduodenal artery (Figure 39.16).

 Carefully expose this artery by picking away the connective and pancreatic tissue along the duodenal border.

Follow the cranial mesenteric artery as it subdivides into numerous **intestinal arteries**, supplying the jejunum and ileum (Figure 39.16). The next branch, the **ileocolic artery**, supplies the terminal portion of the ileum, the cecum, the ascending colon, and proximal portions of the transverse colon. Occasionally a separate **right colic artery**, branching from the cranial mesenteric artery, supplies the ascending and proximal portions of the transverse colon. The last branch of the cranial mesenteric artery is the **middle colic artery**, supplying the transverse and descending colon

(Figure 39.16). Continuity among these colic vessels is the result of anastomoses.

The third major abdominal artery, the **caudal mesenteric artery**, is the smallest and least complexly branched. This vessel can be masked by the mesentery and usually must be exposed by picking away this tissue. The branches of the caudal mesenteric include the **left colic artery**, supplying blood to the distal portion of the descending colon, and the **cranial hemorrhoidal artery**, supplying the proximal portion of the rectum (Figure 39.13C). Notice the anastomosis that may occur between the left and middle colic arteries.

Drainage of the viscera, including the intestines, stomach, and spleen, is accomplished by the hepatic portal system. Therefore, substances, such as breakdown products of digestion, toxins, vitamins, or any chemical that is absorbed across the walls of the digestive system, are transported to the liver. Because the hepatic portal system has capillaries on either end, it requires a special injection of yellow latex directly into the system to be able to dissect and identify the vessels described here. As with all vessels, you must carefully remove associated tissue to expose this system. Be careful not to destroy nearby vessels.

Blood is drained from the large intestine by the **caudal mesenteric vein**. The portion of the caudal mesenteric that drains the rectum and distal descending colon is the **cranial hemorrhoidal vein**, and the proximal portion of the descending colon is drained by the **left colic vein**. The **cranial mesenteric vein** drains the transverse and ascending colon, cecum, small intestine, and head of the pancreas by way of the following branches. A **middle colic vein**, draining the transverse and often the distal portion of the ascending colon, joins the caudal mesenteric vein. If your cat has a right colic artery, it more than likely will also possess a **right colic vein**, draining the ascending colon. Drainage of the cecum and ileum is by way of the **ileocolic vein**. Numerous **intestinal veins** drain the coils of the small intestine. The **posterior pancreaticoduodenal vein** drains blood from the head of the pancreas and the distal portion of the duodenum. Note that the caudal mesenteric vein joins the cranial mesenteric vein, in contrast to the arterial circuit, where each branches independently from the aorta (Figure 39.13 and Figure 39.16).

The stomach, the spleen, and the pancreas are drained by the **gastrosplenic vein**. Locate the **anterior splenic** and **posterior splenic veins** draining their respective regions of the spleen. Find the **left gastroepiploic vein**, paralleling the artery of the same name, draining the greater curvature of the stomach and greater omentum and joining the posterior splenic vein (Figure 39.15).

A variable number of **gastric veins** drain the body of the stomach and also join the posterior splenic vein (Figure 39.15). The pancreas may be drained by one or more veins that empty into the gastrosplenic vein before it joins the cranial mesenteric vein to form the **hepatic portal vein**. In most cats, the **gastroduodenal vein**, formed by the juncture of the **right gastroepiploic** and **anterior pancreaticoduodenal veins,** joins the hepatic portal vein shortly after its formation. The right gastroepiploic vein, draining the right end of the greater curvature of the stomach, and anterior pancreaticoduodenal vein, draining the tail of the pancreas and the proximal portion of the duodenum, may join the hepatic

portal vein, independently, shortly after its formation. The **coronary vein,** draining the lesser curvature of the stomach, joins the hepatic portal independently. A small pyloric vein, draining the pyloric portion of the stomach and emptying into the gastroduodenal vein, also may be present. Follow the hepatic portal vein into the liver (Figure 39.15). The branching pattern of the hepatic portal system varies among mammals.

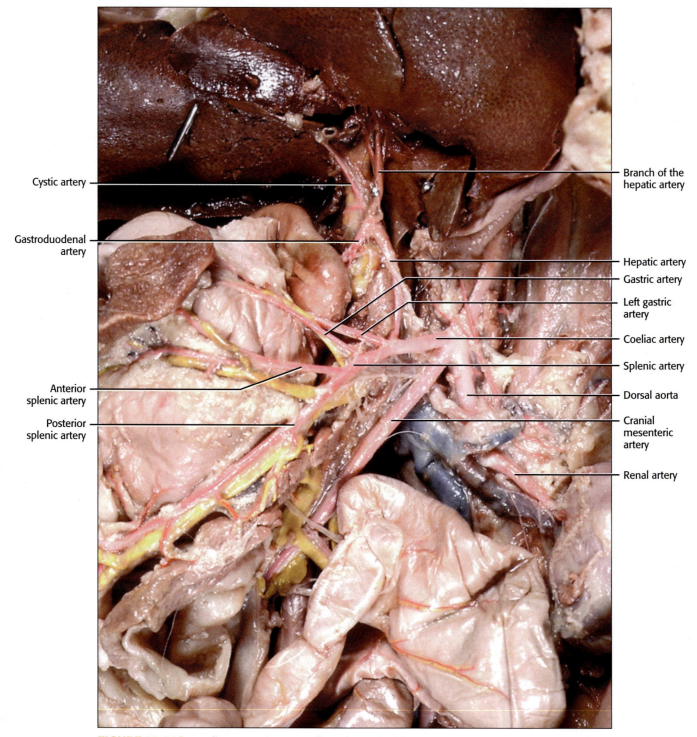

FIGURE 39.14A Coeliac artery (organs reflected to cat's right).

In addition to the unpaired arteries are several paired vessels that typically occur at various levels along the length of the abdominal aorta. **Adrenolumbar arteries** arise on either side of the body a few centimeters posterior to the diaphragm and just posterior to the **cranial mesenteric artery** (Figure 39.16). Their name stems from the destination of their branches. A **posterior phrenic artery** services the diaphragm, and an **adrenal artery** takes blood to the adrenal gland. The adrenolumbar artery then continues laterally to supply the dorsal body wall. A number of variants can occur in a large class. Find the complementary veins, usually lying in close proximity to the artery.

A pair of **renal arteries** emanates from the aorta at the level of the kidneys. Because the right kidney is somewhat

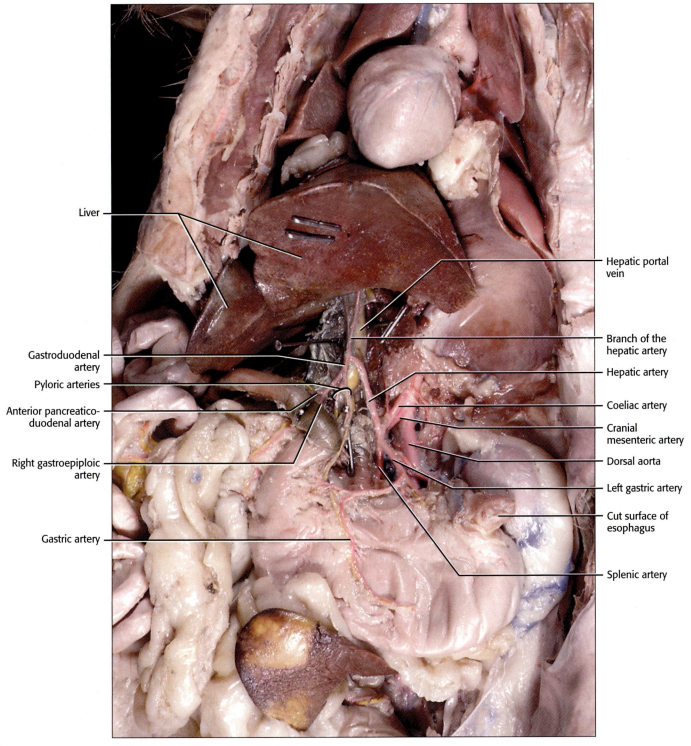

Liver

Gastroduodenal artery

Pyloric arteries

Anterior pancreatico-duodenal artery

Right gastroepiploic artery

Gastric artery

Hepatic portal vein

Branch of the hepatic artery

Hepatic artery

Coeliac artery

Cranial mesenteric artery

Dorsal aorta

Left gastric artery

Cut surface of esophagus

Splenic artery

FIGURE 39.14B Branches of the coeliac artery.

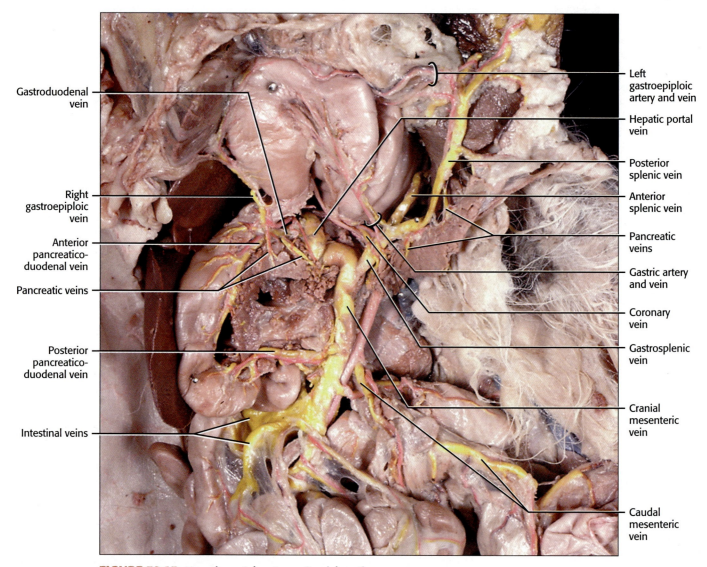

FIGURE 39.15 Hepatic portal system: Cranial portion.

anterior to the left kidney, its renal artery is also slightly anterior to the left kidney. One of the variations mentioned in the preceding paragraph may be the origin of the adrenal artery from these vessels. Identify the renal veins that run parallel to the renal arteries (Figure 39.17).

As we follow the aorta caudally, the next paired arteries that are given off supply the primary sex organs of the cat. In the male, small-diameter **internal spermatic arteries** leave the aorta and extend toward the internal inguinal ring, where they accompany the vas deferens, nerves, lymphatic vessels, and **spermatic veins**, all of which are wrapped in connective tissue known as the **spermatic cord** and lead to the testes (Figure 39.18). The configuration of these vessels may vary in males. Not only may the right artery originate more posteriorly, but also the left may originate from the left renal artery.

The female **ovarian arteries** are of larger diameter, usually originating closer to the level of the position of the

ovary and then extending laterally and supplying blood to the ovary, uterine tube, or oviduct and uterine horns (Figure 39.19). An anastomosis between the branches of the uterine horns and the uterine branch of the middle hemorrhoidal artery is common.

Several pairs of **lumbar arteries** originate from the aorta in the lower back region. They supply blood to the dorsal muscles.

The final conspicuous pair of abdominal vessels originating from the aorta just posterior to the caudal mesenteric artery consists of the **deep ilial circumflex arteries**. Each extends laterally across the iliopsoas muscles taking blood to the dorsal body wall (Figure 39.20).

At about the level of the sacrum, the aorta gives off a large **external iliac artery**, which extends toward the hindlimb on either side of the body. In contrast to the matching arterial and venous pattern of the pelvic and hindlimb region, a major exception should be noted: There is no common iliac artery, but there is a **common iliac vein** (Figure 39.22). Just before the external iliac artery emerges through the body wall from the body cavity, it gives off a **deep femoral artery**. Generally, three branches emanate from the deep femoral artery (Figure 39.21).

1. The **caudal epigastric artery**, which extends along the abdominal surface of the rectus abdominis muscle and anastomoses with a branch of the internal mammary artery (sometimes called the cranial epigastric artery).
2. The **external pudendal artery**, embedded in inguinal fat and carrying blood to the bladder and external genitalia.
3. A deep vessel, the **medial femoral circumflex artery**, carrying blood to the "hamstring" muscles (the biceps femoris, the semitendinosus, and the semimembranosus), as well as the adductor femoris and tenuissimus muscles.

You may notice a small branch, the external spermatic artery, joining the spermatic cord as it makes its journey to the scrotum in the male.

When the external iliac artery appears outside the body wall on the medial surface of the leg, it now is called the **femoral artery**. The first of several branches, the **lateral femoral circumflex artery**, emerges laterally and carries blood to the quadriceps complex (rectus femoris, vastus

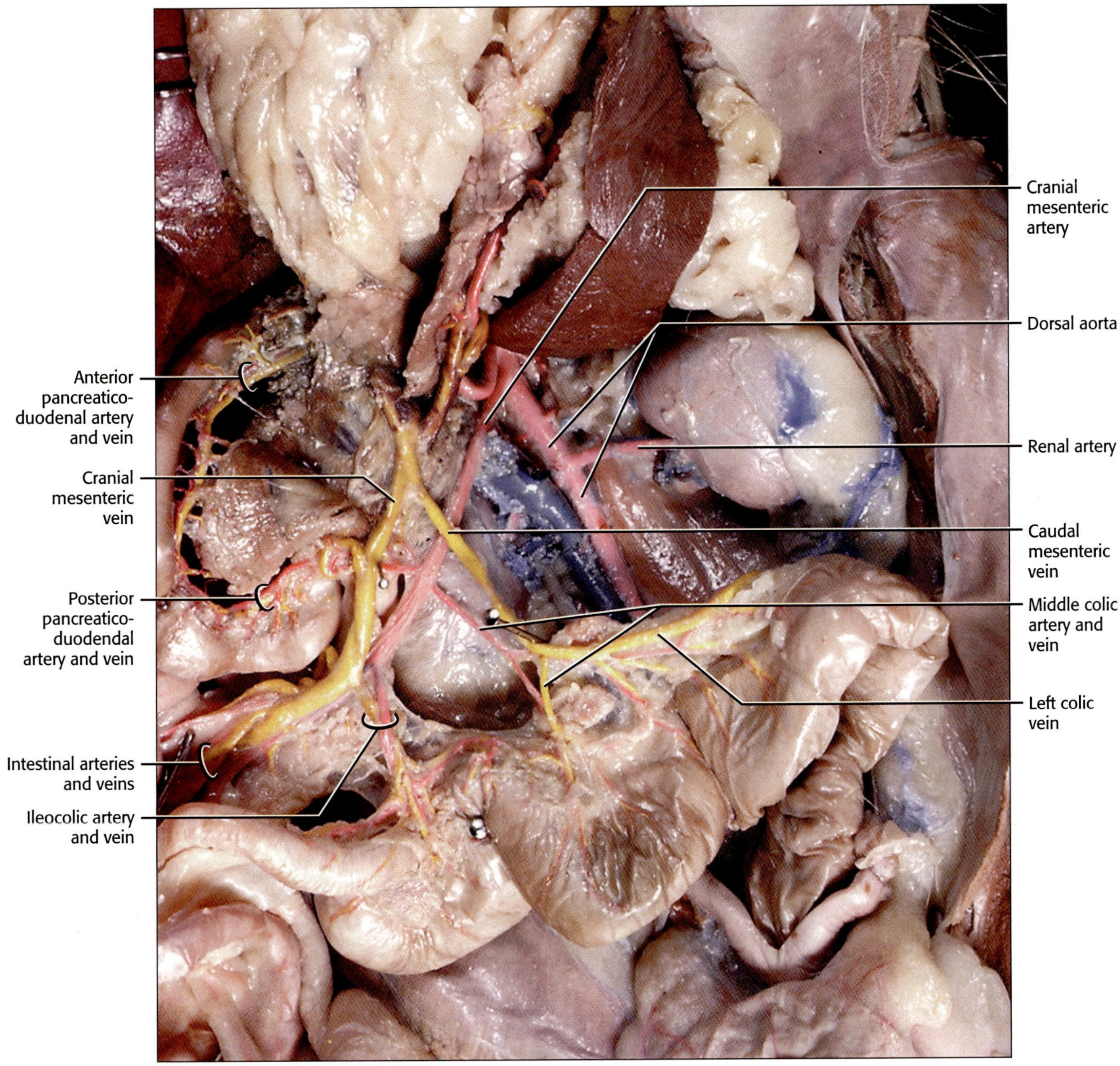
Cranial mesenteric artery
Dorsal aorta
Renal artery
Caudal mesenteric vein
Middle colic artery and vein
Left colic vein
Anterior pancreaticoduodenal artery and vein
Cranial mesenteric vein
Posterior pancreaticoduodenal artery and vein
Intestinal arteries and veins
Ileocolic artery and vein

FIGURE 39.16 Cranial mesenteric artery.

lateralis, vastus medialis, and vastus intermedius), as well as the sartorius and tensor fascia latae muscles. A somewhat more distal medial branch, the **muscular artery**, supplies blood to the adductors, gracilis, and semimembranosus muscles.

As the femoral artery approaches the knee, it gives off the **articular artery** (Figure 39.21), which transports blood to the gracilis, semimembranosus, and vastus lateralis muscles. At this point or slightly distal, the superficial **saphenous artery** originates from the femoral and courses over the medial surface of the thigh and shank. In the region of the knee, the femoral artery continues between the vastus medialis and semimembranosus, becoming the **popliteal artery.** A number of branches of this artery supply

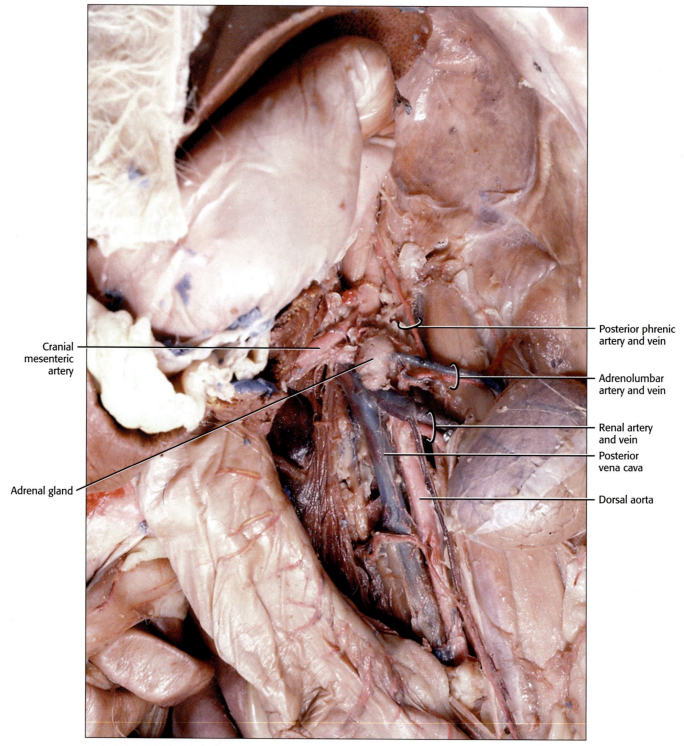

Cranial mesenteric artery

Adrenal gland

Posterior phrenic artery and vein

Adrenolumbar artery and vein

Renal artery and vein

Posterior vena cava

Dorsal aorta

FIGURE 39.17 Renal and adrenolumbar vessels (left side).

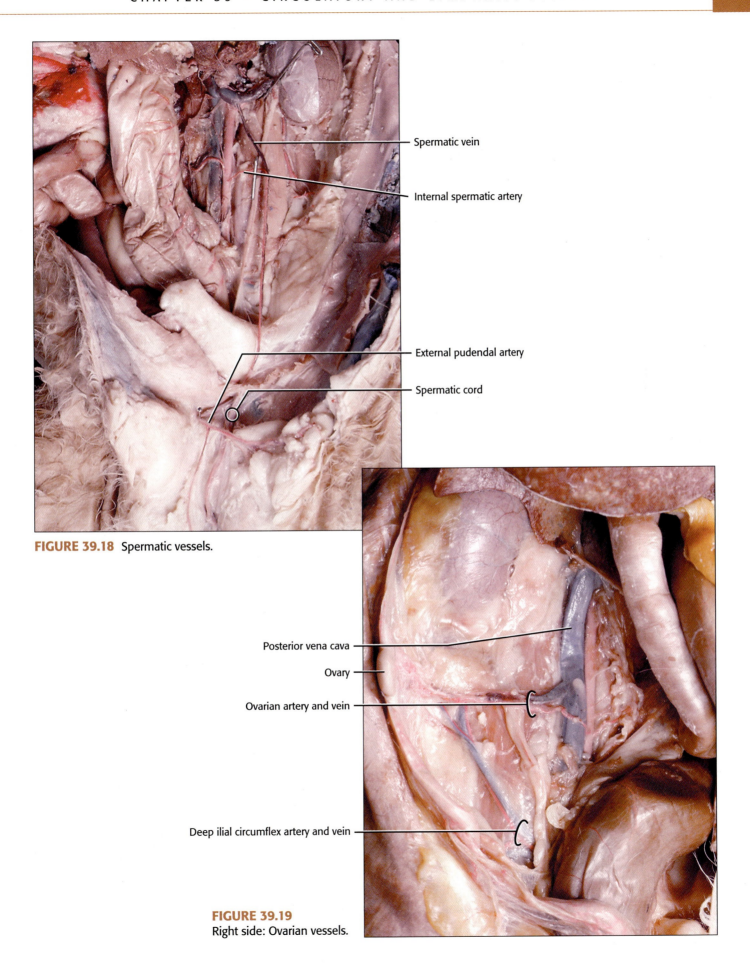

Spermatic vein

Internal spermatic artery

External pudendal artery

Spermatic cord

FIGURE 39.18 Spermatic vessels.

Posterior vena cava

Ovary

Ovarian artery and vein

Deep ilial circumflex artery and vein

FIGURE 39.19
Right side: Ovarian vessels.

blood to muscles of the shank, foot, and thigh (Figure 39.21).

Now return to the point at which the external iliac arteries originated and note that the aorta persists as a smaller version of its abdominal self and dives deeply into the dorsal portion of the pelvic cavity where it is obscured by fat and connective tissue. When you dissected the reproductive system and separated the two os coxae, you undoubtedly observed some of the distal branches of this vessel. After reading the text and checking Figure 39.21 and Figure 39.22 to observe these vessels, you will have to perform your usual careful job of slowly picking away the fat and connective tissue. When you have done so, you will see that almost immediately a pair of **internal iliac arteries**

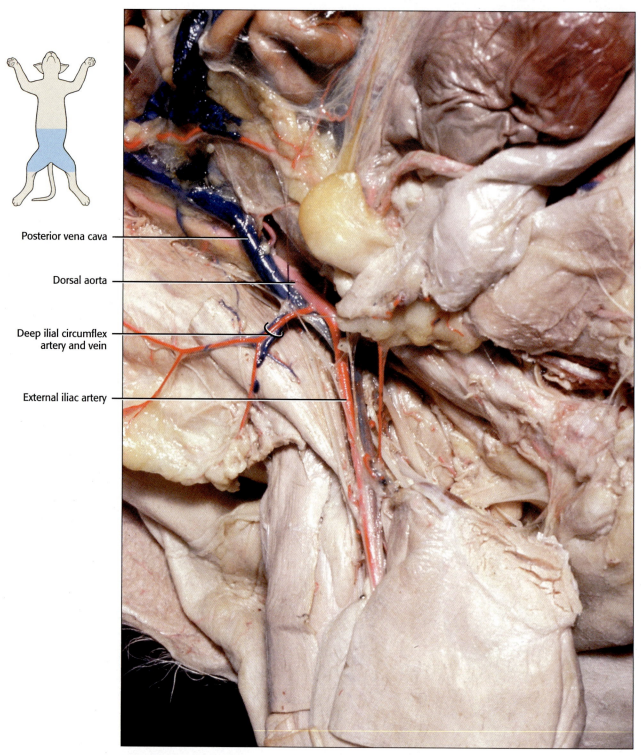

Posterior vena cava

Dorsal aorta

Deep ilial circumflex artery and vein

External iliac artery

FIGURE 39.20 Right side: Deep iliocircumflex artery.

is given off, and then the aorta persists as the small **median sacral artery** continuing along the dorsal aspect of the pelvic cavity, and then along the ventral region of the tail as the **caudal artery**.

The first branch of the internal iliac artery, branching almost immediately, is the **umbilical artery**, extending laterally, entering the fat of the lateral vesical ligament of the bladder and supplying blood to the urinary bladder. This artery represents the remnant of the umbilical arteries that carry blood to the placenta during fetal development (Figure 39.22).

The next branch is the **cranial gluteal artery**, which turns lateral and caudal, taking blood to hip muscles such as the gluteals, pyriformis, and some thigh muscles.

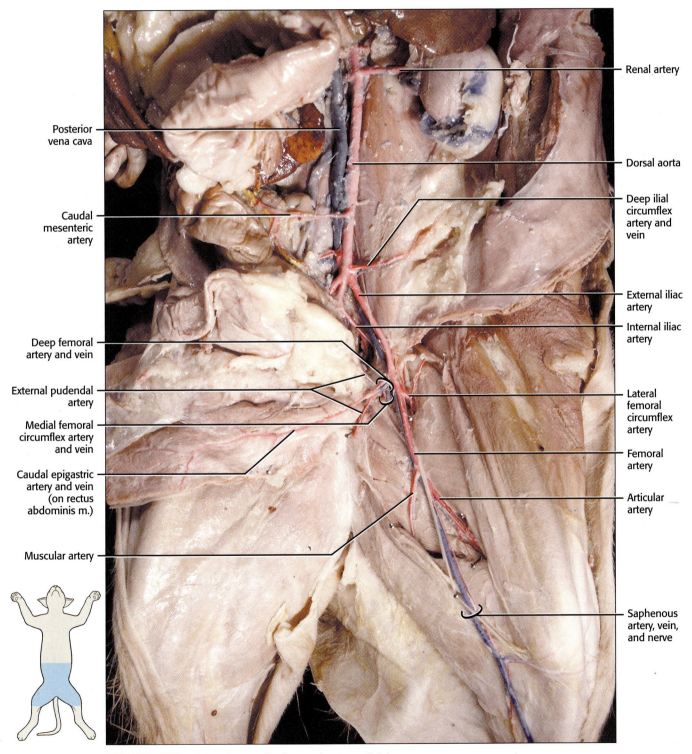

Posterior vena cava

Caudal mesenteric artery

Deep femoral artery and vein

External pudendal artery

Medial femoral circumflex artery and vein

Caudal epigastric artery and vein (on rectus abdominis m.)

Muscular artery

Renal artery

Dorsal aorta

Deep ilial circumflex artery and vein

External iliac artery

Internal iliac artery

Lateral femoral circumflex artery

Femoral artery

Articular artery

Saphenous artery, vein, and nerve

FIGURE 39.21 Left side: Hindlimb vessels: Superficial.

The next vessel, the **middle hemorrhoidal artery**, extends medially toward various structures of the reproductive system, urinary bladder, and rectum. In females, particularly those that were pregnant or had been pregnant recently, a conspicuous sub-branch, the **uterine artery**, extends cranially along the uterus and anastomoses with uterine branches of the ovarian artery.

Finally, the **caudal gluteal artery** represents the end of the iliac artery. It supplies blood to gluteal and pyriformis muscles, as well as the region at the base of the tail (Figure 39.22).

Identify the branches of the **external** and **internal iliac veins** and observe that they merge into the **common iliac vein**. The two common iliac veins join to form the **posterior vena cava** (Figure 39.22). As the posterior vena cava progresses cranially, note that the veins draining adjacent organs and tissues merge with it, adding their blood to its volume. These veins should have been identified already, along with the arteries of the same name. In its passage dorsal to the liver, it is joined by a pair of **hepatic veins** draining the liver and contributing blood from the abdominal viscera that was carried by the hepatic portal vein to the liver (Figure 39.23). To observe the hepatic veins, you will have to carefully scrape tissue from the cranial surface of the liver. Follow the posterior vena cava into the right atrium.

The posterior vena cava in mammals has a complex embryological origin and is a composite of some homologous vessels seen in *Necturus* in addition to some new vessels seen only in embryonic mammals. Included in the posterior vena cava of mammals are portions of the posterior cardinals, renal portals, subcardinals, and hepatics, found in *Necturus*, and supracardinals, appearing for the first time among mammals.

The renal portal system is absent in adult mammals. The posterior portion of the body and kidneys drain directly into the posterior vena cava. In addition, adult mammals do not possess a ventral abdominal vein, but in the embryo the ventral abdominal vein is the source of the fetal umbilical vein(s). The umbilical vein(s) is the venous conduit leading from the placenta associated with the female reproductive lining to the fetus. This vein carries oxygen- and nutrient-rich blood from female to fetus.

LYMPHATIC SYSTEM

Because specimens with injected lymphatic vessels are difficult to obtain, treatment of this system will be confined largely to a discussion. Lymphatic nodes, glands, and organs can be identified, however. Although not discussed previously, other vertebrates have lymphatic systems. They are even more difficult to detect than that of the mammal.

The architecture of the lymphatic system consists of a network of thin-walled, highly permeable vessels located throughout the body. Small capillaries called lacteals project into the finger-like villi of the small intestine and are involved in fat absorption. The lymphatic fluid is returned to the circulatory system by way of a number of smaller vessels that coalesce into the larger thoracic duct opening into the venous system near the left subclavian vein.

A large number of small organs, lymph nodes, embedded in connective tissue and often not easily identified, are clustered along the vessels. Larger lymph nodes that are more noticeable are identifiable in the cervical, intestinal, and groin regions. As a matter of fact, students often misidentify the cervical nodes as salivary glands. Those found in the mesenteries of the intestines are particularly obvious and readily identified.

Large, prominent, lymphoid organs that may have been identified previously during the study of other systems include the palatine tonsils, the thymus gland, and the spleen. Humans have a similar architecture with some modifications such as Peyer's patches, a cluster of lymph nodules associated with the intestines.

The functions of this system are several. At the capillary-tissue level in the body, fluid pressure differences at the arteriole end cause fluids to move from the vessels into the tissues. Most of the fluid diffuses back into the circulatory system at the venule end again because of pressure differences, but a small volume remains in the tissues and is returned to the circulatory system by way of lymphatic vessels. Following fat absorption through the lacteals, these lipid molecules are transported to the blood in lymphatic vessels.

Lymph nodes and nodules are part of the internal "body sanitation and defense department." They are responsible for filtering and destroying foreign proteins and particles such as viruses and bacteria. These also are the sites of lymphocyte production involved in the immune response to foreign matter.

Of the lymphoid organs, the tonsils function in a manner similar to the lymph nodes, eliminating possible foreign invaders. The thymus gland, whose size is influenced by age in mammals—young mammals have much larger thymuses than adults do—is the site of production of lymphocytes that migrate to lymph nodes, providing immune defense functions.

The largest of the lymphoid organs, the spleen is the site of immune surveillance and production of leukocytes. It is the site of aged erythrocyte destruction with release of the components into the circulatory system for production of new blood cells. A number of other organs also play an important role in the life cycle of erythrocytes. In the fetus, the spleen is a major site of erythrocyte production. This usually ceases after birth but can be reactivated in adults under conditions of erythrocyte decimation. The spleen further serves as a reservoir of blood platelets.

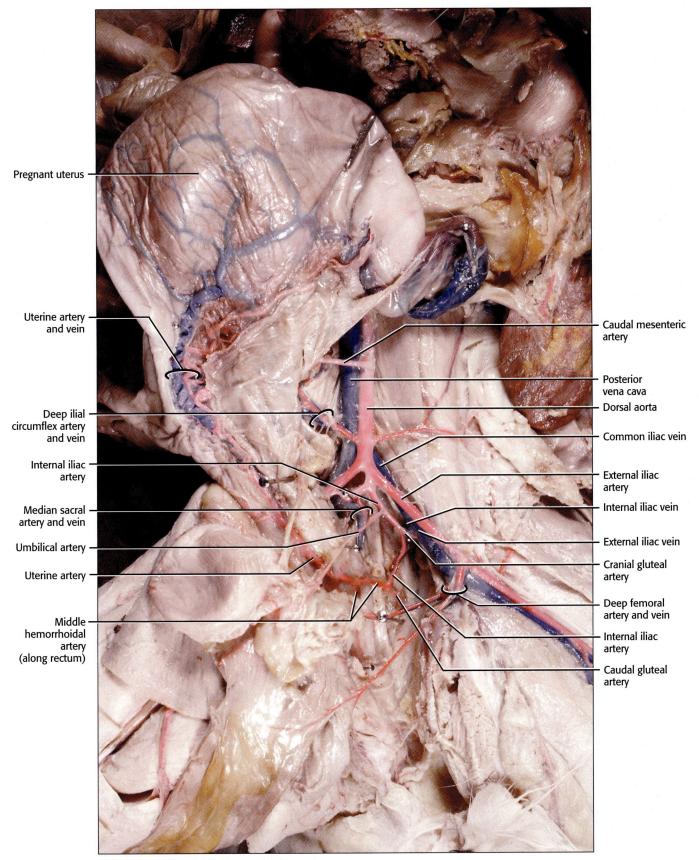

Pregnant uterus

Uterine artery
and vein

Deep ilial
circumflex artery
and vein

Internal iliac
artery

Median sacral
artery and vein

Umbilical artery

Uterine artery

Middle
hemorrhoidal
artery
(along rectum)

Caudal mesenteric
artery

Posterior
vena cava

Dorsal aorta

Common iliac vein

External iliac
artery

Internal iliac vein

External iliac vein

Cranial gluteal
artery

Deep femoral
artery and vein

Internal iliac
artery

Caudal gluteal
artery

FIGURE 39.22 Hindlimb vessels: Deep.

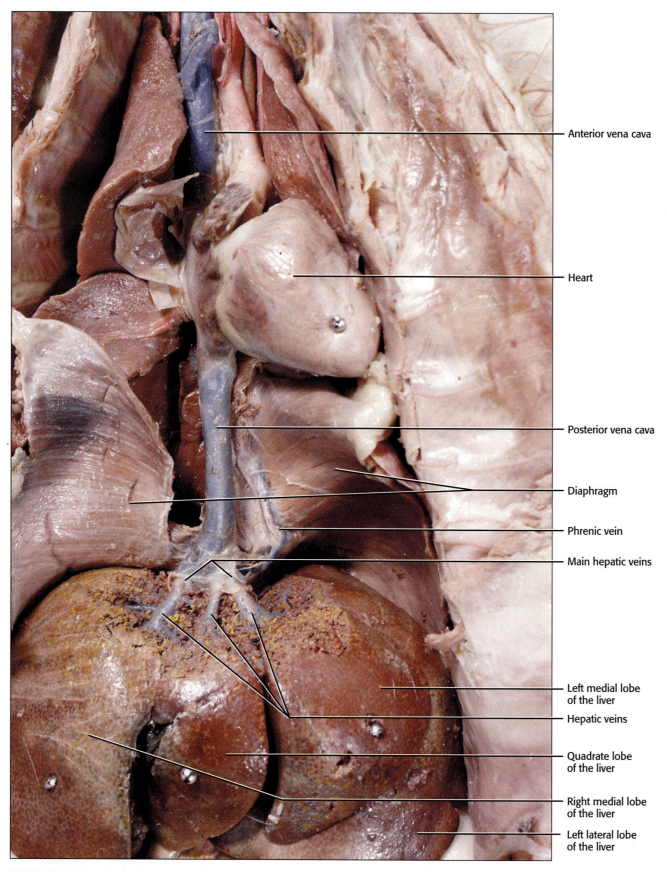

— Anterior vena cava

— Heart

— Posterior vena cava

— Diaphragm

— Phrenic vein

— Main hepatic veins

— Left medial lobe
of the liver

— Hepatic veins

— Quadrate lobe
of the liver

— Right medial lobe
of the liver

— Left lateral lobe
of the liver

FIGURE 39.23 Circulation from the liver to the right atrium.

CHAPTER FORTY

Cat — Nervous System and Sense Organs

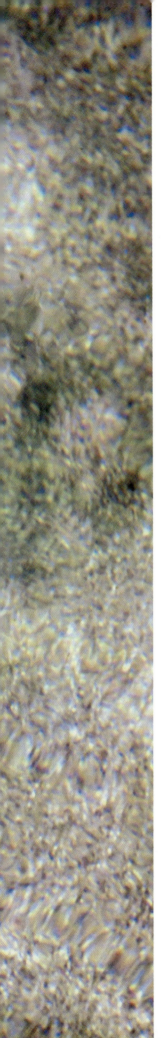

The nervous systems of vertebrates exhibit many similarities at various evolutionary levels. During early development of the nervous system of vertebrates, three primitive embryological divisions of the brain form the enlarged anterior end of the neural tube: the anterior prosencephalon or forebrain and the posterior rhombencephalon or hindbrain, separated by the mesencephalon or midbrain. As differentiation progresses, the mesencephalon remains intact but the prosencephalon subdivides into the telencephalon and the diencephalon, and the rhombencephalon divides into the metencephalon and myelencephalon, yielding a five-part brain.

These gross similarities exist in vertebrates, but differences are apparent in the nervous system of amniotes and anamniotes. The brain of anamniotes is basically linear. Beginning at the anterior end, it is arranged in the following order, similar to the shark and *Necturus*:

1. telencephalon,

2. diencephalon,

3. mesencephalon,

4. metencephalon, and

5. myelencephalon.

The amniote brain undergoes flexure, resulting in a brain in which the telencephalon and diencephalon are angled with respect to the rest of the brain. Cranial nerves XI (the spinal accessory) and XII (the glossopharyngeal) arise within the amniotes and are present in mammals. In addition, the nervous circuitry has some distinct differences. For example, the more sophisticated wiring in mammals permits them to respond to stimuli with more complex and refined behavior that is less apt to be stereotypic.

BRAIN

The mammalian brain is unique among vertebrates in many ways. The most prominent portions are the cerebrum and the cerebellum, both of which have evolved in a similar fashion to accommodate the immense increase in neurons and supporting cells of this part of the central nervous system. The surfaces of both exhibit numerous convolutions (gyri) and crevices (sulci) as a means to increase the surface area substantially, with the result that many more cells can fit into these areas while still conserving space in the cranium. Without this design, mammal heads more than likely would be even larger than they are.

Further, in contrast to nonmammalian vertebrates, the cerebrum and cerebellum of mammals possess a thin outer layer of gray matter—the cortex—consisting mainly of nerve cell bodies and unmyelinated nerve fibers overlying the main mass of white matter consisting primarily of myelinated nerve fibers. The prominent paired optic lobes of the mes-encephalon—the major integration and coordination center of nonmammalian vertebrates—appear in mammals as two paired lobes, the corpora quadrigemina, which are centers for regulation of auditory, visual, and other reflex activities. The centers of integration and coordination now reside in the cerebrum and the cerebellum.

Because accessing the intact brain of the cat is most difficult and often results in a poor specimen, we suggest using a sheep brain to represent the typical mammalian pattern of anatomy. We have found that brains with intact cranial nerves and pituitary glands and with the meninges removed are the best specimens for dissection. We also recommend that your instructor, using a sharp kitchen knife, cut each brain into two equal halves so the internal sagittal anatomy can be studied.

Meninges and Ventricles of the Brain

The central nervous system is isolated from the rest of the body by three protective layers known as meninges. The meninges of mammals are much better and more distinctly developed than those of the shark and the mudpuppy.

1. The innermost layer, the **pia mater**, remains in intimate contact with the brain surface and can be recognized by its abundant vascularity.

2. The outermost whitish, tough, fibrous layer, the **dura mater**, often remains adherent to the area surrounding the pituitary gland (hypophysis) even after the meninges have been removed.

3. The middle layer, the **arachnoid** membrane, is delicate, weblike, and extends between the pia and dura maters.

Similar to other vertebrates, the mammalian brain contains ventricles. Ventricles I and II lie in the left and right cerebral hemispheres, respectively. Each is connected by a small canal, the foramen of Monro, to ventricle III, which occurs in the diencephalon. Ventricle III is connected by a canal called the cerebral aqueduct in the mesencephalon to ventricle IV, lying in a V-shaped space of the medulla oblongata. The cerebral aqueduct is the remnant of the larger ventricle occurring in the mesencephalon of the shark and mudpuppy.

Each of these ventricles is roofed by a complex of tissues called a choroid plexus. A choroid plexus consists of the inner lining of the brain, the ependyma, and a highly vascularized portion of the pia mater. This is the site of the "blood-brain barrier," which is effective in preventing contamination of the central nervous system by potentially damaging invaders such as bacteria and viruses. The choroid

plexi produce a specialized secretion, the cerebrospinal fluid. Cerebrospinal fluid circulates within the ventricles of the brain, the central canal of the spinal cord, and subarachnoid space between the pia mater and arachnoid membrane in mammals. It functions as a lubricant, prevents damage from mechanical shock, and also acts as a buoyant fluid, allowing the brain to float and be perceived as a fraction of its actual weight.

External Anatomy

Telencephalon

From the dorsal view, observe the cerebrum, consisting of paired **cerebral hemispheres** separated from each other by the deep **longitudinal cerebral fissure**. Notice the numerous **gyri** and **sulci** sculpting the surface (Figure 40.1).

On the ventromedial aspect of the cerebrum are the **olfactory bulbs**. The bulbs lie above the cribriform plate of the ethmoid bone and receive fibers from neurons in the olfactory epithelia lining the nasal cavity. Those fibers represent part of the **olfactory nerve (I)**, which carries sensory information to the cerebrum. Two flat bands, the more conspicuous **lateral olfactory band** and the **medial olfactory band**, lead from the olfactory bulbs toward the olfactory portion of the brain, the **pyriform lobe**, that is separated from the rest of the cerebrum by a somewhat

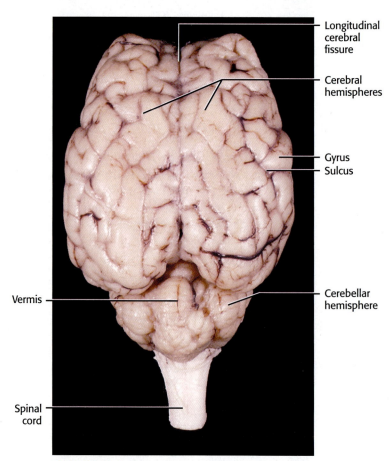

FIGURE 40.1 Brain: Dorsal view.

indistinct **rhinal sulcus**. A triangular area between the two bands is often recognized as the **olfactory trigone** (Figure 40.2).

Diencephalon

Because the cerebrum has developed to such an extraordinary extent in mammals and has attained a size that brings it into contact with the second largest portion of the brain, the cerebellum, a great deal of the **diencephalon** can be best seen in the sagittal section and in the dissection of the brainstem (to be done later). Three areas—the floor or **hypothalamus**, the sides or **thalamus**, and the thin roof, the **epithalamus**—enclose the third ventricle of the diencephalon.

A major landmark, the **optic chiasma**, an X-shaped structure, demarcates the cranial end of the hypothalamus on the ventral surface of the brain. The two stout bands of tissue at the cranial end of the chiasma are the **optic nerves** (II), consisting of nerve fibers carrying sensory information from the eyes. At the chiasma, some fibers from each eye cross over to the opposite side, similar to sharks and *Necturus*; other fibers pass straight through to their respective sides by way of the **optic tracts** (Figure 40.2, Figure 40.3A, and Figure 40.3B). This "straight-through" circuitry allows mammals to see life three dimensionally.

A delicate, slender tube, the **infundibulum**, extends from the hypothalamus to the **pituitary gland** (**hypophysis**). The slightly convex area of the hypothalamus just posterior to the infundibulum is the **tuber cinereum**. Just caudal to the tuber cinereum are the paired **mammilary bodies**, marking the caudal end of the hypothalamus (Figure 40.3A, Figure 40.3B and Figure 40.4).

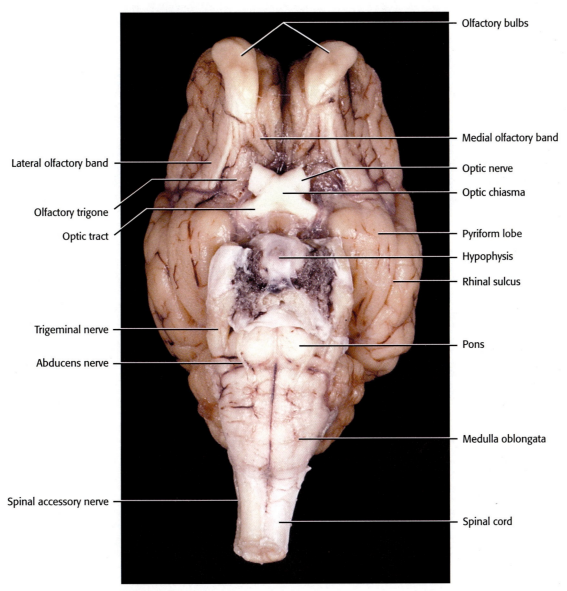

Olfactory bulbs
Medial olfactory band
Optic nerve
Optic chiasma
Pyriform lobe
Hypophysis
Rhinal sulcus
Pons
Medulla oblongata
Spinal cord

Lateral olfactory band
Olfactory trigone
Optic tract
Trigeminal nerve
Abducens nerve
Spinal accessory nerve

FIGURE 40.2 Brain: Ventral view.

Mesencephalon

While gently spreading the cerebrum and cerebellum apart, observe the bulges representing the **corpora quadrigemina**. These bodies consist of a larger, more prominent, pair of **superior colliculi** resting on a smaller, less conspicuous pair of **inferior colliculi**. On the ventral surface, observe elongated paired longitudinal **cerebral peduncles** representing bundles of nerve fibers extending between anterior and posterior areas of the brain (Figure 40.3A and Figure 40.3B). The yellowish flat bands emanating from the ventral surface of the cerebral peduncles and often adhering to the dura mater attached in the region of the hypophysis are the **oculomotor nerves (III)**. The **trochlear nerve (IV)** is unique because it is the only cranial nerve to emerge from the dorsal

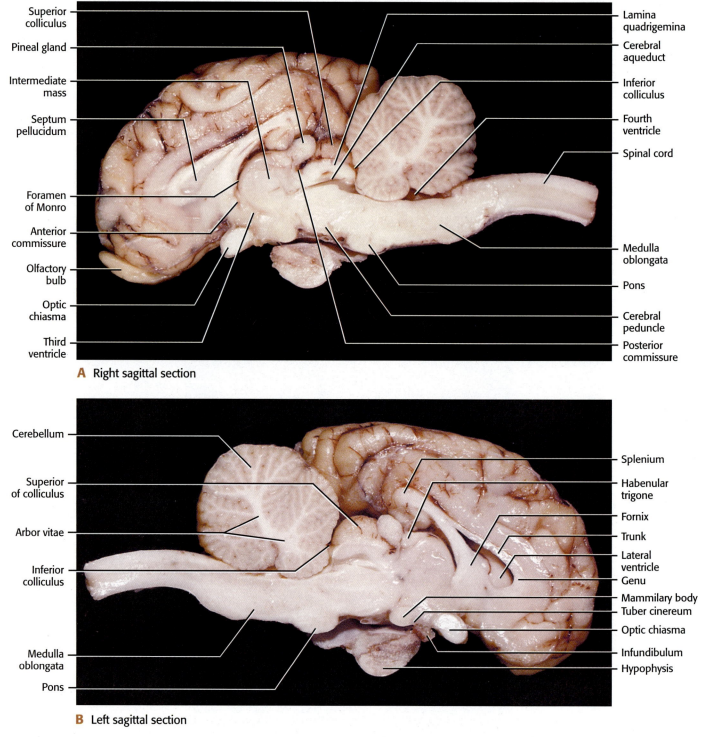

A ▪ Right sagittal section

B ▪ Left sagittal section

FIGURE 40.3 Brain.

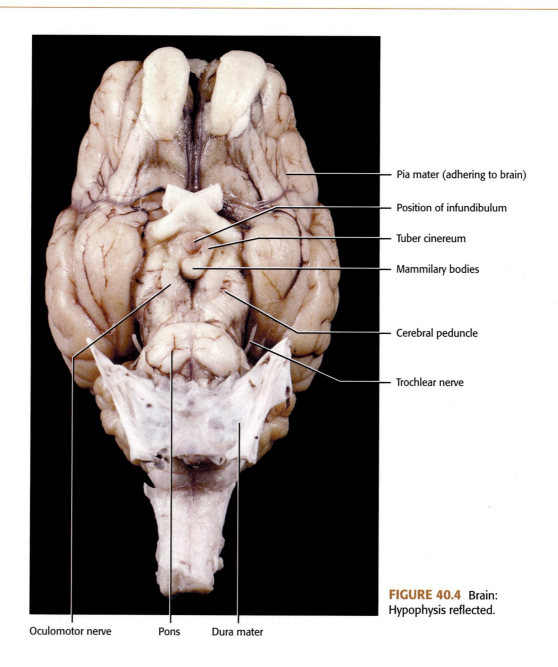

Pia mater (adhering to brain)

Position of infundibulum

Tuber cinereum

Mammilary bodies

Cerebral peduncle

Trochlear nerve

FIGURE 40.4 Brain: Hypophysis reflected.

Oculomotor nerve   Pons   Dura mater

surface of the mesencephalon just anterior to the pons (Figure 40.4).

Metencephalon

The **pons**, a unique mammalian region of the metencephalon consists of a ventral bulging band of transverse fibers located just posterior to the cerebral peduncles. By far, the larger portion of the metencephalon consists of the dorsal **cere-bellum**. Notice the medial **vermis**, flanked by paired **cere-bellar hemispheres** (Figure 40.1, Figure 40.2, and Figure 40.3B).

Myelencephalon

The **medulla oblongata** contains fiber tracts permitting communication between the brain and the spinal cord (Figure 40.2, Figure 40.3A, Figure 40.3B, and Figure 40.5).

Associated with the medulla oblongata are cranial nerves V, the **trigeminal;** VI, the **abducens;** VII, the **facial;** VIII, the **vestibulocochlear;** IX, the **glossopharyngeal;** X, the **vagus;** XI, the **spinal accessory;** and XII, the **hypoglossal.** The spinal accessory, as its name implies, also receives fibers from the anterior end of the spinal cord.

Internal Anatomy: Sagittal Section of Brain

A careful sagittal section usually results in two halves that are similar. The external features just discussed, in many cases, will be seen in this dissection, sometimes in greater detail. The overlying relationship of the cerebrum to the diencephalon and mesencephalon is easier to see. The thin layer of gray matter (actually tan in preserved specimens)

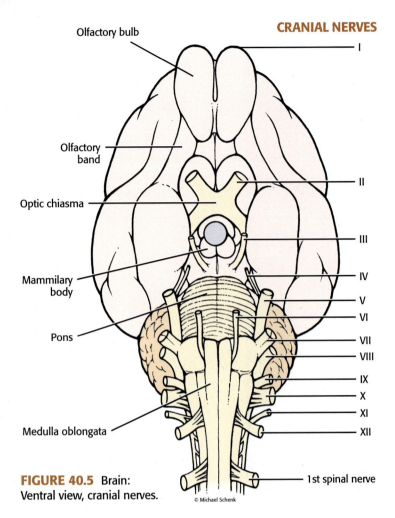

Olfactory bulb

Olfactory band

Optic chiasma

Mammilary body

Pons

Medulla oblongata

CRANIAL NERVES

I
II
III
IV
V
VI
VII
VIII
IX
X
XI
XII

1st spinal nerve

FIGURE 40.5 Brain: Ventral view, cranial nerves.

© Michael Schenk

on the outer surface of the cerebrum and cerebellum is obvious in this section. In the following discussion, refer to Figure 40.3A, Figure 40.3B, and Figure 40.6.

Within the three groups of mammals living today—the monotremes or egg-laying mammals, the marsupials, and the placental mammals—only in the placental mammals has a **corpus callosum** evolved. It consists of transverse nerve fibers that permit transmission of nervous impulses between the cerebral hemispheres. The cranial curve of the corpus callosum is known as the **genu** and the caudal end is called the **splenium,** with the **trunk** extending between them.

Another paired bundle of fibers, the **fornix,** more extensive than can be seen in this section, appears ventral to the corpus callosum. Connecting the corpus callosum and the fornix is the thin wall of tissue called the **septum pellucidum,** consisting of a double sheet of gray and white matter separating the two lateral ventricles of the cerebrum. Often, the sagittal cut will have left the septum entirely in one of the brain halves. If that is true, you will easily observe one of the lateral ventricles, on the side devoid of the septum.

Now direct your attention to the area almost directly ventral to the regions just discussed. If the sagittal cut was

successful in approximating the midsagittal plane, a rather prominent circular mass, the **intermediate mass (massa intermedia),** will be distinguishable, residing within a shallow irregular space, the **third ventricle.** Each of the cerebral hemispheres contains a **lateral ventricle** that communicates with the third ventricle by way of the **foramen of Monro.** Recall that the third ventricle is enclosed within the walls of the diencephalon and the **cerebral aqueduct** sits within the confines of the mesencephalon, leading to the **fourth ventricle** resting within the medulla oblongata.

The **intermediate mass** is a central connective bridge between the left and the right **thalamus** that make up the two lateral walls of the diencephalon. Actually, all of the tissue lateral to the third ventricle in this section, with the exception of the single layered ependyma lining the ventricle, is thalamus. The floor and ventral walls of the diencephalon constitute the **hypothalamus.**

Observe the relationships of the **optic chiasma, infundibulum, hypophysis,** and **mammilary bodies,** all associated with the hypothalamus and seen in external view. The thin **lamina terminalis** forms the cranial wall of the diencephalon. Notice the small thickening, the **anterior commissure,** in the dorsal portion of the lamina terminalis. The **epithalamus** is the very thin roof of the diencephalon. The most prominent structure here is the **pineal gland** projecting from its surface. Just cranial to it is the **habenular trigone** and ventral to it is the **posterior commissure.**

Features of the mesencephalon include the dorsally located **corpora quadrigemina,** consisting of a pair of **superior colliculi,** controlling visual reflexes, and a pair of **inferior colliculi,** controlling auditory reflexes and resting on the **lamina quadrigemina.** The thick floor consists of the **cerebral peduncles,** bundles of nerve fibers extending between the cerebrum and other brain areas. The usually narrow space enclosed in the mesencephalon is the **cerebral aqueduct,** connecting the third and fourth ventricles.

The rounded protuberance posterior to the cerebral peduncles is the **pons,** and the somewhat elongated, flattened region just caudal to the pons is the **medulla oblongata.** Caudal to the medulla oblongata, the central nervous system continues as the **spinal cord** whose cavity is the **central canal.**

The large, deeply grooved body lying dorsally over the fourth ventricle and pons and medulla oblongata is the **cerebellum.** Note the similarity in tissue arrangement between the cerebrum and the cerebellum, with the outer layer of gray matter lying over the white matter. In the sagittal section, this relationship of the tissues of the cerebellum suggested a treelike construction to early anatomists, and they named this arrangement of white matter in the cerebellum the **arbor vitae** (tree of life).

Brainstem

The brainstem of mammals consists of the mesencephalon, pons, and medulla oblongata. The brainstem is the site of a number of relay and vital reflex centers. The specimen in Figure 40.7A includes portions of the diencephalon as well.

 To produce a dissection similar to Figure 40.7A, grasp the cerebrum of one of your sagittal sections and carefully separate it from the rest of the section. Now slice through the thalamus of the diencephalon to create a similar specimen. Carefully lift the cerebellum and remove it from the brainstem, using a sharp knife or scalpel, while watching its connection to the underlying tissue so tears do not occur in the area. After removal, if necessary, use a sharp scalpel to make a smooth slice through the anterior, middle, and posterior cerebellar peduncles similar to Figure 40.7A.

Beginning with structures associated with the diencephalon that are not part of the brainstem, locate the **thalamus**, the **intermediate mass**, the **habenular trigone**, and the **pineal gland**. Notice the two subtle bulges on the lateral wall of the thalamus. The more anterior and less obvious of these diencephalic structures, the **lateral geniculate body**, site of the visual impulse relay, is associated with the **superior colliculus**. The posterior and more obvious **medial geniculate body**, associated with the **inferior colliculus**, serves as a relay of auditory impulses (Figure 40.7A and Figure 40.7B).

Locate the **anterior, middle,** and **posterior cerebellar peduncles,** representing bundles of fibers connecting the cerebellum to the brainstem. The space between the Y-shaped halves of the medulla oblongata is the **fourth ventricle.** This Y-shaped configuration consists of fiber tracts transmitting sensory impulses between the spinal cord and the medulla oblongata. Note the shallow central dorsal groove, the **dorsal median sulcus.** The two halves of the Y consist of

		Neuron Type		
Name	**Number**	**Sensory**	**Motor**	**Distribution**
Terminal	O	X		Neurosensory cells of nasal epithelium
Olfactory	I	X		Neurosensory cells of nasal epithelium
Optic	II	X		Sensory fibers of the retina
Oculomotor	III	†	X	Innervates dorsal rectus, ventral rectus, medial rectus and ventral oblique muscles; innervates retractor bulbi, levator palpebrae superioris, and intrinsic ciliary muscles
Trochlear	IV	†	X	Innervates dorsal oblique muscle
Trigeminal	V	X	X	Nerve consists of three branches, all associated with the facial and jaw regions 1. Ophthalmic branch: innervates the skin in region of eye and nose, upper eyelid, eyeball, lacrimal glands 2. Maxillary branch: innervates the palate, upper lip, upper teeth, vibrissae, skin of upper jaw, and vicinity 3. Mandibular branch: innervates the lower lip, lower teeth, pinna, skin of lower jaw, and cheek; muscles of mastication—masseter, temporal and pterygoid
Abducens	VI	†	X	Innervates lateral rectus
Facial	VII	X	X	Innervates facial and digastric muscles; sensory innervation of the tastebuds of anterior two-thirds of tongue; mandibular, sublingual, and lacrimal glands
Vestibulocochlear	VIII	X		Two branches: 1. Vestibular branch innervates organs of equilibrium, saccule, utricle, and semicircular canals. 2. Cochlear branch innervates the acoustic organs, organs of Corti, in cochlea
Glossopharyngeal	IX	X	X	Innervates pharyngeal muscles; sensory innervation of pharynx and tastebuds of posterior one third of tongue; parotid gland
Vagus	X	X	X	Innervates pharynx, larynx, esophagus, lungs, heart, abdominal viscera
Accessory	XI	†	X	Innervates cleidomastoid, sternomastoid, trapezius muscles
Hypoglossal	XII	†	X	Innervates styloglossus, hyoglossus, genioglossus, and intrinsic muscles of the tongue

TABLE OF CRANIAL NERVES

† Although cranial nerves III, IV, VI, XI, and XII are considered motor nerves, they carry proprioceptive fibers, which are sensory and transmit information concerning muscle status.

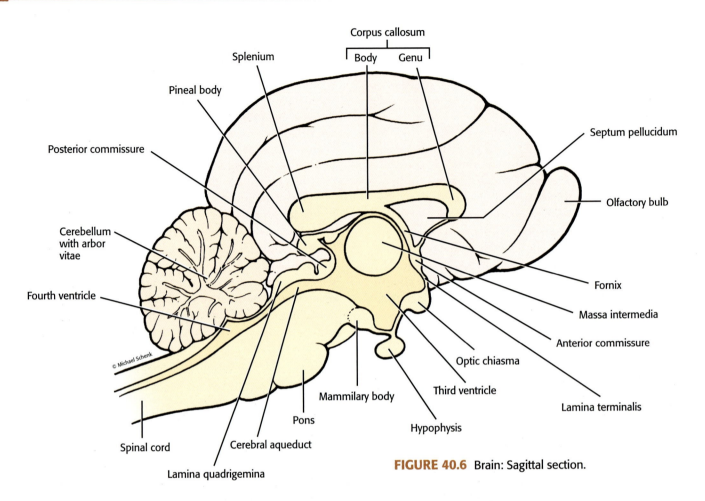

FIGURE 40.6 Brain: Sagittal section.

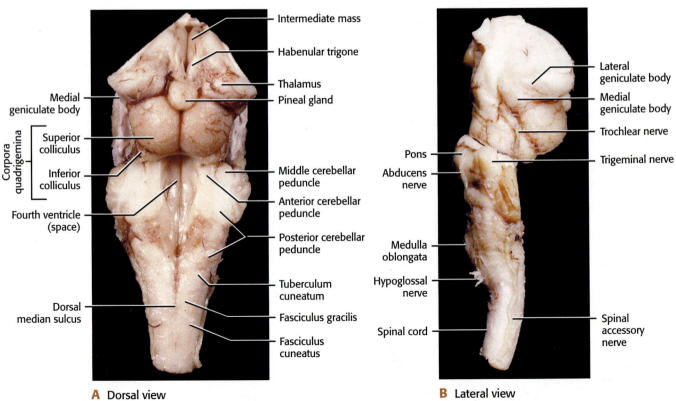

A Dorsal view

FIGURE 40.7 Brainstem.

B Lateral view

the **fasciculus gracilis**. Lateral to the fasiculus gracilis, locate the **fasciculus cuneatus** and the **tuberculum cuneatum** (Figure 40.7A).

SPINAL CORD

Note that the more or less cylindrical spinal cord is continuous with and posterior to the **medulla oblongata.** Like the central nervous system of the shark and *Necturus*, white matter surrounds the gray matter. This portion of the central nervous system is the site of interconnecting synapses between sensory, association, and motor neurons involved in various reflex actions. Recall that the nerve cord in vertebrates is defined as dorsal and hollow.

 Examine the cut end of the spinal cord of the brain specimen and note the small **central canal** that constitutes the "hollow" part of the definition.

The central canal is continuous with the **fourth ventricle;** therefore, the cerebrospinal fluid circulates not only within the brain ventricles but also within the central canal of the spinal cord. Passage of the fluid through foramina in the fourth ventricle into the subarachnoid space allows circulation of the cerebrospinal fluid, thereby bathing the outer surface of the central nervous system.

In the body, the spinal cord lies within the vertebral canal formed by the articulated vertebrae. The spinal cord is not uniform in diameter but has two conspicuous enlargements—the anterior cervical, which supplies nerves to the forelimb, and the posterior lumbosacral, which supplies nerves to the hindlimb. It continues posteriorly, ending in the tail as a slender thread of tissue called the filum terminalis.

Spinal Nerves

The cat has 38–39 pairs of spinal nerves: 8 cervical, 13 thoracic, 7 lumbar, 3 sacral, and 7 or 8 caudal nerves. Each spinal nerve generally carries both sensory and motor fibers. Spinal nerves at their spinal cord ends have a dorsal root carrying the sensory neurons and a ventral root carrying the motor neurons. The roots combine to form the spinal nerve that emerges between adjacent vertebrae and then subdivides into dorsal and ventral rami (branches) supplying their respective body segments.

Complex interrelationships of the ventral rami in certain body regions are identified as plexes. Commonly, three of these—the cervical in the neck region, the brachial in the shoulder and forelimb region, and the lumbosacral in the hip and region of the hindlimb—are discussed. In our treatment here, the cervical and brachial are combined and discussed as the brachial plexus. In other treatments, spinal nerves 1–5 (cervical nerves 1–5) are referred to as the cervical plexus. Cervical nerves 6–8 along with the first thoracic nerve are referred to as the brachial plexus.

Brachial Plexus

During dissection of the muscular and circulatory systems, many of the nerves of this plexus were exposed. To do an orderly dissection of this complex, you will begin at its anterior end and work posteriorly. For the following dissection, see Figure 40.8. As reference points, locate the **hypoglossal nerve (XII)**, which overlies the hyoglossus muscle discussed previously during dissection of the neck muscles, and the **spinal accessory nerve (XI).**

 To expose the spinal accessory nerve, carefully cut the sternomastoid and cleidomastoid muscles at their insertion ends. Because these muscles are innervated by this nerve, exercise great caution.

Find a pair of swellings, the **superior cervical** and **nodose ganglia,** which occur just ventral to the hypoglossal nerve. Find the **vagus nerve (X)**, extending posteriorly from the nodose ganglion and running along the common carotid artery in a sheath closely applied to the trachea.

Although the vagus, spinal accessory, and hypoglossal are cranial nerves, they are important as references to a complete and orderly dissection. The nerves of the brachial plexus often have more than one point of origin; e.g., the first subscapular nerve arises from the sixth and seventh cervical nerves. In a plexus, interconnecting nerves carry fibers of the ventral rami from one spinal nerve to another, thereby contributing to the multiple origin of certain nerves in these complexes.

The **first cervical nerve,** quite small and difficult to recognize because of copious amounts of connective tissue in the area, passes medially over the ventral surface of the longus colli muscle to supply ventral muscles of the neck.

Locate the **second, third, fourth** and **fifth cervical nerves,** somewhat obscured by connective tissue and supplying the tissues of the shoulder and neck area.

Fibers of the **fifth** and **sixth cervical nerves** contribute to formation of the **phrenic nerve,** which travels lateral to and in close association with the vagus nerve through the thoracic region and innervates the diaphragm.

 Hint: It is easier to locate the phrenic nerve on the surface of the diaphragm and trace it cranially to its origin.

The **sixth cervical nerve** extends to the area of the shoulder joint, where the prominent **suprascapular nerve** originates and passes between the supraspinatus and subscapularis muscles, supplying the supraspinatus and infraspinatus muscles. Notice that a small branch to the skin of the shoulder originates at the same point as the suprascapular nerve.

The **sixth** and **seventh cervical nerves** give rise to the **first subscapular nerve,** which innervates the subscapularis muscle and travels with the subscapular vessels. The **axillary nerve** originates from these two cervical nerves and travels with the posterior humeral circumflex vessels beneath the posterior edge of the biceps brachii muscle to innervate lateral shoulder muscles such as the teres major, teres minor, deltoids, and cleidobrachialis muscles. Likewise, the **musculocutaneous nerve** originates from the sixth and seventh cervical nerves and branches near the biceps brachii muscle, innervating the biceps brachii, the brachialis, and the coracobrachialis muscles, as well as the skin of the forelimb.

The **second subscapular nerve** originates from the seventh cervical nerve and innervates the teres major muscle.

The **anterior ventral thoracic nerve** arises from the seventh cervical nerve, travels with the ventral thoracic vessels, and innervates the pectoral muscles.

The largest nerve in the plexus is the **radial nerve**, originating from cervical nerves 6, 7, 8, and the first thoracic nerve. The radial nerve travels with the deep brachial vessels, and innervates the triceps brachii, epitroclearis, and extensor muscles of the forelimb.

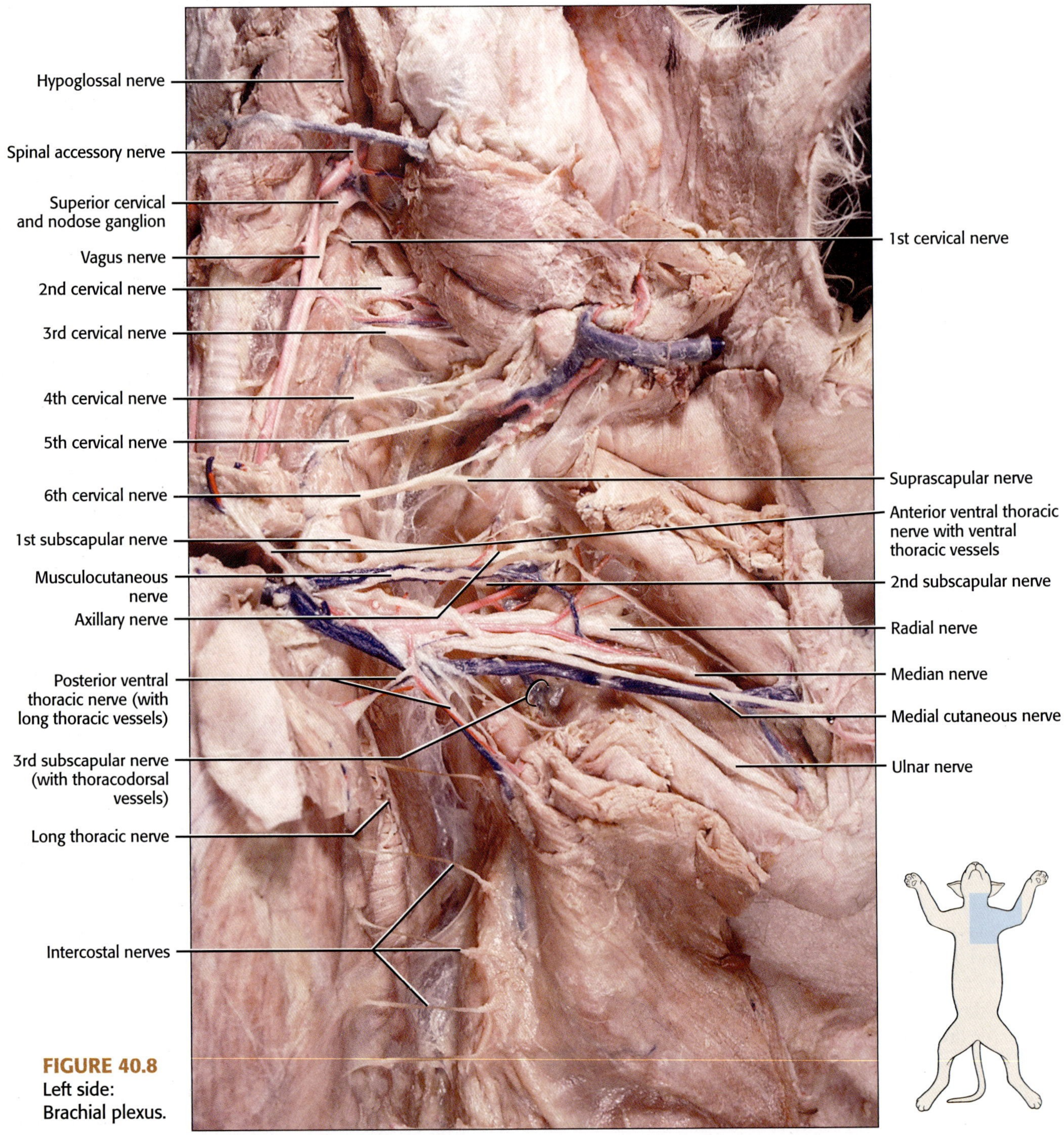
Hypoglossal nerve
Spinal accessory nerve
Superior cervical and nodose ganglion
Vagus nerve
2nd cervical nerve
3rd cervical nerve
4th cervical nerve
5th cervical nerve
6th cervical nerve
1st subscapular nerve
Musculocutaneous nerve
Axillary nerve
Posterior ventral thoracic nerve (with long thoracic vessels)
3rd subscapular nerve (with thoracodorsal vessels)
Long thoracic nerve
Intercostal nerves
1st cervical nerve
Suprascapular nerve
Anterior ventral thoracic nerve with ventral thoracic vessels
2nd subscapular nerve
Radial nerve
Median nerve
Medial cutaneous nerve
Ulnar nerve

FIGURE 40.8
Left side: Brachial plexus.

The **median nerve** arises from cervical nerves 7, 8, and the first thoracic nerve, travels with the brachial artery, and passes through the supracondyloid foramen to innervate some of the flexor muscles of the forelimb.

The **posterior ventral thoracic nerve** originates from the eighth cervical and first thoracic nerves, travels with the long thoracic blood vessels and innervates the pectoral muscles.

The **ulnar nerve** arises from the eighth cervical and first thoracic nerves, travels parallel with the median nerve, crosses the medial epicondyle of the humerus, and innervates flexor muscles of the forelimb.

The **medial cutaneous nerve** originates from the first thoracic nerve and innervates the skin of the forelimb.

The **third subscapular nerve** originates from cervical nerves 7 and 8, travels with the **thoracodorsal vessels**, and innervates the latissimus dorsi muscle.

The **long thoracic nerve** lying along the lateral surface of and supplying the serratus ventralis muscle, originates from the seventh cervical nerve.

With the exception of the first thoracic nerve, thoracic nerves are referred to as the **intercostal nerves**. They supply intercostal muscles, lateral thoracic muscles, some abdominal muscles, back muscles, and skin.

Lumbosacral Plexus (Ventral View)

The nerves that supply the lumbar and sacral regions and the hindlimb consist of a series of paired nerves—seven lumbar and three sacral—passing between adjacent lumbar and sacral vertebrae. Not surprisingly, the basic construction of nerves in this area mirrors those in the brachial region, in that they possess dorsal and ventral rami. The dorsal rami supply dorsal muscles and structures of the skin. The anatomical architecture of the ventral rami of the first three lumbar nerves consists of medial and lateral branches that innervate muscles and structures in this region. The basic structural plan of the lumbosacral plexus consists of the ventral rami of lumbar nerves 4–7, sacral nerves 1–3, and communicating branches between these regions.

To study the architecture of the neural anatomy of these areas, we will begin by finding the **medial branch of the third lumbar nerve** as it emerges from beneath the illiopsoas muscle and passes lateral to it (Figure 40.9).

 Trace the third lumbar nerve anteriorly to its source by carefully removing the anterior portion of the illiopsoas muscle. Exercise care removing it, as other lumbar nerves lie beneath it.

Find the **lateral branch of the third lumbar nerve**. By removing the illiopsoas muscle, you should be able to find the **second** and **first lumbar nerves** with their medial and lateral branches (Figure 40.9).

The remainder of the dissection involves the **lumbosacral plexus** (refer to Figure 40.9). One of the distinctive, but delicate and fragile, nerves of this complex is the **genitofemoral nerve**, which originates as the medial branch of the fourth lumbar nerve. An inconspicuous lateral branch of the fourth lumbar nerve joins the fifth lumbar nerve. First, locate the genitofemoral nerve on the surface of the external iliac blood vessels. Follow it anteriorly as it passes beneath the dorsal aorta and posterior vena cava to its source. In good dissections, particularly in males, you should be able to distinguish a genital branch innervating structures in the pelvic region.

The **lateral femoral cutaneous nerve** originates primarily from the fifth lumbar nerve, with minor contributions from the fourth lumbar nerve. It emerges from beneath the psoas minor muscle, passes over the illiopsoas muscle, travels with the illiolumbar vessels, and supplies the lateral surfaces of the hip and thigh regions.

Most of the fibers of lumbar nerve 6, in company with a branch of the fifth, form the prominent **femoral nerve**. This nerve passes between the illiopsoas and psoas minor muscles, penetrating the abdominal wall, and subdivides into three branches, one of which is the **saphenous nerve**, running parallel to the femoral and saphenous vessels. One of the other branches innervates the quadriceps muscular complex, and the third innervates the sartorius muscle.

The **obturator nerve** originates from the sixth nerve, with contributions from the fifth and seventh lumbar nerves. It passes into the pelvic region through the obturator foramen, where it branches to supply the thigh adductors, pectineus, gracilis, and obturator externus muscles.

The **lumbosacral trunk**, or cord, appears medial to the obturator nerve and is formed by fibers of the sixth and seventh lumbar nerves. As the cord passes caudally, it receives fibers from the sacral nerves.

 Carefully expose the cord and the three sacral nerves and note their interconnections.

Autonomic Nervous System

Although the **sympathetic trunks** are present from the head to the tail, this is the region in which they may be viewed best. They appear as two delicate strands of nervous tissue medial to the lumbar and sacral nerves, appearing to lie almost on the ventral surface of the vertebral column. In each body segment, paired **ganglia** appear in the trunks, with connecting rami to the spinal nerves.

The parasympathetic division of the autonomic nervous system is associated with cranial nerves III, VII, IX, X and three sacral nerves. These are difficult to distinguish.

Lumbosacral Plexus (Dorsal View)

The following dissection will expose the nerves of the lumbosacral plexus, nerves originating from the lumbosacral cord,

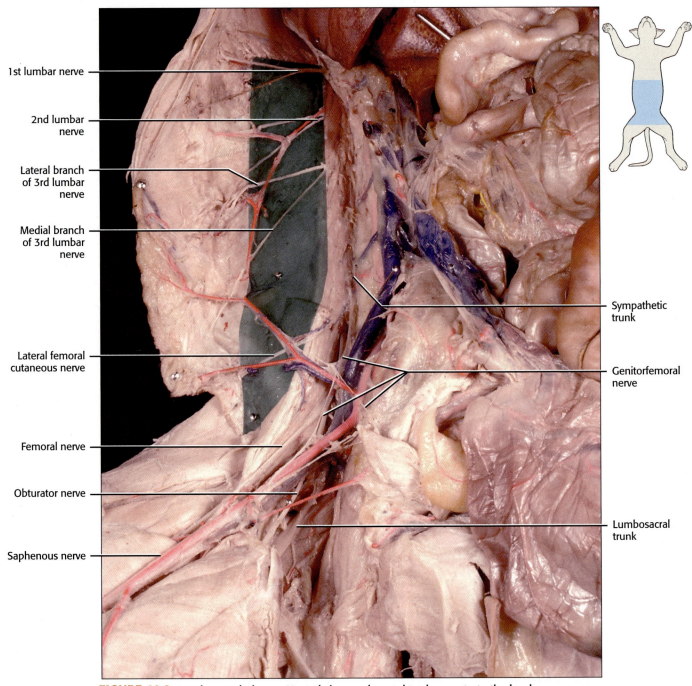

1st lumbar nerve

2nd lumbar nerve

Lateral branch of 3rd lumbar nerve

Medial branch of 3rd lumbar nerve

Lateral femoral cutaneous nerve

Femoral nerve

Obturator nerve

Saphenous nerve

Sympathetic trunk

Genitorfemoral nerve

Lumbosacral trunk

FIGURE 40.9 Lumbosacral plexus: Ventral view—enhanced to demonstrate the lumbar nerves.

and branches of the sacral nerves, and appear in Figure 40.9 and Figure 40.10. During dissection of the deep hip muscles, many of these nerves were partially exposed.

The largest and most readily observed of these nerves is the **ischiadic (sciatic) nerve,** which extends from the lumbosacral cord. It courses over the lateral muscles of the thigh, giving off a **muscular branch,** which divides to supply the biceps femoris, the semimembranosus, semitendinosus, tenuissimus, and other thigh flexor muscles. As it continues distally, the ischiadic nerve gives off a delicate **sural nerve,**

which passes across the gastrocnemius muscle to the ankle. As it nears the knee, the ischiadic nerve divides into the **common peroneal nerve** and **tibial nerve.** The common peroneal nerve pierces the lateral head of the gastrocnemius muscle to supply the peroneal muscles, the tibialis anterior, and extensor digitorum longus muscles, eventually innervating muscles of the digits. The tibial nerve extends between the heads of the gastrocnemius muscle and innervates the gastrocnemius, plantaris, soleus, and other muscles associated with extension and flexion of the foot.

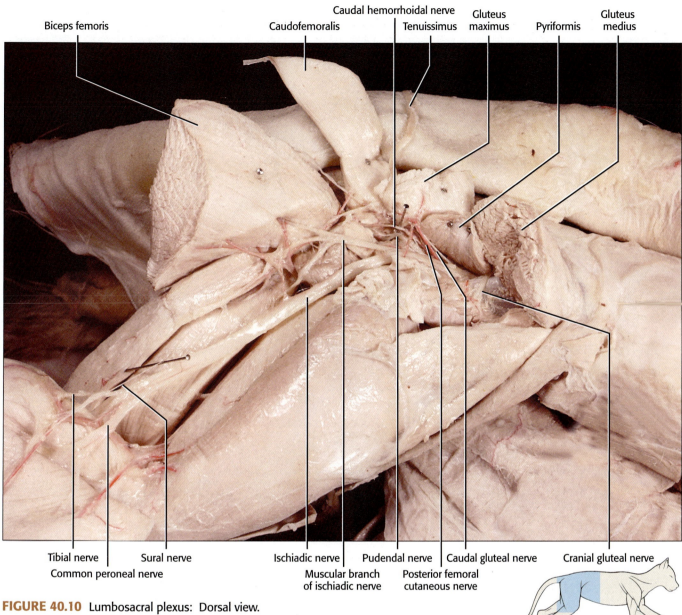

FIGURE 40.10 Lumbosacral plexus: Dorsal view.

The **cranial gluteal nerve**, a branch originating from the lumbosacral cord, passes over the dorsal aspect of the ilium and beneath the gluteus minimus muscle, innervating the gluteus minimus, gluteus medius, gemellus cranialis, and tensor fascia latae muscles. The **caudal gluteal nerve**, also originating from the lumbosacral cord, passes posteriorly to innervate the caudofemoralis and the gluteus maximus muscles (Figure 40.10 and Figure 40.11).

The **posterior femoral cutaneous nerve** arises from the second and third sacral nerves and travels with the caudal gluteal blood vessels. It arches over to the biceps femoris

muscle and extends over the surface of that muscle, innervating the skin at the base of the tail and over the biceps femoris muscle.

The **pudendal nerve** arises from the second and third sacral nerves, appearing as a forked nerve that innervates the anus and genital structures in both sexes.

The **caudal hemorrhoidal nerve**, deepest of these hip nerves, originates from sacral nerves 2 and 3, passes deep, and supplies the rectum and the urinary bladder. Usually, this nerve is difficult to find, as it lies in connective tissue that must be removed carefully.

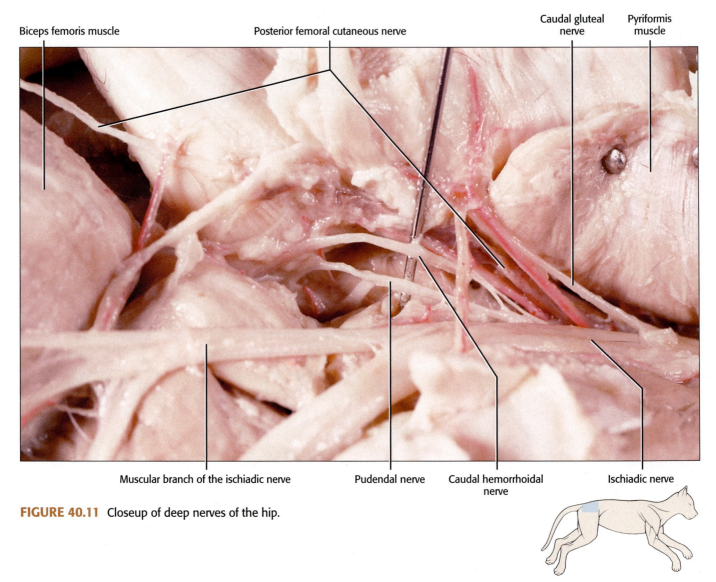

FIGURE 40.11 Closeup of deep nerves of the hip.

SENSE ORGANS

In mammals, like other vertebrates, the major organs associated with detection of environmental stimuli—eyes, auditory apparatus, olfactory apparatus, and tongue—are concentrated in the head. Imbedded in these organs are cells of the nervous system specialized for sensing changes in external conditions. The sensory information is transmitted to the appropriate areas of the brain, where it is interpreted, evoking sophisticated perception and appropriate action.

Tongue

In most mammals the tongue actually is employed in a variety of activities—feeding, drinking, grooming—in addition to gustation (tasting). The receptors associated with gustation are located within papillae scattered over the surface of the tongue. Recall that the various papillae and their distribution were included in the earlier discussion of the digestive system.

Olfactory Apparatus (Nose)

The nose functions during respiration and olfaction (smelling). The sense receptors involved with detecting odors are located within the epithelium of the mucous membranes lining the nasal cavities.

Auditory and Equilibrium Apparatus

The ear of mammals is difficult to dissect because it is imbedded within heavy bone of the skull. The ear consists of external, middle, and inner regions (see Figure 40.12A, Figure 40.12B, and Figure 40.12C). The external ear, also known as the **pinna**, is unique and easily identifiable in most mammals and functions as a collecting funnel of sound waves. This sound energy passes through the external acoustic meatus and causes the **tympanic membrane** (eardrum) to vibrate. The vibration is transmitted sequentially to the inner ear through a series of three ear ossicles— the **malleus**, the **incus**, and the **stapes**—located in the **tympanic cavity** of the middle ear.

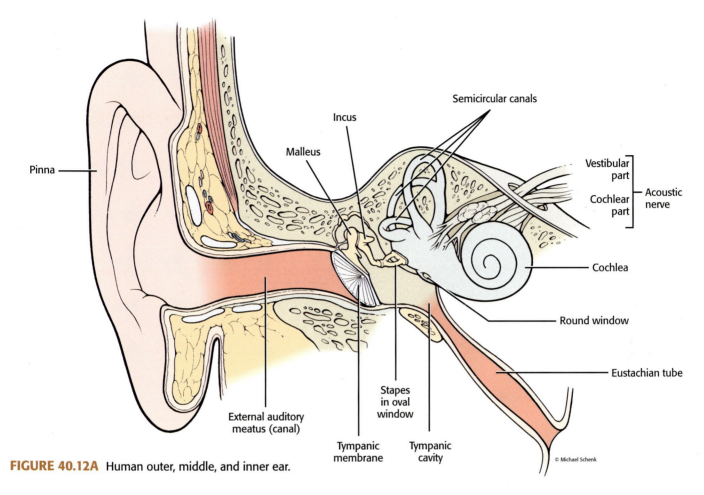

Pinna

Incus

Malleus

Semicircular canals

Vestibular part

Cochlear part

Acoustic nerve

Cochlea

Round window

Eustachian tube

External auditory meatus (canal)

Stapes in oval window

Tympanic membrane

Tympanic cavity

© Michael Schenk

FIGURE 40.12A Human outer, middle, and inner ear.

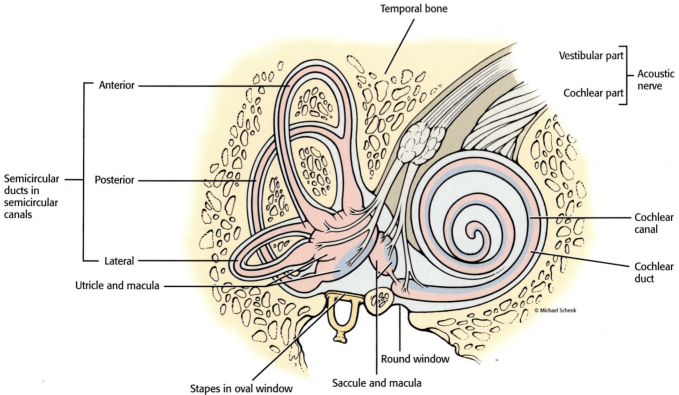

Temporal bone

Vestibular part

Cochlear part

Acoustic nerve

Anterior

Semicircular ducts in semicircular canals

Posterior

Lateral

Utricle and macula

Cochlear canal

Cochlear duct

© Michael Schenk

Stapes in oval window

Saccule and macula

Round window

FIGURE 40.12B Membranous and bony labyrinth of human inner ear.

COCHLEA

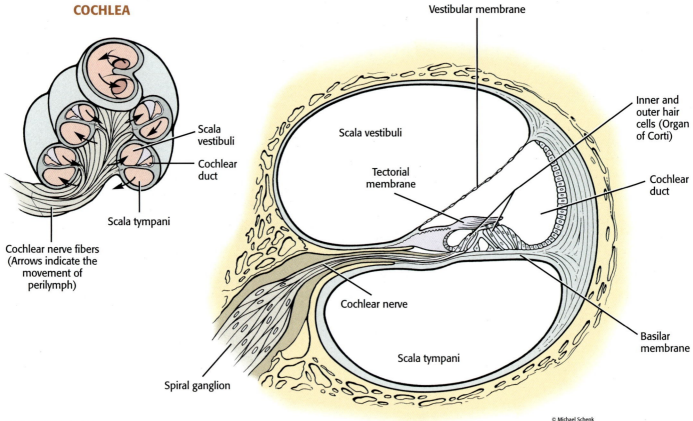

Scala vestibuli

Cochlear duct

Scala tympani

Cochlear nerve fibers (Arrows indicate the movement of perilymph)

Vestibular membrane

Scala vestibuli

Tectorial membrane

Inner and outer hair cells (Organ of Corti)

Cochlear duct

Cochlear nerve

Spiral ganglion

Scala tympani

Basilar membrane

© Michael Schenk

FIGURE 40.12C Section through human cochlea.

The footplate of the stapes fits into the oval window, transmitting vibrations to the fluid-filled cochlea. The membranous labyrinth consists of a blind tube, the **cochlear duct,** shaped like a snail shell, that is continuous with the organ of equilibrium, consisting of the **saccule, utricle,** and three **semicircular canals**. These organs rest in the bony labyrinth of the petromastoid portion of the skull. Within the membranous labyrinth circulates the endolymph, and within the bony labyrinth circulates the perilymph.

Located within the cochlear duct is the **organ of Corti**. Waves of fluid within the cochlea cause the microvilli of the hair cells of the organ of Corti to bend and produce nerve impulses that are transmitted by way of cochlear nerves to the brain for interpretation of sound. The saccule, utricle, and semicircular canals are involved with equilibrium.

Eye

External Structures

The eyeball of the cat is difficult to dissect because it must be removed from the head. An excellent substitute, a cow or sheep eye, is readily available. It is advantageous to dissect two eyes—one to study the extrinsic eye muscles that position the eyeball within the orbit and the other to view the internal eye structures.

 To facilitate study of the eye muscles, carefully remove the white connective tissue and fat, taking care not to remove any pink, tan, or gray structures, which may be muscles or glands associated with the eyeball.

The **optic nerve,** which is white, and located on the back and near the center of the eyeball, must be conserved (Figure 40.13).

All mammals possess the typical six extrinsic eye muscles of vertebrates: a ventral oblique, a dorsal oblique, a medial rectus, a lateral rectus, a dorsal rectus, and a ventral rectus (Figure 40.13). To identify these muscles correctly, it is imperative to determine whether the eyeball is left or right. The key to this determination is to identify the **ventral oblique muscle**. This muscle is found on the bottom of the eyeball and is the only muscle that naturally wraps around the circumference of the eyeball. Its insertion is on the lateral aspect of the eyeball very near the insertion of the **lateral rectus muscle**. The cut end of the ventral oblique muscle represents the medial aspect of the eyeball.

You now should be able to identify the dorsal, ventral, medial, and lateral aspects of the eyeball. Identify the **ventral rectus muscle**, over which the ventral oblique muscle

Lacrimal gland   Levator palpebrae superioris   Dorsal rectus   Dorsal oblique

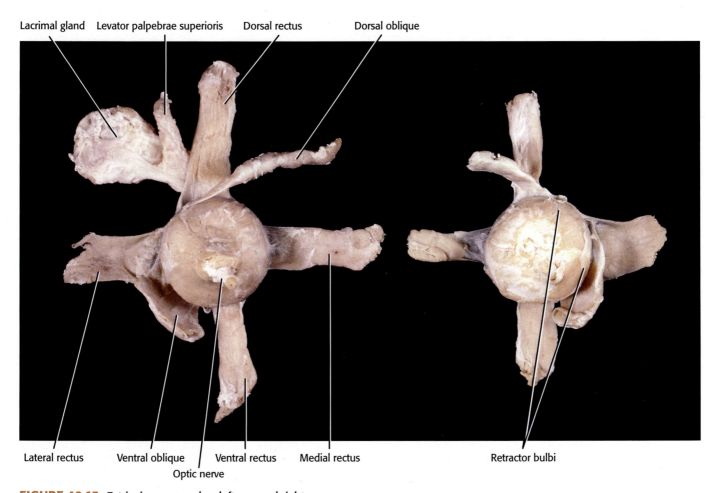

Lateral rectus   Ventral oblique   Ventral rectus   Medial rectus   Retractor bulbi
Optic nerve

FIGURE 40.13 Extrinsic eye muscles; left eye and right eye.

lies. Next identify the **medial rectus muscle**, appearing approximately opposite the **lateral rectus muscle**. Continue around the eyeball dorsally and identify the **dorsal rectus muscle**, which inserts approximately opposite the ventral rectus muscle. Finally, identify the **dorsal oblique muscle**, whose insertion and muscle direction is, as its name implies, somewhat oblique, and opposes the ventral oblique muscle.

On the posterior surface of the eyeball and surrounding the projecting optic nerve is a four-part muscle, the **retractor bulbi** (Figure 40.13). This muscle retracts the eyeball. Another muscle, the **levator palpebrae superioris**, attached to the upper eyelid and found on the anteriodorsal surface of the eyeball, lifts the eyelid. You may see yet another muscle, the **orbicularis oculi**, arranged in a circular sphincterlike pattern in and around the eyelids. It functions during forced closure of the eyelids (e.g., when the cat is being disciplined).

Eyeball moisture for most terrestrial vertebrates affords a well-lubricated surface to assure that the surface remains clean and can function as part of the lens system. The **lacrimal gland**, located on the dorsal lateral aspect of the eyeball, secretes a solution known as tears (Figure 40.13).

This secretion moves across the eyeball, keeping it moist, then drains into the nasolacrimal duct and eventually into the nasopharynx. A transparent fold, the **nictitating membrane**, located in the medial angle of the eye, acts as a protective mechanism in many mammals (Figure 40.14).

Internal Structures

If you received two eyeballs to dissect, choose the one not used for external structures. When only one eyeball is available, use caution during the following dissection so you can preserve muscles and other tissues and organs previously identified.

 To prepare a specimen that is superior for the study of internal structures, insert the tip of a sharp scalpel into the eyeball approximately 1 cm from the posterior edge of the cornea. Continue to cut around the circumference of the eyeball to separate it into anterior and posterior parts. Store the anterior and posterior parts of the eyeball in the open air at room temperature for 24–48 hours or until the jellylike substance has dried out.

Nictitating membrane

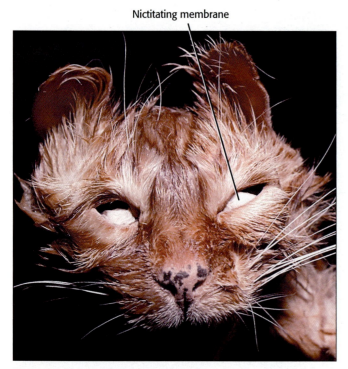

FIGURE 40.14 Nictitating membrane.

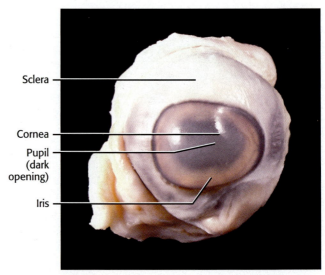

Sclera

Cornea

Pupil (dark opening)

Iris

FIGURE 40.15 Anterior eyeball: Exterior.

Retina   Sclera   Choroid

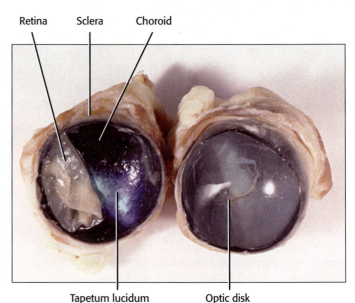

Tapetum lucidum   Optic disk

FIGURE 40.16 Tunics of the eyeball.

This gelatinous substance, the **vitreous humor**, occurs between the lens and the retina. It functions in holding the retina and lens in place and serves as a refractive medium as part of the lens system of the eye. Between the cornea and the ciliary body is the **aqueous humor**, a fluid secreted continuously by the ciliary body into this space. The aqueous humor also is removed continuously through a venous sinus surrounding the eye. Therefore, under normal conditions its volume remains constant. It functions in keeping the structures of the eye in position, as well as serving as part of the lens system of the eye. Further, it supplies the structures in this area with nutrients and oxygen while removing metabolites.

Three distinct concentric tunics or layers form the eyeball. In the posterior end of your preparation, identify the three layers (Figure 40.15 and Figure 40.16).

1. The outer layer is a tough, white, fibrous coat, the opaque **sclera**. Over the anterior surface of the eyeball the sclera is modified as a transparent window called the **cornea** (Figure 40.15).

2. The middle layer, the **choroid**, is heavily pigmented and appears black.

3. The **retina**, the innermost tunic, consists of an outer pigmented layer abutting the choroid and an inner neural layer abutting the vitreous humor. The neural layer will appear creamy, folded, and displaced from its normal position, because of preservation, while the pigmented layer generally associates with the choroid.

Because of this displacement, an iridescent bluish-green area, the **tapetum lucidum** is obvious (Figure 40.16). This is common in vertebrates that are active in subdued light. It functions to enhance light reflecting from dimly illuminated objects onto the retina. Have you noticed that cat eyes, when suddenly illuminated by incidental light in the dark, appear yellow or green? This is light reflecting from the surface of their tapetum lucidum.

 Gently move the retina to one side to see its point of attachment at the back of the eyeball.

This point of attachment is the **optic disk**. Through this area pass the neurons of the optic nerve and nutrient blood vessels. The optic disk is the area known as the blind spot, as no photoreceptors, rods and cones, are located there.

Anteriorly, the choroid is modified as the **iris**. It consists of smooth muscle arranged radially and circularly around an opening known as the **pupil** and regulates its diameter depending upon incident light (Figure 40.15 and Figure 40.17). Pigment of the iris influences eye color. The iris is continuous with a second modification of the choroid, the **ciliary body**. The ciliary body consists of **ciliary muscles** and the **ciliary process.**

The delicate **zonule fibers** of the suspensory ligament extend from the ciliary body to the lens. The ciliary muscles are smooth muscles that regulate lens shape. The folded ciliary process secretes the aqueous humor. The scalloped junction along the posterior aspect of the ciliary body is known as the **ora serrata** (Figure 40.17). It marks the anterior margin of the neural portion of the retina.

The **lens** is a biconvex transparent structure responsible for fine-focusing the ocular system (Figure 40.17 and Figure 40.18). In your specimen, the lens will appear opaque as a result of the preservation process.

The anatomy of the human eye is virtually identical to that of other mammals. Figure 40.19 represents a cross-section of a human eye.

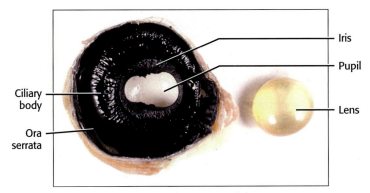

FIGURE 40.17 Internal anatomy of the eye.

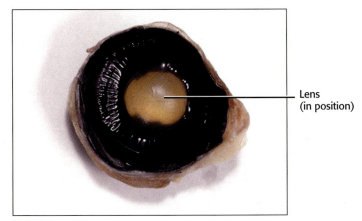

FIGURE 40.18 Lens in natural position.

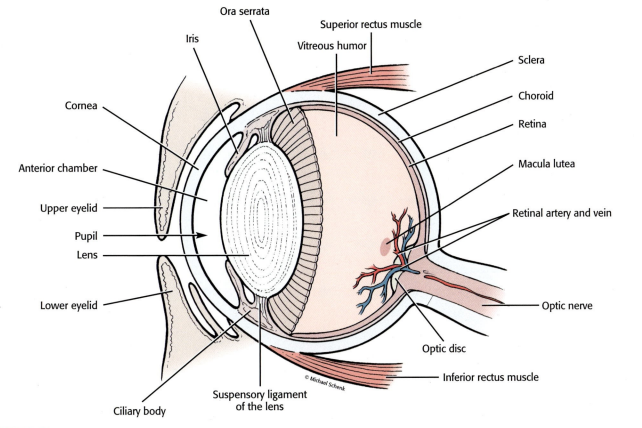

FIGURE 40.19 Sagittal section of human eye.

CHAPTER FORTY-ONE

Cat — Endocrine System

T he endocrine system of mammals is similar to that of the shark and *Necturus*.

PITUITARY GLAND OR HYPOPHYSIS

The **pituitary gland** (**hypophysis**) is connected to the ventral surface of the hypothalamus of the brain by a stalk, the **infundibulum,** and rests in the sella turcica of the sphenoid (Figure 41.1 and Figure

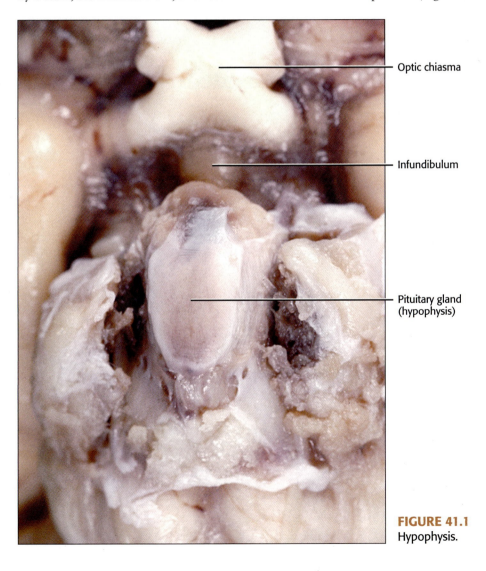

Optic chiasma

Infundibulum

Pituitary gland
(hypophysis)

FIGURE 41.1
Hypophysis.

41.2). It produces a myriad of hormones, many of which control secretions of other glands.

PINEAL GLAND

The **pineal gland** projects from the dorsal surface of the brain (Figure 41.2 and Figure 41.3). The hormone this gland produces may play a role in the onset of puberty.

THYROID GLAND

The **thyroid gland** is paired and lies on either side of the **trachea** just posterior to the larynx (Figure 41.4). Commonly, the lobes of this gland are connected by a delicate, narrow band of tissue called the isthmus. Hormones produced by this endocrine tissue are involved in a wide range of metabolic activities, tissue maturation, sexual maturation,

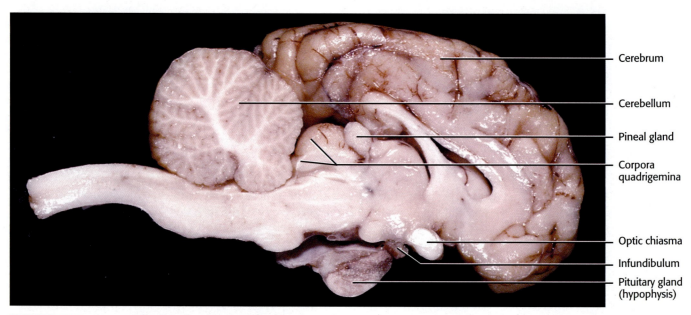

Cerebrum

Cerebellum

Pineal gland

Corpora quadrigemina

Optic chiasma

Infundibulum

Pituitary gland (hypophysis)

FIGURE 41.2 Brain: Sagittal section.

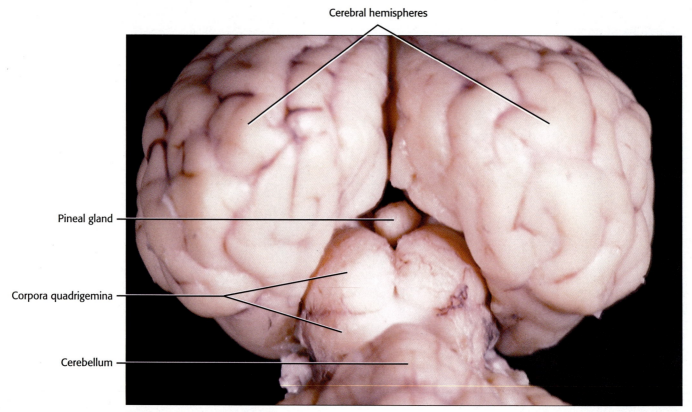

Cerebral hemispheres

Pineal gland

Corpora quadrigemina

Cerebellum

FIGURE 41.3 Pineal gland.

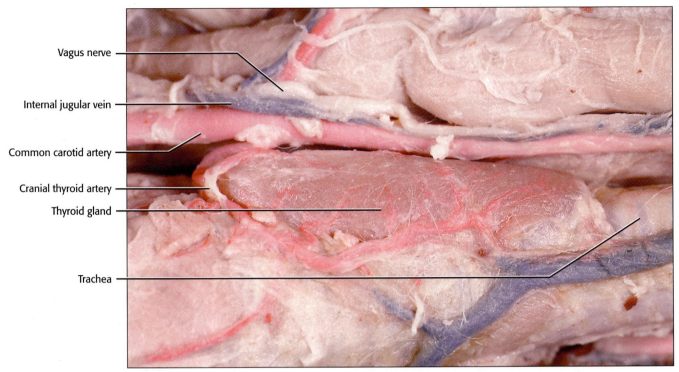

Vagus nerve

Internal jugular vein

Common carotid artery

Cranial thyroid artery

Thyroid gland

Trachea

FIGURE 41.4 Thyroid gland.

and other essential cellular activities such as energy utilization.

PARATHYROID GLAND

The four, very small, parathyroid glands are imbedded in the dorsal surfaces of the thyroid glands but cannot be seen without visual aids. Their hormone secretion is critical to calcium balance.

THYMUS GLAND

The **thymus** is a lobular gland that is very obvious in young mammals but may be difficult to find in older cats (Figure 41.5). It lies on the ventral aspect of the trachea and may extend over the surface of the heart. During the juvenile period of mammalian life, the thymus functions as a source of specialized leukocytes that migrate to other tissues to participate in immune activities.

ADRENAL GLANDS

The **adrenal glands** do not cap the kidneys, as their name implies, but are located a short distance anterio-medial to them (Figure 41.6). This organ is constructed of two areas—the outer cortex and the inner medulla. The hormones secreted by the cortex regulate a number of metabolic activities, and those of the medulla

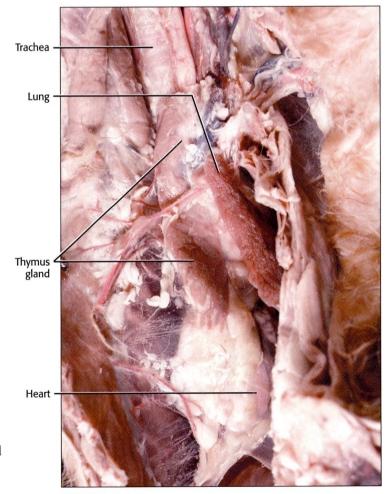

Trachea

Lung

Thymus gland

Heart

FIGURE 41.5 Thymus gland.

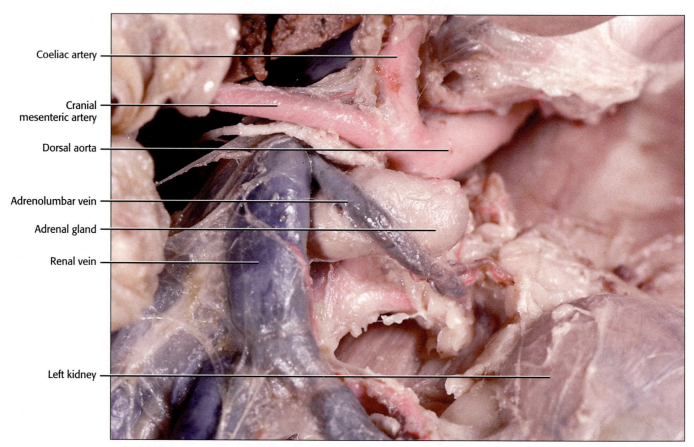

Coeliac artery

Cranial mesenteric artery

Dorsal aorta

Adrenolumbar vein

Adrenal gland

Renal vein

Left kidney

FIGURE 41.6 Adrenal gland (left side).

intensify and prolong the characteristic syndrome stimulated by the sympathetic portion of the autonomic nervous system.

PANCREAS

The **pancreas** is a lobular gland located in the mesoduodenum and part of the gastrosplenic portion of the greater omentum (Figure 41.7). This gland consists of two types of tissue—one that functions as endocrine tissue and the other that secretes chemicals involved in digestion. Hormones produced by the endocrine portion affect carbohydrate metabolism.

OVARIES

Two **ovaries**—small, oval organs of the reproductive system —are located within the peritoneal cavity and are suspended

from the dorsal body wall (Figure 41.8). The endocrine portion of the ovary produces female sex hormones, which affect sexual and reproductive behavior.

TESTES

The **testes**, the male reproductive organs, located in the **scrotum**, produce sex hormones affecting sexual and reproductive behavior (Figure 41.9).

OTHER ENDOCRINE TISSUES

Tissues in the kidney and digestive organs produce hormones whose activities tend to be more localized and confined to the functions of the organs in which they are produced.

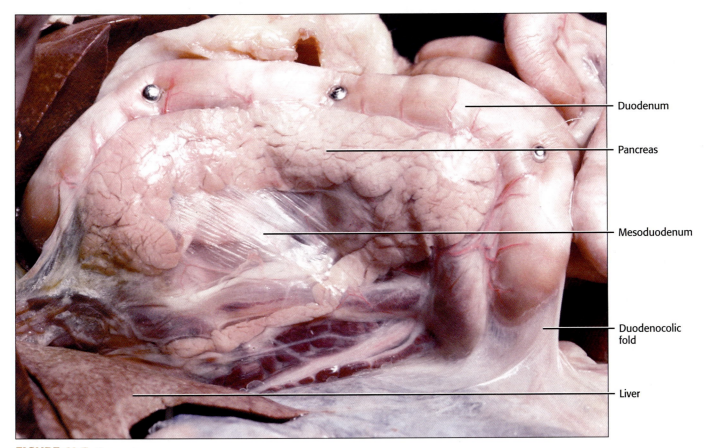

FIGURE 41.7 Pancreas.

Duodenum

Pancreas

Mesoduodenum

Duodenocolic fold

Liver

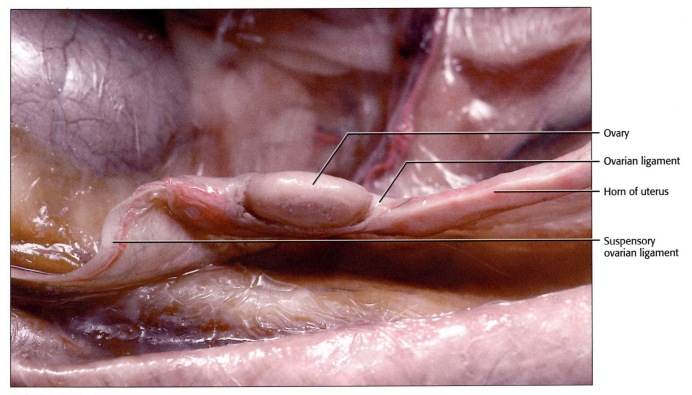

FIGURE 41.8 Ovary.

Ovary

Ovarian ligament

Horn of uterus

Suspensory ovarian ligament

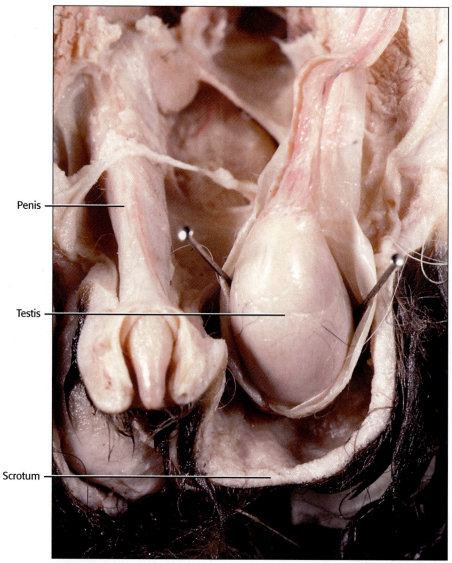

Penis

Testis

Scrotum

FIGURE 41.9 Testis.

References

Ahlberg, P. E., and A. R. Milner. 1994. The origin and early diversification of tetrapods. *Nature* 368: 507–514.

Barrington, E. J. W. 1945. The supposed pancreatic organs of *Petromyzon fluviatilis* and *Myxine glutinosa*. *Quart. J. of Microscop. Sci.* 85: 391–417.

Blackburn, D. G. 1991. Evolutionary origins of the mammary gland. *Mamm. Rev.* 21: 81–96.

Brusca, R. C., and G. J. Brusca. 1990. *Invertebrates*. Sunderland, MA, Sinauer Associates.

Carroll, R. L. 1997. *Patterns and Processes of Vertebrate Evolution*. Cambridge, MA, Cambridge University Press.

Cerny, R., P. L. Wigle, R. Ericsson, D. Meulemans, H. Epperlein, and M. Bronner-Fraser. 2004. *Developmental Origins and Evolution of Jaws: New Interpretation of "Maxillary" and "Mandibular."* Develop. Bio. 276(1) 225–236.

Chiasson, R. B. 1960. *Laboratory Anatomy of Necturus*. Dubuque, IA, Wm. C. Brown.

Clark, J A. 2005. Getting a leg up on land. *Sci. Amer.* 293 (6): 100–107.

Clemente, C. D. 1985. *Gray's Anatomy of the Human Body*. Philadelphia, Lea and Febiger.

Coates, M. I., and J. A. Clack. 1991. Fish-like gills and breathing in the earliest known tetrapod. *Nature* 352: 234–235.

Colbert, E. H., and M. Morales. 1991. *Evolution of the Vertebrates*, 4th ed. New York, John Wiley and Sons.

Daescher, E. B., Shubin, N. H., and F. A. Jenkins. 2006. A Devonian tetrapod-like fish and the evolution of the tetrapod body plan. *Nature* 440: 757–763.

Demski, L. S. 1993. Terminal nerve complex. *Acta Anat.* 148: 81–95.

Duellman, W. E., and L. Trueb. 1994. *Biology of Amphibians*, 2d ed. Baltimore, Johns Hopkins University Press.

Eddy, S., and J. C. Underhill. 1974. *Northern Fishes*, 3d ed. Minneapolis, University of Minnesota Press.

Feduccia, A., and E. McGrady. 1991. *Torrey's Morphogenesis of the Vertebrates*, 5th ed. New York, John Wiley and Sons.

Francis, E. T. 1934. *The Anatomy of the Salamander*. London, Oxford University Press.

Gilbert, S. G. 1987. *Pictorial Anatomy of the Cat*, 2d ed. Seattle, University of Washington Press.

Goodrich, E. S. 1918. On the development of the segments of the head in *Scyllium*. *Quart. J. of Microsc. Sci.* 63: 1–30.

Goodrich, E. S. 1930. *Studies on the Structure and Development of Vertebrates*. London, Macmillan. (reprinted 1996 University of Chicago Press, Chicago).

Gould, S. J. 1977. *Ontogeny and Phylogeny*. Cambridge, MA, Harvard University Press.

Guimond, R. W., and V. H. Hutchinson. 1972. Pulmonary, branchial, and cutaneous gas exchange in the mudpuppy, *Necturus maculosus* maculosus (Rafinesque). *Comp. Biochem. Physiol.* 42 (A): 367–392.

Hildebrand, M., and G. E. Goslow, Jr. 2001. *Analysis of Vertebtrate Structure*, 5th ed., New York, NY, John Wiley and Sons.

Homberger, D. G., and W. F. Walker. 2004. *Vertebrate Dissection*. 9th ed. Belmont CA, Thomson-Brooks/Cole.

Hubbs, C. L., and C. F. Lagler. 1958. *Fishes of the Great Lakes Region*. Ann Arbor, University of Michigan Press.

Janvier, P. 1996. *Early Vertebrates*. Oxford, Clarendon Press.

Ji, Q., P. J. Currie, M. A. Norell, and S. A. Ji. 1998. Two feathered dinosaurs from northeastern China. *Nature* 393: 753–761.

Kardong, K. V., and E. J. Zalisko. 1998. *Comparative Vertebrate Anatomy*. New York, WCB/McGraw-Hill.

Kent, G. C., and R. C. Carr. 2001. *Comparative Anatomy of the Vertebrates*, 9th ed., New York, NY, McGraw-Hill.

Liem, K. F., W. E. Bemis, W. F. Walker, and L. Grande. 2001, 3rd ed. *Functional Anatomy of the Vertebrates*. Belmont, CA, Thomson-Brooks/Cole.

Maderson, P. F. A. symposium organizer, 1972: The vertebrate integument. *Amer. Zool.* 12: 171.

Moore, J. A., Ed. 1964. *Physiology of the Amphibia*. New York, Academic Press.

Moss, S. A. 1984. *Sharks: An Introduction for the Amateur Naturalist*. Englewood Cliffs, NJ, Prentice Hall.

Moyle, P. B., and J. J. Cech, Jr. 2000. *Fishes: An Introduction to Ichthyology*, 4th ed. Saddle River, NJ, Prentice Hall.

Moy-Thomas, J. A., and R. S. Miles. 1971. *Palaeozoic Fishes*, 2d ed. London, Chapman and Hall Ltd.

Noble, G. K. 1931. *The Biology of the Amphibia*. New York, McGraw-Hill.

Northcutt, R. G. 1990. Ontogeny and phylogeny: A reevaluation of conceptual relationships and some applications. *Brain, Behavior and Evolution* 36: 116–140.

Northcutt, R. G. 1993. A reassessment of Goodrich's model of cranial nerve phylogeny. *Acta Anat.* 148: 71–80.

Ota, K. G. and S. Kuratani. 2007. *Cyclostome embryology and early evolutionary history of vertebrates.* Integrative and Comparative Biology, Vol. 47(3): 329–337.

Padian, K. 1998. When is a bird not a bird? *Nature* 393: 729–730.

Pough, F. H., C. M. Janis, and J. B. Heiser. 2005. *Vertebrate Life*, 7th ed. Upper Saddle River, NJ, Prentice Hall.

Putnam, J. L. and J. F. Dunn. 1978. Septation in the ventricle of the heart of *Necturus maculosus*, *J. Herpetol.* 34:292–297.

Reighard, J., and H. S. Jennings. 1935. *Anatomy of the Cat*, 3d ed., revised by R. Elliot. New York, Holt, Rinehart and Winston.

Robert, K. A., and M. B. Thompson. 2001. Viviparous lizard selects sex of embryos. *Nature* 412: 698–699.

Romer, A. S., and T. S. Parsons. 1986. *The Vertebrate Body*, 6th ed. New York, Saunders College Publishing.

Scott, W. B., and E. J. Crossman. 1979. *Freshwater Fishes of Canada*. Ottawa, Fisheries Research Board of Canada, Bulletin 184.

Shubin, N. H., E. B. Daeschler, and F. A Jenkins. 2006. The pectoral fin of *Tiktaalik roseae* and the origin of the tetrapod limb. *Nature* 440: 764–771.

Trautman, M. 1981. *The Fishes of Ohio*. Columbus, Ohio State University.

Ulmer, M. J., R. E. Haupt, and E. A. Hicks. 1971. *Anatomy of the Cat (Atlas and Dissection Guide)*. New York, Harper and Row, Publishers.

Wake, M. L. Ed. 1979. *Hyman's Comparative Vertebrate Anatomy*, 3d ed. Chicago, University of Chicago Press.

Walker, W. F., and D. G Homberger. 1993. *A Study of the Cat (with reference to human beings)*, 5th ed. Orlando, FL, Saunders College Publishing.

Weichert, C. K. 1970. *Anatomy of the Chordates*, 4th ed. New York, McGraw-Hill.

Wischnitzer, S. 1993. *Atlas and Dissection Guide of Comparative Anatomy*, 5th ed. New York, W. H. Freeman and Co.

Young, J. Z. 1981. *The Life of Vertebrates*, 3d ed. London, Oxford University Press.

Glossary

A

abdominopelvic cavity Caudal subdivision of the coelom created by the diaphragm; can be arbitrarily subdivided into abdominal and pelvic spaces.

abduct Describes muscle action that pulls a body part away from midline of body.

acoelous Type of vertebra having a centrum with essentially flat surfaces at both ends.

acrodont Dentition in which the teeth are set on the crest or lingual side of the jaw.

Actinopterygii Ray-finned fishes.

action potential Electrical disturbance triggered by a threshold stimulus causing the flow of ions across the membrane of a neuron or muscle fiber.

adduct muscle action to pull a body part toward midline of body.

adhesive organs Organs of attachment on the anterior end of metamorphosing tunicates.

afferent Designates a direction toward a reference point (*e.g.*, toward an organ such as the heart).

Agnatha Non-monophyletic fish group lacking jaws; includes "ostracoderms" and cyclostomes.

alimentary canal Digestive tract, extending from cranial mouth to caudal anus.

ammonotelic Describes an animal that excretes ammonia (*e.g.*, some fishes).

amniotes Vertebrate group possessing an amnion during embryogenesis (*e.g.*, reptiles, birds, mammals).

amphiarthrosis Joint in which there is slight movement (*e.g.*, between vertebrae).

amphicoelous Type of vertebra having a biconcave centrum.

ampullae of Lorenzini Highly specialized parts of a shark's lateral line system that permit detection of minute electrical currents such as those generated by neurons innervating resting muscle; enables sharks to detect prey, even though they might be camouflaged (*e.g.*, covered by sand).

analogous Describes structures having similar functions.

analogy Term denoting homoplastic structures with similar functions that may or may not resemble one another (*e.g.*, wing of a bird and wing of a bat, gills of a fish and lungs of a salamander); the structures do not have a common ancestry so are not homologous.

anamniotes Vertebrate group lacking an amnion (*e.g.*, fishes and amphibians).

anapsid Tetrapod skull that does not have the temporal openings found in early reptiles.

anterior Directional term meaning *ahead* or *before*.

Anura Tailless amphibians or frogs and toads.

anus Terminal end of digestive system through which undigested material (often called *feces*) passes.

aorta Major artery that carries blood away from heart.

apomorphy A derived characteristic (*e.g.*, hooves in some mammals); if the characteristic is shared, it is known as a *synapomorphic trait*.

appendicular division Portion of skeleton that includes bones or skeletal elements of anterior appendage, pectoral girdle, pelvic appendage, and pelvic girdle.

arboreal Describes habitat in which animal spends most of its life in trees.

archenteron Primitive gut formed during gastrulation.

archinephric duct Embryonic urogenital tube that in adult male vertebrates often becomes the vas deferens or remains as a urinary tube.

arteries Blood vessels that carry blood away from the heart.

artiodactyl Hoofed mammal having an even number of toes (*e.g.*, cow, antelope, giraffe).

astrocytes Cells in central nervous system that provide support and transfer nutrition to cells.

atlas First cervical vertebra modified to allow nodding motion between it and the occipital condyles of the skull.

atriopore Opening from atrium to environment.

atrium Space surrounding pharynx in urochordates and cephalochordates; water passing through pharyngeal gill slits travels to the outside via this space.

audition The act of hearing.

auricle Earlike flap attached to atria of heart in mammals that increases their volume somewhat.

autonomic nervous system (ANS) Branch of peripheral nervous system that can be subdivided into sympathetic and parasympathetic divisions.

aves Taxon of birds.

axial division Portion of skeleton that encompasses the skull, mandible, hyoid, sternum, ribs, sacrum, and tail.

axis Second cervical vertebra modified to articulate with atlas, permitting rotation of the head.

B

basapophysis Transverse process located at the base of the centrum for the articulation of subperitoneal ribs in fish.

benthic Deep layer of ocean.

bichir African member of the chondrostei that includes sturgeon and paddlefish.

bicipital Rib possessing two heads, a tuberculum and a capitulum.

blastocoel Cavity of the blastula.

blastopore Opening into the archenteron created during gastrulation of embryo.

blastula Sphere of cells resulting from cleavage of zygote and generally containing a cavity, the blastocoele.

bolus Mass of chewed food, saliva, and enzymes produced in the oral cavity.

brainstem Consists of mesencephalon, pons, and medulla oblongata, collectively, where reflex centers control activities such as vomiting, breathing, and heart functions, among others.

branchial Pertaining to the gills.

branchial arches Embryological gill supports 3–7, found in fishes.

bronchi Subdivisions of trachea that increase surface area of respiratory system in lungs.

buccal Referring to the mouth region.

buffer Solution that controls the pH of another solution.

C

caecilians Limbless, tropical, fossorial amphibians.

canal Usually tubelike passage with rounded or oval opening in bones through which nerves and blood vessels may pass.

capillaries Small-diameter blood vessels that generally connect arteries and veins; primary site of gaseous exchange, fluid and electrolyte exchanges, and other functions.

capitulum or **head** Proximal end of rib that articulates with demifacets between adjacent vertebral bodies.

carnivory Description of an animal that eats other animals.

caudal Directional term that means toward the butt or tail.

caudal vertebra Vertebrae of the tail.

cecum Blind pouch attached to intestine.

central nervous system (CNS) Consists of brain and spinal cord.

Cephalochordata Taxon including amphioxus.

ceratotrichia Fibrous fin rays that stiffen the fin.

cervical vertebra Vertebra found in neck region.

cervix Distal doughnut-shaped portion of uterine body that projects into vagina.

characins Group in the Ostariophysii that includes the piranhas.

Chimaera Group of cartilaginous fish related to sharks and rays, such as the ratfish.

choana Internal nares.

Chondrichthyes Group of fish possessing cartilaginous skeletons.

chondrocranium Cartilaginous braincase found in vertebrates.

chordate Animal possessing a notochord, dorsal hollow nerve cord, pharyngeal slits, a subpharyngeal organ that binds iodine, and a postanal tail at some time during the life cycle.

choroid Layer of highly vascularized tissue lying between the sclera and the retina.

chyme Mass of processed food in stomach.

circadian cycles 24-hour rhythms involving metabolic and/or behavioral activities.

circannual cycles Annual rhythm involving metabolic and/or behavioral activites.

cladogram Diagram constructed by phylogenetic systematists to illustrate possible relationships among organisms.

cleavage Early mitotic divisions of zygote.

cleidoic egg Shelled egg found among amniotes.

clitoris Erectile female sex organ, part of which is homologous with penis; found in a depression in ventral wall of urogenital sinus.

cloaca Space into which open the urinary, reproductive, and digestive systems.

coccyx Fused caudal vertebrae of humans and tailess apes.

coelom Body cavity further divisible into compartments (*e.g.*, thoracic, pericardial, abdominopelvic); outer wall of coelom is lined by a parietal serous membrane that is continuous with a visceral serous layer covering all organs that project into the cavity; mesenteries are double layers of membrane that suspend the organs.

collecting ducts Tubes draining the nephrons and transporting urine to renal pelvis.

colliculus(i) *See* corpora quadrigemina.

colloid Chemical compound secreted by thyroid gland; essential in binding iodine.

colonial Group of similar organisms living in close proximity.

condyle Smooth, rounded projection at end of articulating bones or skeletal elements (*e.g.*, medial and lateral condyles of femur).

contralateral Describes an action or structure on opposite side of reference.

convergence Term given to homoplastic structures in two remotely related organisms that occupy similar environments, leading to similar adaptations (*e.g.*, fins and streamlined, teardrop-shaped bodies of fish and whales); similarities are not homologous.

cornea Clear portion of sclera at anterior part of the eye through which light passes.

coronary sinus Main venous space of mammalian heart, carrying coronary venous blood to right atrium; opens into right atrium.

corpora quadrigemina Consists of two pairs of bodies known as superior and inferior colliculi, making up roof of mesencephalon of mammals; superior colliculi are centers for visual reflexes and inferior colliculi are centers for auditory reflexes.

cosmoid Thick scale with bony layers covered by cosmine, a dentine-like compound found among sarcopterygian fishes, the ancestral group of tetrapods.

cranial directional term meaning toward the head of an animal.

Craniata Group of chordates possessing a skull (cranium).

cranium Skull.

cremasteric pouch Internal sac derived from peritoneal tissues that suspends testis within outer scrotum.

ctenoid Thin, bony scale with tooth-like spines on caudal end; found among the more advanced teleosts such as basses, perches, etc.

cursorial Describes running habit of an animal.

cycloid Thin bony, rounded scale found among more primitive teleosts such as minnows, herrings, carp, etc.

cyclostomes Group of jawless fishes, probably not having a common origin (*e.g.*, lampreys and hagfishes).

D

demi-facet Smooth articulating surface on body of two adjacent vertebrae where capitulum of a rib articulates.

denticles Minute spines projecting from surface of scales of early fishes that appear to have given rise to placoid scales of sharks and true teeth of vertebrates.

dentine Bone-like compound that forms the major portion of a vertebrate tooth or scales; covered by enamel.

dermal bone Bone developing in membranes and representing remnants of primitive external bony armor.

dermis Inner layer of skin derived from mesoderm.

deuterostome Description of echinoderms and chordates in which the blastopore becomes the anus and the mouth forms from the stomodeum.

diaphragm Muscular partition between thoracic and abdominopelvic cavities of mammals; integral, essential component of respiratory system.

diapophysis Transverse process on a vertebra with which tuberculum of tetrapod rib articulates.

diapsid Tetrapod skull having two temporal openings; found in reptiles such as dinosaurs and most modern reptiles.

diarthrosis Joint in which there is free movement (*e.g.*, between femur and os coxa).

diencephalon Portion of prosencephalon from which develops the thalamic regions of the brain.

digitigrade Foot posture in which animal walks on toes.

dioecious Organisms with separate sexes.

diphyodont Limited tooth replacement found in mammals, resulting in two sets of teeth ("milk" and permanent).

Dipnoi Lungfish.

distal Refers to some distance from point of attachment or origin of a structure.

dorsal Directional term meaning toward the back of an animal.

E

ecdysis Process of shedding epidermis in squamate reptiles, amphisbaenians, lizards, and snakes.

ectoderm Embryological germ layer giving rise to epidermis, most of nervous system, etc.

ectotherms Animals whose internal temperature is determined by external heat sources and often actively attained through the animal's behavior.

edentulous Without teeth.

efferent Refers to a direction away from a reference point (*e.g.*, away from gills).

embryogenesis Developmental processes involved in formation of structures, organs, and organisms.

enamel Hardest bony material found in vertebrate body; covers surface of teeth.

endochondral bone Bone developing within cartilaginous models.

endocrine gland Ductless gland that secretes and/or releases chemical compounds (hormones) into circulatory system; these chemicals affect target tissues throughout the body to control metabolism, secretion, growth, etc.

endoderm Embryological germ layer giving rise to lining of digestive and repiratory systems; organs such as liver and bulk of pancreas are derivatives of the lining.

endolymph Fluid found in membranous labyrinth of inner ear.

endostyle Subpharyngeal organ found in invertebrate chordates that secretes mucus and binds iodine; may be homolog of thyroid gland of vertebrates.

endotherms Animals that internally maintain a more or less constant body temperature.

epaxial Refers to muscle mass dorsal to horizontal septum.

ependyma Epithelium lining the central nervous system.

epicardium Visceral epithelium covering the heart.

epicondyle Projection above or dorsal to condyle (*e.g.*, lateral epicondyle of humerus); muscles originate from these projections.

epidermis Outer layer of skin, derived from ectoderm.

epididymis Convoluted tubules occurring on dorsal surface of testes but connected to seminiferous tubules of testes, where sperm maturation occurs.

epithalamus Roof of diencephalon giving rise to pineal gland and parietal eye.

esophagus Upper portion of digestive tract between pharynx and stomach; or between pharynx and intestine, as in lampreys, where stomach is absent.

euryapsid Tetrapod skull having a single dorsal temporal opening; found in reptiles such as the large marine plesiosaurs.

Eutheria Group of vertebrates including the marsupial and true placental mammals.

exocrine glands Gland that have a duct (*e.g.*, salivary glands).

extension Muscle action causing an increase in angle of a joint and movement of a distal body part away from a proximal one (*e.g.*, extension of forearm away from upper arm by Triceps brachii).

extraembryonic membranes Four membranes exterior to embryo of amniotes, consisting of amnion, chorion, allantois, and yolk sac.

F

facet Smooth, concave surface on lateral surface of vertebral transverse process that articulates with tuberculum of a rib.

false vocal cords Anterior pair of folds at cranial end of the larynx; not involved in sound production.

fenestra Opening in bone, generally large and in the skull.

fimbria(e) Movable, fingerlike projections around edge of infundibulum that sweep the surface of ovary, making it likely that oocytes released from ovary will move into oviduct.

fissure Often elongated opening with jagged edges in bone or cartilage through which nerves and blood vessels may pass.

flexion Muscle action causing reduction of angle at a joint and movement of a distal body part toward a proximal one (*e.g.*, flexion of forearm toward upper arm by Biceps brachii).

foramen Short, smooth, usually round or oval opening in bone or cartilage through which nerves and blood vessels may pass.

fossa Depression in an organ in which another organ may sit (*e.g.*, cerebellar fossa of skull that houses cerebellum of brain).

fossa ovalis In mammals, a depression in interatrial septum, best seen in right atrial wall; marks position of fetal opening between right and left atrium, allowing oxygenated blood returning from placenta to be shunted into the "left heart" to be pumped to anterior portion of fetal body.

fossorial Describes animal specialized for digging (*e.g.*, mole).

frenulum Fold of tissues that connect lip to gum (labial) or tongue to floor of oral cavity (lingual).

frontal section Section parallel to horizontal plane of body or organ.

G

gametes Sex cells.

ganglion Collection of nerve cell bodies occurring in peripheral nervous system and outside central nervous system.

ganoid Thick scale consisting of layers of bone overlain with ganoin, an enamel-like compound found in early bony fishes and modern gars.

gastralia Ventral and abdominal bony riblike structures in some reptiles.

gastrula Embryological process in which blastula is converted into double-layered stage, resulting in establishment of germ layers ectoderm, mesoderm, and endoderm with obliteration of blastocoele and formation of primitive gut, the archenteron or gastrocoele.

geotaxis Movement toward (+) or away from (−) substratum.

germ layers Ectoderm, mesoderm, and endoderm established during gastrulation and from which the tissues and organs of the body are formed.

gills Respiratory organs found among Anamniotes.

glomerular kidney Filtering kidney occurring only in vertebrates; functional unit of kidney consists of a specialized arterial capillary, the glomerulus, associated with cup-shaped Bowman's capsule of renal tubule.

glottis Slit leading into trachea.

glycogen Polysaccharide storage product found in liver and muscles.

gnathostome Vertebrate possessing jaws.

gonads Sex organs producing gametes and sex hormones.

gonopodium Modified anal fins in some male bony fish to transfer sperm.

gustation Sense of taste.

gyrus Convolution of brain surface.

H

Hemichordata Taxon including acorn worms.

hemipenis Structure of penis unique to reptiles in which penis is bifurcated and only half is active during copulation.

herbivory Description of an animal that eats plants.

hermaphroditic Description of an individual producing both eggs and sperm.

heterochrony Development of a body part or parts at a different time than in the ancestor.

heterocoelous Type of vertebra with a centrum having saddle-shaped caudal ends with cranial end shaped to allow articulation with caudal end of an adjacent vertebra.

heterodont Dentition in which tooth shape relates to function (*e.g.,* incisors, canines, and molars of mammals).

Holocephali Taxon among chondrichthyes including chimaera.

holonephros Describes possible primitive kidney of vertebrates consisting of a series of renal tubules.

homeostasis A state in which internal environment of an organism is maintained within a narrow range in response to changes in external environment.

homodont Dentition in which all teeth are similar in shape.

homologous Describes structures or organs inherited through a common ancestry but may or may not resemble one another.

homoplastic Describes structures that may resemble one another and/or may be used in a similar fashion but have evolved independently and are not homologous.

horizontal septum Sheet of connective tissue dividing trunk musculature into dorsal and ventral masses.

hormones Chemical compounds secreted by tissues or organs that regulate functions and growth in their own or other tissues and organs.

hydroxyapatite Pyramidal crystalline form consisting of calcium salts occurring in bone; hydroxyapatite crystals generally are aligned parallel to stress lines in bone and lend strength to their structure.

hypaxial Describes muscle mass ventral to horizontal septum.

hyperosmotic Describes the relationship of two solutions in which one contains more osmotically active particles than the one to which it is compared.

hypophyseal portal system Complex of veins extending between hypothalamus and adenohypophyis that transports inhibitory and releasing hormones to secretory cells in the pituitary.

hypophysis (pituitary gland) Complex organ associated with floor of diencephalon; anterior portion (adenohypophysis) develops as an evagination of roof of pharynx, and posterior portion (neurohypophysis) develops as an evagination of floor of diencephalon; several essential hormones are released from hypophysis.

hyposmotic Describes the relationship of two solutions in which one contains fewer osmotically active particles than the one to which it is compared.

hypothalamus Floor of diencephalon containing regulatory centers of visceral activity.

hypoxia Condition of reduced oxygen concentration.

I

imbricate Describes overlapping cycloid or ctenoid scales.

infundibulum Hood-like funnel attached to proximal end of oviduct that envelops ovary and receives oocytes as they are released from ovary, thereby conveying them to oviduct; also, stalk that connects hypophysis to hypothalamus.

insertion Generally, the more movable end of a muscle, causing movement of a body part.

interstitial Term meaning between cells (*e.g.,* interstitial fluid).

intestine Portion of digestive tract between stomach and anus or cloaca where most of digestion and absorption occur.

ipselateral Describes an action or structure on the same side of reference.

isosmotic Describes two solutions that contain equal amounts of osmotically active particles.

K

keratin Protein found in epidermis and its derivatives of vertebrates (*e.g.,* feathers, hairs, horns, hooves).

keratinization Keratin production creating structures, such as nails and the epidermis.

L

labial Pertaining to lips.

larynx Boxlike organ whose wall is reinforced by a number of cartilages; vocal cords may be associated with it and glottis opens into it.

lateral Toward outer surface of a structure or side of animal.

lateral line system Primitive sensory system found in fish, allowing them to detect fluid pressure changes, thereby enabling them to determine position of object, size of object, movement of object, etc; vertebrate equilibrium and inner-ear organs seem to have evolved from some portions of anterior part of lateral line system.

Latimeria Genus of coelacanths, a sarcopterygian fish, possessing a combination of characteristics of the group as well as a number of specializations; rediscovered in the 1930s after thought to be extinct for about 65 million years.

Lepidosiren South American lungfish.

ligamentum arteriosum Ligamentous band between pulmonary trunk and aortic arch in mammals; in fetus, ductus arteriosus, a short vessel that allows most "right heart" blood to be shunted into lower body circulation away from lungs, which are nonfunctional in fetus.

linea alba White line of connective tissue that extends down midline of belly.

lingual Pertaining to tongue.

liver Large gland-like organ involved in bile secretion, metabolic activities, detoxification of potentially harmful substances, etc.

longitudinal Dimension running the length of a structure or body of an animal.

lophophore Ciliated, armlike feeding organs found in some hemichordates and echinoderms.

lumbar vertebra Vertebra found in lower back between thoracic and sacral vertebrae.

lungs Respiratory organs originating as evaginations of embryonic pharynx and found in tetrapods and some fish.

M

manus Wrist and hand.

mastication The act of chewing.

meatus Passageway or canal (*e.g.*, external auditory meatus leading to eardrum or tympanic membrane).

medial Toward inner surface of a structure or midline of animal.

mediastinum Potential space in central part of chest occupied by pericardial cavity and its contents: trachea, esophagus, nerves, and blood vessels; connective tissue occuring in mediastinum that is important in securing these structures in this area.

membranous labyrinth Membranous inner ear.

meninges (sing., mennix) Investing membranes associated with brain and spinal cord.

mesencephalon Midbrain.

mesenchyme Embryonic tissue that may give rise to most adult structures other than tissues derived from epithelium.

mesoderm Embryological germ layer giving rise to most of skeltal, circulatory, musclar systems, etc.

metanephros Functional kidney of amniotes, reptiles, birds, and mammals.

metencephalon Anterior part of rhombencephalon that gives rise to cerebellum and pons in birds and mammals.

microglia Small cells in central nervous system acting as phagocytes; responsible for eliminating cellular debris.

midline Imaginary line that extends down middle of dorsal and ventral surfaces of animal.

midsagittal Section that passes through exact midline of body or organ (there is only one of these).

monophyletic Description of a taxon that is derived from a common ancestor.

monophyodont Mammalian dention consisting of only one set, such as molars.

monotremes Group of egg-laying mammals (*e.g.*, platypus and echidna).

mouth Anterior or cranial opening into digestive tract.

mucous glands Glands that secrete mucin, which, when it comes into contact with water, forms mucus.

myelencephalon Posterior part of rhombencephalon; gives rise to medulla oblongata.

myelin Fatty insulating layer surrounding myelinated neuron axons; promotes rapid action potential transmission.

myocardium Muscular wall of heart.

myomere Blocklike mass of muscle occurring in invertebrate chordates and in some primitive adult vertebrates.

myosepta Sheets of connective tissue separating myomeres.

N

nares Synonymous with external nostrils (internal nares open into roof of mouth).

Neoceratodus Australian lungfish.

neoteny A form of heterochrony involving a retardation of the development of some features of a vertebrate, resulting in the retention of larval characteristics and seen in several salamander taxons.

nephric ridge Mesodermal tissue that gives rise to urinary and reproductive organs.

nephron Structural and functional unit of vertebrate kidney including glomerulus within Bowman's capsule and a network of tubules.

neural crest Ectodermal tissue derived during formation of neural tube; gives rise to tooth dentin, pigment cells, neurons of autonomic nervous system, tooth etc.

neuromast Sensory organ that is part of lateral line system, which detects water currents

neuron Functional unit of nervous system.

neurotransmitters Chemicals released from telodendria into a synapse; may be inhibitory or excitatory (*e.g.*, acetylcholine and serotonin).

nictitating membrane Membrane found in eye that moistens and cleans.

notochord Longitudinal skeletal rod of tissue derived from mesoderm; characteristic found in all members of phylum Chordata some time during their life cycle.

nucleus Site of concentration of nerve cell bodies within central nervous system.

O

olfaction Detection of odoriferous chemicals.

oligodendrocyte Cells found in central nervous system that create myelin sheaths around axons.

omentum Specialized mesenteries occurring in abdominal cavity; greater omentum is a double-layered, apron-like organ that hangs over viscera; lesser omentum extends between liver and stomach.

omnivory Description of an animal consuming a wide variety of food, including animals and plants.

oocyte Stage in the development of the egg.

opisthocoelous Type of vertebra having a centrum with a convex anterior end and a concave posterior end.

opisthonephric Functional kidney of anamniotes, fish, and amphibians.

optic chiasma X-shaped area of optic neural circuit on ventral aspect of diencephalon; some optic fibers pass straight through, and others cross over as they course to respective sides of brain.

oral cavity Space between gums and pharynx.

oral vestibule Space between lips or cheeks and gums.

origin Generally, the more stable and less movable end of a muscle.

Ostaryophysii A monophyletic group of primitive bony fishes with a number of specializations (*e.g.*, minnows, carp, catfish, characins).

Osteichthyes Group of fish with bony endoskeketon (internal skeleton) and bony scales covering the body.

ostium tubae Opening of infundibulum of oviduct.

ostracoderms Non-monophyletic Paleozoic group of jawless fishes that were characterized by the presence of bone in the skin

ovary Primary reproductive organ in the female from which primary oocytes or ova are released and in which important hormones controlling reproductive functions and influencing secondary sex characteristics are synthesized and secreted.

oviduct Reproductive tube leading from peritoneal cavity to uterine horns; fertilization often occurs here; also known as uterine tubes.

oviparity Egg-laying.

P

paedomorphosis Presence of juvenile characteristics in the adult.

palate Hard palate delimits roof of mouth; soft palate in mammals extends from posterior rim of hard palate into nasopharynx..

pallium Outer thin layer of grey matter of cerebrum, particularly well developed in birds and mammals.

pancreas Organ that secretes digestive enzymes and hormones associated with regulation of glucose.

papilla Small, nipplelike structure.

parallelism Term denoting homoplastic structures of two somewhat related organisms that occupy similar environments leading to similar adaptations (*e.g.*, long legs and long tails of jerboas of the Old World and kangaroo rats of the New World; both of these rodents are hopping mammals that use their front feet for food manipulation; similarities result from similar selection pressures in similar environments, and not homologies).

parapophysis Transverse process on a vertebra with which capitulum of tetrapod rib articulates.

parietal Associated with wall of a body cavity.

parietal eye A medial, light-sensitive eye occurring in some fish and reptiles.

parturition Process of giving birth.

pelagic Living in open marine waters.

penis Erectile copulatory organ in males that functions as an avenue to transfer sperm to female reproductive tract and also to allow elimination of urine.

pericardial cavity Space in which the heart lies.

pericardium Serous membrane associated with pericardial cavity; parietal layer lines pericardial sac, and visceral layer covers heart and is synonymous with epicardium of heart.

perilymph Fluid in inner ear that surrounds membranous labyrinth.

peripheral nervous system Portion of nervous system that includes cranial and spinal nerves.

perissodactyl Hoofed animal possessing odd number of toes (*e.g.*, horse, zebra, tapir, rhinoceros).

peritoneum Serous membrane associated with abdominopelvic cavity; parietal layer lines the cavity and visceral layer covers organs that project into the cavity.

pes Ankle and foot.

phagocytes Cells capable of eliminating foreign proteins and cellular debris.

pharynx Cavity shared by digestive and respiratory systems.

pheromones Chemical compounds secreted to facilitate species recognition and social reaction such sexual attraction.

phototaxis Movement toward (+) or away from (–) light.

phylogenetic system *See* cladistic system (if added as a term)

physoclist Fish whose swim bladder is independent of the gut.

physostome Fish having a swim bladder with a persistent pneumatic duct connected to the gut.

pineal organ (eye) Evagination of roof of diencephalon occurring in vertebrates; seems to function as a visible light monitor, and in some vertebrates produces a hormone involved in development of sexual maturity; in some animals, a second organ, the parietal eye, develops near the pineal and also functions as a photoperiod-controlling structure; some biologists believe these two organs originated as a pair of dorsal eyes that functioned to warn early vertebrates of overhead danger.

placenta A complex of maternal uterine and fetal extraembryonic tissues that grows during gestation across which exchange materials (nutrients, waste products, gases, hormones, toxins, etc.) between mother and developing young. Placental development occurs among some fish, some amphibians, some reptiles and many mammals.

placoid Scale found in skin of sharks; a remnant of the ancient dermal armor of the "ostracoderms."

plantigrade Foot posture in which animal walks on entire foot.

plesiomorphy Primitive characteristic (*e.g.*, four limbs among tetrapods); if the characteristic is shared, it is known as a symplesiomorphic trait.

pleura Serous membranes associated with lungs; parietal layer delimits pleural cavity and visceral layer covers lungs that project into pleural cavity.

pleural cavity Subdivision of thoracic cavity that houses the lungs.

pleurodont Dentition in which teeth are set on labial side of jaw.

plexus Network of nerves, blood vessels, or lymphatic vessels.

plica(e) Fold(s) that increases internal surface area of small intestine of some animals.

podocyte Finger-like processes involved in kidney filtration.

polyphyodont Continuous tooth replacement, found in nonmammalian vertebrates.

portal system Network of veins that transports blood from capillaries to capillaries (*e.g.*, hepatic portal system, renal portal system).

posterior Directional term meaning after or behind.

postzygapophysis or **posterior zygapophysis** Posterior projection of neural arch of vertebra whose articulating surface points ventrally and articulates with prezygapophysis of adjacent vertebra posterior to it.

prepuce Fold of skin that envelops tip (glans) of penis; sometimes referred to as foreskin.

prezygapophysis or **anterior zygapophysis** Anterior projection of neural arch of vertebra whose articulating surface points dorsally and articulates with a postzygapophysis of adjacent vertebra anterior to it; this arrangement permits adjacent vertebrae to support each other.

proboscis Snoutlike organ found in hemichordates that functions as a sense organ and burrowing and feeding apparatus.

process Projection of a bone usually distinct from main body of bone.

procoelous Type of vertebra having a centrum with concave anterior end and convex posterior end.

pronation Muscle action that results in palm of hand pointing dorsally.

pronephros Embryonic vertebrate kidney contributing to formation of archinephric duct.

proprioceptor Receptor in muscles, joints, and tendons that relays information concerning the activity in these structures.

prosencephalon Embryonic forebrain including telencephalon and diencephalon.

Protopterus African lungfish.

protostome Invertebrates in which the mouth becomes the mouth.

protraction Muscle action that advances a body part anteriorly in a direction parallel to longitudinal axis (*e.g.*, as the leg swings forward, it is protracted).

proximal Next to or nearest point of attachment or origin of a structure.

pterygiophores Skeletal fin supports; basal pterygiophores are proximal, and radial pterygiophores are distal.

pygostyle Fused caudal vertebrae of birds that supports the tail.

pylorus Sphincter muscle between pyloric stomach and duodenum; regulates passage of chyme into small intestine.

R

raphe Line of connective tissue indicating junction of two muscles (*e.g.*, between two halves of mylohyoid).

receptor Nerve ending specialized to respond to specific stimuli that generate a nerve impulse.

reflex arc Nervous circuit consisting of at least a sensory and motor neuron giving rise to a stereotypic behavior (*e.g.*, withdrawing a finger after touching a hot object).

refractive index Measure of ability of a lens to bend light; lenses with high refractive indexes(ices) have great light-bending qualities.

respiration Process of oxygen/carbon dioxide exchange.

retina Inner layer of eye, containing rods, cones, pigment cells, and neurons.

retraction Muscle action that pulls a body part posteriorly in a direction parallel to longitudinal axis (*e.g.*, as leg swings backward, it is retracted).

retroperitoneal Describes an organ situated beneath or behind parietal peritoneum (*e.g.*, kidney, ureter).

rhombencephalon Embryonic hindbrain including metencephalon and myelencephalon.

rostrum Anterior projecting shelf found in modern sharks.

S

sacral vertebra Vertebra between lumbar and caudal vertebrae.

sacrum Fused sacral vertebrae.

sagittal section Section parallel to midline of animal or organ, of which there are many.

Sarcopterygii Group of bony fish with fleshy fins ancestral to tetrapods.

Schwann cells Cells derived from neural crest material that produce myelin. a fatty insulating material found surrounding many nerve fibers in the peripheral nervous system.

sclera Outer, dense, white, connective tissue layer of eye.

secondary palate Bony and sometimes fleshy shelf that develops ventral to primary palate.

sessile Describes an organism that is attached to the substrate and sedentary.

sinus Irregular space in organ (*e.g.*, frontal sinus in frontal bone).

sinusoids Small vascular spaces having incomplete epithelial linings; found in organs such as the liver.

somatic Pertaining to the body exclusive of viscera.

spiracle Respiratory opening representing first gill slit preceding branchial gill slits.

splanchnocranium Visceral or branchial skeleton consisting of gill supports.

spleen Organ associated with circulatory system acting a reservoir for erythrocytes and a site of blood cell destruction.

stomach Portion of digestive tract between esophagus and intestine.

subunguis Underlying surface of claws, hooves, and nails.

sulcus Groove in surface of brain.

supination Muscle action that results in palm of hand pointing ventrally.

suprarenal gland Adrenal tissue lying on ventral surface of kidney.

suture Immovable joint between skull bones.

swim bladder Gas-filled sac located dorsal to viscera, permitting fish to approximate neutral buoyancy.

symphysis Type of amphiarthrotic joint between bones (*e.g.*, vertebrae or two pubes or ischia).

symplesiomorphy An ancestral characteristic shared by a group of related organisms belonging to a taxon or taxa.

synapomorphy A derived characteristic shared by a group of related organisms belonging to a taxon or taxa.

synapse Space between a presynaptic membrane and a postsynaptic membrane across which chemical transmitters move.

synapsid Tetrapod skull that has one lateral temporal opening; found among mammal-like reptilian ancestors.

synapticulae Tiny bars of cartilage that connect contiguous pharyngeal bars in amphioxus.

synarthrosis Joint in which there is little or no movement (*e.g.*, between skull bones).

synovial joint Freely movable joint containing synovial fluid (*e.g.*, knee).

synsacrum Complex fusion of vertebrae and ribs that articulates with illium and ischium of pelvis in birds.

T

taxon Group of related organisms.

telencephalon Portion of prosencephalon giving rise to cerebrum and associated olfactory regions.

teleost Group of Actinopterygii to which most modern fishes belong.

telodendrion(a) Distal ending on axon that may synapse with muscle fiber or other neuron(s)s; usually there are more than one.

testis (pl., testes) Primary reproductive organ of male where sperm are produced and hormones involved in controlling reproductive activities and influencing secondary sex characteristics are synthesized and secreted.

thalamus Lateral portions of diencephalon in which various relay centers occur.

thecodont Dentition in which teeth are set in sockets in jawbone.

therians Marsupial and eutherian mammals.

thoracic cavity Cranial subdivision of coelom created by diaphragm; compartmentalized into pleural and pericardial cavities.

thoracic vertebra Vertebra found in thorax or chest with which ribs articulate.

trabecula Small bar or rod-shaped structure usually associated with the bony tissues.

transformation Conversion of an organ found in an earlier group of animals to an often unrecognizable homolog in a later group (*e.g.,* hyomandibula of fishes, a jaw articulation element, becomes the stapes, a sound transmission element in tetrapods).

transverse or cross-section Section that passes perpendicular to longitudinal axis of body or organ.

transverse process Lateral vertebral projection.

true vocal cords In mammals, a pair of folds posterior to false vocal cords at cranial end of larynx; involved in sound production.

tuberculum Smooth, concave projection distal to capitulum of rib that articulates with transverse process of a vertebra.

tuberosity Roughened projection on the bone surface where muscles insert.

tunic Layer of tissue.

turbinates Scroll-like cartilage or bony processes occurring on ethmoid, maxilla, and nasal bones; increase nasal surface area to aid in regulating moisture and temperature of respiratory air.

U

unguis Upper surface of claws, hooves, and nails.

unguligrade Foot posture in which animal walks on toe tips.

ureotelic Describes animal that excretes urea (*e.g.,* mammals).

ureters Tubes transporting urine from renal pelvis to urinary bladder in amniotes.

uricotelic Describes animal that excretes uric acid (*e.g.,* some reptiles and birds).

Urochordata Taxon including the tunicates or sea squirts.

urodeles Tailed amphibians or salamanders.

urogenital Combined urinary and reproductive systems.

urogenital sinus Space in female mammal reproductive tract distal to vagina where vagina (vaginal orifice) and urethra (urethral orifice) open; continues to the outside, where it opens by way of urogenital aperture.

uropygial gland Oil-producing gland in birds on dorsal surface of fleshy portion of tail.

urostyle Skeletal rod consisting of fused caudal vertebrae found in anurans.

uterine horns Enlarged distal portions of oviduct where developing fetuses are carried.

uterus Muscular organ resulting from fusion of distal portions of uterine horns where developing fetuses are carried.

V

vagina Portion of female mammalian reproductive tract that receives penis of male during copulation.

vas deferens Tube leading from epididymis and conveying sperm to urethra.

veins Blood vessels that carry blood toward heart.

velum Membrane in amphioxus and larval lampreys involved in feeding and/or respiration.

vena cava Major veins that deliver blood directly to right atrium.

ventral Directional term meaning toward belly of animal.

vestibule Cavity of oral hood of amphioxus; also, space between lips and teeth in oral cavity of mammals.

vestigial Refers to organ that is a reduced version of a fully developed structure in a related organism (*e.g.,* coccyx [reduced tail] of humans and tail of baboon).

vibrissae Tactile coarse hairs occurring on body of mammals.

villi (villus) Fingerlike projections that increase internal surface area of small intestine in some animals.

visceral Associated with surface of organs projecting into a body cavity.

visceral arches Embryological pharyngeal supports occurring in vertebrates; in jawed vertebrates, 1 and 2 form the jaws and hyomandibula, respectively; in fishes 3–7 support the gills.

viviparity Reproductive process in which female gives birth to living young.

vomeronasal organ A pocket found as an invagination in roof of oral cavity in most tetrapods; associated with chemoreception (*e.g.,* sex pheromones, tracks of prey).

W

Weberian ossicles Series of small bones extending from swim bladder to inner ear of ostariophysian fishes; transmit pressure changes in swim bladder caused by sound vibrations in the environment and interpreted as sound by the fish.

Index

acanthodians, 37, 39, 41, 119
acetylcholine, 80
acorn worms, 9 (Fig. 1.1), 10 (Figs. 1.2, 1.3)
actin, 52
action
  muscle, 53
  potential, 80
adenohypophysis, 95, 96
adrenal (suprarenal)
  cortex, 97
  glands, 97, 433
  medulla, 88
  mudpuppy (*Necturus*), 274, 275 (Fig. 32.5)
adrenocorticotropic hormone (ACTH), 95
agnatha, 99
  chondrocranium, 36
  notochord, 42
aldosterone, 97
alimentary canal
  cat, 355, 357
  dogfish shark, 162 (Fig. 19.8)
allantois, 63
alligator, brain, 81 (Fig. 12.3)
amino acids, 61, 97. *See also* protein digestion
ammocetes, 116, 117
ammonia, 61
amniotes
  adrenal medulla, 97
  autonomic nervous system, 88
  brain, 83, 84, 411
  fertilization, 66, 68
  intestine, 59
  kidney, 62
  motion, 54
  skin (Fig. 4.3), 30
  spinal nerves , 85
  sternum, 46
  urinary bladder, 63
  veins, 77
  vertebrae, 45
amphibians, 211. *See also* salamanders
  atlas in, 45
  brain and meninges, 82, 83
  circulatory system, 72, 75
  fertilization in, 65
  girdle, 49 (Fig. 5.18)
  glands, 33
  lateral line system in, 90, 124, 199
  lymph system in, 78
  muscle, 225
  respiration in, 58
  skin / epidermis, 30, 102, 214, 215–216 (Figs. 24.5–24.6), 246
  sound transmission, 92
  teeth and diet, 40
  urinary bladder, 63
  vertebrae, 42
amphioxus, 21, 22 (Figs. 3.1, 3.2), 23 (Fig. 3.3), 26 (Fig. 3.9, 3.10)
  circulation, 71
  digestion, 98
  excretory system, 24, 26
  female, 24 (Fig. 3.5)
  hindgut, 24, 26 (Fig. 3.10)
  male (midgut), 23, 25 (Figs. 3.7, 3.8)
  oral hood, 24 (Fig. 3.4)
  pharynx (female), 24 (Fig. 3.6)
ampullae of Lorenzini, 91 (Fig. 12.12)
anamniotes
  brain, 82, 83, 411
  facial nerve in, 87
  fertilization, 65
  kidney, 62
  nerve cord in, 85
androgens, 98
annelids, 1, 55
anterior, definition, 4
antlers, 32 (Fig. 4.5), 33, 282
anurans, 58, 75. *See also* salamanders
anus, 59, 60
  in deuterostome, 2 (Fig. 1.1)
  in mammals, 355
aortic arch, 74 (Fig. 11.6), 75 (Fig. 11.7)
appendicular division, 35
appendix, 59
aqueous
  fluid, 92
  humor, 204, 428
archenteron, 2 (Fig. 1.1)
archinephric duct, 63
archosaurs, 42
arginine vasotocin, 96
armadillos, 68
arteries, 69, 75. *See also* blood vessels
  ancestral circulation pattern, 76 (Fig. 11.8)
  carotid, 74, 76, 114
arthropods, 1
Ascidiacea, class, 15
astrocytes, 80
atriopore in adult tunicate, 15
audition. *See* hearing
auditory apparatus. *See also* ear
  mammal, 200
autonomic nervous system, 82, 88, 114
Aves class, 4
axial division, 35
axons, 79, 80, 81, 96

Balanoglossus, 9 (Fig. 1.1), 10–11 (Figs. 1.3, 1.4)

baleen, 32 (Fig. 4.5), 33, 40
basapophyses, 136
bats, 67
beaks, 31, 40. *See also* birds
bichir scales, 30
bile, 60, 98
binomial nomenclature, 2
birds
  aortic arch in, 75
  beaks, 31, 40
  brain and meninges, 81 (Fig. 12.4), 83
  bursa of Fabricus, 97
  carrion-eating, 82
  cecum, 59
  claws, 31
  crop, 58, 96
  feathers, 30
  gizzards, 59
  heart, 72
  limb changes in, 49
  renal portal system, 78
  reproduction, 68
  skin, 33
  skull, 42
  sound transmission in, 92
  sternum and synsacrum, 45
  tongues, 58
  uricotelic, 62
  vertebrae, 43
bladder. *Also see* urinary system
blastopore, 2 (Fig. 1.1)
"blind spot," 206
blood, 69, 72. *See also* circulation; heart
  clotting, 69
  fetal, 72, 74
  pressure, 96, 97
blood vessels, 69
  arteries, 76 (Fig. 11.8)
  veins, 76 (Fig. 11.9), 77 (11.10), 88
B lymphocytes, 97
boas, 94
body cavities, 55 (Fig. 7.1)
bone
  endochondral, 35, 99
  facial, mammal, 284
bony fishes, 37, 38, 119
  aortic arches in, 74
  kidneys in, 97
  pectoral girdle in, 48, 54
  ribs and vertebral column, 43, 45
  scales, 30 (Fig. 4.2)
  teeth in, 39, 41 (Fig. 5.6A), 42 (Fig. 5.7)
Bowman's capsule, 62, 371
brain, 79, 83 (Fig. 12.4)
  cavities / ventricles, 82–83

447

evolution / development, 80 (Fig. 12.2)
    mammal, 416 (Fig. 40.5), 418 (Fig. 40.6)
    vertebrate, 130, 191
braincase, 130, 218. *See also* chondrocranium
brainstem, 82, 417
Buccal cirri (amphioxus), 21, 22 (Fig. 3.2), 23
    (Figs. 3.3, 3.4)
budding, 15
buffer solutions, 60

caecilians, 68
calcitonin, 97, 98
canine teeth, 40, 277, 284, 288
capillaries / capillary beds, 71, 78. *See also*
        blood vessels; circulatory system
carnivores
    cecum, 59
    reproduction, 67
    teeth, 40
    tongues, 58
cat, 1, 99
    alimentary canal, 355, 357
    appendicular division, 302
    arterial schematic, 388 (Fig. 39.7A)
    atlas and axis, 283 (Fig. 34.1), 284, 299
        (Figs. 34.22, 34.23)
    carpals, 303
    body cavities and sinuses, 296, 345–346
        (Fig. 36.3)
    chondrocranium, 284
    circulatory system, 381
    clavicle, 304 (Fig. 34.32)
    cranial foramina, 296–297
    cranial nerves (table), 417
    dental formula, 284
    dentary, 298
    digestive tube, 360, 361 (Fig. 37.7)
    esophagus, 361 (Fig. 37.3)
    excretory system, 372 (Fig. 38.1)
    external anatomy / features, 279 (Fig. 33.1)
    female, 371
        reproductive system, 353–354 (Figs. 36.12,
            36.13)
        urogenital system, 374–378 (Fig.
            38.4–38.10), 380
    femur, 306, 307 (Fig. 34.37)
    gallbladder, 364
    head, dissecting, 358–359 (Figs. 37.2, 37.4)
    humerus, 303, 304 (Fig. 34.32)
    hyoid, 297, 299 (Fig. 34.20)
    intestines, 362–363 (Figs. 37.8–37.9)
    kidneys, 371–373 (Figs. 38.1–38.2)
    liver, 361, 363 (Fig. 37.11), 364
    lungs, 364, 369 (Fig. 37.16)
    lymphatic system, 381
    male, 371
        abdominopelvic region, 352 (Fig. 36.11)
        urogenital system, 373 (Fig. 38.3), 374,
            377–379 (38.11–38.13), 380
    mandible, 297, 298 (Fig. 34.19)
    manus, 305 (Fig. 34.35)
    mesenteries, 345–347 (Fig. 36.4), 351–353
        (Figs. 36.9–36.12)
    mouth (oral cavity), 357
    muscles
        abdominal, 312 (Fig. 35.2), 313
        back, 313–314 (Fig. 35.4), 320

dissection, 310
forelimb (brachium and antebrachium),
    325–330 (Figs. 35.16–Fig. 35.20)
gastrocnemius, 336 (Fig. 35.26)
head, 322–324
hip, 332 (Fig. 35.22), 339–341 (Fig. 35.28)
jaw (masseter), 323 (Fig. 35.13), 355
lumbar, 317 (Fig. 35.7), 318
neck, 318–322 (Figs. 35.8–35.12)
pes, 336 (Fig. 35.26), 338 (Fig. 35.27),
    339
plantaris, 338 (Fig. 35.27), 339
quadriceps, 334–336 (Fig. 35.25)
shank, 336–338 (Figs. 35.26–35.27)
shoulder, 314 (Fig. 35.4), 321–322 (Figs.
    35.10–35.12), 324
tail, 340, 342 (Fig. 35.28)
temporalis, 323–324 Fig. 35.14)
teres minor, 325 (Fig. 35.15)
thigh, 331–334 (Fig. 35.21–35.24)
thoracic, 310, 311 (Figs. 35.1), 312, 313
    (Fig. 35.3), 315–316 (Figs. 35.5, 35.6),
    317
nictating membrane, 428 (Fig. 40.14)
nostrils, 279 (Fig. 33.1)
omentum, greater and lesser, 346, 349–350
    (Fig. 36.6, 36.7)
os coxa, 306 (Fig. 34.36)
pancreas, 364, 365–366 (Figs. 28.12, 28.13)
parotid gland, 355
patella, 307 (Fig. 34.37)
pectoral girdle and forelimb, 302
pelvic girdle and hindlimb, 305, 306 (Fig.
    34.36)
pericardia, 348 (Fig. 36.4C)
pes, 307
phalanges, 305 (Fig. 34.35)
pinna (Fig. 33.1)
radius, 304 (Fig. 34.33)
reproduction, 374–379 (Figs. 38.4–38.6)
respiratory system, 364, 366–369 (Figs.
    37.14–37.16)
ribs, 301–302 (Fig. 34.29)
salivary glands, 356 (Fig. 37.1)
scapula, 302, 303 (Fig. 34.30)
sinuses, 296
skeleton, 283 (Fig. 34.1)
skinning, 309–310
skull, 284, 285–288 (Figs. 34.2–34.3)
    basisphenoid, 293 (Fig. 34.15)
    ethmoid, 288, 290 (Fig. 34.8)
    frontal, 291 (Fig. 34.10)
    lacrimal, 288, 289 (Fig. 34.7)
    nasal, 288, 289 (Fig. 34.6)
    occipital, 292 (Fig. 34.14), 293
    palatine, 287 (Fig. 34.3), 288, 289 (Fig.
        34.5)
    parietals / interparietals, 291, 292 (Figs.
        34.12, 34.14)
    phalanges, 308 (Fig. 34.39)
    premaxilla and maxilla, 285–288 (Figs.
        34.2–34.4)
    presphenoid and sphenoid, 294 (Fig.
        34.16), 295 (Fig. 34.18)
    surface features, 296
    temporal, 294, 295 (Fig. 24.17)
    vomer, 290 (Fig. 34.9)
    zygomatic, 291 (Fig. 34.11)

spleen, 364, 366 (Fig. 37.13)
sternum, 301 (Fig. 34.28)
stomach, 361
thoracic vertebrae, 297
tibia, 307, 308 (Fig. 34.38)
tongue, 359 (Figs. 37.5–37.6)
trachea, 367–368 (Figs. 37.14, 37.15)
ulna, 305 (Fig. 34.34)
urinary system, 371
venous schematic, 389 (Fig. 39.7B)
vertebrae / vertebral column, 297–298,
    299–301(Figs. 34.21–3.27)
whiskers, 279
catfish, 58
caudal
    definition, 4
    fins, 46, 47 (Fig. 5.15)
Caudata order, 211
cecum, 59
cellulose, 16
central nervous system, 79–80, 82, 84 (Fig.
    12.5)
centrum, 42–43, 44 (Fig. 5.9)
Cephalochordata / cephalochordates, 2, 21, 42,
    84, 119. *See also* amphioxus
cerebellum, 81 (Fig. 12.3), 83 (Fig. 12.4), 86
    (Fig. 12.7), 269, 411, 412
cerebrospinal fluid, 83–84
cerebrum, 81 (Fig. 12.3), 82, 83, 411, 412
cervix, 68
Cetacea, testes, 65
chameleons, tongue, 58
characins, 39
chimaeras (ratfish), 40
cholecystokinin, 98
chondrichthyes, 37, 39, 58, 61, 62, 65, 68, 70,
    89, 91, 119, 129
chondrocranium, 35, 36 (Fig. 5.1), 37
Chordata phylum / chordates, 1, 2, 9, 16, 21, 99
chromatophores, 29, 33, 216
chromosomes, 65
chyme, 98
cilia, 9, 10, 11, 119
circadian and circannian cycles / rhythms, 82,
    94
circulation; circulatory system, 60, 69, 97. *See
    also* heart
    double pump circuit, 73 (Fig. 11.5)
    hormones in, 207
    single pump circuit, 71 (Fig. 11.4), 114
cladistic system, 3–4
cladogram, 3 (Fig. 1.2), 4
claspers, 68, 122, 123 (Fig. 15.5), 164 (Fig.
    10.13), 171, 174 (Fig. 20.6)
claws, 30, 31 (Fig. 4.4), 32, 282
clitoris, 68
cloaca, 60, 62–63, 68
coccyx, 45
cochlea and cochlear canal, 38, 91, 95 (Fig.
    12.13)
coelacanth, 91
coelom, 24, 55, 68, 109, 113, 153
colon, 59
Columbidae family, 58
copulation, 68. *See also* fertilization
cornea, 92, 94 (Fig. 12.14)
corporora quadrigemina, 82

corpus callosum, 416
corpus luteum, 95
corticoisterone and cortisol, 97
Craniata taxon, 4
cranial nerves, 86–88, 90, 92
craniates
    digestive and respiratory evolution in, 159
    olfaction, 89
    portal system in, 78
cranium, 41, 284. *See also* skull
crocodiles, 30, 42, 59, 68, 72, 78
crop (birds), 58
crotalids, 94
cyclostomes, 68, 70, 84

dendrites, 79 (Fig. 12.1), 80, 81
dental formulae, 40–41
dentary. *See* teeth
denticles, 29 (Fig. 4.1)
dentin, 126, 127 (Fig. 15.13)
dermatocranium, 36, 41 (Fig. 5.6), 42 (Fig. 5.7)
dermis, 29
    of mammals, 31, 33, 281
    shark, 126, 127 (Fig. 15.12)
deuterostomes, 1, 2 (Fig. 1.1)
diencephalon, 80–81 (Figs. 12.2, 12.3), 82, 92, 95, 96
digestive system / digestion, 58, 59, 60, 62, 98
    mammal, 355
    shark, 159, 163
dinosaurs / Dinosauria taxon, 4, 30, 40, 42, 49, 59
dissection tools, 5
distal, definition, 4
diverticulum (amphioxus), 23, 24
dogfish shark, 1, 4, 74, 99, 119
    alimentary canal, 162 (Fig. 19.8)
    ampullae of Lorenzini, 191, 199, 201 (Fig. 22.12)
    anus, 162, 164 (Fig. 19.12)
    archinephric duct, 170
    auditory apparatus, 195, 200, 202
    arteries, 176, 179–186 (Figs. 21.8–21.21)
    autonomic nervous system, 199
    axial and appendicular divisions, 129 (Fig. 16.1), 130, 138
    brain and cranial nerves, 191, 192–196 (Fig. 22.1)
    cartilage, 129, 133 (Fig. 16.10), 134 (Figs. 16.11, 16.12)
    caudal vertebrae, 135 (Figs. 16.13, 16.14)
    chondrocranium, 129, 130–133 (Figs. 16.3–16.8), 191
    cloaca, 162 (Fig. 19.8), 164 (Figs. 19.11, 19.12)
    cranial nerves, 192–196 (Figs. 22.1–22.5), 198 (table), 284
    cutting into, 159 (Fig. 19.1), 160 (Fig. 19.2)
    diencephalon, 191, 192
    duodenum, 98
    endocrine system, 207
    epidermis, 126 (Fig. 15.11), 127 (Fig. 15.13)
    epiphysis, 192
    equilibrium apparatus, 200, 201 (Fig. 22.13)
    esophagus, 161 (Fig. 19.6), 162 (Figs. 19.7, 19.8)
    external anatomy, 121, 122 (Fig. 15.2)

eye, 124, 125 (Fig. 15.9)
female, 122, 123 (Fig. 15.6), 126, 140 (Fig. 16.23), 157 (Fig. 18.8), 169, 170–171 (Figs. 20.2, 20.3)
fins, 121, 122 (Fig. 15.2), 123 (Fig. 15.4), 136, 137 (Figs. 16.17, 16.18), 138 (Fig. 16.19), 139 (Fig. 16.20)
foramina, 131–133 (Figs. 16.4, 16.5, 16.8)
gallbladder, 165 (Fig. 19.14)
gills / gill slits / gill rays, 125 (Fig. 15.10), 132, 133 (Fig. 16.10), 160 (Fig. 19.2), 161, 166 (Fig. 19.16), 167 (Fig. 19.17), 168
girdle
    pectoral, 129 (Fig. 16.1), 138, 139 (Fig. 16.20)
    pelvic, 138, 140 (Figs. 16.22, 16.23)
glands, 208 (Figs. 23.2, 23.3)
glossopharyngeal foramina and nerves, 131, 132 (Fig. 16.7)
heart, 176–178 (Figs. 21.2–21.7)
hyoid arch, 133–135 (Figs. 16.9–16.11)
infundibulum, 207 (Fig. 23.1)
intestine, 161 (Fig. 19.6), 162 (Fig. 19.8), 163 (Fig. 19.10)
keel, 123 (Fig. 15.4)
kidney, 169 (Fig. 20.1)
lateral line system, 199, 200 (Fig. 22.11)
liver, 161 (Fig. 19.6), 162 (Fig. 19.8), 165 (Fig. 19.14)
lymphatic system, 189–190
male, 122, 123 (Fig. 15.5), 126, 140 (Fig. 16.22), 157 (Fig. 18.9), 169, 170, 172–174 (Figs. 20.4–20.6)
mandibular arch, 134–135 (Figs. 16.10–16.11)
medulla oblongata, 134–135 (Figs. 16.10–16.11), 195
meninges, 197–198
mesencephalon, 191, 192
mesenteries, 155, 156 (Figs. 18.6–18.7), 157 (Figs. 18.8–18.9)
metatarsals, tarsals, and phalanges, 308 (Fig. 34.39)
muscles, 141
    branchiomeric, 146–150 (Figs. 17.12–17.16)
    dorsal fin, 143 (Fig. 17.4), 146
    hypobranchial, 150–151 (Fig. 17.16–17.18)
    pectoral and pelvic fins, 143–145 (Figs. 17.5–17.10)
    trunk or axial, 143–144 (Figs. 17.4–17.6)
oculomotor nerve, 192
olfactory apparatus, 199, 200 (Fig. 22.10)
opening for dissection, 153 (Fig. 18.1), 154–155 (Figs. 18.2–18.3)
optic nerve, 206
pancreas, 164 (Fig. 19.13), 165 (Fig. 19.15), 209
pineal gland, 308 (Fig. 23.2)
placoid scales, 121 (Fig. 15.1), 127 (Fig. 15.12)
pleuroperitoneal and pericardial cavities, 155 (Fig. 18.4)
portal system, 186–188 (Fig. 21.24)
rectum, 162 (Fig. 19.8), 163 (Fig. 19.10), 164 (Figs. 19.11, 19.13)

reproductive system / organs, 170, 209, 210 (Figs. 23.5, 23.6)
respiratory system, 166–167 (Figs. 19.16–19.18), 168
rostrum, 130–133 (Figs. 16.3–16.9)
sense organs, 199
skeleton, 129–130 (Figs. 16.1, 16.2)
skin, 126 (Fig. 15.11)
skinning, 141, 142 (Figs. 17.1–l7.3)
spiracle, 124, 125 (Figs. 15.9, 15.10)
splanchnocranium, 129, 132, 133–134 (Figs. 16.9–16.12)
spleen, 161 (Fig. 19.6), 162 (Fig. 19.8), 163 (Fig. 19.10), 164 (Fig. 19.13), 165 (Figs. 19.14, 19.15)
stomach, 161 (Fig. 19.6), 162 (Fig. 19.7), 164 (Fig. 19.13)
teeth, 159, 160 (Figs. 19.2–Fig. 19.4)
telencephalon, 191
tongue, 160 (Fig. 19.3), 161 (Fig. 19.5)
trunk vertebrae, 135, 136 (Figs. 16.15, 16.16)
urogenital system, 169–171, 173 (Fig. 20.5)
uterus, 170 (Fig. 20.2)
veins, 186–190 (Figs. 21.24–21.26)
ventricles, 197 (Fig. 22.8)
vertebrae, 135, 136 (Figs. 16.15, 16.16)
dorsal, definition, 4
duodenum, 98

ear
    bones / ossicles, 4, 424
    cochlea, 426 (Fig. 40.12C)
    human, 425 (Fig. 40.12)
    inner, 92, 131
    in larval lamprey, 117 (Fig. 14.30)
    mammals, 38 (Fig. 5.4), 277, 424
    middle, 38, 58, 92, 424
    parts of, 93 (Figs. 12.12, 12.3)
eardrum. *See* tympanic membrane
ecdysis, 30
echidnas, 91, 277
Echinodermata phylum / echinoderms, 1, 2, 9
elasmobranches / Elasmobranchii, 98, 119
electrolytes and electroreception, 90, 91, 95, 97, 122
endocrine system / tissue / glands / organs, 60, 79, 95, 431, 434
endolymph / endolymphatic ducts, 91, 124 (Fig. 15.7), 125, 202
endomysium, 52
endostyle (amphioxus), 23, 23 (Fig. 3.6)
endotherms, heart, 72
Enteropneusta / enteropneusts, 9 (Fig. 1.1), 11
enzymes, digestive, 59, 60, 98, 355
ependyma, 83
epidermis, 29, 33. *See also* skin
    in mammals, 280–281
    in squamate reptiles, 30
epiglottis, 58
epimysium, 52
epinephrine, 97
epiphysis, 192, 208. *See also* pineal gland
epithalamus, 82, 191, 416
epithelium, 97, 126
    amphioxus, 24
    olfactory, 89 (Fig. 12.9)

equipment, lab, 5
erythrocytes, 69, 99, 166
erythrophores, 33
esophagus, 58, 59
estrogens, 98
Eulamprus tympanum, 65
Eusthenopteron, 42 (Fig. 5.7), 48 (Fig. 5.17)
evolutionary system, 2–3
eye
    of aquatic vertebrates, 92
    human, 94 (Fig. 12.14), 429 (Fig. 40.19)
    lateral, 92, 94, 124
    lizard, 94
    median, 94
    muscles of vertebrates, 426
    optic nerve, 87
eyelids, 94, 124
eyespot, amphioxus, 21, 22 (Figs. 3.1, 3.2), 23
        (Fig. 3.3)

fasiculi, 52
feathers, 30, 33
    tail, 45
feces, 62, 355. *See also* anus
feet, tetrapod, 49, 50 (Fig. 5.20)
female reproductive system, 65, 68
    marsupial, 66 (Fig. 10.2)
    reproductive tract, 66 (Fig. 10.1), 72
femur, 48 (Fig. 5.17)
fertilization, 65, 68, 98. *See also* reproduction
        system
    amphioxus, 26
fibula, 48 (Fig. 5.17)
"fig.ht-or-flight" response, 97
fin(s), 46, 47 (Fig. 5.15), 48 (Fig. 5.17)
    muscle, 53 (Fig. 6.4), 54
fish. *See also* bony fishes
    bottom-feeding, 90
    brain, 82
    cartilaginous, 61, 90, 129
    corneas, 92
    crossopterygian, 40
    freshwater, 61, 63
    jawed, 119
    jawless, 36 (Fig. 5.2A), 38
    lungs in, 57
    marine, 61
    respiratory system, 166
    skin and glands, 33
follicle stimulating hormone (FSH), 95
foramina / foramen, 135
frogs, reproduction in, 68
frontal section, definition, 4
fur, 31. *See also* hair

gallbladder, 98
gametes (amphioxus), 26
ganglia, 80
gars, scales in, 30 (Fig. 4.2)
gastrin, 98
gastrula / gastrulation, 2
genes, 65
genus, 2
Gila Monster, 40
gills / gill slits / gill cartilages, 35
    in agnatha, 36–37
    blood flow through, 71, 72

diffusion through, 61, 62
and respiration, 57, 58
girdles
    musculature in tetrapods, 54
    pelvic and pectoral, 47, 48 (Fig. 5.17), 49
        (Fig. 5.18)
gizzards, 59
glands. *See also* endocrine; hypophysis
    adrenal, 97
    amphibian, 33
    apocrine and exocrine, 33, 282
    ductless, 95, 207
    integumentary, 33
    in mammals, 33
    mucous, 89 (Fig. 12.8), 214
    parathyroid, 98, 274, 433
    scent, 282
    sebaceous, 282
    sweat, 33, 88
    tear, 92
    thymus, 97
    thyroid, 95, 97, 98, 119
glial cells, 80
glomerulus, 62
glottis and epiglottis, 58
glucagon, 60, 98
glucocorticoids, 95, 97
glycogen / glucose, 60, 97
gnathostomes, 1, 36, 42, 68, 74, 78, 97, 119,
        202, 283
gobies, 47
gonadotropic releasing hormone (GnRH), 87
gonadotropin, 98
gonads, 65
    amphioxus, 26
gray matter, 81–82, 84 (Fig. 12.5)
growth hormone (GH), 96
guanine, 33
gustation, 90–92
gustatory impulses, 88
gut, 60. *See also* digestive system

hagfish, 4, 38, 42, 62, 91, 202
hair, 31, 33, 281
Hatschek's pit, 21, 23 (Fig. 3.3)
HCl, 98
hearing, 87, 91. *See also* ear
heart, 69, 78. *See also* circulation
    fetal, 74
    mammalian, 381
    vertebrate, 70 (Figs. 11.1, 11.2)
Hemichordata / hemichordates, 1, 2, 9, 11
hemoglobin, 99
herbivores, 40, 58, 59
hererochrony, 211
hindlimb, mammalian, 280, 305
holocephalans / Holocephali, 59, 129
holonepros, 62
homeostasis, 82
homologous characteristics, 4
homoplasty, 4
hooves, 282
hormones, 60, 95, 96, 207, 434. *See also* en-
        docrine system
    adrenal, 97
    enteric, 98
    male, 65

pituitary, 95
reproductive / sex, 65, 97, 371
horns, 32 (Fig. 4.5), 33, 282
humerus, 48 (Fig. 5.16)
hyoid, 37, 58
hyostyly, 135
hypoglossal nerve, 88
hypophysis, 95, 96 (Fig. 13.1)
hypothalamus, 82, 95, 96, 191

Ichthyomyzon, 118
immune system, 97
incisors, 40, 284
incus, 38, 93 (Fig. 12.13), 277, 424
inner ear, 91, 93 (Fig. 12.13)
insulin, 60, 98
integumentary system. *See* glands; skin
intercostal spaces, 78
interneurons, 80, 85 (Fig. 12.6)
intestines, 59, 60, 98
    acorn worm, 11 (Fig. 1.5)
    mammal, 355
iodine, 2, 97
iridophores, 33

jaws, 37 (Fig. 5.3), 38 (Fig. 5.4), 99
    mammalian, 284
    shark, 38, 121, 135
    articulation and ear ossicles, 38 (Fig. 5.4)

keel, 122, 123 (Fig. 15.4)
keratinization, 30, 31, 214, 281
kidneys, 61, 62, 63, 78, 97, 99

labyrinthodonts, 40
Lagamorpho, testes, 65
Lampetra, 118
lamprey, 1, 4, 38, 42, 55, 59, 60, 62, 71, 91, 97,
        99, 101
    anus, 110, 111 (Fig. 14.19), 162, 164 (Fig.
        19.12)
    archinephric duct, 111
    body cavities, 237
    brain, 81 (Fig. 12.3), 114, 191
    cartilage, 107 (Figs. 14.11, 14.12), 108
    circulatory system, 113 (Figs. 14.22, 14.23),
        114
    cloaca, 110, 111 (Fig. 14.19)
    cranial nerves, 86
    dermis, 104, 105 (Fig. 14.8)
    digestive system, 109–110 (Fig. 14.17)
    epidermis, 104, 105 (Fig. 14.7)
    equilibrium, 91
    esophagus, 110 (Fig. 14.17), 116
    excretory system, 110–111 (Fig. 14.18)
    external anatomy, 101–102
    eye, 101, 108–109
    gallbladder, 116
    heart, 113, 116, 118 (Fig. 14.31)
    integumentary system, 102, 104, 105–106
        (Figs. 14.7–14.9)
    internal anatomy, 106
    intestines, 110, 116, 117 (Fig. 14.30)
    kidney, 110–111 (Fig. 14.18), 117 (Fig.
        14.30)
    larvae, 101, 114, 115–118 (Figs.
        14.24–14.31)

lateral line system, 101, 102 (Fig. 14.1), 103 (Fig. 14.4), 114
liver, 109 (Fig. 14.15), 110, 116
mouth, 101, 104 (Fig. 14.5), 105
muscular system, 108 (Figs. 14.13, 14.14)
nervous system, 114
notocord, 107 (Figs. 14.11, 14.12), 108 (Fig. 14.14), 109 (Fig. 14.15), 114, 118 (Fig. 14.31)
ova / ovaries, 112 (Figs. 14.20, 14.21)
pigment, 118
portal system, 113
reproductive system, 112 (Fig. 14.20), 114
respiratory system, 110 (Fig. 14.17), 119
semicircular ducts, 202
sense organs, 114
skeletal system, 107 (Figs. 14.11, 14.12), 108
skin, 101, 104, 105 (Figs. 14.7, 14.8)
spinal nerves, 84
sucker, 108, 109
testes, 111 (Fig. 14.18)
tongue, 101–102, 109
Larvacea, class, 15
larynx, 58, 88
lateral line system, 90–91 (Fig. 12.12)
dogfish shark, 199, 200 (Fig. 22.11)
lamprey, 101, 102 (Fig. 14.1), 103 (Fig. 14.4), 114
mudpuppy (*Necturus*), 214 (Fig. 24.4), 271–272 (31.7)
Latimeria, 30, 53
lens (eye), 92, 94 (Fig. 12.14)
Lepidosiren, 72, 75
leukocytes, 69, 408
limbs
with digits, 47, 48 (Fig. 5.16)
evolution of terrestrial, 47
muscle, 53 (Fig. 6.4), 54
pectoral, 48 (Fig. 5.16)
postures, primitive and advanced, 49 (Fig. 5.19)
Linnaeus, Carolus, 2
liver, 58, 72, 74, 78, 98
lizards, 30, 42, 58, 65, 94
tongue, 58
locomotion
cerebellum role in, 82
in mammals, 49
muscle role in, 51, 54
spinal cord and, 84
terrestrial, 220
lophophores, 11, 12 (Fig. 1.6)
lung(s). *See also* respiration
in early fishes, 57
in mammals, 364
terrestrial vertebrate, 243
lungfish (dipnoi), 39, 59, 91
heart and circulation in, 71, 72, 74, 78
pancreas, 98
luteinizing hormone (LH), 95
lymph / lymphatic vessels / system, 69, 78, 97
lymphocytes, 78

malleus, 38, 92, 93 (Fig. 12.13), 277, 424
mammals
brain and brainstem, 411, 416 (Fig. 40.5), 417, 418 (Fig. 40.6)

central nervous system, 84 (Fig. 12.5)
cud-chewing, 59
diaphragm, 225
digestive system, 355
endocrine system, 431
epidermis, 280–281
eye muscles, 426
feet, 50 (Fig. 5.20)
female eutherian, 67 (Fig. 10.3)
glands, 33
hair, 31, 33, 281
heart, 72, 381
herbivorous, 355
integument / skin of, 34, 280–282
jaw articulation and ear ossicles, 38 (Fig. 5.4)
limbs and locomotion, 49
lungs, 364, 408
lymph, 78
meninges, 83, 412
milk production, 58
monotreme, 33, 91, 277, 355, 371
nipples, 280
pectoral girdle and forelimb, 302
placental, 72
portal system, 408
reproduction, 66, 371
sense organs, 424
skull, 284
stomach, 59
tail, 280
teeth, 40, 355
therian, 68
thymus, 408, 433
tongue, 58, 424
unguligrade, 49
urinary system, 62, 371
vertebral column / vertebrae, 45, 298
mammary glands, 33, 96, 282
mandible (Meckel's cartilage) and mandibular arch, 134
nerve branch, 87
marsupials, 66 (Fig. 10.2), 355
mechanoreception, 31, 124
medial, definition, 4
medulla oblongata, 81 (Fig. 12.3)
melanin, 29, 33
melanocyte stimulating hormone (MSH), 96
melanophores, 126
melatonin, 96, 97
meninges, 83, 84 (Fig. 12.5)
mesencephalon, 80 (Fig. 12.2), 81, 82, 83
mesenteries, 55–56, 109, 345
mesoderm, 55, 97
mesonephros, 62
metabolism, 60, 97, 98
mammal, 284
metamorphosis, 97, 118
metanephros, 62
metencephalon, 80 (Fig. 12.3), 81, 269, 411
mineralocorticoids, 95, 97
mitosis, 126
moles, 67
mollusks, 1
molting, 97
monotreme mammals, 33, 91, 277, 355, 371
motor nerves and neurons, 50, 85, 87
mouth

in acorn worm, 10 (Fig. 1.3)
in deuterostome, 2 (Fig. 1.1)
in protostome, 2 (Fig. 1.1)
in tunicate, adult, 15 (Fig. 2.1)
movements, produced by muscles, 52, 53
mudpuppy (*Necturus*), 1, 99, 211
adrenal, 274, 275 (Fig. 32.5)
appendicular division, 217 (Fig. 25.1), 223
archinephric duct, 249 (Fig. 29.1), 253 (Fig. 29.8)
arteries, 258–262 (Figs. 30.5–30.11)
atlas, 220, 221
axial division, 217 (Fig. 25.1) 218–222
bladder, 238 (Fig. 27.3)
brachial plexus (mudpuppy), 270 (Fig. 31.4)
brain, 267, 269–270 (Figs. 31.1–31.3)
chondrocranium, 218 (Fig. 25.3)
circulatory system, 255, 258–266 (Figs. 30.5–30.18)
cloaca, 254 (Fig. 29.9)
cranial nerves (table), 269
dermatocranium, 218–219
digestive system, 243–246 (Figs. 28.1–28.7)
duodenum, 245, 246 (Fig. 28.6)
endocrine system, 273
esophagus, 243 (Fig. 28.1), 244
external anatomy / features, 213–214 (Figs. 24.1–24.3)
eyes, 213, 247 (Fig. 28.9), 272 (Fig. 31.6)
female, 242 (Fig. 27.11), 250–252 (Figs. 29.3–29.4), 254, 275, 276 (Fig. 32.7)
forelimb, 223 (Fig. 25.18)
gallbladder, 245, 246 (Fig. 28.6)
gills / gill slits, 213 (Fig. 24.2), 214, 217 (Fig. 25.1), 220 (Fig. 25.10), 223, 224 (Fig. 25.19), 243 (Fig. 28.1), 244, 246, 247 (Fig. 28.8)
heart, 255, 257–258 (Figs. 30.1–30.4)
hindlimb, 223, 224 (Fig. 25.19), 234 (Fig. 26.13), 235 (Fig. 26.14), 236 (Fig. 26.16)
humerus, 223 (Fig. 25.18)
hypophysis, 273 (Fig. 32.1)
integumentary system, 214, 215–216 (Figs. 24.5, 24.6)
intestine, 244 (Fig. 28.2)
kidney, 249–250 (Figs. 29.1, 29.2)
lateral line, 214 (Fig. 24.4), 271–272 (Fig. 31.7)
ligaments, 238 (Figs. 27.2, 27.3), 240–241 (Figs. 27.2–27.8)
liver, 245, 246 (Fig. 28.6)
lungs, 247, 248 (Fig. 28.13)
male, 242 (Fig. 27.10), 250, 253 (Figs. 29.7, 29.8), 254, 275, 276 (Fig. 32.8)
mandible, 219
medulla oblongata, 270
meninges, 267–268 (Fig. 31.1)
mesenteries, 239–242 (Figs. 27.4–27.11)
muscles
abdominal, epaxial, hypaxial 226, 227 (Figs. 26.3, 26.4)
forelimb, 223 (Fig. 25.18), 233 (Fig. 26.12), 234
head and shoulder, 228 (Figs. 26.5, 26.6), 229
hindlimb, 234 (Fig. 26.13), 235 (Fig. 26.14), 236 (Fig. 26.16)

medial, 232 (Fig. 26.11)
pectoral, 231 (Fig. 26.10), 232
pelvic, 235 (Fig. 26.13)
shoulder, 232 (Fig. 26.11)
subarcuales, 230–231
nervous system, 267
olfactory apparatus, 271, 272 (Fig. 31.6)
opening, 237–238 (Figs. 17.1–27.3)
ovaries, 275, 276 (Fig. 32.7)
pancreas, 244 (Fig. 28.2), 245, 274, 275 (Fig. 32.6)
parathyroid glands, 274
pectoral and pelvic girdles, 223–224 (Figs. 25.18–25.19)
pharynx, 243 (Fig. 28.1), 244
pineal gland, 273 (Fig. 32.2)
portal system, 262–264 (Figs. 30.12, 30.15)
radius bone, 223 (Fig. 25.18)
reproductive system, 250, 251–254 (Figs. 29.3–29.9)
respiratory system, 243, 245–246, 247–248 (Figs. 28.8–28.13)
ribs, 221 (Figs. 25.12, 25, 13), 222, 223 (Fig. 25.17)
sacral, 222 (Fig. 25.15)
scapula, 223 (Fig. 25.18)
sense organs, 271, 272 (Fig. 31.6)
skeletal system, 217 (Fig. 25.1, 25.2)
skin / epidermis / dermis, 214, 215 (Fig. 24.5, 24.6), 216 (Fig. 24.7), 246
skinning, 225, 226 (Fig. 26.1, Fig. 26.2)
skull, 218, 219 (Figs. 25.5, 25.6), 220 (Fig. 25.9)
spinal nerves and spinal cord, 270–271 (Figs. 31.4, 31.5)
splanchnocranium, 218, 219–220
stomach, 244 (Fig. 28.3)
suprarenal (adrenal), 274, 275 (Fig. 32.5)
teeth, 219, 220 (Fig. 25.7–Fig. 25.9)
testes, 242, 250, 253 (Fig. 29.7, 29.8), 275, 276 (Fig. 32.8)
throat, 229 (Fig. 26.7), 230 (Fig. 26.8)
thyroid and thymus glands, 274 (Figs. 32.3, 32.4)
tongue, 243 (Fig. 28.1), 244
ulna, 223 (Fig. 25.18)
urinary bladder / system, 238 (Fig. 27.3), 249–250 (Figs. 29.1–29.2)
veins, 262–266 (Figs. 30.12–30.18)
vertebral column / regions, 220, 221–222 (Figs. 25.12–25.16)
visceral arches, 220
mud-skippers, 47
muscle(s)
actions, 53
amphibian, 225
anatomy / architecture, 52 (Figs. 6.2, 6.3)
branchiomeric, 54
cardiac, 52
eye, of vertebrates, 202, 426, 427 (Fig. 40.13)
striated / skeletal, 51 (Fig. 6.1), 52
of throat and head region, 54
tongue, 54
types of, 51 (Fig. 6.1)
musculature, epaxial and hypoaxial, 54, 53
myelencephalon, 80 (Fig. 12.2), 81
myelin, 80

myomeres / myosepta / myosin, 52, 53, 84, 108 (Fig. 14.13), 116 (Fig. 14.28) 118, 143
Myxini subphylum, 4

nails, 31 (Fig. 4.4), 282. See also claws
neck, evolution of, in fish, 48
Necturus, 58, 72, 211. See also mudpuppy
aortic arch, 74 (Fig. 11.6), 75
brain, 81 (Fig. 12.3)
central nervous system, 84 (Fig. 12.5)
gills, 58, 75, 125, 126 (Fig. 15.10)
lateral line system in, 90
Neoceratodus, 72
neoteny, 211
nephrons, 62, 371. See also kidneys
nerve cord, hollow, 2, 82, 99
amphioxus, 21, 22 (Figs. 3.3, 3.4)
shark, 191
tunicate, 18 (Fig. 2.6)
nerves
cranial (trigeminal, trochlear), 86–90
facial, 87
glossopharyngeal, 88, 90
olfactory, 87
vagus, 94
vestibulocochlear, 87, 202
nervous system / cells, 79, 85. See also central nervous system
sympathetic division, 86 (Fig. 12.7)
neurohypophysis, 96
neuromast, 90 (Fig. 12.11)
neurons / neurotransmitters, 79 (Fig. 12.1), 80, 82, 96
nictating membrane (cat), 428 (Fig. 40.14)
norepinephrine, 80, 97
nose. See olfaction
notochord, 99
in vertebrates, 35, 36 (Figs. 5.1, 5.2A), 42, 43, 135

ocellus, 18 (Fig. 2.6), 19–20 (Figs. 2.8, 2.9)
odontoblasts, 126
olfaction, 89
olfactory lobes and bulbs, 81 (Fig. 12.3), 82, 83 (Fig. 12.4), 87, 191
oligodendroglia, 80
omnivores, 40
optic lobes and nerves, 82, 87, 94 (Fig. 12.14), 426
oral
cavity, 58
hood (amphioxus), 21, 23 (Fig. 3.4)
osmolarity / osmotic pressure, 62, 69
osmoregulation, 102, 162, 250
ossicles, 91. See also ear
ostariophysii, 91
osteichthytes, 119. See also bony fishes
jaw, 37 (Fig. 5.3C)
nostrils, 89
skeleton, 35
ostracoderms, 99, 119
lateral line system, 199
skeleton / vertebrae, 35, 36, 42
ova / ovaries / oviducts, 65, 98
amphioxus, 24 (Fig. 3.6), 26
in chondrichthytes, 65
in cyclostomes, 68

mammalian, 371
shark, 209, 210 (Fig. 23.5)
oval window (ear), 92
oviducts, 65, 66
oxygenation, 114
oxytocin (OXY), 96

paedomorphosis, 211
palate, 41–42 (Fig. 5.7)
Paleozoic era, 119
pancreas, 60, 98
pars distalis, 95, 96
pectoral limb, 47, 48 (Figs. 5.16, 5.17)
pelagic, definition, 15
penis, 66, 68, 371
pepsin / pepsinogen, 98
pericardium, 381
perilymph, 91
perimysium, 52
peripheral nervous system, 79, 80, 84, 88
peritoneum, 55 (Fig. 7.1)
Petromyzon marinus, 101
pH, 98
pharyngeal
pores, 11 (Fig. 1.4)
slits, 2, 9, 11, 16, 23, 99. See also pharynx
pharynx, 58, 59, 88, 97
amphioxus, 21, 23, 24 (Fig. 3.5)
in chordates, 57
nerves to, 88
and taste receptors, 58
pheromones, 90
photoreceptors, 92, 94
amphioxus, 22 (Fig. 3.2), 23 (Fig. 3.3)
physoclists / physotomes, 57, 58
"pigeon milk," 58, 96
pigment / pigmentation, 33–34, 96
pineal eye / organ, 94, 101, 103 (Fig. 14.3), 114
pineal gland, 96–97
pinna, 93 (Fig. 12.13), 279 (Fig. 33.1), 424
piranha, 39
pituitary, 95, 273, 431. See also hypophysis
pit vipers, 94
placenta, 68, 72, 98
placoderms, 119
jaws, 37, 38–39
skeleton, 35
platelets, 69
platypus, 91, 277
plesiomorphies, 3 (Fig. 1.2)
Plethodontidae, 246
pons, 82, 83 (Fig. 12.4), 417
portal system, 78
hypophyseal, 96 (Fig. 13.1)
lamprey, 113
mammal, 408
posterior, definition, 4
primates
claws / nails in, 32
reproduction in, 68
sweat glands in, 33
tails, 280
primitive gut, 2, 9
proboscis, 9 (Fig. 1.1), 10 (Fig. 1.2), 12 (Fig. 1.6)
progesterones, 98
prolactin, 58, 96
pronephros, 62, 117